COMPSTAT 1974

Proceedings in Computational Statistics

Edited by
Gerhart Bruckmann
Franz Ferschl
Leopold Schmetterer

Physica Verlag Wien 1974
ISBN 3 7908 0148 8

Printed in Germany by: repro-druck "Journalfranz", Arnulf Liebing, Würzburg
ISBN 3 7908 0148 8

P R E F A C E

Perhaps neither mathematicians specialized in probability theory or statistics nor experts in electronic data processing will look on computational statistics as a serious scientific subject. Nevertheless, we were daring enough to organize a symposium on computational statistics, and that for two reasons: first, the foundation of a computer centre at a university is in most cases connected with the needs of numerical mathematics or physics. At the university of Vienna, however, the connection between computer application and statistics is a good old tradition. And secondly, as 1971 Walter Freiberger and Ulf Grenander wrote in their Course in Computational Probability and Statistics, "we felt the time was ripe for a systematic exploi tation of modern computing techniques in mathematical statistics and applied probability". The field of probability theory and of statistics as well as the standard of computational devices has been growing at a spectacular rate - a rate which as we hope will now result in techniques of model building being "very different today from what was in pre-computer days". If we succeed in making statisticians aware of the great possibilities of modern computing facilities, which at any rate go beyond simple numerical computation, the Symposium serves its purpose. Selection of topics is - particularly in a new field - to some extent a matter of personal preferences and prejudices. We hope, however, we will have the benefit of advice and criticism of colleagues and participants in the Symposium.

The Symposium could not be held without the help of many persons and institutions. It is a pleasure for us to acknowledge our gratitude to the Austrian Ministry for Science and Research for financial and moral backing, particularly to Dr. Hertha Firnberg, Minister for Science and Research. We are grateful to the Lord Mayor of

Vienna,Leopold Gratz,for the same reason.Valuable support has been rendered by the Austrian Association for Statistics and Data Processing.Our thanks are also due to the computer firms Control Data Corporation,Honeywell-Bull,and IBM for giving financial and practical aid.Finally,we would like to thank cordially all our colleagues whose energetic assistance enabled the Symposium to be held.

Vienna,September 1974 J.Gordesch,P.P.Sint

Organizing Committee

H.Abele (Univ. of Fribourg), G.H. Fischer (Univ. of Vienna), J. Gordesch (Free Univ. of Berlin), P. Sint (Austrian Academy of Science, Vienna), G. Vinek (Univ. of Linz)

CONTENTS

SUBJECT GROUP A

Computational Probability

SUBJECT GROUP B

Automatic Classification

SUBJECT GROUP C

Numerical and Algorithmic Aspects of Statistical Methods

SUBJECT GROUP D

Simulation and Stochastic Processes

SUBJECT GROUP A

Computational Probability

On the Distribution of the Parameter of Curvature k of the Fisher-Tippett Distributions under the Hypothesis k = 0

Konrad Cehak, Vienna

Summary

For discrimination between the three types of Fisher-Tippett extreme value distributions the frequency distribution of the parameter of curvature k as dependent on the sample size n is investigated. On account of the difficulties of an analytical determination of this frequency distribution series of random numbers were generated, which were distributed according to a Fisher-Tippet I distribution. From these series the parameters of the Fisher-Tippett type II or III distributions were calculated and the frequency distribution of the parameter of curvature k was determined. It could be shown that k is normally distributed. The mean value is 0, the standard deviation depends on the sample size n according to the expression $\sigma_k = 1.163\, n^{-0.5962}$.

Über die Verteilung des Krümmungsparameters k in den Fisher-Tippet Verteilungen unter der Hypothese k = 0.

Zusammenfassung

Zur Unterscheidung der drei Typen der Fisher-Tippett Extremwertverteilungen wurde die Verteilung des Krümmungsparameters k in Abhängigkeit vom Stichprobenumfang n untersucht. Wegen der Schwierigkeit einer analytischen Bestimmung dieser Verteilung wurden Zufallsreihen generiert, die nach Fisher-Tippett I verteilt waren. Aus ihnen wurden die Parameter der Fisher-Tippett II und III Kurven berechnet und die Verteilung des Krümmungsparameters k bestimmt. Es zeigt sich, daß k normal verteilt ist. Der Mittelwert ist 0, die Streuung hängt gemäß $\sigma_k = 1{,}163\ n^{-0{,}5962}$ vom Stichprobenumfang n ab.

1. INTRODUCTION

The three possible forms of the asymptotic extreme value distributions, which are called Fisher-Tippett distributions I to III after the authors of the first investigations of extreme value distributions [FISHER and TIPPETT, 1928], may according to JENKINSON [1955]be written in a common form by which they are connected to a reduced variate y:

$$x = x_o + a\,\frac{1 - \exp\,(-ky)}{k} \quad . \tag{1}$$

In expression (1) x is the observed variate, x_o is the value of this variate for y = 0, a is the slope of the x,z-curve in the point $(x_o|0)$ and k is a parameter of curvature. Expression (1) may be transformed in a way that the meaning of the coefficients comes out more clearly:

$$x = x_g - b\,\exp\,(-ky) \quad . \tag{2}$$

In expression (2) x_g is the limiting value of the variate x, which is transgressed with probability zero. The reduced variate is distributed according to the double-exponential distribution function

$$F(y) = \exp\,\{-\,\exp\,(-y)\} \quad , \tag{3}$$

which is also known as Gumbel's distribution. In a probability diagram, in which one basic axis is the linear y-axis renumbered according to (3), whereas the other basic axis is the linear axis of the variate x, the Fisher-Tippett I distribution is a straight line (k = 0), type II (k < 0) is convex to the y-axis and type III (k > 0) is concave to the y-axis. It is essential that in type II the variate x is limited towards small values, in type III towards large values, while in type I it is unlimited in both directions.

In applications one will wish to use the Fisher-Tippett I distribution, as it has the simplest form and causes the least troubles in fitting the observations. Usually, it is even applied in cases, in which it is known that the variate has to be bound towards large or small values.

Especially in using extreme value distributions one meets the case that the sample values near to the ends of the distribution function deviate relatively widely from the theoretically expected values, so the goodness of fit in the extreme value probability diagram is not very strong. Although one could e.g. by means of control curves [SNEYERS, 1963] prove the validity of the fitting by means of the Fisher-Tippett I curve to the sample distribution it seems desireable to derive a parametric test for the opposite problem, namely, having fitted a Fisher-Tippett II or III curve for deciding whether it had been necessary to do so or whether it had been sufficient to use a Fisher-Tippett I curve.

Such a test could be based on the parameter of curvature, k, by testing the hypothesis $k = 0$ with the sample values. For being able to do so one has to know the distribution of k.

2. THE DISTRIBUTION OF THE PARAMETER OF CURVATURE k

Starting from the maximum-likelihood-principle the calculation of the parameters x_o, a and k affords the solution of a set of three transcendental equations, which is done by an iteration method, for which one has to know a set of approximate solutions as a first guess (see e.g. CEHAK [1971]). This shows that the calculation of k, and therefore, of its distribution, too, in a purely analytical way will lead to large difficulties. Out of this reason the following numerical way has been chosen.

For various sample sizes (n = 1o, 2o, 3o, 6o, 1oo and 3oo) series of random numbers, which are distributed according to Fisher-Tippett I, have been generated. For this purpose, pseudo-random numbers have been generated by means of the program RANDU from the IBM Scientific Subroutine Package on the IBM 113o computer of the Zentralanstalt für Meteorologie und Geodynamik. This program produces random numbers, which are equally distributed in the interval [0, 1] , by means of the power residue method. These random numbers were transformed using expression (3) into a series of numbers being distributed according to Fisher-Tippett I. Because of the properties of the random number generator it is ascertained that the samples used consist of "observations" being independent from each other and they themselves are independent from each other, too, as the repeating period of the pseudo-random numbers has not yet been reached.

For each of the quoted sample sizes approximately 1oo series were generated and from them the necessary parameters for expression (2) were calculated using the method described by CEHAK [1971] . For each sample size the distribution of the parameter k was determined. The result, which is presented in figure 1, shows that, for each sample size, the distribution of k is very near to a normal one. The same may be derived from the sample moments, which are contained in table 1. The sample means are very near to zero, therefore, taking into consideration the sizes of the standard deviations one may assume that the population means of the k distributions are zero independent of the sample size. The deviations of the sample means from zero is mostly less than the standard deviation, only in the case for n = 3oo it lies between two and three times the standard deviation. Also the parameter of skewness only once, for n = 6o, deviates from zero more than twice the standard deviation. If for a test given by Fisher (quoted by SNEYERS [1973]) one calculates the quantity X^2, which is χ^2-distributed with twelve degrees of freedom (twice the sample size, which in this case equals six)

$$X^2 = -2 \sum_{i=1}^{6} \ln \{\Phi(z_i)\} \tag{4}$$

one gets

$$X^2 = 5{,}82 \tag{5}$$

if one uses the parameter of skewness standardized by its standard deviation

$$z_i = \frac{\gamma_{1,i}}{\sigma_\gamma} \tag{6}$$

as the variates z_i in expression (4). With twelve degrees of freedom (5) has a probability for being transgressed by chance of more than 9o %, therefore, there is no reason for refusing the hypothesis that k is normally distributed for all sample sizes n on account of one significant deviation from zero in six cases.

The standard deviations of the normals distributions of k evidently depend on the sample size n. Figure 2, presenting σ_k

versus n in a double-logarithmic scale, shows that the σ_k values depend on the sample size according to the following equation:

$$\lg \sigma_k = \lg p + q \lg n \quad . \qquad (7)$$

Calculating the parameters lg p and q using the method of least squares one gets the following values:

$\lg p = 0.0659$

$p = 1.163$

$q = 0.5962$

Hence, the dependence of σ_k on n is given by

$$\sigma_k = \frac{1.163}{n^{0.5962}} \quad . \qquad (8)$$

The proximity of the exponent of n in expression (8) to the value 0.5 suggests to use a simplified expression, in which only the nominator has to be determined by a least squares scheme. This yields

$$\sigma_k = \frac{0.8037}{n^{0.5}} \quad . \qquad (9)$$

Both approximating functions are scetched in figure 2. Testing the goodness of fit by calculating the average deviations of the sample values from the approximating curves shows that expression (9) yields a mean deviation (0.012) which is nearly two times that from expression (8) (0.0066). Therefore, the approximation by (9) has to be preferred to that by (8).

In the application, having calculated a value of k one has to calculate its standard deviation σ_k, which corresponds to the sample size, from expression (8). If the sample value of k deviates more than the three-fold standard deviation σ_k from zero the use of a Fisher-Tippett II or III distribution is statistically justified, otherwise the fitting of a simple Fisher-Tippett I distribution would suffice.

REFERENCES

Cehak, K.: Der Jahresgang der monatlichen höchsten Windgeschwindigkeiten in der Darstellung durch Fisher-Tippett III-Verteilungen. Archiv Meteor. Geoph. Biokl. B 19, 165-182 (1971)

Fisher, R.A., and H.C. Tippett: Limiting Forms of the Frequency Distributions of the Largest or Smallest Member of a Sample. Proc. Cambridge Phil. Soc. 24, 180 (1928).

Jenkinson, A.F.: The Frequency Distribution of the Annual Maximum (or Minimum) Values of Meteorological Elements. Quart. J. Roy. Meteor. Soc. 87, 158-171 (1955).

Sneyers, R.: Du test de validité d'un ajustement basé sur les fonctions de l'ordre des observations. Inst. Roy. Météor. de Belgique, Publ. B 39 (1963).

Sneyers, R.: Unpublished manuscript for a WMO Technical Note on Statistics in Climatology (1973).

Table 1. Statistical parameters of the distributions of k in the six sample series.

Sample size n	1o	2o	3o	6o	1oo	3oo
Number of series	1o1	1o1	1o9	1o9	1o9	114
Mean value $\overline{k}$	o.o27	o.o25	-o.oo2	o.oo5	o.oo1	-o.oo9
Standard deviation σ_k	o.285	o.185	o.165	o.1o6	o.o76	o.o37
Skewness γ_1	o.o4	-o.o1	o.22	o.91	o.62	-o.16
Kurtosis γ_2	-1.o8	o.26	-o.1o	o.93	o.7o	-o.2o
Standard deviation of skewness	o.24	o.24	o.23	o.23	o.23	o.23

Adress of the author: Univ.Prof. Dr. Konrad Cehak, Zentralanstalt für Meteorologie und Geodynamik. Hohe Warte 38, A-119o Wien. Austria.

Figures:

Fig. 1. Frequency distribution of the k-values in the six sample series in a Gaussian probability diagram.

Fig. 2. Dependence of σ_k on n in double-logarithmic paper. Solid line: fitting by expression (8). Broken line: fitting by expression (9).

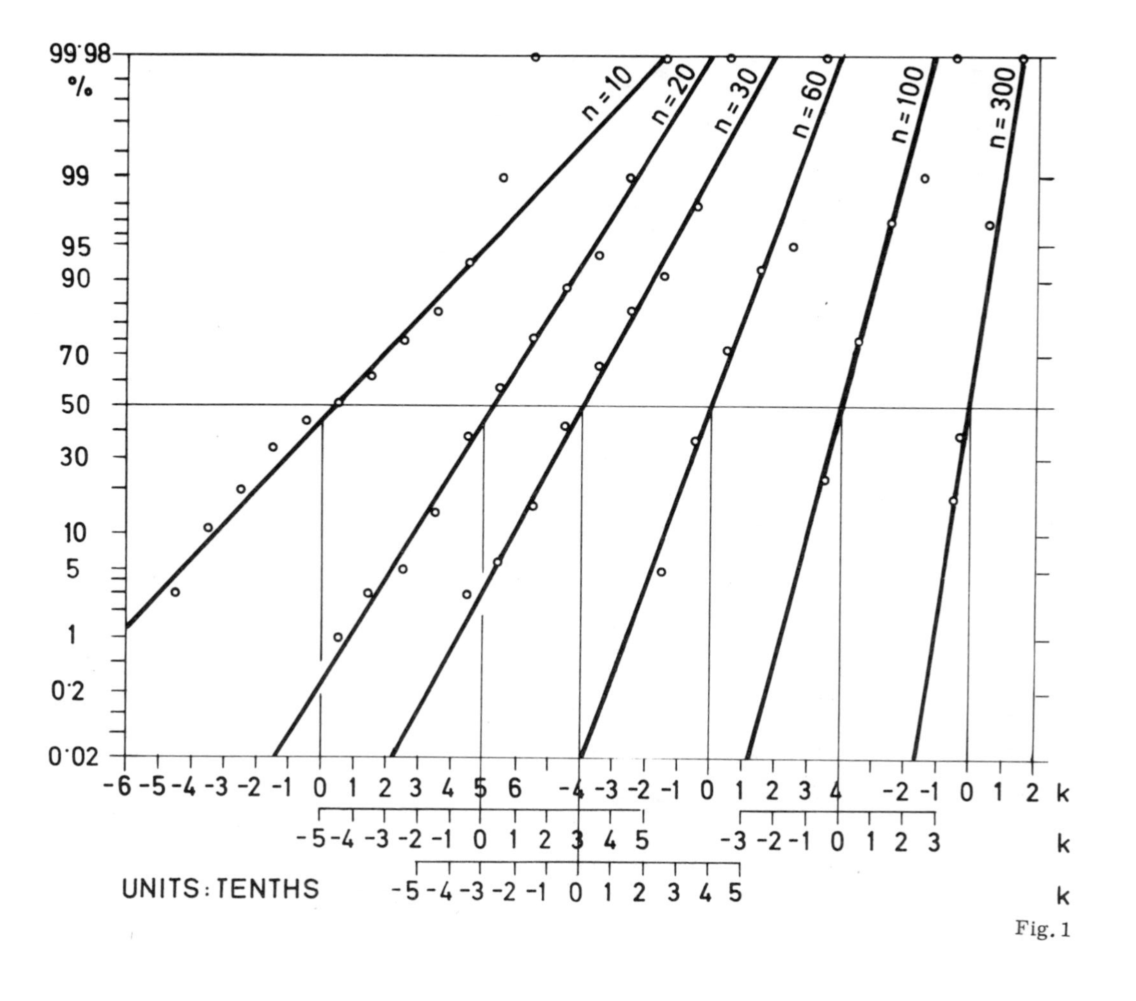

99·98
%
99
95
90
70
50
30
10
5
1
0·2
0·02
n = 10
n = 20
n = 30
n = 60
n = 100
n = 300
-6 -5 -4 -3 -2 -1 0 1 2 3 4 5 6 -4 -3 -2 -1 0 1 2 3 4 -2 -1 0 1 2 k
-5 -4 -3 -2 -1 0 1 2 3 4 5 -3 -2 -1 0 1 2 3 k
UNITS: TENTHS
-5 -4 -3 -2 -1 0 1 2 3 4 5 k

Fig. 1

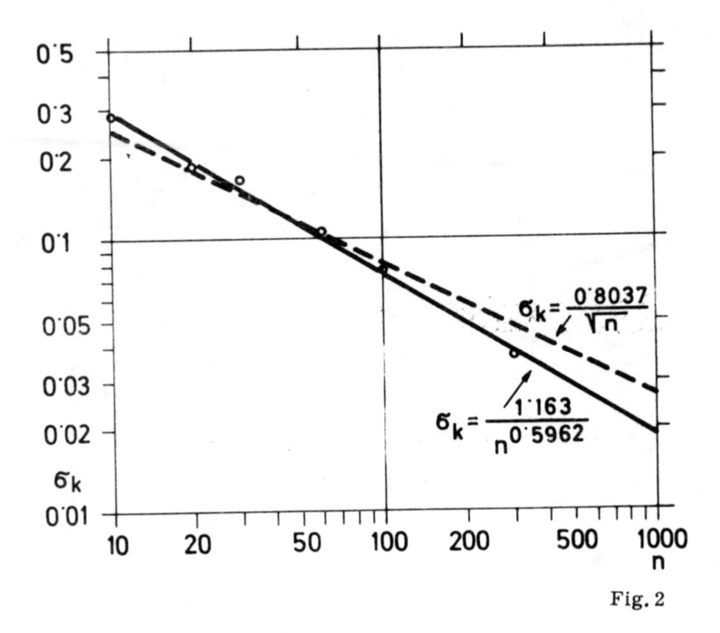

0·5
0·3
0·2
0·1
0·05
0·03
0·02
σk
0·01
10
20
50
100
200
500
1000
n
σk = 0·8037/√n
σk = 1·163/n^0·5962

Fig. 2

A Fast Generator for Gamma-Distributed Random Variables

Arthur J. Greenwood, New York

1. JÖHNK'S ALGORITHM

JÖHNK (1964) has given a mathematically sound and easily programmable algorithm for generating random variables with the gamma density function

$$x^{p-1}e^{-x}/\Gamma(p). \tag{1}$$

Premising that a source of independent random variables r uniformly distributed on the interval

$$0 \leq r \leq 1 \tag{2}$$

is available, and that an algorithm for exponentially distributed random variables y has been constructed ($y = -\log_e r$ will serve, but see Section 3 below), Jöhnk proceeds as follows.

1. If p is an integer n, generate x as the sum of n independent exponential variables.

2. If p is not an integer, set $p = n+f$, where $0 < f < 1$. If $n = 0$, set $x_0 = 0$ and generate the exponential variable x_1; otherwise, generate x_0 as the sum of n independent exponential variables and x_1 as an additional exponential variable.

3. Let r_1 and r_2 be two independent uniform variables. Compute

$$w_1 = r_1{}^{1/f} \text{ and } w_2 = r_2{}^{1/(1-f)}.$$

If $w_1+w_2 > 1$, reject r_1 and r_2 and take a new pair; otherwise, generate $x_2 = w_1/(w_1+w_2)$. Then x_2 has the beta density

$$x^{f-1}(1-x)^{-f}/[\Gamma(f)\Gamma(1-f)];$$

x_1x_2 has the gamma density

$$x^{f-1}e^{-x}/\Gamma(f);$$

and $x = x_0+x_1x_2$ has the gamma density of eq. (1).

2. DIFFICULTIES WITH JÖHNK'S ALGORITHM

It is convenient to state here the only safe procedure for generating the sum of n exponential variables: generate the n variables separately, by whatever algorithm for exponential random variables you have programmed; then sum them. Do not program the seductive short-cut (JÖHNK 1964, p. 12) which has crept into some handbooks of simulation, ex.gr. NAYLOR et al. (1966, pp. 88-89), and even into the careful textbook of KNUTH (1969, p. 115), employing the identity

$$\Sigma \log r_i = \log \Pi r_i;$$

this is a mathematical, not a computational, identity. Taking the floating-point hardware of the UNIVAC 1100, for example, the product of 4 non-zero r_i can underflow; the product of 90 non-zero r_i will underflow with frequency exceeding 50%. Not only does the underflow destroy the latent value of $\Sigma \log r_i$, but also, depending on the operating system in use, the underflow or the subsequent futile attempt to compute the logarithm of zero can print diagnostic messages or abort the computation altogether.

The principal disadvantage of Jöhnk's algorithm is the large number of uniform random variables used: p for integer p and approximately $p+3$ otherwise. If random variables are to hand, this is only a disadvantage in time; the computation of n logarithms becomes onerous for $n \gtrsim 1000$, say. If, as is usual in modern computing, pseudo-random variables, generated by a subroutine on demand, are used instead of random variables, the provision of independent r_i poses a difficult problem. Published uniform generators of good quality have passed tests for independence in pairs and triples; the distributional properties of these generators for blocks of n, $n > 3$, are almost unknown. Defining a robust random-variable algorithm as one that delivers random variables with the distribution demanded when uniform random numbers are input, and, when the input is switched to high-quality pseudo-random numbers, delivers pseudo-random variables of quality not much worse than the input, we say that for $p > 3$ Jöhnk's algorithm is not robust.

The range [0,1] of eq. (2) for uniform random numbers, quoted from Jöhnk, is not the most convenient range either to generate or to use. Assuming a source of random decimal digits: to approximate a uniform variable on [0,1), take a string of s digits and prefix a decimal point; on (0,1], compute $r' = 1-r$, where r is a uniform variable on [0,1); on (0,1), prefix a decimal point to a string of $(s-1)$ digits and suffix a 5. The modifications to accommodate a source of binary digits are obvious. Use of the open interval (0,1) as range ensures that any transformed random variable will assume neither of the values $\pm\infty$.

3. EXPONENTIAL GENERATORS

Exponentially distributed random variables are of direct use in simulation of queuing problems, and are a natural intermediate in generating random variables with the normal, gamma, Gumbel, and Weibull distributions; so a robust exponential generator is a desideratum in any laboratory that pretends to make Monte Carlo calculations. J. v. Neumann's method (JÖHNK 1969, p. 12) requires at least 3 successive r_i to be independent to guarantee a correct exponential distribution on [0,1); and $(2n+1)$ successive r_i to be independent for a correct distribution on $[0,n)$. The direct method

$$y = -\log_e r$$

makes no demands on independence of the r_i. It suffers not from lack of robustness, but from discretization error: first, a test is necessary to circumvent log 0 (KNUTH 1969, p. 114); second, the tail of the distribution of y is coarse-grained. Specifically, assume that the r_i are presented to the logarithm routine as strings of s decimal digits preceded by a decimal point; then values of r for which $10^{-s} < r < 2\times10^{-s}$ are absent from the input, and hence y for which

$$s \log_e 10 - \log_e 2 < y < s \log_e 10$$

are absent from the output. (Again, the modifications for binary digits in the input are obvious). If $s = 8$, y is confined to the range [0,18.41), and random normal deviates computed as $2y^{\frac{1}{2}}\cos r$ and $2y^{\frac{1}{2}}\sin r$ to the range (-6.08,+6.08). A personal communication from Professor David Durand suggests that this last restriction, while possibly quite innocuous for a Monte Carlo study of the median, say, in samples from a normal distribution, could adversely affect a simulation of the range in such samples.

The program REXP shown in Appendix I requires an average of 1000/999 r_i for each exponential variable generated. If RANDEM yields random numbers, the tail of the exponential distribution will be correct out to 2^{20} or beyond, the exact upper bound depending on the details of floating-point addition in the machine used; if RANDEM yields pseudo-random numbers that are independent in blocks of n, the tail will be correct out to $3n \times \log_e 10$.

4. THE PROPOSED ALGORITHM

A well-known approximation, due originally to WILSON & HILFERTY (1931), asserts that, if y is a standardized normal variable with the density

$$(2\pi)^{-\frac{1}{2}}\exp(-\tfrac{1}{2}y^2), \tag{3}$$

then

$$z = f[1 - 2/9f + (2/9f)^{\frac{1}{2}}y]^3 \tag{4}$$

has approximately a chi-squared distribution with f degrees of freedom. The argument leading to eq. (4) rests on equalization of asymptotic moments, and does not constrain f to be an integer; so that

$$x = p[1-(9p)^{-1}+(9p)^{-\frac{1}{2}}y]^3 \tag{5}$$

has approximately the distribution (1). To generate exactly gamma-distributed variables it is natural to start with (5) and adjust the distribution by rejection. Thus it is required to represent (1) as the product

$$x^{p-1}e^{-x}/\Gamma(x) = a\varphi(x)g(x), \tag{6}$$

where $\varphi(x)$ is the density function of x from (5) and $g(x) \leq 1$. A little manipulation yields

$$dx/dy = p^{\frac{1}{2}}\{1-(9p)^{-1}+(9p)^{-\frac{1}{2}}y\}^2 = p^{-1/6}x^{2/3},$$

$$\varphi(x) = (2\pi)^{-\frac{1}{2}}\exp(-\tfrac{1}{2}y^2)\,dy/dx = (2\pi)^{-\frac{1}{2}}p^{1/6}x^{-2/3}\exp(-\tfrac{1}{2}y^2),$$

$$\log_e[ag(x)] = \tfrac{1}{2}\log_e(2\pi)-1/6\log_e p+(p-1/3)\log_e x+\tfrac{1}{2}y^2-x. \tag{7}$$

The expression (7) is found to have a maximum at

and so

$$y_0 = -3^{\frac{1}{2}}+(9p)^{-\frac{1}{2}},\quad x_0 = p[1-(3p)^{-\frac{1}{2}}]^3,\quad p \geq 1/3; \tag{8}$$

$$a = (2\pi)^{\frac{1}{2}}p^{p-\frac{1}{2}}[1-(3p)^{-\frac{1}{2}}]^{3p-1}\exp[-p+(3p)^{\frac{1}{2}}+\tfrac{1}{2}+4/3(3p)^{-\frac{1}{2}}+(18p)^{-1}].$$

The approach of a to its asymptotic value 1 is rapid: for $p = 1/3$, $a = 2.189$; for $p = \frac{1}{2}$, $a = 1.337$. The rejection rule can now be stated: generate a standardized normal variable y, and compute x by (5); generate a random number r independent of y. If

$$r \leq (x-x_0)^{p-1/3}\exp(\tfrac{1}{2}y^2-x-\tfrac{1}{2}y_0^2+x_0) \tag{9}$$

accept x, otherwise start over. On the average, a pairs (y,r) are needed to yield an x; and the accepted x have the distribution (1). It is convenient to rewrite (9) in terms of an exponentially distributed $z = -\log_e r$; if

$$0 \leq z+(p-1/3)(\log_e x-\log_e x_0)+\tfrac{1}{2}(y^2-y_0^2)-x+x_0,$$

accept x.

5. NOTES ON THE FORTRAN PROGRAMS

The programs in Appendix I conform to the ISO FORTRAN recommendations. With the exception of RANDEM, they should be independent of word size.

Statement 11 in REXP is needed if RANDEM produces true random numbers. If RANDEM generates pseudo-random numbers, the equality cannot occur and the test may be omitted.

Function REXP2 is used to protect against the possibility that if statement 21 of RGAMMA were to read

```
21     RGAMMA = REXP(JA)+REXP(JA)
```

some sophisticated compiler would compile *one* call to REXP and add the result to itself.

Function RNORM makes use of the identities

$$\sin 2w = 2\tan w\ /\ (1 + \tan^2 w),$$
$$\cos 2w = (1 - \tan^2 w)/(1 + \tan^2 w).$$

Function TAN (the circular tangent), not required by ISO, is supplied in most manufacturers' libraries. If it is not supplied, the version in Appendix I can be used.

In function RGAMMA, ALPHA denotes the exponent called p in the discussion above. The program implements Jöhnk's algorithm for

$p < .5$ and the algorithm of Section 4 for $p > .5$. Shortcuts are programmed for $p = .5$, 1.0, 1.5, 2.0 (statements 15, 17, 19, 21). A negative return denotes failure to generate a gamma variable: specifically,

RGAMMA = -1. $p < 0$; gamma distribution not defined.

RGAMMA = -2. Small p, where Jöhnk's algorithm is defeated by underflow.

RGAMMA = -3. Large p, where x computed by (5) will agree with p to 5 figures or more.

A pseudo-random number routine free from superfluous and time-consuming code must be written in machine-dependent language; some compiler languages that exploit hardware masking and shift operations are suitable. The program RANDEM in Appendix I is not fast: each pseudo-random number generated requires 8 double-precision multiplications and 4 double-precision divisions. The program is so far machine-independent that it will yield the same numbers on any computer that accommodates positive integers of length 31 bits or more, and allots 48 bits or more to the mantissa of double-precision numbers. For suggesting that the use of double-precision arithmetic can yield machine-independent programs, I am greatly indebted to Professor L. H. Herbach.

The numbers generated by RANDEM are independent in blocks of 17 and uncorrelated in blocks of 34. Statement 13 counts the number of calls to the routine and can be deleted if the count is not wanted. To run the gamma generator using RANDEM as printed in Appendix I, the calling program must contain the array declarator

```
      DIMENSION JA(19)
```

This declarator must be inserted in REXP, REXP2, RNORM, and RGAMMA. The FUNCTION statement at the head of RGAMMA must be changed to

```
      FUNCTION RGAMMA(ALPHA,JA)
```

RANDEM as printed in Appendix I is not self-starting; the calling program should initialize JA by storing a random integer (9 decimal digits) in each of locations JA(1) through JA(18) .

APPENDIX I. FORTRAN PROGRAMS

```
      FUNCTION REXP(JA)
C     EXPONENTIAL PSEUDO-RANDOM NUMBER GENERATOR
```

```
      DATA AA/6.9077553/
C     AA = LOGE(1000)
      REXP = 0.
10    AC = RANDEM(JA)
C     RANDEM IS A USER-SUPPLIED GENERATOR FOR RANDOM OR PSEUDO-
C     RANDOM REAL NUMBERS, 0. .LE. R, R .LT. 1.
      IF (AC .GT. .001) GO TO 12
      AC = REXP
      REXP = REXP+AA
11    IF (REXP .EQ. AC) RETURN
      GO TO 10
12    REXP = REXP-ALOG(AC)
      RETURN
         END

      FUNCTION REXP2(JA)
      REXP2 = REXP(JA)
      RETURN
         END

      FUNCTION RNORM(JA)
C     NORMAL PSEUDO-RANDOM NUMBER GENERATOR
      DATA IA/0/,PI/3.14159265/
      IF (IA .NE. 0) GO TO 10
      IA = 1
      AB = SQRT(2.*REXP(JA))
      RNORM = TAN(PI*RANDEM(JA))
C     RANDEM IS A USER-SUPPLIED GENERATOR FOR RANDOM OR PSEUDO-
C     RANDOM REAL NUMBERS, 0. .LE. R, R .LT. 1.
      AC = RNORM**2
      AD = 1.+AC
      AC = 1.-AC
      RNORM = 2.*AB*RNORM/AD
      RETURN
10    IA = 0
      RNORM = AB*AC/AD
      RETURN
         END

      FUNCTION RGAMMA(ALPHA)
C     GAMMA PSEUDO-RANDOM NUMBER GENERATOR, CUBED NORMAL METHOD
      DATA AA/.333333333/,AC/1.73205081/,AL/-1./
C     AC = SQRT(3.)
      BL = ALPHA
      IF (BL .GT. 0) GO TO 10
      RGAMMA = -1.
      RETURN
10    IF (BL .EQ. AL) GO TO 30
      IF (BL .LT. 8E-9) GO TO 22
      IF (BL .LT. 1E12) GO TO 12
      RGAMMA = -3.
      RETURN
12    IF (BL-.5) 13,15,16
13    BC = 1./BL
      AD = 1./(1.-BL)
```

```
14    RGAMMA = RANDEM(JA)**BC
C     RANDEM IS A USER-SUPPLIED GENERATOR FOR RANDOM OR PSEUDO-
C     RANDOM REAL NUMBERS, 0. .LE. R, R .LT. 1.
      IF (RGAMMA .EQ. 0.) RETURN
      AF = RANDEM(JA)**AD+RGAMMA
      IF (AF .GT. 1.) GO TO 14
      RGAMMA = REXP(JA)*RGAMMA/AF
      RETURN
15    RGAMMA = .5*RNORM(JA)**2
      RETURN
16    IF (BL-1.) 29,17,18
17    RGAMMA = REXP(JA)
      RETURN
18    IF (BL-1.5) 29,19,20
19    RGAMMA = REXP(JA)+.5*RNORM(JA)**2
      RETURN
20    IF (BL-2.) 29,21,29
21    RGAMMA = REXP(JA)+REXP2(JA)
      RETURN
22    RGAMMA = -2.
      RETURN
29    AL = BL
      AD = BL-AA
      AE = AA/SQRT(AL)
      AG = AL*(1.-AC*AE)**3
      AG = AD*ALOG(AG)-AG+.5*(AE-AC)**2
30    AX = RNORM(JA)
      RGAMMA = 1.+AE*(AX-AE)
      IF (RGAMMA .LE. 0.) GO TO 30
      BL = REXP(JA)-AG
      RGAMMA = AL*RGAMMA**3
      IF (RGAMMA .GT. BL+AD*ALOG(RGAMMA)+.5*AX*AX) GO TO 30
      RETURN
         END

      FUNCTION TAN(X)
C     CIRCULAR TANGENT
C     NOT A GENERAL-PURPOSE ROUTINE
C     VALID ONLY FOR X IN RANGE ZERO TO PI
      POL(Y) = (((((.0095168091*Y+.002900525)*Y+.0245650893)*Y+
     A.0533740603)*Y+.1333923995)*Y+.3333314036)*Y+1.
      AA = X
      IA = 1+INT(AA*4./3.14159265)
      GO TO (10,11,11,12,13),IA
10    TAN = AA*POL(AA*AA)
      RETURN
11    AA = 1.5707963267949D0-AA
      TAN = 1./(AA*POL(AA*AA))
      RETURN
12    AA = AA-3.14159265358979D0
      GO TO 10
13    TAN = 0.
      RETURN
         END
```

```
      FUNCTION RANDEM(JA)
C     UNIFORM FLOATING-POINT PSEUDO-RANDOM NUMBER GENERATOR
      DOUBLE PRECISION AA,AB
      DIMENSION AA(17),JA(19),JB(17),JC(4)
      DATA AA,JB/32773D0,32781D0,32789D0,32797D0,32805D0,32813D0,
     A32821D0,32829D0,32837D0,32845D0,32853D0,32861D0,32869D0,
     B32877D0,32885D0,32893D0,32901D0,0,1,3,7,9,15,21,27,31,37,45,
     C55,57,61,69,75,87/
      DO 12 IB = 1,4
        IA = MOD(IABS(JA(18)),17)+1
        JA(18) = IA
        AB = IABS(JA(IA))
10      AB = DMOD((AB+1D0)*AA(IA),2147483648D0)
        JC(IB) = AB
        IF (MOD(JC(IB),8388608).LT. JB(IA)) GO TO 10
        JA(IA) = JC(IB)
        JC(IB) = JC(IB)/8388608
12      CONTINUE
      JC(1) = JC(1)/2
      RANDEM = ((JC(1)*256+JC(2))*256+JC(3))*256+JC(4)
      RANDEM = RANDEM/(32768.*65536.)
13    JA(19) = JA(19)+1
      RETURN
        END
```

REFERENCES

JÖHNK, M. D. Erzeugung von betaverteilten und gammaverteilten Zufallszahlen. *Metrika 8:* 5–15 (1964)

KNUTH, D. E. *The art of computer programming, volume 2/ Seminumerical algorithms.* Addison-Wesley: Reading (Massachusetts) (1969)

NAYLOR, T. H.; J. L. BALINTFY; D. S. BURDICK; K. CHU. *Computer simulation techniques.* John Wiley & Sons: New York (1966)

WILSON, E. B.; M. M. HILFERTY. The distribution of chi-square. *Proc.Nat.Acad.Sci.U.S.A. 17:* 684–688 (1931)

A Long-Period Random Number Generator with Application to Permutations

Robert Salfi, New York

1. RANDOM PERMUTATION ALGORITHMS

The best publicized algorithm for random permutations is that of DURSTENFELD (1964), reproduced in Appendix I. Durstenfeld's algorithm is exact, in the sense that if *random* delivers random real numbers continuously and uniformly distributed on [0,1), *SHUFFLE* delivers equidistributed permutations. If, however, *random* delivers strings of s decimal digits preceded by a decimal point, then the best that *SHUFFLE* can deliver is permutations of n letters with frequency varying from

$$m^{-1}\prod_q 10^{-s}[10^s/q] \text{ to } m^{-1}\prod_q 10^{-s}(1+[10^s/q]),$$

where m is the product of integers less than n of the form 2^a5^b, q runs over the integers from 3 to n and not of the form 2^a5^b, and [] denote the function *entier*. These deviations from equidistribution are negligible for practical purposes: but they suffice to vitiate any chi-square test of the permutations (BERKSON 1938, p. 527).

We omit the rather fussy code required to make Durstenfeld's algorithm exact for discrete input, since an algorithm given by MOSES & OAKFORD (1963, pp. 4-5) and formalized as PERMUT in Appendix II accepts strings (usually short) of random binary digits and delivers equidistributed random permutations without discretization error. KNUTH (1969, p. 125) fails to distinguish between the algorithms of Moses & Oakford and of Durstenfeld.

2. CONGRUENTIAL GENERATORS AND PERMUTATIONS

While it is known that the numbers output by a congruential generator fall mainly in the planes (MARSAGLIA 1968), the rather obvious deduction that such a generator is suspect as a source of random permutations seems to have escaped publication. VERDIER (1969, p. 41) states that, in practice, a congruential pseudo-random integer

> generator is used to make a choice between some number B of alternatives. For this purpose, we form the quantity $[BX_j/P]$, where the square brackets denote integer part.

Verdier undertakes to show

> that relatively short sequences of such quantities may be very highly nonrandom, in the sense that if such a sequence is taken to represent a unit hypercube in a hypercube of side B, many of the possible unit hypercubes can never be selected. That is to say, knowledge of the last k choices produced by the generator gives us considerable information about what the next choice will be, for distressingly small k.

In like manner, for distressingly small B, many permutations of B letters fail to appear.

MAGHSOODLOO (1971), working on an IBM 7040, used a generator RANNUM, not further described, with Durstenfeld's algorithm to yield pseudo-random permutations of n letters, $11 \leq n \leq 20$. It is a reasonable conjecture that RANNUM had a period of length 2^{35}. Now $12! < 2^{35} < 13!$, and so, for $13 \leq n \leq 20$, Maghsoodloo sampled a population of 2^{35} (at most) distinct permutations of n letters. In the light of Verdier's work, that population may be expected to be badly non-random.

To illustrate non-uniform distributions of permutations without having to compute 2^{35} permutations, we ran Durstenfeld's algorithm for permutations of 3, 4, 5, 6, 7, 8 letters over a full cycle of 65536 permutations, using the generator

```
KB = MOD(625*KB+625,65536)                    (1)
```

The results are summarized in Table 1; we conjecture that a generator of period c yields satisfactory permutations of n letters so long as

$$n! < \sqrt{c}.$$

TABLE 1

Frequencies of permutations of n letters: generator of eq. (1)

n	*Expected frequency*	*Observed frequency minimum*	*maximum*
3	10922.67	10910	10936
4	2730.67	2706	2758
5	546.13	512	576
6	91.02	79	101
7	13.00	2	26
8	1.62	0	9

3. THE PROPOSED GENERATOR

The algorithm LEMUR, given in detail in Appendix II, combines two known techniques for improving randomness; the combination, however, appears to be new:

1) Since the low-order bits from a generator using a congruence modulo a power of 2 have short periods (KNUTH 1969, p. 12), we retain only the high-order bits.
2) A long incoherent sequence can be obtained by combining two (or more) shorter sequences with relatively prime lengths. This technique dates from 1918 (KAHN 1967, pp. 397-398) and has been applied to yield pseudo-random numbers (DAHLQUIST 1954, p. 34).

A simple modification of the sieve of Eratosthenes yields, instead of prime numbers, relatively prime numbers. Construct a table of integers a_i, $i = 1, \ldots,$ IB, where IB (say 256) is the number of sequences to be combined; $a_1 = 0$, $a_2 = 1$, and each subsequent a_i is the smallest integer for which

$$2^{27}-a_1,\ 2^{27}-a_2,\ \ldots,\ 2^{27}-a_{i-1},\ 2^{27}-a_i \tag{2}$$

are relatively prime. Run IB congruential generators, each with modulus (and period) 2^{34}, and with different multipliers (to ensure that the output of the generators will not lock in step in the short run). At each call to LEMUR an index variable, cycling from 1 to IB, determines which generator is used. Represent the output of generator i as $2^{27}b_i+c_i$, $0 \leq b_i < 2^7$, $0 \leq c_i < 2^{27}$. If $c_i \geq a_i$, deliver b_i as the output of LEMUR; otherwise, continue to run generator i until $c_i \geq a_i$. The output sequence of 7-bit pseudo-random numbers has period

$$\mathrm{IB}\prod_{i=1}^{\mathrm{IB}}(2^{27}-a_i) > 2^{27\times \mathrm{IB}}. \tag{3}$$

Already for IB = 2 the period $2^{35}(2^{27}-1)$ is so long as to make the notion of a full cycle illusory: a computer running LEMUR with IB = 2 without intermission and delivering a 7-bit number every microsecond would first repeat after 146138 years. The reason for using a large value of IB (we have a working version of LEMUR with IB = 1024) is to guarantee that any specified permutation will occur somewhere in the cycle.

4. NOTES ON THE FORTRAN PROGRAMS

The programs in Appendix II conform to the ISO FORTRAN recommendations; PERMUT and TIGER are independent of word size. TIGER exploits the same algorithm as PERMUT, but accepts pseudo-random integers of fixed length 7 bits. LEMUR is so far machine-independent that it will yield the same numbers on any computer that accommodates positive integers of length 27 bits or more, and allots 53 bits or more to the mantissa of double-precision numbers.

The program LEMUR was converted, following Professor Herbach's valuable suggestion (GREENWOOD 1974), from a program originally written in machine-dependent FORTRAN for the UNIVAC 1108. That program, in which arrays AA and AB were of type INTEGER, exploited the following properties of UNIVAC 1100 computers: a single word holds a 35-bit positive integer; when the result of integer multiplication esceeds 35 bits, the low-order 35 bits of the product are correctly stored; remaindering modulo 2^p is effected by applying a mask with the p low-order bits set and the $36-p$ high-order bits cleared; division by 2^p is effected by shifting. Working versions of LEMUR should be written in machine-dependent language; in particular, if using a machine with word size 32 bits, it is preferable to generate 30-bit pseudo-random numbers and partition them into 7 high-order bits and 23 low-order bits.

To increase IB beyond 256 change statement 14 of LEMUR to the new IB; change the dimensions of AA, AB, JA in statements 10, 11, 12 of LEMUR and statement 10 of TIGER to the new IB; change ID in statement 16 of TIGER to an integer such that $(\mathrm{ID}!)^2 < 2^{27\times \mathrm{IB}}$; change the dimensions of JB, JC in statement 10 of LEMUR and in statements 10 and 15 of TIGER to the new ID; add DATA statements for arrays containing additional integers a_i, defined in eq. (2);

add EQUIVALENCE statements that will fill array JA with the integers a_i.

LEMUR must be initialized by the statement

```
IG = LEMUR(0)
```

It is advised to initialize AA by storing a random number (11 decimal digits) in each element. These random numbers should be saved if a subsequent computation requires that LEMUR produce the same sequence of pseudo-random numbers.

By concatenating the results of p successive calls to LEMUR, $(7p)$-bit pseudo-random integers can be generated; they should be excellent for $p \leq$ IB and satisfactory for $p \leq 2\times$IB.

Before generating permutations, initialize TIGER by the call

```
CALL TIGER(1)
```

This call initializes arrays JC of TIGER and AB of LEMUR, and incidentally, if array AA had been cleared to zeroes when loading the programs and has not been subsequently initialized, operates each congruential generator in LEMUR once. Then store the N letters to be permuted in the first N locations of JB. On return from TIGER, location IE contains N if a permutation was generated and zero if no permutation was generated.

5. TEST OF THE PERMUTATIONS PRODUCED

For the reason given at the end of Section 3, to attempt a comprehensive test of LEMUR would be to give substance to the canard denied by KNUTH (1969, p. 66) „that more computer time has been spent testing random numbers than using them in applications!" Since the guaranty of the permutations of TIGER is the independence of successive calls to LEMUR, it suffices to run a short test of TIGER to verify that the contents of JB are being permuted. For the sake of comparison with the short-cycle generator of Table 1, 65536 permutations each of 3, 4, 5, 6, 7, 8 letters were generated; the results are summarized in Table 2.

TABLE 2

Frequencies of permutations of *n* letters: Long-period generator

n	*Expected frequency*	*Observed frequency minimum*	*maximum*	2	*d.f.*
3	10922.67	10832	11018	1.96	5
4	2730.67	2661	2854	19.35	23
5	546.13	481	620	130.65	119
6	91.02	65	127	715.85	719
7	13.00	3	29	4945.12	5039
8	1.63	0	9	40569.78	40319

APPENDIX I. DURSTENFELD'S ALGORITHM

```
procedure SHUFFLE(a, n, random);
  value n; integer n; real procedure random; integer array a;
begin
  integer i, j; real b;
  for i := n step -1 until 2 do
    begin j := entier(i×random+1);
          b := a[i]; a[i] := a[j]; a[j] := b
    end   loop i
end SHUFFLE
```

APPENDIX II. FORTRAN PROGRAMS

```
      SUBROUTINE PERMUT (N1,N2,ARRAY)
      DIMENSION ARRAY(512),JA(512)
      DATA JA/0,1,2*2,4*3,8*4,16*5,32*6,64*7,128*8,256*9/
      IA = N1
      N2 = 0
      IF (IA .LT. 2) RETURN
      IF (IA .GT. 512) RETURN
      DO 12 IB = 2,IA
10      IC = IBIS(JA(IB))
C       FUNCTION IBIS(N) IS A USER-SUPPLIED ROUTINE THAT DELIVERS
C       RANDOM OR PSEUDO-RANDOM INTEGERS OF LENGTH N BITS
        IF (IC-IB+1) 11,12,10
11      AA = ARRAY(IB)
        ARRAY(IB) = ARRAY(IC+1)
        ARRAY(IC+1) = AA
12      CONTINUE
      N2 = IA
      RETURN
        END

      FUNCTION LEMUR(I)
C     SEVEN-BIT LONG-CYCLE PSEUDO-RANDOM INTEGER GENERATOR
      DOUBLE PRECISION AA,AB
10    COMMON /ZOO/ IA,AA(256),JB(465),IE
11    DIMENSION AB(256),JA(256)
12    DIMENSION KA(256)
13    EQUIVALENCE (KA,JA)
14    DATA IB/256/
```

```
      DATA KA/1039,1069,1077,1081,1095,1099,1105,1119,1125,1131,0,
     A1141,1155,1161,1179,1185,1225,1237,1255,1257,1267,1281,1171,
     B1309,1335,1339,1357,1365,1369,1377,1381,1395,1467,1479,1489,
     C1491,1497,1501,1509,1537,1539,1549,1551,1561,1587,1599,1605,
     D1609,1615,1617,1627,1651,1669,1675,1677,1687,1689,1707,1717,
     E1719,1729,1741,1795,1797,1839,1845,1869,1879,1881,1885,1887,
     F1897,1909,1911,1915,1927,1951,1977,1981,2001,2005,2007,2025,
     G2035,2047,2085,2095,2121,2139,2145,2149,2151,2155,2169,2175,
     H2187,2209,2211,2221,2245,2265,2271,2281,2287,2299,2307,2331,
     I2341,2349,2365,2379,2415,2419,2421,2455,2467,2475,2487,2497,
     J2499,2529,2587,2607,2617,2635,2671,2679,2685,2709,2715,2719,
     K2721,2727,2739,2749,2755,2761,2797,2821,2835,2839,2841,2851,
     L2865,2877,2895,2901,2919,2931,2937,2947,2961,2965,2967,2979,
     M105,111,115,129,135,145,165,175,181,187,199,201,207,217,219,
     N231,235,241,259,261,277,279,289,291,319,325,327,357,361,367,
     O375,397,405,427,429,451,471,481,507,517,525,529,549,555,559,
     P565,571,577,597,615,621,625,639,649,651,657,679,681,685,727,
     Q741,751,781,787,789,795,817,829,847,859,861,867,889,891,895,
     R901,921,927,937,945,951,955,969,991,999,19,27,31,39,45,49,1,
     S61,67,69,79,81,91,3,7,9/
      IF (I .NE. 0) GO TO 21
      DO 20 IC = 1,IB
        AB(IC) = 8*IC+262141
20      CONTINUE
21    IA = MOD(IABS(IA),IB)+1
22    AA(IA) = DMOD(1D0+DABS(AA(IA))*AB(IA),17179869184D0)
      IC = DMOD(AA(IA),134217728D0)
      IF (IC .LT. JA(IA)) GO TO 22
      LEMUR = AA(IA)/134217728D0
      RETURN
         END

      SUBROUTINE TIGER(N)
C     PSEUDO-RANDOM PERMUTATION GENERATOR
      DOUBLE PRECISION AA
10    COMMON /ZOO/ IA,AA(256),JB(465),IE
C     JB IS THE ARRAY TO BE PERMUTED
15    DIMENSION JC(465)
16    DATA ID/465/
      IG = N
      IE = 0
      IF (IG .LE. 0) RETURN
      IF (IG .GT. ID) RETURN
      IF (IG .NE. 1) GO TO 25
      DO 23 IG = 2,128
        JC(IG) = MOD(128,IG)
23      CONTINUE
      DO 24 IG = 129,ID
        JC(IG) = MOD(16384,IG)
24      CONTINUE
      IG = LEMUR(0)
C     LEMUR IS A 7-BIT LONG-CYCLE PSEUDO-RANDOM INTEGER GENERATOR
      IG = 3
25    IE = IG
      DO 27 IG = 2,IE
```

```
26        IH = LEMUR(IG)
          IF (IG .GT. 128) IH = 128*IH+LEMUR(IG)
          IF (IH .LT. JC(IG)) GO TO 26
          IH = MOD(IH,IG)
          II = JB(IH+1)
          JB(IH+1) = JB(IG)
          JB(IG) = II
27        CONTINUE
        RETURN
          END
```

REFERENCES

BERKSON, J. Some difficulties of interpretation encountered in the application of the chi-square test. *J.Amer.Statist.Assoc. 33,* 526-536 (1938)

DAHLQUIST, G. Monte Carlo-metoden. *Nordisk Mat.Tidskr. 2,* 27-43 (1954)

DURSTENFELD, R. Algorithm 235: random permutations. *Comm. ACM 7,* 420 (1964)

GREENWOOD, J. A. A fast generator for gamma-distributed random numbers. Contributed to this symposium (1974)

KAHN, D. *The codebreakers.* Macmillan: New York (1967)

KNUTH, D. E. *The art of computer programming, volume 2/ Seminumerical algorithms.* Addison-Wesley: Reading (Massachusetts) (1969)

MAGHSOODLOO, S. An investigation by high speed sampling of the frequency distribution of rank correlation. *Computing 8,* 1-12 (1971)

MARSAGLIA, G. Random numbers fall mainly in the planes. *Proc.Nat. Acad.Sci.U.S.A. 61,* 25-28 (1968)

MOSES, L. E.; R. V. OAKFORD. *Tables of random permutations.* Stanford Univ. Press: Stanford (California) (1963)

VERDIER, P. H. Relations within sequences of congruential pseudo-random numbers. *J.Res.Nat.Bur.Standards 73B,* 41-44 (1969)

Testing Simple Hypotheses by a Monte Carlo Method with Sequential Decision Procedure

M. Wall and A. Meystre, Basle

1. INTRODUCTION

Let R be a sample space with probability measure μ , t a real statistic on R and let V be a critical region of a test of the simple hypothesis: $\mu = \mu_0$. V is supposed to be of the form:

$V = \{r \in R: t(r) < t_\varepsilon\}$, t_ε a real defined by the desired significance-level .

Often, Monte Carlo methods are applied to determine t_ε . It is possible, however, to perform an approximate test with the aid of a Monte Carlo method without determining t_ε . The following procedure has been proposed by M. Dwass [1957] : in order to find out whether a given sample $r' \in R$ is an element of V, generate a number of elements r in R distributed according to μ_0. Acceptance or rejection of the hypothesis can then be done based on the number NLESS of elements for which $t(r) < t(r')$.

The number of random samples ensuring a sufficiently accurate decision can be lowered considerably, if a sequential decision procedure is applied: After each sample generation, examine if the data up to that point permit a decision. If not, generate the next sample. In what follows, this test will be called "Sequential Monte Carlo Test" and the corresponding test based on exact knowledge of t_ε will be called "exact test".

The present paper concerns with this kind of hypothesis testing. As a decision procedure, a truncated modification of the Sequential Probability Ratio Test of A. Wald [1947] has been applied. For some sets of decision-parameters, the operating characteristic and the expected number of samples necessary for a decision (average sample number function, ASN function) were determined as a function of

$$p = \mu_0(\{r \in R: t(r) < t(r')\}) \quad (1)$$

Some of the calculations could be performed by known approximations, but for the present paper, exact methods appeared preferable.

The operating characteristic informs about the probability for obtaining a "wrong result" when applying the Sequential Monte Carlo Test, i.e. a result opposite to that of the exact test.

2. THE DECISION PROCEDURE

Let K be the number of random samples r generated and NLESS the number of random samples generated for which $t(r) < t(r')$. The Sequential Probability Ratio Test of A. Wald [1947] works as follows.

if $NLESS - c \cdot k \leq -a$	accept the hypothesis
if $NLESS - c \cdot k \geq b$	reject the hypothesis
if $-a < NLESS - c \cdot k < b$	generate the next random sample.

a, b and c are preassigned positive real numbers. The number of steps necessary for a decision is finite with probability = 1 WALD, 1947]. If we want to reach a decision after at most NTRUNC steps, we are lead to the following truncated version:

if $K < NTRUNC$, proceed as in the non-truncated procedure
if $K = NTRUNC$ and $NLESS - c \cdot k \leq 0$, accept the hypothesis
if $K = NTRUNC$ and $NLESS - c\ k > 0$, reject the hypothesis.

For the present paper, computations are done only for the truncated version. For the non-truncated version ($NTRUNC = \infty$), exact computations are possible [WALD, 1947] . If NTRUNC is large, however, the influence of truncation can be neglected. In the following, a, b, c and NTRUNC will be called "decision-parameters".

3. SOME CHARACTERISTICS OF THE SEQUENTIAL MONTE CARLO TEST

For given sets of decision-parameters a, b, c and NTRUNC, we shall consider the following characteristics of the Sequential Monte Carlo Test:

- $L(p)$ the operating characteristic
- $ASN(p)$ the average sample number function
- ε the level of significance
- D_1 and D_2 two measures of the difference between the Sequential Monte Carlo Test and the corresponding exact test.

For a given p (being defined as in section 1., L(p) denotes the probability to accept the hypothesis (the operating characteristic of the exact test equals one ore zero always).

The operating characteristic is a decreasing function of p [LEHMANN, 1955] . Further, L(p) is easily seen to be continuous.

$1-\varepsilon$ is the probability to accept the hypothesis if the hypothesis is true. Let F be the cumulative distribution function of the statistic t under the hypothesis (thus, $p = F(t(r'))$). It is easily verified that

$$1-\varepsilon = \int_0^1 L(p)\,dp \tag{2}$$

if F is continuous,

$$1-\varepsilon' = F(t_1) + \sum_{i=1}^{m-1} L(F(t_i))\{F(t_{i+1})-F(t_i)\} \tag{3}$$

if $t_1, t_2 \ldots . t_s$ are the only possible values of t. Notice that

$$\int_0^1 L(p)\,dp \leq F(t_1) + \sum_{i=1}^{m-1} L(F(t_i))\{F(t_{i+1})-F(t_i)\} \tag{4}$$

since L(p) is decreasing. Thus, if the same decision parameters are applied, the level of significance in the discrete case will generally be lower than in the continuous case. In practice, $F(t_i)$ will be unknown (if not, the exact test can be applied). Therefore, (4) must be used as an approximation in the discrete case. The test is conservative then. The approximation is good if the probabilities $F(t_{i+1})-F(t_i)$ are small. Generally, this will be the case if the number m of possible t-values is large.

The measures D_1 and D_2 of the difference between the Sequential Monte Carlo Test and the corresponding exact test with equal level of significance are defined as follows: Let

$$d(p) = \begin{cases} 1-L(p) & \text{if } p < 1-\varepsilon \\ L(p) & \text{if } p > 1-\varepsilon \end{cases} \tag{5}$$

For given p, d(p) is the probability to obtain a result opposite to that of the exact test, if the Sequential Monte Carlo Test is applied. Let

$$D_1 = \int_0^1 d(p) \tag{6}$$

$$D_2 = \int_0^1 d(p) \cdot |1-\varepsilon-p|\,dp \tag{7}$$

D_1 is the expectation of d(p). In D_2 , the "wrong decisions" are weighted in an obvious manner (see fig. 1).

4. COMPUTATIONS

Values of the characteristics considered in the previous section were computed for the following sets of decision-parameters:

1. $a = 10.83 \quad b = 11.53 \quad c = 0.95$
2. $a = 21.73 \quad b = 22.43 \quad c = 0.95$
3. $a = 11.245 \quad b = 12.065 \quad c = 0.99$

with several values of NTRUNC.

For sufficiently large values of NTRUNC, these parameter-sets provide Sequential Monte Carlo Tests with the following properties:

$$\int_0^1 L(p)\,dp = \begin{cases} 0.95 \text{ for sets 1. and 2} \\ 0.99 \text{ for set 3.} \end{cases} \tag{8}$$

$$L(0.94) = 0.99 \text{ for set 1.} \tag{9}$$

$$L(0.945) = 0.99 \text{ for set 2.} \tag{10}$$

$$L(0.948) = 0.99 \text{ for set 3.} \tag{11}$$

Thus, both set one and two provide tests with a significance level of 0.05, but the second test yields results closer to those of the corresponding exact test than the first one.

The computations were performed by means of a simple recursive procedure. Consider the following situations:

(ACC,K) : The hypothesis is accepted after generation of at most K random samples ($K \leq$ NTRUNC).

(REJ,K) : The hypothesis is rejected after generation of at most K random samples.

(ℓ,K) : NLESS = ℓ after generation of K random samples ($-a < \ell - c \cdot K < b$).

Denote by Pr(ACC,K), Pr(REJ,K), Pr(ℓ,K) the corresponding probabilities, supposed to be known for a given p = probability that $t(r) < t(r')$. Then Pr(ACC,K+1), Pr(REJ,K+1), Pr(ℓ,K+1) ($-a < \ell - c \cdot k < b$) can be computed in an obvious manner. Thus, recursive computations can be done, beginning at K=1 . From the results, the characteristics considered in the previous section are easily obtained.

This procedure is simple, but rather time-consuming if NTRUNC is large.

Suppose, however, that the decision parameter c is rational:

$$c = \frac{L}{M}$$

where L and $M > 1$ are integers. It is possible then, to calculate the recursion in steps $K \rightarrow K+M$ instead of $K \rightarrow K+1$, which saves considerable computational work. Here, the principle will only be outlined.

Let ℓ_i (i=1,2 ...s) be the s possible values of NLESS after generation of K random samples (see Fig. 2). Denote the corresponding probabilities by $Pr(\ell_i, K)$. After generation of K+M random samples, the possible values of NLESS are $\ell_i + \Delta$ (i=1,2 ...s), where $\Delta = M \cdot c$. Denote the vector $[Pr(\ell_1, K), \ldots, Pr(\ell_s, K)]$ by p(K) and the vector $[Pr(\ell_1 + \Delta, K+M), \ldots, Pr(\ell_s + \Delta, K+M)]$ by p(K+M). Then

$$p(K+M) = W_M \cdot p(K)$$

where W_M is a (sxs)-matrix. The elements of W_M are conditional probabilities, whose exact meaning is readily understood. Thus, p(K+M) can be computed directly from p(K), if W_M is known. The essential point is that the same matrix W_M can be applied in all recursion steps:

$$p(K+i \cdot M) = W_M^{\,i} \cdot p(K) \tag{12}$$

Further,

$$Pr(ACC, K+M) = Pr(ACC, K) + u_M \cdot p(K)^T \tag{13}$$

$$Pr(REJ, K+M) = Pr(REJ, K) + v_M \cdot p(K)^T \tag{14}$$

where $p(K)^T$ is the transposed of p(K) and u_M, v_M are row-vectors whose elements are also conditional probabilities. Here again, the same vectors u_M, v_M can be applied in each recursion step.

The elements of W_K, u_K and v_K must be computed by single-step recursion $K \rightarrow K+1$. For each element, M steps are necessary.

The expected number of necessary random samples can be computed in a similar, although somewhat more complicated way. The formulas needed for the calculation can be deduced in a straightforward manner and will not be cited here.

This procedure may be extended in the following manner: Compute W_M, u_M, v_M by single-step recursion $K \rightarrow K+1$. With these results, compute $W_{N \cdot M}$, $u_{N \cdot M}$ and $v_{N \cdot M}$ by recursion $K \rightarrow K+M$, whereby N is an optional integer (N recursion-steps are necessary for each element). Finally, using the new results, compute the acceptance and rejection probabilities and the ASN function by recursion steps $K \rightarrow K+N \cdot M$.

The computations were performed on an IBM 370/155 by means of a programm written in FORTRAN. For most steps, double precision was applied. The operating characteristic L(p) and the ASN function were determined for equidistant p-values and several NTRUNC-values. The step-width of the p-values was 0.002 for decision-parameter set 1, 0.001 for set 2 and 0.0004 for set 3. The integrations to get the values of $1-\varepsilon$, D_1 and D_2 were calculated with Simpson's rule. Some of the results are listed in Tab. 1-4.

5. CONCLUDING REMARKS

The Sequential Monte Carlo Test makes possible a decision with correct level of significance without any need of knowledge about the distribution of the test statistic. Clearly, the method is only useful if the exact test cannot be applied. This will be the case if the assumptions to use tables of critical values are not clearly fulfilled (e.g. to many ties in the case of rank order tests; not enough experimental data if only the asymptotic distribution of the statistic is tabulated; missing values in the case of multivariate comparisons), or if tables are not available (e.g. tests based on permutations of the original values; ad hoc-constructed tests for special problems).

Practically, the use of the method is limited by the possibly large amount of computation time necessary for a decision. On the one hand, the computation time depends on the time needed to generate a random sample which is given by the structure of the sample space and can be lowered by appropriate programming. On the other hand it depends on the choice of decision-parameters and, to a large extent, on the actual value of p. The expected number of random samples to be generated (i.e. the ASN function) is large if 1-p is close to ε, the level of significance (which is given by the decision-parameters).

The random generator needed for the sample generation should be selected carefully. In our applications, we have used the subroutine RANDU of the Scientific Subroutine Package of IBM [IBM,H20-0205-3] , which has well known properties [IBM, GC20-8011-0] .

The special decision procedure considered in this paper, i.e. the Sequential Probability Ratio Test of A. Wald, has been applied, because it is simple and well known. There is one case in which this test is optimal (relative to the expected number of random samples), namely if a decision is to be made for one of the two hypothesis H_0: $p = p_0$ and H_1 : $p = p_1$ [WALD, 1947 and Lehmann, 1959].

But, obviously, we are concerned with a different set of hypotheses in our application. It is certainly possible, therefore, to find a sequential decision procedure more appropriate to a Sequential Monte Carlo Test. A sequential decision optimal relative to some optimality criteria of practical interest should be based on some apriori-assumptions or-knowledge about p.

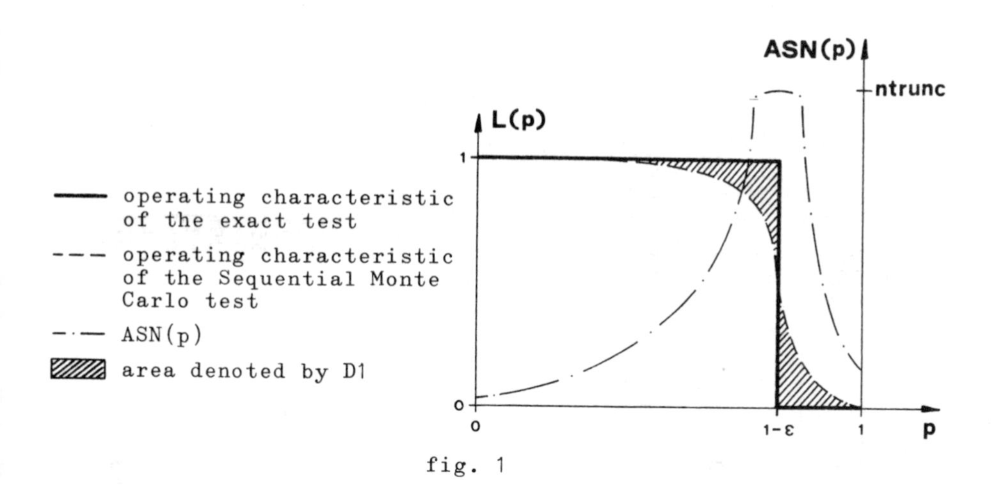
ASN(p)
ntrunc
L(p)
1
0
0
1-ε
1
p
operating characteristic of the exact test
operating characteristic of the Sequential Monte Carlo test
ASN(p)
area denoted by D1

fig. 1

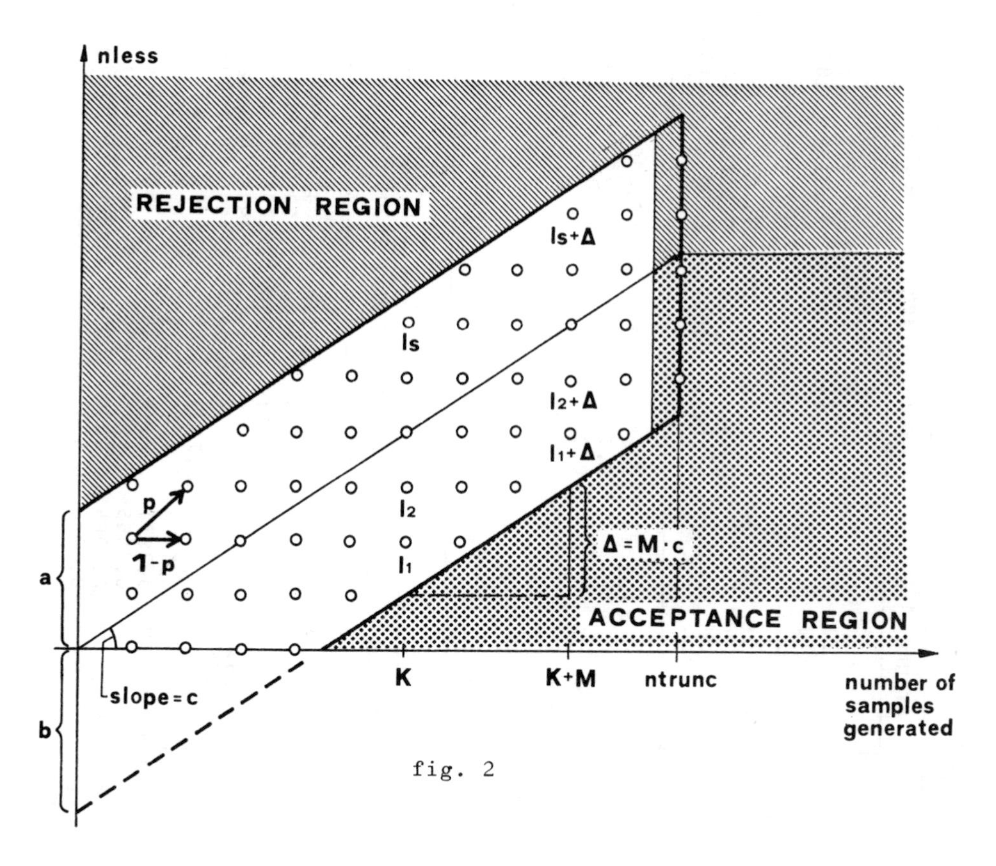
nless
REJECTION REGION
ACCEPTANCE REGION
Is+Δ
Is
I2+Δ
I1+Δ
I2
I1
Δ = M · c
p
1-p
a
b
slope = c
K
K+M
ntrunc
number of samples generated

fig. 2

TAB 1

	NTRUNC	500	1000	2000	5000	10000	30000	100000
SET 1	$100 \cdot \int_0^1 L(p) \cdot dp$	95.0099	95.0049	95.0025	95.0000	95.0000	94.9999	94.9999
	$100 \cdot D_1$	0.77417	0.54906	0.39655	0.30360	0.29031	0.28949	0.28949
	$100 \cdot D_2$	0.00472	0.00237	0.00124	0.00077	0.00072	0.00072	0.00072
SET 2	$100 \cdot \int_0^1 L(p) \cdot dp$	95.0100	95.0045	95.0024	95.0011	95.0006	95.0001	95.0000
	$100 \cdot D_1$	0.77407	0.54869	0.38845	0.24654	0.18253	0.14812	0.14771
	$100 \cdot D_2$	0.00472	0.00237	0.00119	0.00048	0.00027	0.00019	0.00019
SET 3	$100 \cdot \int_0^1 L(p) \cdot dp$	99.0022	99.0010	99.0005	99.0002	99.0003	99.0003	99.0003
	$100 \cdot D_1$	0.34851	0.24878	0.17670	0.11211	0.08073	0.05970	0.05791
	$100 \cdot D_2$	0.00098	0.00049	0.00025	0.00010	0.00005	0.00003	0.00003

TAB 2 DECISION-PARAMETER SET 1

	L(P)						ASN(P)					
P \ NTRUNC	500	1000	2000	5000	10000	≥25000	500	1000	2000	5000	10000	≥25000
0.5	1.0000	1.0000	1.0000	1.0000	1.0000	1.0000	25	25	25	25	25	25
0.9	1.0000	1.0000	1.0000	1.0000	1.0000	1.0000	223	224	224	224	224	224
0.92	0.9967	0.9999	1.0000	1.0000	1.0000	1.0000	344	372	373	373	373	373
0.93	0.9720	0.9960	0.9998	0.9998	0.9998	0.9998	418	536	558	558	558	558
0.94	0.8506	0.9220	0.9731	0.9897	0.9900	0.9900	4[illegible]4	779	1015	1092	1093	1093
0.946	0.6816	0.7313	0.7949	0.8590	0.8705	0.8711	490	899	1431	1958	2052	2057
0.95	0.5287	0.5203	0.5146	0.5102	0.5090	0.5088	495	937	1585	2424	2682	2711
0.954	0.3635	0.2931	0.2145	0.1411	0.1288	0.1282	497	932	1551	2062	2156	2160
0.96	0.1522	0.0665	0.0176	0.0065	0.0064	0.0064	493	841	1083	1140	1140	1140
0.97	0.0101	0.0004	0.0000	0.0000	0.0000	0.0000	460	571	577	577	577	577
1.0	0.0000	0.0000	0.0000	0.0000	0.0000	0.0000	231	231	231	231	231	231

TAB 3 DECISION-PARAMETER SET 2

NTRUNC / P	L(P) 1000	2000	5000	10000	30000	90000	ASN(P) 1000	2000	5000	10000	30000	90000
0.5	1.0000	1.0000	1.0000	1.0000	1.0000	1.0000	49	49	49	49	49	49
0.9	1.0000	1.0000	1.0000	1.0000	1.0000	1.0000	442	442	442	442	442	442
0.93	0.9960	0.9999	1.0000	1.0000	1.0000	1.0000	889	1092	1134	1134	1134	1134
0.94	0.9221	0.9757	0.9989	0.9999	0.9999	0.9999	984	1691	2181	2205	2205	2205
0.945	0.7747	0.8488	0.9442	0.9826	0.9899	0.9899	997	1907	3496	4189	4321	4322
0.948	0.6321	0.6705	0.7460	0.8122	0.8646	0.8668	999	1965	4258	6349	7993	8064
0.95	0.5203	0.5143	0.5091	0.5066	0.5047	0.5045	1000	1980	4472	7129	10114	10422
0.952	0.4040	0.3521	0.2635	0.1909	0.1353	0.1330	1000	1979	4351	6519	8198	8266
0.955	0.2424	0.1528	0.0497	0.0138	0.0082	0.0082	1000	1949	3636	4311	4417	4417
0.96	0.0663	0.0153	0.0003	0.0000	0.0000	0.0000	997	1775	2234	2245	2245	2245
1.0	0.0000	0.0000	0.0000	0.0000	0.0000	0.0000	449	449	449	449	449	449

TAB 4 DECISION-PARAMETER SET 3

NTRUNC / P	L(P) 1000	2000	5000	10000	30000	120000	ASN(P) 1000	2000	5000	10000	30000	120000
0.5	1.0000	1.0000	1.0000	1.0000	1.0000	1.0000	24	24	24	24	24	24
0.98	0.9953	0.9998	1.0000	1.0000	1.0000	1.0000	895	1141	1162	1162	1162	1162
0.984	0.9579	0.9911	0.9999	1.0000	1.0000	1.0000	975	1608	1928	1934	1934	1934
0.986	0.8923	0.9533	0.9951	0.9997	0.9998	0.9998	992	1821	2763	2894	2898	2898
0.988	0.7592	0.8213	0.9168	0.9706	0.9899	0.9900	998	1948	3949	5220	5673	5675
0.9892	0.6385	0.6649	0.7263	0.7903	0.8640	0.8717	999	1982	4525	7272	10363	10684
0.99	0.5427	0.5302	0.5192	0.5142	0.5110	0.5104	1000	1992	4710	8034	13059	14117
0.9908	0.4390	0.3846	0.2961	0.2184	0.1368	0.1290	1000	1996	4696	7676	10959	11275
0.992	0.2829	0.1869	0.0697	0.0190	0.0065	0.0065	1000	1995	4286	5616	5958	5959
0.994	0.0833	0.0209	0.0005	0.0000	0.0000	0.0000	1000	1965	2973	3017	3017	3017
1.0	0.0000	0.0000	0.0000	0.0000	0.0000	0.0000	1000	1207	1207	1207	1207	1207

REFERENCES

Dwass, M.: Modified Randomization Tests
for Nonparametric Hypothesis.
Ann. Math. Stat. 28, 181-187, 1957

Random Number Generation and Testing.
IBM Data Processing Techniques, GC20-8011-0

System /360 Scientific Soubroutine Package, Version III.
IBM Application Programm, H20-0205-3

Lehmann, E.L.: Ordered Families of Distributions.
Ann. Math. Stat. 26, 399-419, 1955

Lehmann, E.L.: Testing Statistical Hypothesis.
John Wiley and Sons, Inc., New York, 1959

Wald, A.: Sequential Analysis.
John Wiley and Sons, Inc., New York, 1947

Monte Carlo Evaluation of the Distributions of the First Three Moments of Small Samples for Some Well-Known Distributions

James R. Wallis, Pisa, N. C. Matalas, and J. R. Slack, Reston

ABSTRACT

A frequent problem in hydrology is to estimate the first three moments of hydrologic records (the mean, standard deviation and coefficient of skewness of flood sequences or annual streamflows, for example) based upon only short observed sequences that often appear to be non-Gaussian. The small sample properties of the moments of most distributions are analytically intractable; however, Monte Carlo simulation on a high speed computer allows one to obtain approximations of these properties which are probably sufficiently accurate for most operational purposes.

In this study sample sizes of 10(10)90 were considered and the distributions used were the Normal, Gumbel, Log-normal, Pearson Type III, Weibull, and Pareto Type I. Values of the coefficient of skewness used in the sample were in the range [0.0, 15.0]. Pronounced skews, biases and constraints were observed, and comparison of these Monte Carlo results with the observed records from 2,000 U.S. streamflow stations show that the observed set of records are unlikely to have resulted from any of the above pure distributions.

Introduction:

Decision variables pertaining to the design of water resource systems are functions of various parameters, including those that characterize the stochastic properties of hydrologic inputs to the systems. Because hydrologic sequences are of finite lengths, only estimates of the hydrologic parameter values can be obtained, and as a result, uncertainty in the design decisions is in part attributed to hydrologic uncertainty. The particular set of hydrologic parameters to be estimated depends upon the purposes and objectives underlying the proposed development of a water resource system. In general, the set is likely to include those parameters which are defined in terms of the low order moments, namely the mean, standard deviation and coefficient of skewness.

Generally a hydrologic sequence of length n is assumed to be a sample on n identically distributed random variables, where each random variable has a finite mean, μ, standard deviation, σ, and coefficient of skewness, γ. Moreover, the n random variables ordered in time is assumed to be a stationary stochastic process. Consequently, estimates of μ, σ, and γ, denoted by $\overline{X}$, S, and G, may be defined in terms of time averages of the n observations forming the hydrologic sequence. The sampling properties of $\overline{X}$, S, and G as functions of n depend upon 1) the method of estimation, 2) the marginal probability distribution function, and 3) the type of generating mechanism of the stochastic process.

To gain some insight as to the sampling properties of $\overline{X}$, S, and G, Monte Carlo experiments were carried out on the basis of the method of moments for estimating μ, σ, and γ, under the assumption that the stochastic process is purely independent, such that the n random variables are independent as well as identical. The experiments were performed on each of several distribution functions for various values of n and γ. From these experiments, the probability distribution functions of $\overline{X}$, S, and G were defined emperically, and values of the mean, standard deviation, coefficient of skewness, and coefficient of kurtosis for $\overline{X}$, S, and G were calculated. In this paper only a brief discussion of the full Monte Carlo results are presented and the interested reader is referred to the paper by Wallis, Matalas, and Slack (1974) for fuller details.

Experimental design:

Six probability distribution functions were considered: 1) Normal (N), 2) 3-Parameter Log-Normal (LN), 3) 3-Parameter Pearson Type III (PIII), 4) Gumbel, Extreme Value Type I (GI), 5) Weibull (W), and 6) Pareto (PIV). For each distribution function, F(y), 100,000 samples of size n = 10(10)90 were generated with $\mu = 0$, $\sigma = 1$, and γ equal to the values indicated by an asterisk in Table 1.

Table 1

Values of γ for distribution function F

F(y) \ γ	0	1/4	1/2	$\sqrt{1/2}$	1	1.14	$\sqrt{2}$	2	3	4	5	10	15
N	*												
LN		*	*	*	*	*	*	*	*	*	*	*	*
P III		*	*	*	*	*	*	*	*	*	*		
G I						*							
W		*	*	*	*	*	*	*	*	*	*	*	*
P IV									*	*	*	*	*

In general, let Y_{ij} denote the i-th "observation" for the j-th sequence of length n. For the j-th sequence, the mean $\overline{Y}_j$, standard deviation S_j, and coefficient of skewness, G_j, were defined as follows:

$$\overline{Y}_j = \sum_{i=1}^{n} Y_{ij} / n \tag{1}$$

$$S_j = [\sum_{i=1}^{n} Y_{ij}^2 / n - \overline{Y}_j^2]^{\frac{1}{2}} \tag{2}$$

$$G_j = [\sum_{i=1}^{n} Y_{ij}^3 / n - 3 \overline{Y}_j S_j^2 - Y_j^3] / S_j^3 \tag{3}$$

Let X_j, denoting any one of the above three statistics, be an observation on the random variable X having mean $\mu(X)$, standard deviation $\sigma(X)$, and coefficient of skewness $\gamma(X)$.

$$\mu(X) = E(X) \quad (4)$$

$$\sigma(X) = \{E[X - E(X)]^2\}^{\frac{1}{2}} \quad (5)$$

$$\gamma(X) = E[X - E(X)]^3\} / \sigma^3(X) \quad (6)$$

From the M = 100,000 "observations" on X, the values of $\mu(X)$, $\sigma(X)$, and $\gamma(X)$, were approximated by

$$\tilde{\mu}(X) = \sum_{j=1}^{M} X_j / M \quad (7)$$

$$\tilde{\sigma}(X) = [\sum_{j=1}^{M} X_j^2 / M - \tilde{\mu}^2(X)]^{\frac{1}{2}} \quad (8)$$

$$\tilde{\gamma}(X) = \sum_{j=1}^{M} X_j^3 / M - 3\tilde{\mu}(X) \quad \tilde{\sigma}^2(X) - \tilde{\mu}^3(X)] / \tilde{\sigma}^3(X) \quad (9)$$

The distribution function $F(x) = P[X < x]$ was defined as follows. For any one of the statistics X, a range $R \equiv [R_X^-, R_X^+]$ was selected as needed, such that $P[R_X^- \leq x \leq R_X^+] < 10^{-4}$. The range R was divided into 400 uniform intervals

$$\{I_k = [\frac{k-1}{400} R_X^+ + \frac{401-k}{400} R_X^-, \frac{k}{400} R_X^+ + \frac{400-k}{400} R_X^-) \mid k=1,\ldots,400\} \quad (10)$$

For each I_k a count C_k was determined of the occurrences of the statistic X_{ij} falling in that interval, whereby, the distribution function $F(x) = P[X < x]$ was defined as $F(x) = \sum_{k|I_k' < x} C_k/100{,}000$ where I_k' is the lower end-point of I_k.

<u>Sequence generating algorithms</u>:

To generate samples from each of the distributions noted above, it is necessary to devise algorithms for the generation of appropriately distributed pseudo-random numbers. The generation of such numbers is at best an art. It whould be noted that a number in a digital computer has a very specific form. Both the range of magnitude and the degree of precision (or "discreteness") of the representation are circumscribed by the computer hardware and software. Some care was taken in this study to see that the computer algorithms selected were indeed performing as expected. For the basic numbers, a well-tested uniform pseudo-random number generator (Lewis et al.: 1969) and a carefully programmed Box-Muller transform (Box and Muller: 1957) for generating normally distributed pseudo-random numbers were used.

For each of the six probability distributions, their density functions, f(y), and their distribution functions F(y), in those cases where F(y) has a closed form representation, are given in Table 2, as well as the algorithms for generating the basic numbers, y, as functions of pseudo-random normal, η, and uniform, u, numbers.

For the algorithm used to generate the basic Pearson Type III numbers given in Table 2, B is distributed as Beta, $B \sim (b - [b], 1-b + [b])$, where [b] denotes the greatest interger equal to or less than b. The algorithm for B is as

Table 2

Distribution functions and algorithms for generating basic numbers

Probability Distribution	f(y)	F(y)	z	Range of y	y
N	$\frac{1}{a\sqrt{2\pi}} e^{-\frac{1}{2}z^2}$	-	$\frac{y-m}{a}$	$(-\infty,\infty)$	η
LN	$\frac{1}{a\sqrt{2\Gamma}(y-c)} e^{-\frac{1}{2}z^2}$	-	$\frac{\ell n(y-c)-m}{a}$	(c,∞)	$c + \exp[m + a\eta]$
GI	$\frac{1}{a} e^{-z + e^{-z}}$	$e^{-e^{-z}}$	$\frac{y-m}{a}$	$(-\infty,\infty)$	$m + a\{-\ell n[-\ell n\ u]\}$
W	$\frac{c}{a} z^{c-1} e^{-z^c}$	$1-e^{-z^c}$	$\frac{y-m}{a}$	(m,∞)	$m + a[-\ell n(1-u)]^{1/c}$
PIII	$\frac{1}{a\Gamma(b)} z^{b-1} e^{-z}$	-	$\frac{y-m}{a}$	(m,∞)	$\{m + a[-m \prod_{k=1}^{[b]} u_k - B\ \ell n\ u]\}$
PIV	$\frac{b}{a} z^{b+1}$	$1-z^b$	$\frac{a}{y}$	(a,∞) $a>0,\ b>0$	$a(1-u)^{-1/b}$

follows: 1) set $r = b - [b]$, $s = 1 - r = 1 - b + [b]$; 2) generate $u_1, u_2 \sim [0,1]$; 3) set $\zeta = u_1^{1/r}$, $\xi = u_2^{1/s}$; 4) if $\zeta + \xi > 1$, return to 2, otherwise proceed to 5; 5) set $B = \zeta / (\zeta + \xi)$. Note, if b is integral, then B = 0 and if $b < 1$, then $[b] = 0$, (Berman: 1971, Jöhnk: 1964).

To generate the various sets of psuedo-random numbers, the values of the parameters appearing in the algorithms were derived by solving the equations relating $\mu(Y)$, $\sigma(Y)$, and $\gamma(Y)$ to those parameters such that $\mu(Y) = 0$, $\sigma(Y) = 1$, and $\gamma(Y)$ equal the particular values shown in Table 1. The equations relating $\mu(Y)$, $\sigma(Y)$, and $\gamma(Y)$ to the various distribution parameters can be found in Wallis, Matalas, and Slack (1974) as well as in standard statistical texts.

Monte Carlo Experimental Results

The derived distribution functions for $\overline{X}$, S, and G over the experimental ranges of sequence length n, coefficient of skewness, $\gamma(Y)$, and probability density function, f(y), are available as paper copy or in microfiche form through the National Technical Information Service, U.S. Department of Commerce, Springfield, Va., 22151. The values of $\tilde{\mu}(X)$, $\tilde{\sigma}(X)$, and $\tilde{\gamma}(X)$, are presented on each figure. It should be noted that for skews greater than five, 100,000 trials do not provide sufficient resolution of the tails of the various sample distribution. In these cases the tails outside the probability range (0.1, 99.9) are of questionable accuracy. Figure 1 represents one of the 369 graphs that are available.

From these results, it can be seen that the distributions of $\overline{X}$, S, and G are functions of n, $\gamma(Y)$, and f(y). As expected $\overline{X}$ is an unbiased estimate of $\mu(Y)$, whereas, S and G are biased estimates of $\sigma(Y)$ and $\gamma(Y)$. In particular the distributions of G observed suggest that G is bounded. Kirby (1974) has shown that bounds are given by

$$|G| \le \frac{n-2}{\sqrt{n-1}} \tag{11}$$

and do not depend upon $\gamma(Y)$ or f(y). The bound on G, shown graphically in Fig. 1, is illustrated by the distribution of G for n = 10 and $\gamma(Y)$ = 3 in Fig. 2.

The mean and standard deviation of G for a 3-parameter log-normal distribution and a sample size of 10 are given in Table 3, from which it can be seen that the biases in G can be quite extreme.

Table 3

Mean and Standard Deviation of G versus γ for a 3-parameter log-normal distribution and n=10

G \ γ	1/4	1/2	1/2	1	2	2	3	4	5	10	15
Mean	0.131	0.255	0.352	0.476	0.626	0.791	0.978	1.099	1.181	1.380	1.465
Standard deviation	0.581	0.588	0.596	0.609	0.628	0.648	0.672	0.682	0.688	0.694	0.693

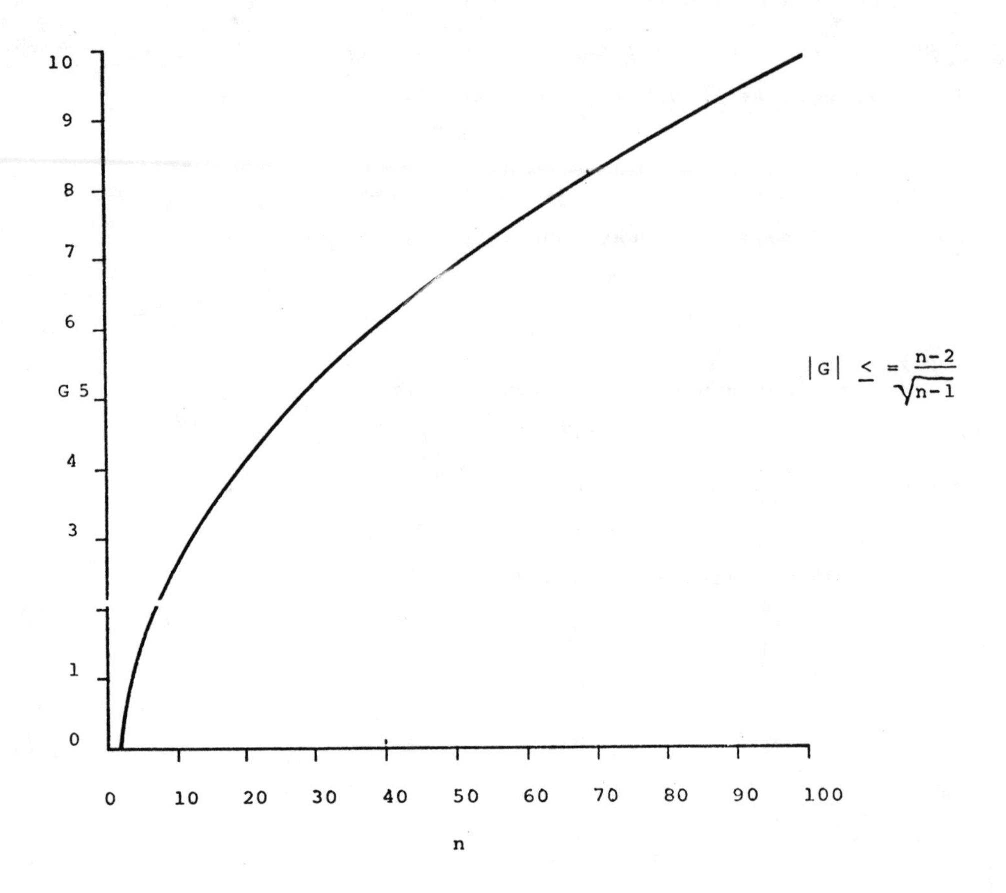

Figure 1 -- Bound on coefficient skewness.

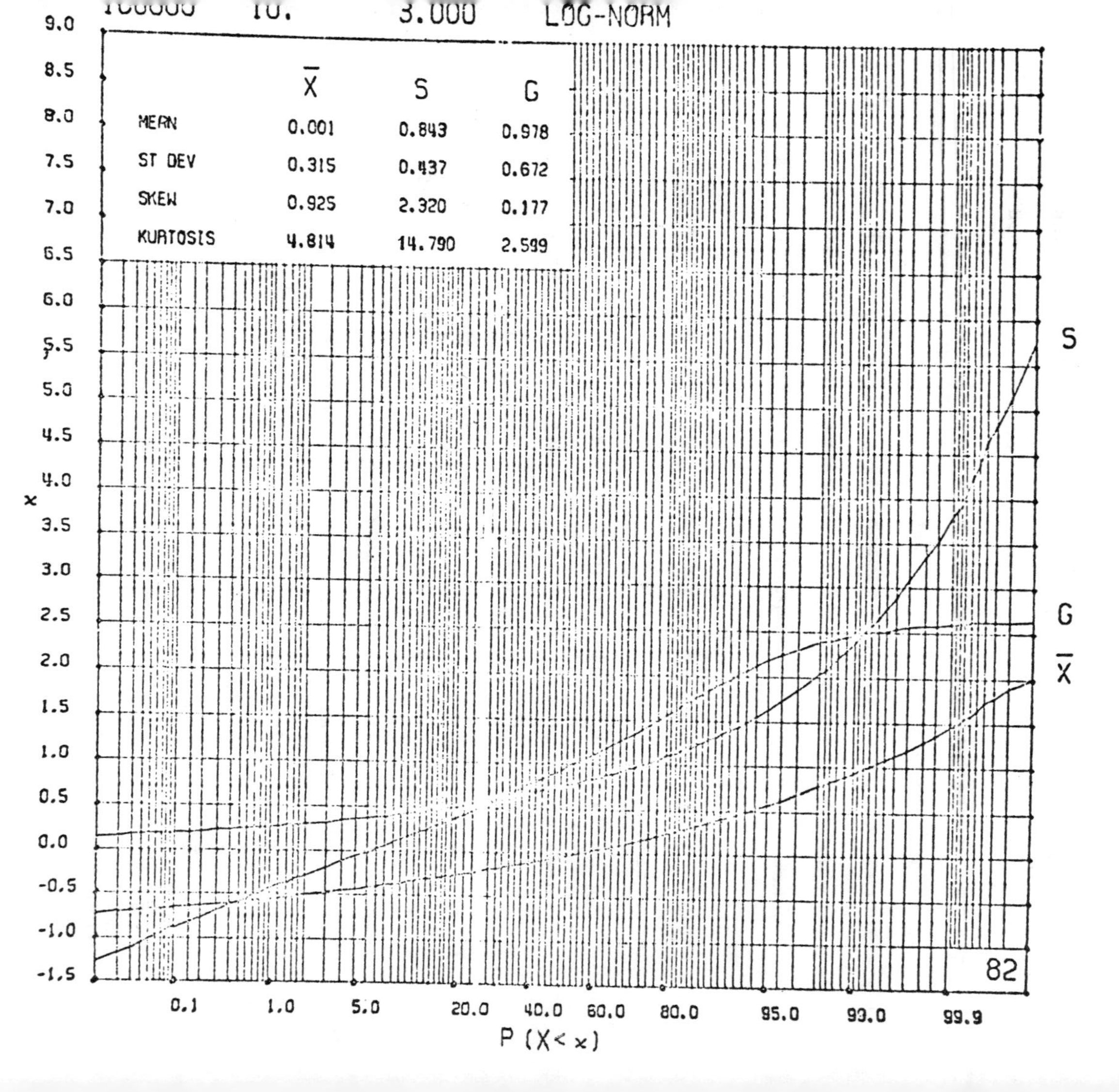

10. 3.000 LOG-NORM
X̄ S G
MEAN 0.001 0.843 0.978
ST DEV 0.315 0.437 0.672
SKEW 0.925 2.320 0.177
KURTOSIS 4.814 14.790 2.599
S
G
X̄
82
x
P (X < x)

Figure 2

Experimental results from some real world data

From the approximately 10,000 streamgaging stations in the United States, some 2,000, having records of good quality and minimal regulation, were considered. From the records of annual highest daily flows for each of the selected stations, the coefficient of skewness, G, was determined for periods of 10(10)70 years. These values of G are a basis for attempting to derive regional estimates of γ. Some preliminary results for those stations contained in the Northeast region of the U.S. are given in Table 4.

From Table 4, it is noted that the mean value of G increases monotonically with n. The reversal in trend at n = 70 is attributed to sampling error as only one 70-year period was available for determining the mean value of G. The trend itself is due perhaps to the lessening of the Kirby constraint as n increases. The general trend was observed for all other regions in the U.S. It was noted that the sampling properties of estimates of γ for each of the six probability distributions considered did not agree with those for the real world data. In particular, in matching the mean values of G derived from the Monte Carlo experiments with those of the real world, it was noted that the standard deviation of G derived from the Monte Carlo experiments was less than that for the real world data. This pattern was very consistent over all regions. Ongoing studies seek to explain this behavior by means of mixed distributions.

Table 4

Estimates of streamflow skewness, G, for the Northeastern United States

Period length	10	20	30	40	50	60	70
Number of periods	682	283	178	89	28	5	1
Mean G	0.928	1.449	1.690	2.024	2.155	2.939	1.082
Standard deviation G	0.743	0.913	1.040	1.091	1.211	1.843	-

Summary

Monte Carlo experiments were carried out to derive sampling properties of estimates of the mean, standard deviation, and coefficient of skewness. The basic aim of these experiments was to provide "data" for assessing the sampling properties of real world floods. The experiments provide a measure of the biases and variabilities associated with estimates of the mean, standard deviation, and coefficient of skewness as functions of the underlying distribution function, sample size, and population value of the coefficient of skewness. The sampling properties of estimates of the coefficient of skewness derived from the Monte Carlo experiments give poor agreement with those for real world flood data. Regional estimates of the coefficient of skewness must take into account that the values of the estimates are bounded as a function of n and that the values tend to increase as n increases. On-going studies seek to determine if mixed distributions will yield better agreement between the Monte Carlo experiments and real world flood data with respect to the sampling properties of estimates of the coefficient of skewness.

REFERENCES

Berman, M. B., Generating gamma distributed variates for computer simulation models, Report R-641-PR, Rand, Santa Monica, Calif., 43 p., Feb. 1971, (AD 720-801).

Box, G.E.P., and M. E. Muller, A note on the generation of random normal deviates, Ann. Math. Statist. 29, 610-611, 1958.

Jöhnk, M. D., Erzeugung von betaverteillen and gammaverteilten Zufallshanlen, Metrika, 8(1), 5-15, 1964.

Kirby, W., Algebraic boundedness of sample statistics, Water Resour. Res. , 10(2), 220-222, 1974.

Lewis, P.A.W., A. S. Goodman, and J. M. Miller, A pseudo-random number generator for the system/360, IBM Syst. J., 8(2), 136-146, 1969.

A Monte Carlo Estimator of the Gradient

Ryszard Zieliński, Warsaw

1. INTRODUCTION. The problem of estimation of the gradient of a function using only the values of the function (i.e. without differentiation) is considered.

Let $f: R^k \to R^1$ be a given function and let $f(x)$ denotes its value at the point $x = (x_1,\dots,x_k)$. Let e_i be the k-dimensional unit vector with i-th coordinate equal to 1. It is obvious that the estimation of the i-th coordinate of the gradient by the difference quotient $(2h)^{-1}(f(x + he_i) - f(x - he_i))$ yields the error $O(h^2)$. It can be proved that if h is a suitable random variable then the expected value of the above difference quotient (after a suitable normalization) equals to the i-th coordinate of the gradient of the function f at the point x, i.e. that a normalized difference quotient is an unbiased estimator of the first partial derivative. More precisely: in the paper [Zieliński 1974b] the following result for a function $g: R^1 \to R^1$ is presented. Suppose that g has all derivatives $g^{(i)}(x)$, $i = 1, 2, \cdots$, and that $g(x+h) = \sum_{i=0}^{\infty} (h^i/i!) g^{(i)}(x)$ for $x, h \in R^1$. Let $U = (u_j)_{j=0}^{\infty}$ and $V = (v_j)_{j=0}^{\infty}$ be real sequences such that

$$\sum_{j=0}^{\infty} v_j u_j = \frac{1}{2} \quad , \quad \sum_{j=0}^{\infty} v_j u_j^{2i+1} = 0 \quad \text{for all } i = 1, 2, \cdots$$

Let J be a random variable such that $P(J = j) = q_j > 0$ for $j = 0, 1, \cdots$. If U, V and g satisfy the condition

$$\sum_{j=0}^{\infty} \sum_{i=0}^{\infty} \frac{v_j u_j^{2i+1}}{(2i+1)!} g^{(2i+1)}(x) = \sum_{i=0}^{\infty} \sum_{j=0}^{\infty} \frac{v_j u_j^{2i+1}}{(2i+1)!} g^{(2i+1)}(x)$$

then $v_J(g(x + u_J) - g(x - u_J))/q_J$ is an unbiased estimator of the first derivative $g'(x)$ of the function g at the point x. An example of U, V, g satisfying the above conditions is given in [Zieliński 1974a]. Using the above estimator for every coordinate of the gradient separately we obtain an unbiased estimator of the gradient. This is a 2k-point estimator in the sense that for estimation of the gradient the evaluation of the function at 2k points is needed.

In what follows we shall present a (k+1)-point estimator of the gradient of the function f. It seems that k+1 is the minumal number of points needed for estimation of the gradient. The idea of estimator is similar to that used in the paper [Zieliński 1973] for construction of a (k+1)-point estimator which yields the expected error $O(h^2)$.

2. THE IDEA. Let S_k be the unit sphere in E^k with the centre at the point x. Let $\xi_0, \xi_1, \dots, \xi_k$, $\xi_j = (\xi_{1j}, \dots, \xi_{kj})$, be points which are equidistributed on the surface of S_k, i.e. ξ_j are vertices of a regular simplex so that $\|\xi_j - x\|^2 = \sum_{i=1}^{k} (\xi_{ij} - x_i)^2 = 1$ and $(\xi_j - x, \xi_l - x) = \sum_{i=1}^{k} (\xi_{ij} - x_i)(\xi_{il} - x_i) = \text{const}$ (const being equal to $(k+1)/k$). The points ξ_j will be considered as fixed ones. We shall construct real sequences $V = (v_s)_{s=0}^{\infty}$, $U = (u_s)_{s=0}^{\infty}$, $u_s > 0$ and a probability distribution on the set of all non-

negative integers: $P(S = s) = q_s > 0$, $s = 0, 1, \cdots$ and then we shall apply the following algorithm: 1) sample the index S according to the distribution $Q = (q_s)_{s=0}^{\infty}$; 2) calculate the values $f(x + u_S \xi_j)$ for $j = 0, 1, \cdots, k$; 3) calculate the estimator $(a_1, \cdots, a_k)$ of the gradient by the formula

$$a_i = \frac{k}{k+1} \cdot \frac{v_S}{q_S} \cdot \sum_{j=0}^{k} \xi_{ij} f(x + u_S \xi_j)$$

It will be proved that $(a_1, \cdots, a_k)$ is an unbiased estimator of the gradient of the function f at the point x. Without loss of generality we assume that $x = 0$. The class $\mathcal{F}$ of functions for which the above result holds is to be defined below; it is defined in such a way that the first partial derivatives of f can be presented as a linear combination of the values of the function at suitable points and that some technical conditions which guarantee existence of the expected values are satisfied.

3. THE RESULT. Let $\mathcal{F}$ be a class of functions $f: R^k \to R^1$ such that (i) all partial derivatives $\partial^{(r)} f / \partial x_1^{\alpha_1} \ldots \partial x_k^{\alpha_k}$

where $\alpha_j \geq 0$ and $\sum \alpha_j = r$ exist and

$$f(x+y) = \sum_{r=0}^{\infty} \sum_{\alpha_1+\dots+\alpha_k=r} \frac{y_1^{\alpha_1} \cdots y_k^{\alpha_k}}{r\,!} \cdot \frac{\partial^{(r)} f(x)}{x_1^{\alpha_1} \cdots x_k^{\alpha_k}} \tag{1}$$

for all $x, y \in R^k$; (ii) there exists a sequence $(\gamma_r)_{r=0}^{\infty}$ of positive numbers and a number $\beta \in (0,1)$ such that $\gamma_r > \beta^r$ and the series

$$\sum_{r=0}^{\infty} \sum_{\alpha_1+\dots+\alpha_k=r} \gamma_r \frac{\xi_{1j}^{\alpha_1} \cdots \xi_{kj}^{\alpha_k}}{r!} \cdot \frac{\partial^{(r)} f(0)}{x_1^{\alpha_1} \cdots x_k^{\alpha_k}} \tag{2}$$

converges absolutely.

<u>Theorem.</u> Let $f \in \mathcal{F}$. Let $U = (u_s)_{s=0}^{\infty}$ and $V = (v_s)_{s=0}^{\infty}$ be real sequences such that $u_s > 0$, $\sum_{s=0}^{\infty} v_s < \infty$ and

$$\sum_{s=0}^{\infty} v_s u_s^r = \delta_r^1 \quad , \quad \sum_{s=0}^{\infty} |v_s u_s^r| \leq \gamma_r \quad , \; r = 1, 2, \cdots \tag{3}$$

where δ_r^s is the Kronecker symbol. Let S be a random variable such that $P(S = s) = q_s > 0$ for $s = 0, 1, 2, \cdots$. Define the estimator $(a_1, \cdots, a_k)$ of the gradient of the function f at the origin by the formula

$$a_i = \frac{k}{k+1} \cdot \frac{v_S}{q_S} \cdot \sum_{j=0}^{k} \xi_{ij} f(u_S \xi_j) \tag{4}$$

Then Ea_i is equal to the first partial derivative of the function f (with respect to x_i) at the origin, i.e. $(a_1, \cdots, a_k)$ is unbiased estimator of the gradient.

Proof. By (4) and (1) we obtain

$$Ea_i = \sum_{s=0}^{\infty} q_s E(a_i \mid S=s) = \frac{k}{k+1} \sum_{s=0}^{\infty} v_s \sum_{j=0}^{k} \xi_{ij} f(u_s \xi_j) =$$

$$= \frac{k}{k+1} \sum_{s=0}^{\infty} \sum_{j=0}^{k} \sum_{r=0}^{\infty} \sum_{\alpha_1+\dots+\alpha_k=r} \frac{v_s \xi_{ij} u_s^r \xi_{1j}^{\alpha_1} \cdots \xi_{kj}^{\alpha_k}}{r!} \cdot \frac{\partial^{(r)} f(0)}{\partial x_1^{\alpha_1} \cdots \partial x_k^{\alpha_k}}$$

By the second condition of (3) and utilizing the absolute convergence of the series (2) we can change the order of summation and then we get

$$Ea_i = \frac{k}{k+1} \sum_{r=0}^{\infty} \sum_{j=0}^{k} \left(\sum_{s=0}^{\infty} v_s u_s^r \right) \sum_{\alpha_1+\dots+\alpha_k=r} \frac{\xi_{ij} \xi_{1j}^{\alpha_1} \cdots \xi_{kj}^{\alpha_k}}{r!} \cdot \frac{\partial^{(r)} f(0)}{\partial x_1^{\alpha_1} \cdots \partial x_k^{\alpha_k}}$$

Now by the first condition of (3) all terms except that for $r = 0$ and $r = 1$ vanish and

$$Ea_i = \frac{k}{k+1}\left(\sum_{j=0}^{k}\sum_{s=0}^{\infty} v_s \xi_{ij} f(0) + \sum_{j=0}^{k}\sum_{l=1}^{k} \xi_{ij}\xi_{lj} \cdot \frac{\partial f(0)}{\partial x_l}\right)$$

Because $\xi_0, \xi_1, \ldots, \xi_k$ are vertices of a unit simplex centered at the origin we have

$$\sum_{j=0}^{k} \xi_{ij} = 0 \quad \text{and} \quad \sum_{j=0}^{k} \xi_{ij}\xi_{lj} = \frac{k+1}{k}\delta_i^l$$

so that $Ea_i = \partial f(0)/\partial x_i$, q.e.d.

Remarks. (i) As a simple example of a function $f \in \mathcal{F}$ we can take any function f with r-th partial derivatives bounded by c^r, c being any positive constant ; then (2) holds with e.g. $\gamma_r \equiv 1$.

(ii) Let (γ_r) be a given sequence of positive numbers such that $\gamma_r > \beta^r$ for a number $\beta \in (0,1)$. As the sequence U we can take any strictly increasing sequence of positive numbers such that $u_0 = \beta$ and $u_s \to \infty$. Then there exists a sequence

V satisfying (3). The proof is a slight modification of that given in [Zieliński 1974a] and is based on the proof of the well-known Polya theorem (see for example [Cooke 1950], th. 2.5, I) ; the proof consists in construction of the sequence V when sequences U and (γ_s) are given. Another method of construction of sequences U and V is discussed in [Zieliński 1974b].

4. OPTIMAL UNBIASED ESTIMATION OF THE GRADIENT. The conditions (3) for the sequences U and V are rather loose in the sense that there are infinitely many sequences satisfying (3). Moreover the results hold for every distribution $Q = (q_s)_{s=0}^{\infty}$. The problem arises how to choose U, V and Q so as to get the "optimal" estimator, for example the estimator with minimal variance. The problem will be considered in a separate paper.

REFERENCES

Cooke, R.G. (1950): Infinite matrices and sequence spaces. London

Zieliński, R. (1973): A randomized finite-differential estimator of the gradient. Algorytmy, X, No 18, 21–30

Zieliński, R. (1974a): Unbiased estimation of the derivative of a regression function. Symposium to honour Jerzy Neyman. Warsaw, April 3–10, 1974

Zieliński, R. (1974b): An unbiased finite-differential estimator of the gradient (will appear in Algorytmy)

SUBJECT GROUP B

Automatic Classification

A Split Clustering Procedure in Taxonomy

Antonio Bellacicco and Enrico Nervegna, Rome

1-Introduction.

The content of this paper is an outline of some computational aspects of a clustering procedure on binary data,called "SPLIT".

Generally speaking,it is well known that cluster analysis is a splitting device operating on a given set of objects,S,performed through a maximization of the similarity of the objects in each subgroup and/or a maximization of the dissimilarity among the subgroups,/ 1/ .

Two main procedures are at disposal of the researchers: the hierarchical clustering and the aggregating clustering procedures. We will not spend time on general remarks on cluster analysis and we will refer to / 1 / and / 2 / as general references.

"SPLIT" clustering procedure is a kind of synthesis of the two mentioned procedures,just because it is based on the binary splitting of the set S in two subgroups,S' and S",mutually exclusive, through the clustering of the objects around two "Nucleus of Clustering",(abb. NC), via the maximization of the dissimilarity of the two growing subgroups.

Binary data are quite common in many psycological and sociological researches as basic protocols.SPLIT was expecially developed for the purpose of building new complex concepts, in an operational way,superimposed on the polarized subgroups,S' and S".

Let us consider a given population of objects,$S = (S_1, S_2, \ldots, S_N)$ and a set of binary predicates ,$P = (P_1, P_2, \ldots, P_M)$.It is easy to see that 2^M is the number of binary partitions of P,that is the set of all the possible sequences of predicates

$$P^j = (P_1) * (P_2) * \ldots * (P_M) \qquad J = 1,2,\ldots,2^M$$

where (P_i) means either the predicate P_i or his negation, for $i = 1,2,\ldots,M$, and " * " means the logical conjunction "and".

In case of binary clustering of S, we have,at the same time,not only the splitting of S in two subgroups, S' and S" , but also the splitting of P such that the two subgroups of predicates ,P' and P", are logically complementary and correspond to two sequences ,P^j and P^i, for $(i,j) = 1,2,\ldots,2^M$.

The two sequences are logically complementary too and the positive attributes are the necessary,but not sufficient conditions, for the identification of P^j and P^i. The two subgroups ,S' and S"" are the "extensional" size of the abstract concepts, labelled P^j and P^i,/ 3 /.

It is possible to give a factorial interpretation to the splitting procedures just considering P^i and P^j as the opposite directions of a factorial axis ,that is the most explicative factor of the whole set of attributes,P.

2 - The Intragroup Similarity and the Intergroup Dissimilarity.

In order to clarify SPLIT clustering procedure , we consider a basic distance function. The technical detail can be found in 4.

Let us start just considering an object-predicate table, X , labelled the Matrix of Basic Data,(abb.MBD):

X	P_1, P_2, ...,	P_j	, ..., P_M
S_1		⋮	
S_2		⋮	
⋮		⋮	
S_i	....	x_{ij}	....
⋮		⋮	
S_M		⋮	

where $x_{ij} = \begin{cases} 1 & \text{iff } P_j(S_i) \text{ is true} \\ 0 & \text{otherwise} \end{cases}$

and $P_j(S_i)$ means that the object S_i enjoies the property P_j.

As a general rule we will suppose that all the rows and the columns of the matrix X are not equal, but they differ at least for one element, x_{ij}.

We consider a normalized distance function

$$0 \leq d_{ij} = \sum_{h=1}^{M} \left| x_{ih} - x_{jh} \right| / M \leq 1$$

between two objects whatsoever S_i and S_j, $(i,j) = 1,2,\ldots,M$.

We can consider the index of Intragroup Similarity

$$V(S_1,S_2,\ldots,S_k) = \sum_{i=1}^{j} \sum_{j=2}^{N} d_{ij}$$

and his normalization

$$v(S_1;S_2,\ldots,S_k) = \frac{V(S_1,S_2,\ldots,S_k)}{\max V(S)_k}$$

It is possible to show that:

i) if $N = 2K$ then $\max V(S)_k = M K^2$

ii) if $N = 2K + 1$ then $\max V(S)_k = M K (K + 1)$

More general formulas can be written in case of many copies of the same object and can be seen in / 4 /.

The Intergroup Dissimilarity Index is

$$B(S',S'') = \sum_{h=1}^{M} \sum_{i=1}^{N'} \sum_{j=1}^{N''} \left| x_{jh} - x_{ih} \right|$$

where $S' = (S'_1,S'_2;\ldots,S'_{N'})$ and $S'' = (S''_1,S''_2,\ldots,S''_{N''})$ and $N'+N''=N$.

As for $V(S)_k$, it is possible to write down a general formula for $\max B(S',S'')$; in short we write :

i) for $N' = 2K'$; $N'' = 2K''$; $N' < N''$ then

$$\max B(S',S'') = M(N' + N'(K''- 1) + K')$$

ii) for $N' = 2K'$; $N'' = 2K''+1$; $N' < N''$ then

$$\max B(S',S'') = M N'(K'' + 1)$$

iii) for $N' = 2K'+1$; $N'' = 2K'' + 1$; $N' < N''$ then

$$\max B(S',S'') = M N'(K'' + 1)$$

iv) for $N' = 2K'+1$; $N'' = 2K''$; $N' < N''$ then

$$\max B(S',S'') = M (N' + N'(K'' - 1) + K' + 1 - M^{-1}).$$

It is possible to write down more general formulas for the case of many copies of the same object. See for reference / 4 /.

A general consideration for the understanding of the previous formulas is based on the simple fact that in case of binary data, the distance function d_{ij} obeys the simple triangular property:

$$d(S_i,S_j) + d(S_i,S_h) = M$$

at most when $S_J = S_h$ and S_i is the logical complement of the two previous objects. We remember that two objects S_i and S_h are complementary if we can write in the already known notation

$$P^i(S_i) = 1 - P^h(S_h).$$

X is therefore partitioned in two submatrices X' and X" and the normalized distance between them is therefore

$$b(S',S'') = \frac{B(S', S'')}{\max B(\cdot ,\cdot)}$$

3 - The "SPLIT" Clustering Procedure

"SPLIT" is a FORTRAN V computer program runned on the UNIVAC 1110 of the InterDept. Center of the University of Rome.

SPLIT considers two population at a time, S(1) and S(2) and performes a set of comparisons following the above scheme:

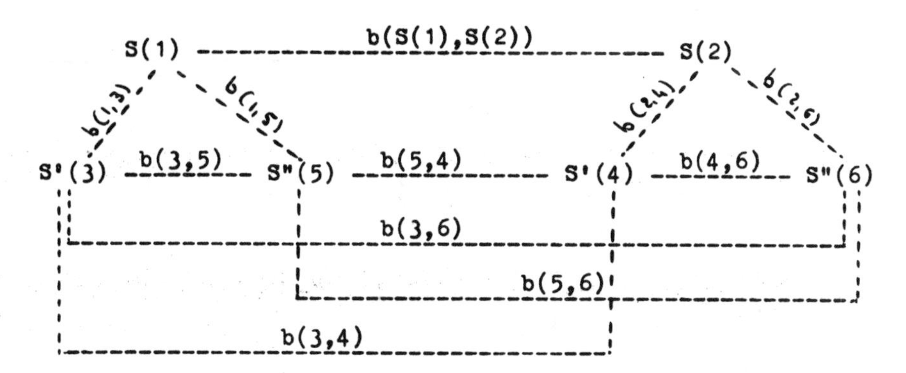

where odd numbers identify the first population S(1) and his subgroups and even numbers identify the second population S(2) and his subgroups.

In case of many populations S(1),S(2),....., S(G), to be considered simultaneously, SPLIT avoids comparisons between populations.

In order to run the program, it is necessary to consider the following parameters:

i) IFS : multiplicative scalar factor when x_{ij} are real numbers, positive and bounded.

ii) KSOG : the maximum number og objects considered in the G populations; max 750.

iii) NN : max 300 objects in the same population;

iv) MM : max 30 attributes;

SPLIT program reads:

a) the format of the SK through which it will read the MBD;

b) the number GG of the available populations;

c) the switch indicating either the SK preselected in the populations or in a special area of data parking;

d) the level u of a threshold when the x_{ij} are real, positive bounded numbers and we like filter the data such that:

$$x'_{ij} = \begin{cases} 1 & \text{if } x_{ij} \geq u \\ 0 & \text{if } x_{ij} < u \end{cases}$$

where x'_{ij} is the transformed data. A subroutine ADGIUST enables the researcher to the desired transformation.

e) the PRINT switch of the MBD of the populations and of the subgroups;

SPLIT procedure works in this simple way. We start choosing two Nucleus of Clustering , $NC'(S_i,S_j)$ and $NC''(S_h,S_k)$ such that:

$$d(S_i,S_j) = \min_{i,j} \quad \text{and} \quad b((S_i,S_j);(S_h,S_k)) = \max_{i,j,h,k}$$

for $(i,j,h,k) = 1,2,\ldots,M$.

As a secondstep we cluster a fifth object, S_r,just enlarging NC' or, arternatively, NC",following the criterion of choosing the min V(∘,∘) in the following table :

objects	NC'	NC"
S_{r1}	V(∘,∘)	V(∘,∘)
S_{r2}	∘	∘
∘	∘	∘
∘		
∘	∘	∘
∘		
$S_{r_{N-4}}$	∘	∘

where V(∘,∘) means the Intragroup Similarity computed on the two nucleous, NC' and NC", enlarged any time with one of the last N-4 elements of the given population.

In case of equality of some V(∘,∘),we choose the new element S_r such that we have

either $\max b((NC',S_r);NC")$ or $\max b((NC");(NC',S_r))$.

SPLIT program stops when all the objects are allocated and the two subgroups are completely built.

The output of the SPLIT procedure is the following one with switches on PRINT:

1) The MBD of the NN objects and MM attributes.
2) The percentage of the presence of each attribute in MBD, called the Profile Vector (abb. PV) of MBD.
3) The name of the elements of NC' and NC", with V(NC') and b(NC',NC").
4) The MBD of each NC' and NC" of each population.
5) The V(S') and V(S") of each population.
6) The PV of each MBD;
7) The b(S',S") , labelled in the program as B max.
8) The sequence of attributes whose percentage of occurrence is greather than 70%, corresponding to the rejection of the null hypothesis 50% of occurrence.
9) Graphical subroutine of the percentages.

4 - A Didactic Example.

We give a simple illustrative example of a splitting procedure.

X	P_1	P_2	P_3	P_4
S1	0	0	1	0
S2	1	0	1	1
S3	0	1	1	0
S4	0	0	1	1
S5	1	0	1	0
S6	1	1	1	1
S7	0	0	0	0

The Matrix of Basic Data

Let us calculate the Matrix $D_2(S)$ of the distances d_{ij} on S.

	S_1	S_2	S_3	S_4	S_5	S_6	S_7
S_1	0	2/4	1/4	1/4	1/4	3/4	1/4
S_2	-	-	3/4	1/4	1/4	1/4	3/4
S_3	-	-	-	2/4	2/4	2/4	2/4
S_4	-	-	-	-	2/4	2/4	2/4
S_5	-	-	-	-	-	2/4	2/4
S_6	-	-	-	-	-	-	1
S_7	-	-	-	-	-	-	-

The two nucleous are: NC'(1,3) ; NC"(2,6). The second step:

V(∘,∘)	(1,3)	(2,6)
S_4	0,125	0,180
S_5	0,125	0,125
S_7	0,125	0,250

For the third step we compute

V(·,·)	(1,3,7)	(2,6)
S_4	0,140	0,180
S_5	0,140	0,125

and finally we have

V(·,·)	(1,3,7)	(2,5,6)
S_4	0,14	0,14

and therefore the final allocation of S_4 is indifferent.

- Concluding Remarks.

SPLIT clustering procedure can be improved in two directions:

a) the completion of the whole dendrogram, just as a hierarchical clustering;

b) the generalization of the splitting, from binary to n-ary splitting;

We did not mentioned in this paper any application of the outlined procedure SPLIT, just because it is our purpose to issue the basic results in psycological and sociological journals.

We conclude this presentation just remembering the simplifying role of the binary splitting in many kinds of researches, where we need to enphasize the presence of two polarized groups. As a matter of fact, there is not the perfect pass-partout in cluster analysis and it is a fundamental caution to relate themselves to the actual aim of the research.

REFERENCES

1 - Sokal R.R., Sneath P.H. -PRINCIPLES OF NUMERICAL TAXONOMY
Freeman, San Francisco, 1963

2 - Andenberg M.R. -CLUSTER ANALYSIS FOR APPLICATIONS
Academic Press, New York, 1973

3 - Bellacicco A. -"Empirical Clustering and Theory Construction"
Summer School on Philosophy of Science
Chiavari- 3/15 of Sept. 1973

4 - Bellacicco A. -"Aspetti e Problemi della Classificazione Booleana"
Associazione Italiana di Ricerca Operativa
Giornate di Lavoro, 1973, Padova

Clustering Structures

Johannes Gordesch and Peter Paul Sint, Vienna

1. INTRODUCTION

Traditional cluster methods classify elements characterized by vectors. It appears more appropriate, however, to cluster elements characterized by more complex structures, e. g., family structures represented by graphs.

The concept of category provides a working mathematical tool which clarifies and provides a sufficiently general theory for the concept of strucure. In this paper the notion of clustering is generalized by the use of categorical algebra. The structures which can be described by graphs are also treated in some detail. A certain familiarity with the traditional cluster methods will be assumed.

2. CONCEPTS OF CATEGORICAL ALGEBRA

A category $\underline{C}$ is a class $\underline{C}$ (not necessarily a set) of objects A, B, C, ..., together with two mappings:

(1) a mapping which assigns to each pair of objects $A, B \in \underline{C}$ a set mor(A,B) called the set of morphisms with domain A and codomain B;

(2) a mapping which assigns to each triple of objects A,B,C $\underline{C}$ a law of composition

$$mor(A,B) \circ mor(B,C) \rightarrow mor(A,C)$$

(if $f \in mor(A,B)$ and $g \in mor(A,B)$, then the composite of f and g, written gf or $g \circ f$, is a morphism with domain A and codomain C).

The mappings so defined must, moreover, satisfy the axioms:

C1 Equality: $mor(A_1,B_1)$ and $mor(A_2,B_2)$ are disjoint unless $A_1=A_2$ and $B_1=B_2$.

C2 Associativity: Given $f\in mor(A,B)$, $g\in mor(B,C)$, and $h\in mor(C,D)$, then $(hg)f=h(gf)$.

C3 Identity: To each object $A\in \underline{C}$ there is a morphism $i_A:A\rightarrow A$ such that, for all $f\in mor(A,B)$, $g\in mor(C,A)$, we have

$$fi_A=f,\ i_Ag=g$$

A <u>subcategory</u> $\underline{C}_S$ of a category $\underline{C}$ is defined by the following postulates: the class of objects as well as the set of morphisms of $\underline{C}_S$ are parts of the corresponding class resp. set of $\underline{C}$; and the composition of morphisms leads in $\underline{C}$ and in $\underline{C}_S$ to the same result, $g\circ f = g\circ_S f$.

$\underline{S}$, the category of sets and mappings, results if the class of objects is the class of all sets, and the set of morphisms $mor(A,B)$ is the set of mappings with domain A and codomain B. The identity morphism for A,i_A, is the identity mapping $A\rightarrow A$.

A wellknown example is the category of groups and homomorphisms, or the category of a certain group and its inner automorphism.

Let $G(A,E,\zeta,\eta)$ be a directed graph with arrows (edges)A and nodes E, : A E,η : $A\rightarrow E$.

E_1, the set of starting points, is the image under ζ and E_2, the set of terminal points, is the image under η where $E_1UE_2=E$.

Let $G(A,E,\zeta,\eta)$ and $G'(A',E',\zeta',\eta')$ be two graphs. Then a graph morphism G to G', denoted by f:G G', is defined by the following commutative diagrams:

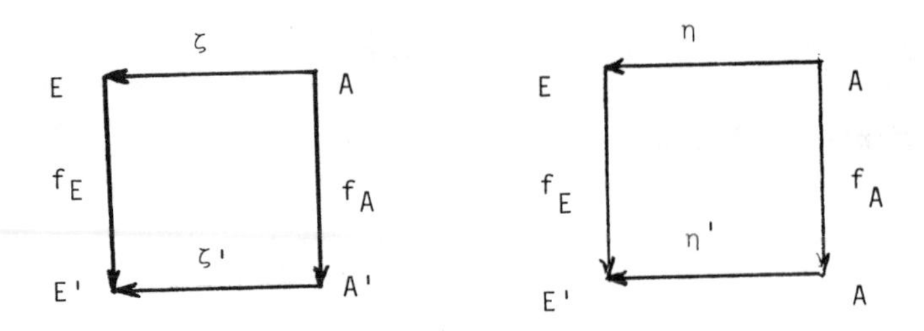

where f_E is a set morphism from E to E', and f_A a set morphism from A to A'.

The class of all directed graphs, together with the above defined graph morphisms and composition by components, from the category of graphs, $\underline{G}$. If, instead, the class of objects consists only of one graph, and f_E resp. f_A are automorphisms, the category of a certain graph, $\underline{H}$, is formed. Similar definitions can be given for undirected graphs, labeled graphs, and multigraphs.

A <u>functor</u> $\underline{f}$ mapping a category $\underline{C}$ to a category $\underline{D}$ consists, by definition, of two mappings:

the object mapping that assigns to each A in $\underline{C}$ an object $\underline{f}(A)$ in $\underline{D}$; and
the mapping that assigns to each morphism $h: A \to B$ of $\underline{C}$ a morphism
$\underline{f}(h): \underline{f}(A) \to \underline{f}(B)$ of $\underline{D}$.

These mappings satisfy the axioms:
F1: For each $A \in \underline{C}$: $\underline{f}(i_A) = i_{f(A)}$, and

F2: For each composite gf defined in $\underline{C}$: $\underline{f}(gh) = \underline{f}(g)\underline{f}(h)$ (covariant functor), resp.

F2': $\underline{f}(gh) = \underline{f}(h)\underline{f}(g)$ (contravariant functor).

A <u>transformation</u> of two functors $\underline{f}$, $\underline{g}$: $\underline{C} \to \underline{D}$ is a mapping which assigns to each object $X \in \underline{C}$ a morphism $t_X: \underline{f}(X) \to \underline{g}(X)$ in $\underline{D}$.

In the case of a *natural transformation* (which is the more usual concept in categorical algebra), every morphism h:X Y of C yields a commutative diagram, i. e.

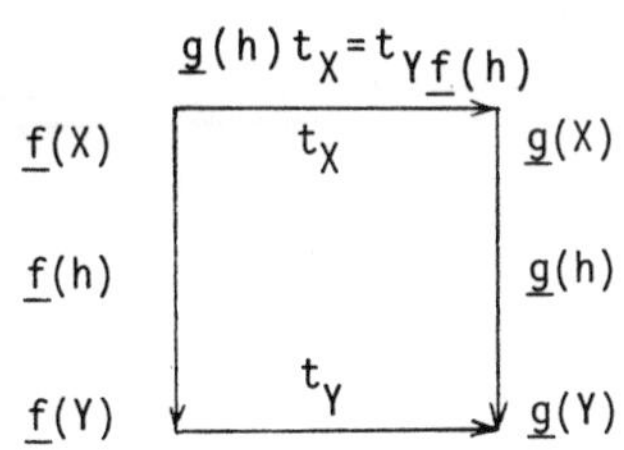

Let C be a category of sets bearing some algebraic structure with morphisms preserving this structure (e. g. groups, modules, topological spaces). Assigning the underlying set to each object of C, we arrive at the so called *forgetful* functor.

3. GENERAL DEFINITION OF CLUSTERING

Clustering is a procedure for classification of some objects. It may be defined formally as a functor mapping the given category C, where the morphisms of this category C and its image are defined by the particular clustering procedure. The composition of morphisms guarantees the continuation of the clustering process.

The concepts of category and functor were introduced into algebra in order to provide a theory which can deal with mappings of ordinary elements (leading to groups, rings, integral domains, etc.) as well as with mappings of structured sets, thus giving a general theory of algebraic structures.

In cluster analysis, the usefulness of these general concepts is twofold: first, the appropriate theory is formulated not only for the unrestricted sets, but also for the clustering of structures; and secondly, the theory covers the classification of infinite sets as well as the classification into an infinite number of classes. Some examples for the classification of infinite sets are the partitioning of the domain of a complex-valued function according to its singularities, and the decomposition of an infinite group into a series of subgroups (e. g. composition series). In the latter example, the clusters overlap, as the intersection of subgroups contains at least the subgroup consisting of the neutral element alone.

As a further instance, the classification into an infinite number of classes is given by the partitioning of the domain of a doubly periodic function according to its period, i. e. the partitioning into period parallelograms.

4. SPECIAL TYPES OF CLUSTER METHODS

Let $\underline{S}$ be the category of sets and mappings. Let the category $\underline{P}$ consist of the set X and all its subsets, and define mor(A,B) as set morphism. Then we construct a family of categories

$\underline{P}_n [X_n$ and its subsets $X^m{}_n$, $mor(X^m{}_n, X_n^{m'})]$,

$X_n \preccurlyeq X, X^m{}_n \preccurlyeq X_n$, n of some index set I, mor defined as above, and form the category

$\underline{T}$ set of all $\underline{P}_n$; $mor(\underline{P}_n, \underline{P}_{n'})$,

$mor(\underline{P}_n, \underline{P}_{n'})$ being defined by the special clustering method (e.g. in hierarchical clustering some ordering relation).

The functor $\underline{P} \to \underline{T}$ is called traditional clustering.

In the case that the clustering functor is a forgetful functor the structure is not retained. Thus general clustering becomes equivalent to traditional methods. If the set $mor(\underline{P}_n, \underline{P}_n)$ is restricted according to some specification sufficient to define an algorithm, a particular cluster method results. If, e. g., the morphisms belonging to $\underline{P}$ and $\underline{P}_n$ define a graph of similarity (by some similarity mapping,Hartigan,1967),we speak of a graph theoretic cluster method. The clustering of graphs, which is a specialization of the general procedure for the case of the category of graph, should not be confused with the graph theoretic cluster methods. The set of all mappings of a set A to a particular structured (e. g. ordered) set B (set of valuations) is called a fuzzy set, denoted by B^A (Kaufman,1973),and may also be formulated in terms of categorical algebra. In the case that B is the real closed interval 0,1 and the mappings obey the axioms of probability calculus, one is lead to the notion of a probability space.

The set A may already consist of mappings.

If the objects in the above definitions are no longer sets but fuzzy sets, we then speak of clustering of fuzzy sets, and if the objects are sets but the morphisms form a fuzzy set, we speak of fuzzy clustering. A combination of these two clustering methods is also possible. In practice, these notions are not always differentiated. Of course, too much fuzziness will severely complicate algorithms. Fuzziness is often the result of statistical reasons.

Conditional clustering (Sint,1973)is defined as a transformation from a functor clustering in the domain of predictor variables to a functor clustering in the domain of criterion variables. Optimality conditions partly result in the restriction to a natural transformation. Statistical reasons (refering to regressionanalysis) lead quite naturally to fuzzy conditional clustering, which has been so far the only conditional clustering method of importance.

5. CLUSTER METHODS FOR GRAPHS

To make the notion of structure more concrete and to relate it to the structures yet studied in cluster analysis, we will study categories, the objects of which are graphs. The traditional approaches to clustering assume as starting values elements characterized by vectors, or elements which are connected by a similarity structure. This description often seems to be sufficient to give an adequate picture.

For example, let us classify family structures. We may state the number of relation ships of a certain form (father - son, grandmother - grandson, etc.) in form of a vector, but we can hardly get a good description this way.

On the other hand it is easy to express the relationships with a graph:

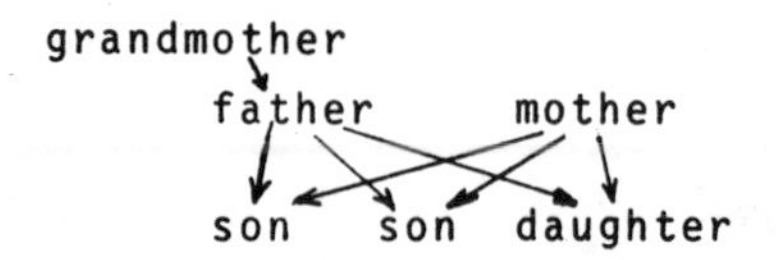

There may be nodes which do not correspond to existing persons (the father may be missing) Other structures may be expressed more appropriately with undirected graphs or labeled multigraphs instead of directed graphs. As always we hope, to put togrther in one cluster the most similar objects by applying cluster methods to these graphs.One method of measuring the similarity of the two graphs could be to use the numbers of nodes and /or edges in common (differing) for the two graphs. For instance we may use a function isotone in the number of nodes and edges in common, and antitone in the number of differing nodes edges.

As a simple example we may take the following figure:

Cluster 1 Cluster 2

In order to realize which way this concept is related to classifying vectors let us assume two binary vectors are given, e. g. a=(1,1,0,0,0)
b=(1,0,1,0,0)

We may find graphs a 0 1 b 0 1

which describe these vectors completely. Because the nodes are the same in both pictures we may determine similarity measure by counting the number of edges in common between the two graphs (plus, possibly, those missing in both graphs). This way we may define the traditional measures of similarity in graph theoretic terms.

a \ b	present	not present
present	3	2
not present	2	3

This way we think of zero and one as symmetric position, and we could confine ourselves to the upper part of the graph alone. But it is also possible to study separately the connections with the two nodes 0 and 1:

	a \ b	connected with 1	connected with 0
connected with	1	1	1
	0	1	2

and to handle these connections differently. This will result in other traditional similarity measures.

This may also serve to clarify another point. Naturally, it is always possible to describe a graph as a vector (e. g. by connecting the rows of the incidence matrix) and thus to return to vector description. But the purpose of introducing graph language is to clarify the relationships between various aspects of the vector and to describe more easily the necessity to handle different parts of the graph in different ways. We may introduce several concepts to handle these problems. If a certain structure, e. g.

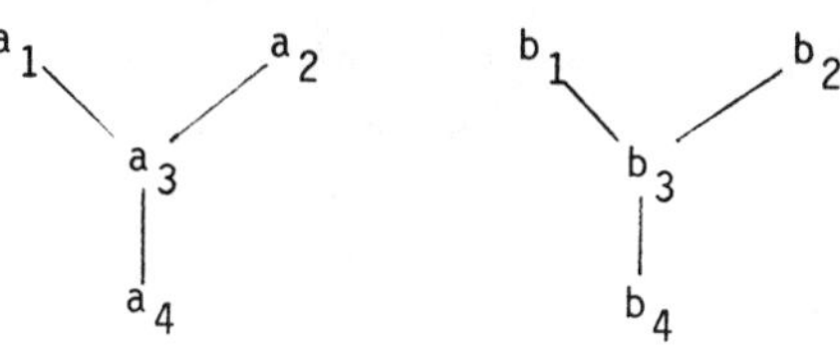

is given, the objects corresponding to a_1, a_2 ... will, in general, not be the same as b_1, b_2 ..., but they will only be similar to a certain degree. If a_1 and b_1 are both horses, they still may differ to a quite large extent. Or, relating to our previous example, it may be possible that the difference between father and stepfather is not expressed.

This variability may be accounted for in at least two different ways: either we think of a_1 belonging to a class C_1 with a certain probability, or to be similar to the objects of C_1 to a certain extent. The same would be true for b_1. That means that the elements in our graph are characterized by the property of belonging to certain cluster.

Or we may assume that the mapping of a a_1 to b_1 is uncertain, and try to give a quantification of that uncertainty by studying pairs ($a_1 \rightarrow b_1$, p) of mappings together with their uncertainties, i. e. a fuzzy mapping.

A special case of this situation would be the problem of homologue elements (variables) studied by Jardine(1967). Rubin (1967)handles the case of non applicable variables, which has its natural place in this context. These are variables (e. g. colour of fur), which have only a meaning if a certain condition (animal with fur) is fulfilled.

In this case we can isolate a subgraph (properties of fur) and define similarities of this subgraph with a class (cluster). The whole subgraph will then be introduced into the graph to be compared (together with some statements of the similarities), e. g. as one node. If the overall similarity of two graphs is high enough it may be necessary to take into account the structure (similarity) of the subgraphs in more detail. That means: the splitting up of the subgraphs will only be necessary in certain cases. Generally, we will speak of a branching analysis. The total graphs will first be simplified by reducing their complexity in combining subgraphs to one element. These elements will be analyzed - a branching into the subgraph will occur - only if these parts of the graphs are similar enough to make the study worthwhile.

We see also that feed-back may occur at this stage: at first we identify the subgraphs in a preliminary way. Then we study the whole graphs, which gives a hint on the cluster to which the subgraph belongs. Now we may use this knowledge to reclassify the subgraph.

Still another aspect would be the possibility to consider the connections within the graphs as fuzzy and uncertain, even if the elements are clearly identified.

This case and the above case, where an element of the graph belongs to a class with a certain probability, could be referred to as clustering fuzzy graphs, while an uncertain mapping ($a_1 \rightarrow b_1$, p) could be named fuzzy clustering.

We will not argue any more on special similarity measures, for they are highly dependent on intuitive and substantial knowledge in the field involved. However we hope that the whole concept of classifying structures presented above will draw explicit attention to the importance of the role of the specialist of an applied science.

REFERENCES

Cormack R.M.,A review of classification,J.Roy.Stat.Soc.A,139, 321-360,1971

Ehrig H.,M.Pfender et al.,Kategorien und Automaten,De Gruyter, Berlin,New York,1972(in German)

Hartigan J.A.,Representation of similarity matrices by trees, J.Am.Statist.Ass.,62,1140-1158,1967

Jardine N.,The concept of homology,Brit.J.Philos.Sci.,18,125-139 1967

Kaufman A.,Introduction à la théorie des sous-ensembles flous, Masson,Paris,1973(in French)

Lerman I.C.,Les bases de la classification automatique,Gauthier--Villars,Paris,1970(in French)

MacLane S.,Categories.For the Working Mathematician,Springer, Berlin,Heidelberg,New York,1972

Rubin J.,Optimal classification into groups: an approach for solving the taxonomy problem,J.Theor.Bio.,15.103-144,1967

Schubert H.,Kategorien I,II (two volumes),Springer,Berlin, Heidelberg,New York,1970 (in German)

Sint P.P.,Conditional Clustering,Contributed Papers of the 39th Session of the International Statistical Institute, Vol. 2, 940-946, 1973

Data Base Reorganization via Clustering

Samuel Gorenstein, New York

1. INTRODUCTION

For large data bases with storage hierarchies one would like to cluster into units of transfer, called blocks, from one level to a higher (faster access) level those records that will be used together. The object is to maximize the probability that a required record will be found in the higher level. However, if the system does not know future requirements and if there is no a priori knowledge of such association among records, we are left with past experience as a basis for predicting simultaneous use.

Thus, we would like to cluster records into blocks on the basis of past experience. However, one of the main difficulties of using clustering algorithms for this purpose, is the number of items to be clustered. Data bases can be very large, hundreds of thousands of records, and the references to them can run into the millions during the course of a week. We therefore require a simple clustering algorithm that can handle a problem of this size. Generally, clustering algorithms are not applied to such large problems, and we find that 1000 items to be clustered is usually considered a large number and most algorithms will have a limit of this order of magnitude on the number of items they can handle (Wishart).

Methods which involve the computation of similarity coefficients between all items to be clustered have to be eliminated since for 300,000 items we would require 4.5×10^{10} calculations and the same number of storage locations. Thus we need a method that does computations on-the-fly, and one whose computations are minimal.

Since the problem is so large, iterative techniques to improve an attained classification into clusters are not permissible and a one-pass algorithm is a requirement. That is, a cluster is to be formed, or an item added to a cluster, each time a pass is made through the data and it is then not to be changed except for adjustments which are not iterative but have been introduced to reduce the problem size or to make equal-sized clusters.

The methods chosen for consideration are 1) a variant of one known as "K-means" since K clusters are required and they are represented by their "mean" during the course of the algorithm; and 2) one known as monothetic division, since the cluster is formed by dividing a larger cluster on the basis of only one attribute (Lance & Williams).

Also, an optimization problem can be formulated and a replacement rule developed which determines the cluster to be removed from the high level storage to make room for the entering cluster. However, this optimization is too large to solve, so heuristics have to be used.

The system studied here is a large administrative system which performs accounting and control functions. The implementation strategy is to use the data references during one week to form clusters (blocks); this is a reorganization of the data base which can be done over a week-end when the system is normally idle or doing batch jobs. It is expected that several hours of CPU time will be devoted to the data base reorganization. This organization will be used during the next week and reference information will again be gathered. The process continues this way with weekly reorganizations.

2. APPLICATION TO THE ADMINISTRATIVE SYSTEM

The administrative system is a terminal system used in the branch offices. The unit of work on this system is a transaction, an administrative function such as collections on accounts receivable, entering cash received, recording orders, etc.

Data management trace tapes are available from which one can determine which records were used during a transaction, i.e., the records used by someone entering a transaction at a terminal.

Thus, one can construct a matrix with 0-1 entries, records on one axis and transactions on the other. An entry of "1" indicates a record was referenced in a transaction and a "0" indicates that it wasn't. Figure 1 shows a typical matrix for m records and n transactions. This will be an extremely sparse matrix. (It will not be represented this way for the clustering algorithm - it is represented by a vector of column numbers and a vector of pointers to where each row

Transactions

	T_1	T_2	T_3	T_4 ...	T_n
R_1	1	1			1
R_2		1		1	
R_3	1		1		1
R_4				1	1
R_5	1		1		
R_6		1		1	
.					
.					
.					
R_m					

Figure 1. The matrix of indicators of records referenced by transactions.

starts. But for discussion it is convenient to refer to it as a matrix.) The object is to put the records into clusters based on their use in the transactions which in the terminology of clustering are considered as attributes of the records with the 0-1 entries indicating abscence or presence of the attributes.

The method of monothetic division starts by considering all the data as one cluster and seeks to divide it into two on the basis of one of the columns, i.e., break off one cluster. The criterion originally considered was the minimum

$$F_i = \frac{(\text{average distance}^2 \text{ from center of cluster } i \text{ to be formed})^{1/2}}{\text{distance between cluster } i \text{ to be formed \& remaining elements of parent cluster}} = \frac{DEV_i}{DIST_i}$$

A column (transaction) defines a potential cluster: the records used in that transaction. F_i is computed for each column i and the minimum over i, i=1,2,...,n, selects the cluster to be formed. A lower bound is set for the number of non-zero elements in a column that is permitted to define a cluster. This lower bound is relaxed if the required number of clusters cannot be formed. If transaction K is selected, the records used in that transaction become the new cluster, and the remaining unclustered elements are now subject to division. The process continues until the required number of clusters are formed.

Some explanation of the computations is in order. Each row (record) is an n-dimensional vector. A cluster is a set of vectors (rows) and its center is the center of gravity, i.e., if $x^i = (x_1^{\ i}, x_2^{\ i}, \ldots, x_n^{\ i})$ is row i and S is the set of row indices in the cluster, $i \in S$, then if we call this set cluster S, the center of gravity of S is given by

$$\frac{1}{|S|} = \sum_{i \in S} x^i$$

where $|S|$ is the number of elements in S.

Any metric can be used for distance, we have used Euclidean distance.

$$d(x,y) = (\sum (x_i - y_i)^2)^{1/2}$$

So DEV_i is the square root of the average distance squared of all rows in a set from its center. $DIST_i$ is the distance between the rows selected and the remaining unclustered rows.

3. REVISED COMPUTATIONS

Because of the size of the data matrix, even the simple computations discussed in the previous sections are more than is practically possible. For example, the center of the parent cluster could have 125,000 elements requiring a minimum of 125×10^3 operations to find the distance between clusters, $DIST_i$. If this were done for 125,000 columns, we have 1.56×10^{10} operations for a cluster, and for 3,500 clusters we have on the order of 10^{13} operations - at a microsecond each we have hundreds of hours of computation time. Therefore the criterion was changed to be min DEV_i. The computation for average distance from the center for a small cluster is not prohibitively time consuming since the vectors are so sparse, having only a small number of non-zero components, and these are the only components that need be considered.

To further reduce computation, the following is done:

A) instead of all columns being considered as candidates for forming clusters, they are sampled, tentatively 5%. Resampling is done when the lower bound is relaxed.

B) The algorithm described, with the DEV criterion only, is applied until the required number of clusters are formed. Then for each unclustered row, the total number of connections to clusters already formed are counted, and the row is joined to the smallest cluster to which it has the maximum average number of connections. If it has no connections to clusters already formed, it is either joined to the smallest cluster or starts a new one. In the latter case, a program to merge clusters is required. Clusters are not permitted to grow beyond the block size.

4. AN ALTERNATE ALGORITHM (K-MEANS).

Using only the $\min_i DEV_i$ criterion still involves substantial computation. To reduce this computation, an algorithm can be used which is essentially only the last step - adjoining the element to the smallest cluster to which it has the maximum value for: no. of connections ÷ cluster size. This normalizing factor was suggested by the formula for a cluster's re-reference probability. A START program starts off the required number of clusters with unrelated (no entries in the same column) elements. Sequentially, new elements are adjoined to those clusters to which they have the maximum connections as defined above. Surprisingly, in the small experimental problems, this gave better clusters, according to the average derivation criterion, than the one that used the DEV criterion directly. We can only speculate on the reason for this. Elements became candidates for a cluster on the basis of one column, but, since the matrix is so sparse, they were only minimally related through other columns and they were thus formed into clusters of relatively poor homogeneity. However, there is no guarantee that the DEV criterion has any real significance. The only true test of a method is the one that minimizes the number of misses, i.e., the number of times one has to access the lowest level of the storage hierarchy.

This method is a variant of the K-means method of clustering (MacQueen).

A refinement can be introduced which would probably improve the clustering, but at a cost of more storage and computation. This is described below, but left for future experimentation.

In addition to the data in the Record-Transaction matrix, we record the sequence numbers of references in a transaction and/or maintain a frequency count of nonsequential references to the same record within a transaction (the frequency count is useful in optimization problems that can be formulated). Presumably, the closer in time sequence the use of records in a transaction the more related they are.

Now, instead of merely counting the number of connections to clusters, we weigh these connections inversely by the differences in sequence numbers, and take the weighted average as a measure of relatedness to a cluster, and, as before, attach the element to the smallest cluster to which it is most related. Storing the sequence numbers would double the storage requirements and would make the experimentation more difficult. But it might be considered in an implementation for reorganization of a data base.

PRELIMINARY RESULTS

Using the simplest of the described algorithms, clustering by maximal average connections (MAC), an experiment was conducted on a very well organized large data base used for administrative purposes, which includes such accounting and control functions as cash receipts, collections, inventory, commissions, etc.

The data base consists of 3.4 billion bytes on some 150 disk packs. The data accesses for one day were recorded on a trace tape. Based on these references, the records were clustered with the MAC algorithm, and the day's misses were counted under simulated conditions, in which the transfer blocks from the lower to higher level storage were constructed as follows:

1) the current organization constructed by applications programmers on the basis of knowledge of the accesses made by the applications programs

2) clustering by MAC

3) random blocks

The block sizes were two cylinders of a 2314 disk (about 280,000) bytes, and the units clustered were tracks, about 7000 bytes, or 40 tracks to a cluster. The results are shown in the accompanying chart, Figure 2. While the MAC was not as good as the current organization, it was superior to a random organization. LRU was used as a replacement algorithm in these simulations. We expect that a replacement rule using transition probabilities would show improved performance but this has not yet been tested. From other tests, we have reason to believe the current organization is a very good one because it is well known to the programmers how the data base is being accessed. We expect that for a data base with many diverse users, each accessing the data for different purposes and in different ways, the clustering algorithms would take advantage of knowledge gained by observing the accesses and reorganize accordingly. It is not expected that the clustering would improve the performance of specialized systems serving homogeneous users.

ACKNOWLEDGMENT

Without the efforts of George Galati who prepared the input to the clustering programs from the trace tapes, and processed the output from the clustering program for input to the efficient stack distance programs of Brian Bennett for counting misses under LRU, this work would not have been possible.

Also, discussions with Richard Hirko of his preliminary studies of the traces, provided understanding of the system's accesses to the data base.

Discussions with Hisashi Kobayashi were very helpful in clarifying ideas about the clustering and he provided encouragement and support throughout the course of this work.

The individuals mentioned above are my colleagues at the IBM Research Division.

REFERENCES

Lance, G. N. and Williams, W. T., Computer Programs for Monothetic Classification ("Association Analysis"). Computer Journal, V. 8, 1965, p. 246-249.

Wishart, D., A General Tripartite Clustering Method and Similarity Generating Function. Statistics 1 Division Report No. R-31, Civil Service Department, Whitehall, London SW1, April 1972.

MacQueen, J., Some Methods for Classification and Analysis of Multivariate Observations. In L. M. LeCam and J. Neyman (eds.), Proc. of the Fifth Berkeley Symposium on Mathematical Statistics and Probability, Vol. 1, pp. 281-297. University of California Press, Berkeley, California (1967).

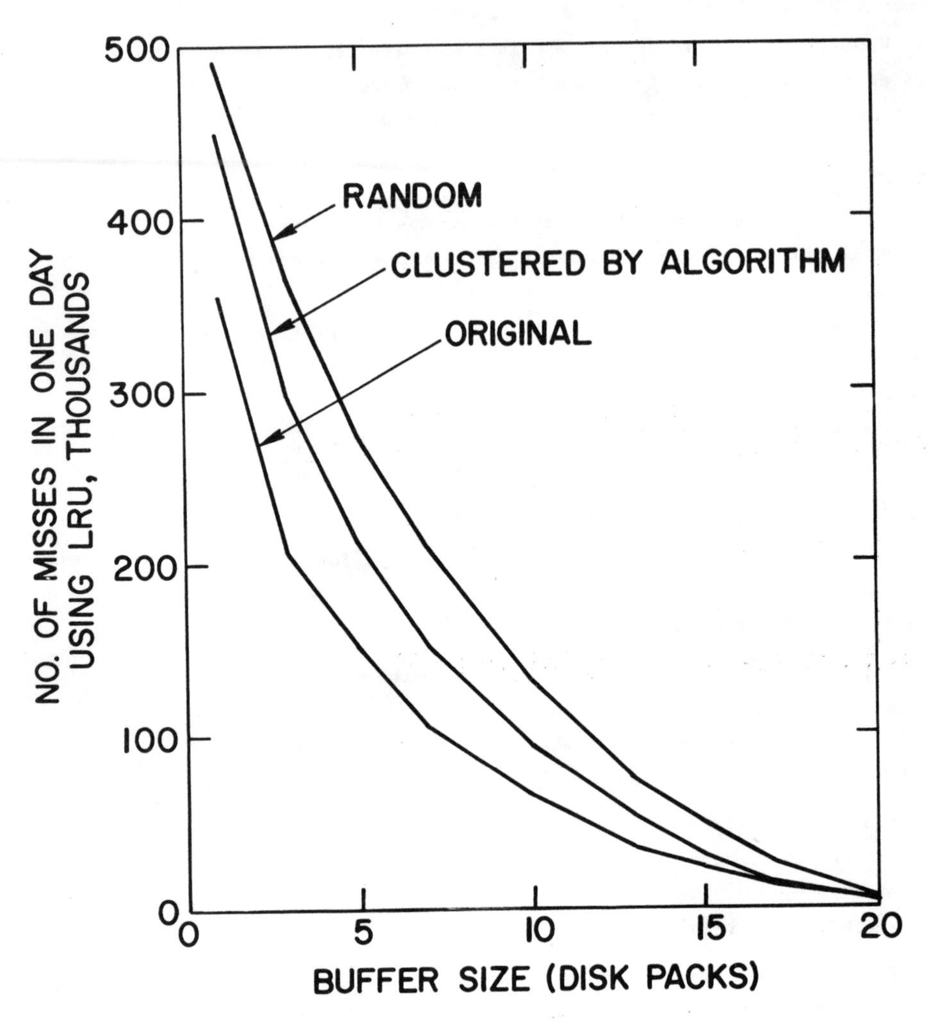

FIGURE 2. MISSES FOR VARIOUS FILE ORGANIZATIONS.

A Stepwise Discriminant Analysis Program Using Density Estimation

J. D. F. Habbema, J. Hermans, and K. van den Broek, Leiden

0. SUMMARY

An outline is given of a discriminant analysis program for continuous variables. The main features of this program are:

- The classification procedure connected with the discriminant analysis, will be based on the criterion of minimization of the expected loss.
- The expected loss will be estimated by a leaving-one-out technique.
- The variables are selected in a forward stepwise way. The selection criterion is based on minimization of expected loss.
- The density of the variables in the populations is estimated with a potential function method.
- The smoothing parameter of the potential function is estimated by a modified maximum likelihood method.

KEYWORDS:
Allocation rule; Density estimation; Discriminant analysis; Expected loss; Forward selection; Leaving-one-out technique; Posterior probability; Smoothing parameter.

1. INTRODUCTION

Discriminant analysis, or allocation theory, is concerned with the problem of allocating an element to one of k populations $A_1, \ldots, A_k$, after observation of variables giving probabilistic information about the population of origin of the element. We consider the situation where all the variables are continuous.

Most computer programs for continuous variables are based on normality assumptions, usually with equal covariance matrices, like the BMD programs for discriminant analysis, DIXON (1971), sometimes with unequal covariance matrices, like VICTOR (1972). However, there can be considerable non-normality in practical data, so a computer program using a general method of density estimation seems worthwhile. The theory of this method is developed in section 2.

Our program can be used for selection of a subset from the total set of variables. Usually it is impossible to investigate all subsets of the original set of variables, so we choose a forward stepwise procedure. The selection criterion is, in accordance with the allocation criterion, based on expected loss. Two programs for linear discriminant analysis using stepwise selection, are the forward selection BMD 07M program, and the backward selection program of MICHAELIS (1972). They use selection criteria that are not directly related to the allocation criteria. MICHAELIS (1972) also describes a quadratic discriminant analysis, where the estimated probability of misclassification is used as the selection criterion. The theory concerning the allocation rule, the selection process, and the estimation of expected loss will be given in sections 3 and 4.

In sections 5 and 6, the options and the CPU time aspects of the program will be discussed, together with a practical example. We conclude with a short section on recommendations for the use of the program.

2. ESTIMATION OF THE PROBABILITY DENSITY

The potential function method is used for estimation of the p-dimensional density in each of the k populations. The general form of this estimator is for a sample $x_{j1}, \ldots, x_{jN_j}$ given by

$$f_j(x) = \frac{1}{\sigma_j^p} \frac{1}{N_j} \sum_{r=1}^{N_j} K\{(x-x_{jr})/\sigma_j\} \qquad j=1,\ldots,k$$

Here $K(\cdot)$ is a potential function, centered at x_{jr}, the r-th sample element, while σ_j is a smoothing parameter. A well-known difficulty is the selection of a value for σ_j. Three steps can be distinghuished in the way we handled this problem. First the variances of the variables within the populations are standardized, by the transformation

$$y_{jti} = x_{jti}/s_{ji} \qquad \begin{array}{l} j = 1, \ldots, k \\ t = 1, \ldots, N_j \\ i = 1, \ldots, p \end{array} \qquad (2)$$

with s_{ji} the sample standard deviation.

We then have in each of the k populations p variables y, each with sample variance equal to 1. So if we choose a Gaussian potential function, it is appropriate to take $\Sigma_j = I$ as its covariance matrix. This gives for the density estimator:

$$h_j(y) = \frac{1}{(\sigma_j\sqrt{2\pi})^p} \frac{1}{N_j} \sum_{r=1}^{N_j} \exp - \left\{ \frac{(y-y_{jr})'(y-y_{jr})}{2\sigma_j^2} \right\} \tag{3}$$

For discussion, references and an application in discriminant analysis see HABBEMA, HERMANS, and van der BURGT (1974).

The second step is the actual calculation of a value for σ_j. A straightforward application of the maximum likelihood (M.L.) method gives the useless result of $\hat{\sigma}_j=0$ as M.L. estimate, for the likelihood function

$$g(\sigma_j) = \prod_{t=1}^{N_j} h_j(y_{jt}) \tag{4}$$

goes to infinity when σ_j goes to zero. This is easily seen by substituting y_{jt} in eq. (3). For the index r=t gives a contribution of 1 to the sum of exponentials, and as a result $h_j(y_{jt}) > c.\sigma_j^{-p}$, and $g(\sigma_j) > (c.\sigma_j^{-p})^{N_j}$. To get rid of this unwanted behaviour we can use a 'leaving-one-out' modification of the M.L. method. The density estimator in the point y_{jt} will be based on all sample elements from population j except element y_{jt}:

$$h_j^{(t)}(y_{jt}) = \frac{1}{(\sigma_j\sqrt{2\pi})^p} \frac{1}{N_j-1} \sum_{\substack{r=1 \\ r\neq t}}^{N_j} \exp\left\{ - \frac{(y_{jt}-y_{jr})'(y_{jt}-y_{jr})}{2\sigma_j^2} \right\} \tag{5}$$

As estimator for the smoothing constant σ_j we take the value $\hat{\sigma}_j$ with

$$g(\hat{\sigma}_j) = \max_{\sigma_j>0} \prod_{t=1}^{N_j} h_j^{(t)}(y_{jt}) \tag{6}$$

The third and final step consists of substitution of the value $\hat{\sigma}_j$ in eq. (3) and the transformation back, from y to x, see eq. (2). We then get for the estimated density in population j for j = 1, ..., k:

$$\hat{f}_j(x) = \frac{1}{(\hat{\sigma}_j\sqrt{2\pi})^p s_{j1}\cdots s_{jp}} \frac{1}{N_j} \sum_{r=1}^{N_j} \exp\left\{ -\frac{1}{2} \sum_{i=1}^{p} \left(\frac{x_i-x_{jri}}{\hat{\sigma}_j s_{ji}} \right)^2 \right\} \tag{7}$$

Eq. (7) is a key formula for the computer program and will be used in each step of the selection process.

3. THE ALLOCATION RULE AND THE ESTIMATION OF THE EXPECTED LOSS

Each allocation or classification procedure can be specified by an allocation function $a(x)$:

$$a(x) = a_i \text{ if } x \text{ is allocated to population } A_i.$$

Let for an element $x \in A_j$ the loss associated with the allocation $a(x) = a_i$ be $L(a_i,A_j)$. In order to determine the allocation function or rule $a(x)$ the Bayes strategy of minimizing the expected loss is chosen in this paper. For the special loss function $L(a_i,A_j) = 0$ if $i=j$ and 1 if $i\neq j$, this criterion is equivalent to minimization of the probability of error classification. Fòr the case of known loss L, prior probabilities $\pi(A_j)$ and densities $f_j(x)$, the optimal rule is e.g. given by RAO (1965). When $f_j(x)$ is unknown we can substitute in the optimal rule the density estimator $\hat{f}_j(x)$, see eq. (7). This leads to the following allocation rule:

$$\left.\begin{aligned} &\text{Let} && b_m(x) = \sum_{j=1}^{k} \pi(A_j)\, L(a_m,A_j)\, \hat{f}_j(x), \\ &\text{then} && a(x)=a_i \text{ if } b_i(x) = \min_{m=1,\ldots,k} b_m(x). \end{aligned}\right\} \qquad (8)$$

The expected loss associated with this rule might be written as

$$\begin{aligned} E(L) &= \sum_{i,j=1}^{k} \pi(A_j)L(a_i,A_j)P(a_i|A_j) \\ \text{with } P(a_i|A_j) &= P\{a(x)=a_i|x\in A_j\}= \int_{\{x|a(x)=a_i\}} \hat{f}_j(x)dx \end{aligned} \qquad (9)$$

For the selection process, described in section 4, it is necessary to calculate the expected loss (9). In fact this means that one has to estimate the allocation probabilities $P(a_i|A_j)$ associated with rule (8). As estimation method we used the leaving-one-out method, see LACHENBRUCH and MICKEY (1968). To estimate $P(a_i|A_j)$ each sample element x_{jr}, $r=1, \ldots, N_j$ is allocated. Not, however, by resubstitution of x_{jr} in rule (8), but in the rule based on all sample elements except x_{jr}. So one does not calculate $b_m(x_{jr})$, see eq. (8), but

$$b_m^{(r)}(x_{jr}) = \sum_{\substack{t=1\\t\neq j}}^{k} \pi(A_t)L(a_m,A_t)\hat{f}_t(x_{jr}) + \pi(A_j)L(a_m,A_j)\hat{f}_j^{(r)}(x_{jr}) \qquad (10)$$

with $\hat{f}_j^{(r)}$ based on (N_j-1) sample elements from A_j: x_{jr} is leaved out.
Now let

$$a(x_{jr}) = a_i \quad \text{if} \quad b_i^{(r)}(x_{jr}) = \min_{m=1,\dots,k} b_m^{(r)}(x_{jr}) \qquad (11)$$

and define for rule (11)

$$m_i^{(j)} = \#\{a(x_{jr}) = a_i \ , \ r = 1,2,\dots,N_j\} \qquad (12)$$

then the relative frequency of sample elements from A_j allocated to A_i by (11), $m_i^{(j)}/N_j$, is the leaving-one-out estimator for $P(a_i|A_j)$.

As pointed out by FUKUNAGA and KESSELL (1971) there is no difference in computation times between the resubstitution and the leaving-one-out estimator for $P(a_i|A_j)$ for the density estimation approach, in contrast to the situation when normal densities are assumed.

Substitution of the derived estimator in eq. (9) gives for the estimated expected loss

$$\hat{EL} = \sum_{i,j} \pi(A_j)L(a_i,A_j) \frac{m_i^{(j)}}{N_j} \qquad (13)$$

Equation (13) plays a central role in the selection process.

4. THE SELECTION PROCESS

The optimality criterion used for the derivation of the allocation rule, was minimizing the expected loss. So an obvious criterion for the selection of variables is also minimizing the (estimated) expected loss. Essentially eq. (13) is used here.

In the first step of our forward stepwise selection procedure, p allocation rules are considered, each based on one of the variables x_i, i=1, ..., p. The estimated expected loss $\hat{EL}(x_i)$ is computed by eq. (13) for these p rules. Variable x_{i_1} is selected with

$$\hat{EL}(x_{i_1}) = \min_{i=1,\dots,p} \hat{EL}(x_i) \qquad (14)$$

In the second step another variable x_{i_2} is selected, such that this new variable, together with x_{i_1}, gives a minimal estimated expected loss for all pairs of variables containing x_{i_1}:

$$\hat{EL}(x_{i_1}, x_{i_2}) = \min_{\substack{i=1,\ldots,p \\ i \neq i_1}} \hat{EL}(x_{i_1}, x_i) \tag{15}$$

This process can be continued p steps; we then arrive at an ordering of the variables $(x_{i_1}, x_{i_2}, \ldots, x_{i_p})$. Usually we like to know whether it is worthwhile to stop the selection process earlier. A stopping criterion, in accordance with the selection criterion, is as follows. Consider some threshold value δ and stop after step r if after this step for the first time

$$\hat{EL}(x_{i_1}, \ldots, x_{i_r}) - \hat{EL}(x_{i_1}, \ldots, x_{i_{r+1}}) \leqslant \delta \tag{16}$$

For an interpretation of eq. (16) it might be usefull to consider δ as the costs of introducing another variable. In the special case of a zero-one loss matrix L the criterion (16) simplifies to

$$\hat{P}_r(\text{error}) - \hat{P}_{r+1}(\text{error}) \leqslant \delta \tag{17}$$

i.e. stop the selection after r steps if the decrease in estimated probability of error classification is less than or equal to δ.

So this process leads finally to a selected set of r variables $(x_{i_1}, x_{i_2}, \ldots, x_{i_r})$. Allocation rule (8) can now be based on this set of variables.

Remark Throughout the whole selection procedure we use values for $(\hat{\sigma}_1, \ldots, \hat{\sigma}_k)$ based on the complete set of variables $(x_1, \ldots, x_p)$ simultaneously. It would be better to use for each computation of a value for $\hat{EL}(x_{i_1}, \ldots, x_{i_r})$ a vector of σ-values based on the set of variables $(x_{i_1}, \ldots, x_{i_r})$. This seems, however, practically infeasible, because in the first step of the selection process one should then p times calculate a vector $(\hat{\sigma}_1, \ldots, \hat{\sigma}_k)$, in the second step (p-1) times, etc.

5. OPTIONS, LIMITATIONS, AND TIME ASPECTS OF THE COMPUTER PROGRAM

Using the computer program, based on the theory of the foregoing sections, a discriminant analysis can be performed with stepwise selection of variables. From the k populations training samples of sizes N_j $(j=1,\ldots,k)$ are assumed. A number of m test groups might be added. The data of these test groups are not taken into account for the density estimation and the selection process.

The prior probabilities $\pi(A_j)$, $j=1,\ldots,k$ can be chosen equal to $1/k$, or proportional to the sample sizes N_j, or can be read. The loss-matrix $L(a_i, A_j)$ can optionally be taken as a zero-one matrix, i.e. $L(a_i,A_j) = 0$ if $i=j$ and 1 if $i \neq j$, or can be read. The smoothing parameters σ_j can be determined by the program by maximizing $g(\sigma_j)$, see eq. (6), or can be read. The threshold loss δ has to be specified in the input. One can also specify a maximal number of steps to be taken in the selection process. In this case the program may stop earlier according to the δ criterion.

The selection process might start by selecting a first variable out of all the p variables. It is also possible to select a subset of q variables in advance. In that case the selection process starts with the selection of the (q+1)-th variable out of the remaining (p-q). By choosing q=p the program can in fact be used in a non-stepwise way.

After each selection step a classification matrix is given. For this matrix each element of the (k+m) groups is allocated to the population A_j, j=1, ..., k, giving the minimal estimated expected loss. In case of a zero-one loss matrix this is equivalent to allocation to the population with the highest posterior probability. The posterior probabilities are given for each element at the end of the program; it is optimal to get them also after each selection step. For an element x of the test-data the posterior probabilities are calculated according to

$$P(A_j|x) = \pi(A_j)\hat{f}_j(x) / \sum_{m=1}^{k} \pi(A_m)\hat{f}_m(x) \qquad ,j=1,\ldots,k \quad (18)$$

For an element x_{jt} of the training-data the posterior probabilities are calculated according to

$$P(A_j|x_{jt}) = \frac{\pi(A_j)\hat{f}_j^{(t)}(x_{jt})}{\sum_{\substack{m=1 \\ m\neq j}}^{k} \pi(A_m)\hat{f}_m(x_{jt}) + \pi(A_j)\hat{f}_j^{(t)}(x_{jt})} \quad (19)$$

A repeated use of the same dataset with different input parameters as well as the use of different datasets is possible in the same job run.

At the moment the following limitations hold for the program:

(1) Number of groups (k+m) < 20,

(2) Number of variables (p) < 25,

(3) N * (k+m) < 5.000, where $N = N_1+\ldots+N_{k+m}$,

(4) N * p < 10.000,

(5) Required memory capacity 136 K.

The program is written in Fortran IV, and run on the IBM 370/158 of the University of Leiden.

The program requires relatively much CPU time, which depends on the number of variables, the number of variables to select, the number of groups, and the number of individuals. The CPU time increases when we have a smaller number of populations for the same total number of individuals. We are not yet able to approximate the required computer time as a function of the quantities involved, but maybe some examples give a good illustration of the orders of magnitude;

Example 1 : Number of variables p=11; number of groups k=2, number of individuals in the groups N_i=30; number of variables to select q:

q	1	2	3	4	5	6	7	8	9	10	11
T (sec)	12	20	25	32	42	53	58	68	70	75	77

Example 2 : p=22, k=2, N_i=30, q is variable :

q	1	2	3	5
T	22	36	53	100

Example 3 : p=5, k=4, N_i=50, q is variable :

q	1	2	3	4
T	56	85	114	145

Example 4 : p=5, k=4, q=3, N_i is variable :

N_i	20	30	40	50
T	19	43	81	114

Example 5 : p=5, N=200, q=3, k is variable :

k	2	4
T	151	114

Example 6 : p=10, k=2, N_i=120, q=5, T $\approx$ 17 minutes

6. EXAMPLE

In genetic counseling the question occurs how to discriminate between normal women and haemofilia A carriers, with the following two variables: $u = {}^{10}\log$(AHF activity) and $v = {}^{10}\log$(AHF-like antigen). Reference (training) data of 45 obligatory carriers and 30 normals were available. Furthermore data on u and v were given for a test group of 23 possible carriers. In the present example the prior probability on carriership equals the genetic chance on carriership. For each of the possible carriers this genetic chance is well-known from the pedigree.

Examination of the distribution of the reference data led to the conclusion that the assumption of 2-dimensional normal distributions with equal covariance matrices was very reasonable. So for the group of possible carriers the posterior probabilities are calculated under these assumptions. On the other hand they are also calculated with the density estimation approach. The results are given in Table 1. This table shows in most of the cases a very close agreement between the two posterior probabilities. The density estimation approach leads to slightly lower posterior probabilities of carriership in the region containing the cases 2, 10, 15, and 17 while slightly higher values occur in the region containing the cases 5, 11, 13, 14, 18, and 23.

TABLE 1. Probabilities on carriership of haemofilia A for 23 possible carriers

Case	Prior proba-bility	Posterior probability (density estimation)	Posterior probability (normal densities)	Case	Prior proba-bility	Posterior probability (density estimation)	Posterior probability (normal densities)
1	.25	.179	.206	13	.50	.248	.062
2	.66	.847	.973	14	.50	.515	.424
3	.33	.038	.013	15	.66	.948	.990
4	.33	.174	.151	16	.66	.129	.050
5	.50	.088	.010	17	.66	.892	.955
6	.25	.024	.009	18	.66	.497	.286
7	.50	.037	.014	19	.50	.935	.996
8	.50	.066	.024	20	.50	.992	.998
9	.50	.002	.002	21	.50	.985	.995
10	.50	.904	.987	22	.66	.960	.971
11	.50	.683	.501	23	.50	.235	.190
12	.50	.813	.792				

In this example the selection of variables is not the central question, but the comparison of posterior probabilities, calculated both under the assumption of normality and under a density estimation approach. In this 2-dimensional case with moderate sample sizes for the training data, it could be checked that the normal model (with equal c.v. matrices) fitted the data very well. It turned out (Table 1) that the approach with the density estimation procedure gives nearly the same results as the approximately optimal normal approach.

7. CONCLUSIONS

As a selection program, this program is first of all suited for rather small-scale problems in k-group discriminant analysis. The particular advantages are direct density estimation, use of the expected loss in the stepwise variable selection procedure, and the use of a leaving-one-out technique in estimating allocation probabilities. The main disadvantage is the amount of computer time required for the selection process, dependent on the number of variables, number of groups, and number of elements per sample. The costs involved with the selection program are, on the other hand, very small relatively to the costs of most biomedical research projects, where the variable selection is required.

As a program for routine use, i.e. for the allocation of new elements the program is also suited for large-scale problems. For the allocation of new elements it is straightforward and requires only minimal computer time: the set of variables is fixed, the value of the vector of the smoothing parameter $\sigma_1,\ldots,\sigma_k$ is known, and allocation has only to be done for a limited set of new elements. The main limitation here is the storage of the training samples of the k populations.

8. REFERENCES

Dixon, W.J. (ed.). BMD Biomedical Computer Programs, Univ. of Calif. Press, 1971

Fukunaga, K., and D.L. Kessell, Estimation of classification error, IEEE transactions on computers, Vol. C-20, 1521-1527, 1971

Habbema, J.D.F., J. Hermans and A.T. van der Burgt, Cases of doubt in allocation problems, Biometrika (61), 1974 (to be published)

Lachenbruch, P.A., and M.R. Mickey, Estimation of error rates in discriminant analysis, Technometrics (10), 1-11, 1968

Michaelis, J., Computerprogramme zur Variabelenreduktion bei der linearen und quadratischen Diskriminanzanalyse für mehrere Gruppen, EDV in Medizin und Biologie (4), 97-102, 1972

Rao, C.R., Linear statistical inference and its applications, Wiley, New-York, 1965

Victor, N., A nonlinear discriminant analysis, Computer Programs in Biomedicine (2), 36-50, 1971

Combinatorial Problems in Non-Parametric Classification Theory

David M. Jackson, Waterloo

1. INTRODUCTION

The purpose of non-parametric classiciation theory is to construct classifications of the objects of a population when there is no known statistical model for the population. For such methods to be adequate it is necessary to confirm that the resulting classification is stable; that is, "small changes" in the data result only in small changes in the classification. The failure of a classification to be stable may be attributed to the inadequacy of the classification algorithm or may indicate that insufficient data has been collected. The question of stability has been discussed elsewhere by Jardine, Jardine and Sibson [1967] and Jackson and White [1971], while discussions of non-parametric classification, with extensive bibliographies, may be found in Dempster [1971] and Cormack [1971]. The details of an error analysis carried out for the case in which the attributes are discrete and errors occur independently in the attributes has been given by Jackson and White [1972] and by Jackson [1972].

The purpose of this paper is to discuss the extent to which the methods of combinatorial enumeration theory can be used effectively to answer computationally some of the difficult combinatorial questions raised by non-parametric classification theory. Many of the problems in classification theory resolve into combinatorial problems because they involve the computation of functions over families of discrete structures. To demonstrate the use of these methods a specific problem is examined. It arose in the investigation of the computational complexity of a method for computing the expectations of statistical functions of generalised distance functions when the attribute states are susceptible independently to error. Jackson and Wadge [1974]. This problem involved the determination of the number of non-negative integer $m \times m$ doubly stochastic matrices with line sum n as a function of m and n. This problem occurs elsewhere in classification theory in connexion with the χ^2 distribution.

Three combinatorial methods will be discussed in the following three sections. These methods involve crude generating functions, group theoretic methods and residue methods. Proofs of the supporting theorems have not been given in the interests of brevity, but appropriate references are cited. Typical results for special cases may be found in Abramson and Moser [1973] and Carlitz [1966].

2. CRUDE GENERATING FUNCTIONS

Let $H_m(n)$ be the number of $m \times m$ matrices over the non-negative integers with row sum n and column sum n. Then

$$H_m(n) = [x_1^n x_2^n \dots x_m^n][y_1^n y_2^n \dots y_m^n] \prod_{1 \le i,j \le m} (1-x_i y_j)^{-1} \qquad (1)$$

where the x_i and y_j are indeterminates and $[x^n]f(x)$ denotes the coefficient of x^n in $f(x)$. The operator is associative, commutative and distributive with respect to addition. This notation will be maintained throughout this paper. Let $\Phi_m(t)$ be the ordinary generating function for the sequence $\{H_m(n)\}$ so

$$\Phi_m(t) = \sum_{n\geq 0} H_m(n)t^n \tag{2}$$

$$\left.\begin{aligned} \text{Then}\quad & \Phi_m(t) = [x_1^0x_2^0\ldots x_m^0][y_1^0y_2^0\ldots y_m^0]\phi(x_1,x_2,\ldots,x_m,y_1,y_2,\ldots,y_m,t) \\ \text{where}\quad & \phi(x_1,\ldots,t) = (1-x_1^{-1}x_2^{-1}\ldots x_m^{-1}t)\prod_{1\leq i,j\leq m}(1-x_iy_j)^{-1} \end{aligned}\right\} \tag{3}$$

The problem of determining $\Phi_m(t)$ is accordingly reduced to one of applying the operator $[x_1^0\ldots x_m^0][y_1^0\ldots y_m^0]$ to the function $\phi(x_1,\ldots,t)$. The latter function is called a crude generating function, following MacMahon [1960], since it devolves into the ordinary generating function after the elimination of the indeterminates.

Methods exist for carrying out the elimination of the indeterminates. Let $<A>$ denote $(1-A)^{-1}$. Then the following are trivial identities:

$$\left.\begin{aligned} <\alpha><\beta> &= <\alpha\beta>(<\alpha>+<\beta>-1) \\ <0> &= 1 \end{aligned}\right\} \tag{4}$$

This identity may be used as the basis for a strategy Elliott, E.B. [1903] for eliminating the indeterminates by selecting α and β such that there is a cancellation of at least one indeterminate in the product $\alpha\beta$. The following example illustrates the method for $\Phi_2(t)$

From Eq.(3), using the notation of Eq.(4), we have

$$\Phi_2(t) = [x_1^0x_2^0][y_1^0y_2^0]<x_1y_1><x_1y_2><x_2y_1><x_2y_2><x_1^{-1}x_2^{-1}y_1^{-1}y_2^{-1}t> \tag{5}$$

Carrying out the first stage of the elimination of x_1 using Eq.(4) we have:

$$\begin{aligned} & [x_1^0]<x_1y_1><x_1y_2><x_1^{-1}x_2^{-1}y_1^{-1}y_2^{-1}t> \\ &= [x_1^0]<x_1y_1><x_2^{-1}y_1^{-1}t>(<x_1y_2>+<x_1^{-1}x_2^{-1}y_1^{-1}y_2^{-1}t>-1) \\ &= <x_2^{-1}y_1^{-1}t>[x_1^0]<x_1y_1><x_1^{-1}x_2^{-1}y_1^{-1}y_2^{-1}t> \quad (\text{since } [x_1^0]<x_1y_1><x_1y_2>=1) \\ &= <x_2^{-1}y_1^{-1}t>[x_1^0]<x_2^{-1}y_2^{-1}t>(<x_1y_1>+<x_1^{-1}x_2^{-1}y_1^{-1}y_2^{-1}t>-1) \\ &= <x_2^{-1}y_1^{-1}t><x_2^{-1}y_2^{-1}t> \end{aligned}$$

The remaining variables may be eliminated in a similar fashion, whence using Eq.(5):

$$\begin{aligned} \Phi_2(t) &= [y_1^0y_2^0][x_2^0]<t>^2(<x_2y_1>+<x_2^{-1}y_1^{-1}t>-1)(<x_2y_2>+<x_2^{-1}y_2^{-1}t>-1) \\ &= <t>^2 = (1-t)^{-2}\ . \end{aligned}$$

Thus $H_2(n) = [t^n]\Phi_2(t) = n+1$.

While this process is easy to apply in small cases, the recursive character of Eq.(4) results in the creation of three new subproblems with each application of the identity. Although some simplification can be obtained by looking for equivalence classes of subproblems under permutation of variables, this is feasible only at the early stages of the computation. It is not easy to construct such equivalence classes, and the decoding of the result of elimination, expressed in the angle-bracket notation, to the standard form also presents problems.

However, some reduction may be obtained by determining the largest values of p and q with which the dexter of

$$F(\underline{a},\underline{b}) \equiv [\lambda^0] \prod_{i=1}^{p} \langle a_i\lambda\rangle \prod_{j=1}^{q} \langle b_j\lambda^{-1}\rangle \tag{6}$$

appears in the elimination process. Intermediate stages will involve special cases of Eq.(6), where some of the a's and b's may be zero. The appropriate elimination formula may be constructed directly from Eq.(6) thereby avoiding repetitive computation of $F(\underline{a},\underline{b})$ for each instance of $\underline{a}$ and $\underline{b}$ as they arise. The programming of this algorithm was considered to be too cumbersome to attempt in full.

3. GROUP THEORETIC METHOD

The matrices may be interpreted as incidence matrices of bichromatic graphs with restricted valencies. The cycle index polynomial of the automorphism group of these graphs may be utilised in the superposition theorem Redfield [1927], Read [1959 and 1963] to determine an expression for the result. The method rests on the following theorem:

THEOREM 1: $H_m(n) = N(h_n^m * h_n^m)$

where i) h_n is the cycle index polynomial in the indeterminates $x_1, x_2, \ldots, x_n$ for the symmetric group S_n

ii) If $A(\underline{x})$, $B(\underline{x})$ are multivariate polynomials and

$A(\underline{x}) = \sum_{(\underline{i})} a_{\underline{i}} \underline{x}^{\underline{i}}$, $B(\underline{x}) = \sum_{(\underline{i})} b_{\underline{i}} \underline{x}^{\underline{i}}$

where the summation is taken over all partitions $(\underline{i}) = 1^{i_1} 2^{i_2} \ldots n^{i_n}$ of n , the inner product $A*B$ is defined by

$$A(\underline{x})*B(\underline{x}) = \sum_{(\underline{i})} a_{\underline{i}} b_{\underline{i}} g(\underline{i}) \underline{x}^{\underline{i}}$$

where $g(\underline{i}) = 1^{i_1} 2^{i_2} \ldots n^{i_n} i_1! i_2! \ldots i_n!$

iii) $N(A(\underline{x})) = A(\underline{x})\big|_{\underline{x}=(1,1,\ldots,1)}$

PROOF: See Read [1963].

For small values of n , Theorem 1 provides useful results. For example, if $n = 2$, then $h_2 = \frac{1}{2}(x_1^2+x_2)$ whence

$$H_m(2) = N(h_2^m * h_2^m) = \frac{(m!)^2}{2^m} \sum_{i=0}^{m} \frac{2^i}{(m-i)!} \binom{2i}{i} \qquad (7)$$

For $m>2$, the multinomial theorem must be applied to exponentiate the cycle index polynomial and this leads to problems when direct computation is to be carried out. The computation is simplified by means of Schur functions, utilising the orthogonality relation that these functions possess. Let $\{\lambda\}$ be the Schur function associated with a partition (λ) of n. Then $h_n = \{n\}$ gives the representation of the cycle index polynomial for S_n in terms of Schur functions. The following lemmas are needed.

LEMMA 1: $h_n^m = \sum_{(\lambda)} \alpha_\lambda \{\lambda\}$

where the summation is over all partitions of mn, and α_λ is rational.

PROOF: h_n^m is a symmetric function.

LEMMA 2: $N(\{\lambda\}*\{\mu\}) = \begin{cases} 1 & \text{if } (\lambda) = (\mu) \\ 0 & \text{if otherwise} \end{cases}$.

PROOF: See Littlewood [1950].

THEOREM 2: $H_m(n) = \sum_{(\lambda)} \alpha_\lambda^2$.

PROOF: Direct from Theorem 1 and Lemmas 1, 2.

The problem of computing $H_m(n)$ is accordingly reduced to the problem of decomposing h_n^m as a linear combination of Schur functions. This may be carried out using the following product scheme Littlewood [1950], Read [1968].

Let $\{\lambda\}$ be the Schur function corresponding to a partition (λ) of r. To evaluate the product $\{\lambda\}\{n\}$ construct the Young diagram for (λ) using "asterisks" and add to this diagram n "dots" in all possible ways, such that

i) the resulting diagram is a Young diagram in the two symbols
ii) no two "dots" lie in the same vertical line.

The successive products may be nested, and the product scheme then may be expressed as a problem of enumerating partitions subject to restrictions imposed by the two rules. Related problems have been discussed by Miller [1974]

Further simplification may be obtained provided tables of plethysms of symmetric groups [Ibrahim] or tables of characters of the symmetric group are available. In general, neither of these is available for large enough values of mn.

The following example illustrates the use of Theorem 2 and the product scheme for the case $H_3(2)$.

Now, with obvious notation:

$$\{2\} = [**] \quad \text{so} \quad \{2\}^2 = [{**}\cdot\cdot] + \left[\begin{smallmatrix}**\cdot \\ \cdot\end{smallmatrix}\right] + \left[\begin{smallmatrix}** \\ \cdot\cdot\end{smallmatrix}\right]$$

whence $\{2\}^2 = \{4\} + \{3,1\} + \{2^2\}$.

Finally $\{2\}^3 = \{4\}\{2\} + \{3,1\}\{2\} + \{2^2\}\{2\}$

$$= \{6\} + 2\{5,1\} + 3\{4,2\} + 2\{3,2,1\} + \{4,1^2\} + \{3^2\} + \{2^3\}.$$

Thus, from Theorem 2:

$$H_3(2) = 1^2 + 2^2 + 3^2 + 2^2 + 1^2 + 1^2 + 1^2 = 21,$$ which agrees with Eq.(7).

Prior information about the form of $H_m(n)$ permits it to be determined by polynomial interpolation, while a symmetry relation reduces the number of values of n at which $H_m(n)$ need be computed. The following theorem is needed:

THEOREM 3: (Conjectured by Anand, Dumir and Gupta [1966]).

i) $H_m(n)$ is a polynomial in n of degree $(m-1)^2$

ii) $H_m(n) = (-1)^{m-1} H_m(-m-n)$.

PROOF: See Stanley [1973].

The combined use of Theorems 2 and 3 has been used by Jackson and van Rees [1974] to extend the known set of $\{H_m(n)\}$ given in Sloane [1973] to $m = 5$ and $m = 6$. The cases $m \geq 7$ require prohibitive amounts of computation by this method and have not been attempted.

4. RESIDUE METHODS

The method employing the identity given in Eq.(4) involves the successive elimination of particular variables by distributing appropriate terms across some of the factors. This is clearly seen in the example of Section 2. This process is analogous to the computation of residues and suggests replacement of the operator $[x_1^n x_2^n \dots x_m^n][y_1^n y_2^n \dots y_m^n]$ by the appropriate contour integration. The exponential and ordinary generating functions, $\Psi_m(t)$ and $\Phi_m(t)$ respectively, are given by the following theorems.

THEOREM 4:

$$\Psi_m(t) = \sum_{\substack{p_1,p_2,\dots,p_m \\ p_1+\dots+p_m=m}} \begin{bmatrix} m \\ p_1,p_2,\dots,p_m \end{bmatrix} (-1)^{p_1+2p_2+\dots+mp_m} \frac{\partial^{m-1}}{\partial t^{m-1}} I_m(\underline{p},t)$$

where

i) $$I_m(\underline{p},t) = \frac{1}{(2\pi i)^m} \oint_C \frac{\exp(x_1^{p_1-1} x_2^{p_2-1} \dots x_m^{p_m-1} t)}{\prod_{1\leq j<k\leq m} (x_j - x_k)^{p_j+p_k}} \cdot (x_1 x_2 \dots x_m)^{m-2} dx_1 \dots dx_m$$

ii) C is the polydisc $\bigcup_{i=1}^{m} C_i$ and C_i is the circle

$$C_i : |x_u| = \rho_i \quad \text{with} \quad 0<\rho_1<\rho_2<\dots<\rho_m<1 ,$$

PROOF: Omitted for brevity.

THEOREM 5:

$$\Phi_m(t) = \sum_{\substack{p_1,p_2,\ldots,p_m \\ p_1+\ldots+p_m=m}} \begin{bmatrix} m \\ p_1,p_2,\ldots,p_m \end{bmatrix} (-1)^{p_1+2p_2+\ldots+mp_m} J_m(\underline{p},t)$$

where

i) $$J_m(\underline{p},t) = \frac{1}{(2\pi i)^m} \oint_C \frac{x_1^{p_1(m-1)} \ldots x_m^{p_m(m-1)} . dx_1 dx_2 \ldots dx_m}{\prod_{1 \le j<k \le m} (x_j - x_k)^{p_j+p_k} (x_1 x_2 \ldots x_m - x_1^{p_1} x_2^{p_2} \ldots x_m^{p_m} t)}$$

ii) C is as in note (ii) of Theorem 4.

PROOF: Omitted for brevity.

The following theorem gives the forms of $\Phi_m(t)$ and $\Psi_m(t)$.

THEOREM 6:

i) $\Psi_m(t) = e^t p_m(t)$ where $p_m(t)$ is a polynomial of degree $(m-1)^2+1$

ii) $\Phi_m(t) = \dfrac{q_m(t)}{(1-t)^{(m-1)^2+1}}$ where $q_m(t)$ is a polynomial of degree $(m-1)(m-2)$ such that $q_m(t) = t^{(m-1)(m-2)} q_m(t^{-1})$.

PROOF: Omitted for brevity (c.f. Theorem 3).

The computation of $\Psi_m(t)$ or $\Phi_m(t)$ requires the evaluation of the integrals $I_m(\underline{p},t)$ or $J_m(\underline{p},t)$ for each $\underline{p}$ such that $p_1+p_2+\ldots+p_m = m$. However a considerable number of these are known to be zero, by the following lemma:

LEMMA 2: If $p_1 \neq 0$ or $p_m = 0$ then

i) $I_m(\underline{p},t) = 0$

ii) $J_m(\underline{p},t) = 0$.

PROOF: If $p_1 \neq 0$ then the integrands are analytic on the interior of C_1 .
If $p_m = 0$ then the integrands are analytic on the exterior of C_m .

Further reduction is offered by the following conjecture:

CONJECTURE Jackson and Johnson [1974].

$$\exists_{j<m} \text{ such that } \sum_{i=1}^{j} p_i \ge j \Rightarrow I_m(\underline{p},t) = 0 \text{ and } J_m(\underline{p},t) = 0 .$$

Thus for the case $m = 3$ only 2 of the 9 integrals need be computed, namely those for $\underline{p} = (0,0,3)$ and $\underline{p} = (0,1,2)$. For the case $m = 4$ only 5 of the 35 integrals need be computed, namely those corresponding to $(0,0,0,4)$, $(0,0,1,3)$, $(0,0,2,2)$, $(0,1,0,3)$, $(0,1,1,2)$.

The integration itself may be carried out straightforwardly by determining the poles of the integrand for a specified variable and obtaining the residues by differentiation. Both these processes may be mechanised in a language, for example, ALTRAN, which permits algebraic manipulation. The following lemma carries out the integration with respect to one variable.

<u>LEMMA 3</u>: Let $$f(\underline{x}) = \frac{g(\underline{x})}{\prod_{r\in\sigma} (x_j - x_r)^{k_r}}$$

where i) $\underline{x} = (x_1, x_2, \ldots, x_m)$

ii) $\sigma \subset \{1,2,\ldots,m\}$

iii) $k_r > 0$ and $g(\underline{x})$ is analytic in a region C_j

then $$\oint_{C_j} f(\underline{x})\,dx_j = 2\pi i \sum_{r\in\sigma} \frac{1}{(k_r-1)!} \frac{\partial^{k_r-1}}{\partial x_j^{k_r-1}} g(\underline{x})\Big|_{x_j=x_r} .$$

Although the two generating functions are equivalent, the exponential generating function $\Psi_m(t)$ involves determining at each stage of integration whether there is an essential singularity in the region of integration. The essential singularities may be avoided by integrating on the interior or the exterior of the appropriate contour. The ordinary generating function $\Phi_m(t)$ may be treated in a manner which avoids this problem. Computation with Theorem 5 using ALTRAN 1.9 has been most encouraging and has resulted in substantial reductions in computation time over the interpolatory method of Section 3 for $m=3$ and $m=4$. It is anticipated that the method will provide a means of determining $\Phi_7(t)$.

<u>ACKNOWLEDGMENTS</u>

The author thanks D.Z. Djokovic, S.C. Johnson, P.A. Morris and R.C.Read for fruitful discussions at various times.

<u>REFERENCES</u>

Abramson, M., and W.O.J. Moser, Arrays with Fixed Row and Column Sums, Discrete Math. 6, 1-14, 1973.

Anand, H., V.C. Dumir and H. Gupta, A Combinatorial Distribution Problem, Duke Math. J., 33, 757-770, 1966.

Carlitz, L., Enumeration of Symmetric Arrays, Duke Math. J., 33, 771-782, 1966.

Cormack, R.M., A Review of Classification, J. Roy. Statist. Soc., Ser. A., 134, 321-353, 1971.

Dempster, A.P., An Overview of Multivariate Data Analysis, J. Mult. Analysis, 1, 316-346, 1971.

Elliott, E.B., On Linear Homogeneous Diophantine Equations, Quar. J. Pure and Appl. Math., 35, 348-377, 1903.

Ibrahim, E.M., Tables for the Plethysms of S-functions, Royal Society (Lond.), Depository of Unpublished Tables.

Jackson, D.M., Expectations of Functions of Sequences over Finite Alphabets with given Transition Probabilities by Methods Independent of Sequence Length, SIAM J. Comp., 1, 203-217, 1972.

Jackson, D.M., and S.C. Johnson, Unpublished Material, 1974.

Jackson, D.M., and G.H.J. van Rees, The Enumeration of Generalised Doubly Stochastic Non-negative Integer Square Matrices, (submitted for publication, 1974).

Jackson, D.M., and W.W. Wadge, Normal Form Reduction of Probabilistic Computations in Non-parametric Classification, Comp. J. (to appear).

Jackson, D.M., and L.J. White, The Weakening of Taxonomic Inferences by Homological Errors, Math. Biosci., 10, 63-89, 1971.

Jackson, D.M., and L.J. White, Stability Problems in Non-statistical Classification Theory, Comp. J. 15, 214-221, 1972.

Jardine, C.J., N. Jardine and R. Sibson, The Structure and Construction of Taxonomic Hierarchies, Math. Biosci. 1, 173-179, 1967.

Littlewood, D., The Theory of Group Characters, Clarendon Press, Oxford 1950.

MacMahon, P.A., Combinatory Analysis, Chelsea, Boston, 1960.

Miller, J.C.P., On the Enumeration of Some Partially Ordered Sets, Combinatorial Theory Conference, Aberwystwyth, 1973.

Redfield, J.H., The Theory of Group Reduced Distributions, Am. J. Math., 49, 433-455, 1927.

Read, R.C., The Enumeration of Locally Restricted Graphs (I), J. Lond. Math. Soc., 34, 417-436, 1959.

Read, R.C., The Enumeration of Locally Restricted Graphs (II), J. Lond. Math. Soc., 38, 344-351, 1963.

Read, R.C., The Use of S-functions in Combinatorial Analysis, Can. J. Math., 20, 808-841, 1968.

Sloane, N.J.A., A Handbook of Integer Sequences, Academic Press, New York, 1973.

Stanley, R.P., Linear Homogeneous Diophantine Equations and Magic Labelings of Graphs, Duke. Math. J. 40, 607-632, 1973.

On Fuzzy Thesauri

Leo Reisinger, Vienna

1. PRELIMINARY REMARK. THE FUNCTIONS OF THESAURI

Every Information System is faced with the problem that the signs (symbols) used by the sender of a message correspond only partially to those used by the receiver. This partial correspondence is caused by the vagueness of natural languages and leads to the well known fact that the documents of a data bank which are relevant for a certain problem are only partially retrieved when an according query is formulated. For every Information System strategies have therefore to be developed to minimize "silence" (set of relevant but not retrieved documents) and "noise" (set of retrieved but not relevant documents) [BUNDESMINISTERIUM DER JUSTIZ, 1972, p. 97; REISINGER, 1974] The most important means to achieve an acceptable quality of selection is the construction of a thesaurus.

A thesaurus can be defined in the following way [KOMITEE TERMINOLOGIE UND SPRACHFRAGEN, 1968] : A thesaurus in the field of documentation consists of a set of signs which stand for concepts and are called the descriptors of the thesaurus and a set of relations between these signs. These relations are defined in such a way that
a) a one to one correspondence of descriptors and concepts and
b) a one to one correspondence of relations between descriptors and relations between concepts is attained. Applying the usage of the formal sciences we may call a thesaurus therefore a "relational system". The definitions of the relations of this system depend of course on the specific field of the Information System. However some of the following relations are usually contained in a thesaurus. It is therefore appropriate to call these relations "general semantic relations"

To attain a one to one correspondence of descriptors and concepts homographs have to be removed by marking descriptors accordingly (Example: "Schloß-Gebäude", "Schloß-Verschluß"). The next step is the definition of synonymity. We have to distinguish here between context independent synonymity where two descriptors are exchangeable in every context (Example: "Schornstein", "Rauchfang") and context dependent synonymity where this is not the case (Example: to "schnell" are context dependent synonyms "rasch", "eilig", "geschwind", "hurtig", "flott", "zügig", "flink", "behend", "lebhaft", "stürmisch"). The distiction between context dependent and context independent synonymity is not a priori defined and depends on the subject of the data bank. Most thesauri solve synonymity by distinguishing a preference synonym that is denoted by USE. All other relevant descriptors are denoted by UF (Example: "Rauchfang" USE "Schornstein", "Schornstein" UF "Rauchfang").

To attain a one to one correspondence of relations between descriptors and relations between concepts hierarchical relations and relations of opposition are defined. The most important hierarchical relation is the relation of abstraction (generic relation, hyponymity) between broader term and narrower term. This relation is denoted by BTG or NTG respectively. The narrower terms of the same broader term are related terms and denoted by RT or RTG (Example: "KFZ" NTG "PKW", "KFZ" NTG "LKW" implies "PKW" BTG "KFZ", "LKW" BTG "KFZ", "PKW" RTG "LKW", "LKW" RTG "PKW"). Some thesauri distinguish another hierarchical relation, the partitive relation, from the generic relation [LANG, 1971]. The partitive relation is defined between total term and part term and is denoted by BTP or NTP respectively. Two part terms of the same total term are contiguous terms (RTP) (Example: "BRD" NTP "Bayern", "BRD" NTP "Hessen" implies "Bayern" BTP "BRD", "Hessen" BTP "BRD", "Bayern" RTP "Hessen", "Hessen" RTP "Bayern"). If a thesaurus does not distinguish between generic and partitive relation the hierarchical relation is denoted by BT, NT and RT.

Three different relations of opposition may be distinguished: Incompatibility, complementarity and antonymity. Two descriptors are incompatible if the first implies the negation of the second and vice versa (Example: "rot"/"schwarz"). Two descriptors are complementary if the first is a synonym of the negation of the second (Example: "männlich"/"weiblich"). Two descriptors are antonyms if they are situated at the ends of a scale in such a way that there exists a continuous transition from one extreme to the other. The negation of the first descriptor does not imply the second (Example: "klein"/"groß", "hell"/"dunkel", "jung"/"alt").

The formal properties of the above mentioned relations may be summarized in the following way:

1. Context independent synonymity is an equivalence relation (reflexive, symmetrical, transitive). If the relations USE and UF are distinguished these are converse relations. Both are asymmetrical.
2. Context dependent synonymity is a resemblance (reflexive, symmetrical).
3. Generic and partitive relation are antireflexive and asymmetrical. If BT/BTG/BTP (or NT/NTG/NTP) denote not only the immediate broader term (or narrower term respectively) these relations are also transitive and therefore relations of order. The generated relation RT/RTG/RTP is symmetrical and transitive.
4. Relations of opposition are antireflexive and symmetrical.

2. THE THEORY OF FUZZY SETS. SOME ELEMENTARY DEFINITIONS

An ordinary thesaurus applies the familiar notions of relations as they are developed in the theory of ordinary sets. The following fundamental dichotomy holds in this case: Let R be a semantic relation, x and y two descriptors, then there exists only one of the following cases: either $(x,y) \in R$ or $(x,y) \notin R$, i.e. every semantic relation has "sharp boundaries". This dichotomy seems to be a severe draw back for certain semantic relations, e.g. a relation of association or a context dependent synonymity. In these examples there exist transitions, the semantic relations have no sharp boundaries, they are "fuzzy". A possibility to introduce the concept of "fuzziness" is supplied by the theory of fuzzy sets that has been developed by L.A. Zadeh and others since 1965 [Zadeh, 1965; an extensive bibliography of fuzzy sets theory is found in Kaufmann, 1973] . In this section we shall give some elementary definitions of fuzzy sets. (To distinguish fuzzy sets we shall use the symbol $\sim$ to denote fuzziness.)

Definition 1. Let E be a set of objects (our "universe of discourse"), then a fuzzy set $\underset{\sim}{A}$ in E is charakterized by a characteristic function (membership function) $f_{\underset{\sim}{A}}(x)$ which associates with each object x in E a real number in the interval $[0,1]$ with the value of $f_{\underset{\sim}{A}}(x)$ at x representing the "grade of membership" of x in $\underset{\sim}{A}$. A fuzzy set $\underset{\sim}{A}$ may therefore be defined as

$$\underset{\sim}{A} = \{(x, f_{\underset{\sim}{A}}(x))\} \quad \forall x \in E, \; f_{\underset{\sim}{A}}(x) \in [0,1] \tag{1}$$

When A is an ordinary set, $f_A(x)$ can take on only two values 1 or 0, according as x does or does not belong to A. The notion of fuzzy sets is therefore an extension of the notion of ordinary sets.

Definition 2.1. Two fuzzy sets $\underset{\sim}{A}$ and $\underset{\sim}{B}$ are equal, $\underset{\sim}{A} = \underset{\sim}{B}$, iff

$$f_{\underset{\sim}{A}}(x) = f_{\underset{\sim}{B}}(x) \quad \forall x \in E \tag{2}$$

2. $\underset{\sim}{\bar{A}}$ the complement of a fuzzy set $\underset{\sim}{A}$ is defined by

$$f_{\underset{\sim}{\bar{A}}}(x) = 1 - f_{\underset{\sim}{A}}(x) \quad \forall x \in E \tag{3}$$

3. A fuzzy set $\underset{\sim}{A}$ is a subset of a fuzzy set $\underset{\sim}{B}$, $\underset{\sim}{A} \subset \underset{\sim}{B}$, iff

$$f_{\underset{\sim}{A}}(x) \leqq f_{\underset{\sim}{B}}(x) \quad \forall x \in E \tag{4}$$

4. The union of two fuzzy sets $\underset{\sim}{A}$ and $\underset{\sim}{B}$ is a fuzzy set $\underset{\sim}{C} = \underset{\sim}{A} \cup \underset{\sim}{B}$, defined by

$$f_{\underset{\sim}{C}}(x) = \max (f_{\underset{\sim}{A}}(x), f_{\underset{\sim}{B}}(x)) \quad \forall x \in E \tag{5}$$

5. The intersection of two fuzzy sets $\underset{\sim}{A}$ and $\underset{\sim}{B}$ is a fuzzy set $\underset{\sim}{C} = \underset{\sim}{A} \cap \underset{\sim}{B}$, defined by

$$f_{\underset{\sim}{C}}(x) = \min (f_{\underset{\sim}{A}}(x), f_{\underset{\sim}{B}}(x)) \quad \forall x \in E \tag{6}$$

In the theory of ordinary sets the most important examples of the notion of distance are the Hamming-Distance and the Euclidian Distance. These may be generalized for fuzzy sets in the following way.

Definition 3. Let $E = \{x_1, x_2, \ldots, x_n\}$, $\underset{\sim}{A} \subset E$, $\underset{\sim}{B} \subset E$, then the Hamming-Distance $d(\underset{\sim}{A}, \underset{\sim}{B})$ is defined by

$$d(\underset{\sim}{A}, \underset{\sim}{B}) = \sum_{i=1}^{n} |f_{\underset{\sim}{A}}(x) - f_{\underset{\sim}{B}}(x)| \tag{7}$$

and the Euclidian Distance by

$$e(\underset{\sim}{A}, \underset{\sim}{B}) = \left(\sum_{i=1}^{n} (f_{\underset{\sim}{A}}(x) - f_{\underset{\sim}{B}}(x))^2\right)^{\frac{1}{2}} \tag{8}$$

If a normalization of distance in the interval $[0, 1]$ is required, we arrive at "relative distances" defined by

$$d^*(\underset{\sim}{A}, \underset{\sim}{B}) = \frac{d(\underset{\sim}{A}, \underset{\sim}{B})}{n} \quad (9) \quad \text{and} \quad e^*(\underset{\sim}{A}, \underset{\sim}{B}) = \frac{e(\underset{\sim}{A}, \underset{\sim}{B})}{\sqrt{n}} \quad (10)$$

Definition 4. If $\underset{\sim}{A}$ is a fuzzy set, then the "nearest ordinary set to $\underset{\sim}{A}$", denoted by A°, is defined by

$$\begin{aligned} f_{A^\circ}(x) &= 0 && \text{if } f_{\underset{\sim}{A}}(x) < 0.5 \\ &= 1 && \text{if } f_{\underset{\sim}{A}}(x) > 0.5 \\ &= 0 \text{ or } 1 && \text{if } f_{\underset{\sim}{A}}(x) = 0.5 \end{aligned} \tag{11}$$

Two "indicators of fuzziness" for a fuzzy set $\underset{\sim}{A}$ are defined by

$$In_1(\underset{\sim}{A}) = 2d^*(\underset{\sim}{A}, A^\circ) \quad (12) \quad \text{and} \quad In_2(\underset{\sim}{A}) = 2e^*(\underset{\sim}{A}, A^\circ) \quad (13)$$

with $0 \leq In_1(\underset{\sim}{A}),\ In_2(\underset{\sim}{A}) \leq 1$

Definition 5. If $E \times E$ is a Cartesian product, a fuzzy relation $\underset{\sim}{R} \subset E \times E$ is defined by

$$\underset{\sim}{R} = \{((x,y), f_{\underset{\sim}{R}}(x,y))\} \quad \forall (x,y) \in E \times E, \ f_{\underset{\sim}{R}}(x,y) \in [0,1] \tag{14}$$

The definitions of equality, complement, union, intersection, distance and indicators of fuzziness are defined for relations in analogy to (2) to (13).

Definition 6. If $\underset{\sim}{R}$ is a fuzzy relation,

1. $\underset{\sim}{R}$ is reflexive, iff

$$f_{\underset{\sim}{R}}(x,x) = 1 \quad \forall (x,x) \in E \times E \tag{15}$$

2. $\underset{\sim}{R}$ is antireflexive, iff

$$f_{\underset{\sim}{R}}(x,x) = 0 \quad \forall (x,x) \in E \times E \tag{16}$$

3. $\underset{\sim}{R}$ is symmetrical, iff

$$f_{\underset{\sim}{R}}(x, y) = f_{\underset{\sim}{R}}(y, x) \quad \forall (x, y) \in E \times E \tag{17}$$

4. $\underset{\sim}{R}$ is antisymmetrical, iff $x \neq y$ and

$$f_{\underset{\sim}{R}}(x, y) \neq f_{\underset{\sim}{R}}(y, x) \text{ or } f_{\underset{\sim}{R}}(x, y) = f_{\underset{\sim}{R}}(y, x) = 0 \quad \forall (x, y) \in E \times E \tag{18}$$

5. $\underset{\sim}{R}$ is asymmetrical, iff $x \neq y$ and

$$f_{\underset{\sim}{R}}(x, y) > 0 \text{ implies } f_{\underset{\sim}{R}}(y, x) = 0 \quad \forall (x, y) \in E \times E$$

6. $\underset{\sim}{R}$ is transitive, iff

$$f_{\underset{\sim}{R}}(x, z) \geqq \max_{y} (\min (f_{\underset{\sim}{R}}(x, y), f_{\underset{\sim}{R}}(y, z))) \quad \forall (x, y), (y, z), (x, z) \in E \times E \tag{19}$$

The definitions of fuzzy equivalence, fuzzy resemblance, fuzzy preorder and fuzzy order are formulated in analogy to the definitions of ordinary relations.

Definition 7. Two fuzzy relations $\underset{\sim}{R}$ and $\underset{\sim}{S}$ are converse, iff

$$f_{\underset{\sim}{R}}(x, y) = f_{\underset{\sim}{S}}(y, x) \quad \forall (x, y) \in E \times E \tag{20}$$

The converse relation of $\underset{\sim}{R}$ shall be denoted by $\underset{\sim}{R}^{c}$.

3. FUZZY THESAURI

We shall now apply the theory of fuzzy sets to the construction of thesauri.

Definition 8. Let T be a set of signs, $T_1 = \{r_1, r_2, \ldots, r_k\}$ and T_2 two nonempty subsets of T with $T_1 \cap T_2 = \emptyset$, then a fuzzy thesaurus TH_f is defined by

$$TH_f = (T_1, T_2; \underset{\sim}{R}_1, \underset{\sim}{R}_2, \ldots, \underset{\sim}{R}_k) \tag{21}$$

where T_1 is a set of relators r_i denoting R_i, T_2 a set of descriptors and $\underset{\sim}{R}_i \subset T_2 \times T_2$ is a fuzzy semantic relation defined by

$$f_{\underset{\sim}{R}_i}(x, y) \in [0, 1] \quad \forall (x, y) \in T_2 \times T_2, \quad i \in \{1, 2, \ldots, k\} \tag{22}$$

$f_{\underset{\sim}{R}_i}$ represents the grade of membership of (x, y) in $\underset{\sim}{R}_i$.

As was mentioned in section 2 the notions of "distance" and "indicator of fuzziness" may be applied directly to semantic relations from (7) to (13).

Let us now examine the counterpart of general semantic relations in a fuzzy thesaurus.

1. A fuzzy context independent synonymity is a fuzzy equivalence relation.
2. A fuzzy context dependent synonymity is a fuzzy resemblance.
3. A fuzzy hierarchical relation is a fuzzy order relation.
4. The connection between $\underset{\sim}{BT}$, $\underset{\sim}{NT}$ and $\underset{\sim}{RT}$ is the following:
   Let $x, y, z \in T_2$ and

a) if $(x, z) \in \underset{\sim}{BT}$ and $(y, z) \in \underset{\sim}{BT}$, then $(x, y) \in \underset{\sim}{RT}$ with

$$f_{\underset{\sim}{RT}}(x, y) = \min \left(f_{\underset{\sim}{BT}}(x, z), f_{\underset{\sim}{BT}}(y, z) \right) \tag{23}$$

b) if $(z, x) \in \underset{\sim}{NT}$ and $(z, y) \in \underset{\sim}{NT}$, then $(x, y) \in \underset{\sim}{RT}$ with

$$f_{\underset{\sim}{RT}}(x, y) = \min \left(f_{\underset{\sim}{NT}}(z, x), f_{\underset{\sim}{NT}}(z, y) \right) \tag{24}$$

5. $\underset{\sim}{USE}$ / $\underset{\sim}{UF}$ and $\underset{\sim}{BT}$ / $\underset{\sim}{NT}$ are converse relations.

With respect to the information retrieval the user of an Information System may be interested not only in the question, which descriptors are connected with a given descriptor by a certain fuzzy relation $\underset{\sim}{R}_i$ (this question is answered by the definition of $\underset{\sim}{R}_i$) but also in the problem of determining all descriptors that are connected with a given descriptor by any relation.

Definition 9. Let $\underset{\sim}{R}_1, \underset{\sim}{R}_2, \ldots, \underset{\sim}{R}_k$ be the fuzzy relations of TH_f, then the fuzzy "total relation " $\underset{\sim}{R}_t$ is defined by

$$\underset{\sim}{R}_t = \bigcup_{i=1}^{k} (R_i \cup R_i^c) \tag{25}$$

Let $x \in T_2$ then $\underset{\sim}{S}_a(x)$ the fuzzy "semantic environment of x" (with respect to level a) is defined by

$$\underset{\sim}{S}_a(x) = \{ y \in T_2 \mid f_{\underset{\sim}{R}_t}(x, y) \geqq a \} \text{ with } f_{\underset{\sim}{S}_a(x)}(y) = f_{\underset{\sim}{R}_t}(x, y) \tag{26}$$

By selecting different levels of a we arrive at different semantic environments of a descriptor. The bigger a, the smaller generally the "silence" of the information retrieval and the bigger the "noise". Different a lead therefore to different retrieval strategies.

4. FUZZY THESAURI AND ASSOCIATION FACTORS

The concept of a fuzzy thesaurus presupposes procedures for determining the characteristic function. This may be done by a documentalist according to his knowledge but in this case the interpersonal and intertemporal comparability will be usually insufficient. Therefore it seems advisable to determine the characteristic function by Automated Data Processing. In this section we shall give an example of doing so by means of association factors [BALL, 1965; LANCE-WILLIAMS 1965; RUSPINI 1970; REISINGER 1972/73] .

Let each document in a data bank be characterized by a set of descriptors and n be the number of documents, then for any two descriptors x and y the following contingency table can be computed.

	y	$\bar{y}$	Σ
x	a	b	a+b
$\bar{x}$	c	d	c+d
Σ	a+c	b+d	n

a is the number of documents, that are characterized both by x and y, b the number of documents that are charcterized by x but not by y , etc.

Usually the following properties are required of any association factor K:
1. $-1 \leq K \leq 1$;
2. $K = 0$ if the descriptors are completely independent;
3. $K = 1$ or -1 if the descriptors are completely dependent.

Association factors that fulfill these requirements are e.g.

1. the coefficient of contingency

$$K_1(x,y) = \frac{ad - bc}{\left((a+b)(a+c)(b+d)(c+d)\right)^{\frac{1}{2}}} \tag{27}$$

2. the coefficient of association

$$K_2(x,y) = \frac{ad - bc}{ad + bc} \tag{28}$$

3. the coefficient of colligation (according to Yule)

$$K_3(x,y) = \frac{\sqrt{ad} - \sqrt{bc}}{\sqrt{ad} + \sqrt{bc}} \tag{29}$$

As can be seen easily the sign of $K_i(x,y)$ indicates the "direction" of association. If $K_i(x,y)$ is positive, then x and y are found together, if $K_i(x,y)$ is negative, then x and y exclude each other in the majority of documents. In the former case we may speak of "positive association", in the latter case of "negative association".

By means of association factors the following fuzzy relations of association may be defined:

Definition 10.1. A fuzzy relation of "positive association" $\underset{\sim}{PA} \subset T_2 \times T_2$ is defined by

$$\begin{aligned} f_{\underset{\sim}{PA}}(x,y) &= K_i(x,y) \quad \text{if } K_i(x,y) > 0 \\ &= 0 \quad \text{else} \end{aligned} \tag{30}$$

$\underset{\sim}{PA}$ is reflexive and symmetrical, i.e. a fuzzy resemblance.

2. A fuzzy relation of "negative association" $\underset{\sim}{NA} \subset T_2 \times T_2$ is defined by

$$f_{\underset{\sim}{NA}}(x,y) = -K_i(x,y) \quad \text{if } K_i(x,y) < 0$$
$$= 0 \quad \text{else} \tag{31}$$

$\underset{\sim}{NA}$ is antireflexive and symmetrical.

3. A fuzzy relation of "association" $\underset{\sim}{RA} \subset T_2 \times T_2$ is defined by

$$f_{\underset{\sim}{RA}}(x,y) = |K_i(x,y)| \tag{32}$$

$\underset{\sim}{RA}$ is a fuzzy resemblance.

If the fuzzy relations $\underset{\sim}{PA}$ and $\underset{\sim}{NA}$ determined by means of an association factor K_i are the only relations defined in a thesaurus, then the corresponding $\underset{\sim}{RA}$ is the fuzzy "total relation" of this thesaurus which determines the fuzzy "semantic environments" of the descriptors.

Concluding we remark that the notion of fuzzy relations of association and semantic environments may be used not only for descriptors but also for documents. Let x and y be two documents of the data bank, then the corresponding table of contingency may be read as follows: a is the number of descriptors that characterize both x and y, b the number of descriptors that characterize x but not y, etc. By applying (26) and (30) to (32) we arrive at fuzzy relations of association between documents and the fuzzy semantic environment of a document. Both notions are of special interest for cluster analysis.

REFERENCES

BALL G.H., Data Analysis in the Social Sciences. What about Details ? In: AFIPS, Proc. of the Fall Joint Computer Conference, 533-559, 1965

BUNDESMINISTERIUM DER JUSTIZ (ed.), Das Juristische Informationssystem. Analyse, Planung, Vorschläge. Karlsruhe 1972

DIN 1463 Richtlinien für die Erstellung und Weiterentwicklung deutschsprachiger Thesauri

KAUFMANN A., Introduction à la théorie des sous-ensembles flous. 1. Eléments théoriques de base. Paris 1973

KOMITEE TERMINOLOGIE UND SPRACHFRAGEN IN DER DGD, Fachwörterbuch der Dokumentation (Thesaurus). In: Nachr. Dok. 19 , 268-273, 1968

LANCE G.N. - W.T. WILLIAMS, Computer Programs for Monothetic Classification (Association Analysis). In: The Computer Journal 8, 246-249, 1965

LANG F.H., Dokumentation und Information. Probleme der Bildungsorganisation. In: IBB Bulletin 7, 1971

LYONS J., Introduction to Theoretical Linguistics. Cambridge 1968

REISINGER L., Thesaurusstrukturen juristischer Datenbanken. In: DSWR 14-16, 1972/73

REISINGER L., Probleme der Bewertung und des Vergleiches automatisierter juristischer Informationssysteme. In: DVR 1,1974

RUSPINI E.H., Numerical Methods for Fuzzy Clustering. In: Information Sciences 2,319-350,1970

ZADEH L.A., Fuzzy Sets. In: Information and Control 8,338-353,1965

SUBJECT GROUP C

Numerical and Algorithmic Aspects of Statistical Methods

A Uniform Numerical Method for Linear Estimation from General Gauss-Markoff Models

Åke Björck, Linköping

1. INTRODUCTION

We consider the general Gauss-Markoff model

$$E(y) = X\gamma \quad , \qquad D(y) = \sigma^2 V \tag{1}$$

with the design matrix X and the variance-covariance matrix V known and σ^2 unknown. We take X to be $m*n$ of rank $r \leq n \leq m$, and also allow V to be singular. We thus include the case of linear constraints on the γ parameter.

In a recent paper Rao [1971] has pointed out that, although it is possible to transform the general model (1) to a standard model where X and V have full rank, a uniform algorithm for (1) would be of great value. We state below the main result in Rao [1971] in a form adopted to suit our derivations in section 3. We first make the following two definitions.

<u>Definition 1</u> Let A be an $m\times n$ matrix of arbitrary rank. A generalized inverse (g-inverse) of A is an $n\times m$ matrix denoted by A^-, such that $x = A^-y$ is a solution of $Ax = y$ for any y which makes this system of equations consistent.

It can easily be shown that an equivalent definition is that A^- should satisfy the condition $A A^-A = A$. Note also that a g-inverse always exists but in general is not unique.

<u>Definition 2</u> Let q be an $n\times s$ matrix. Then the linear function $q'\gamma$ is an estimable function of γ if it admits an unbiased estimator linear in Y .

Rao shows that the problem of inference from the general model can be reduced to the computation of a generalized inverse of a certain symmetric matrix.

Theorem 1 [C.R. Rao] Assume that in the model (1) y satisfies possible consistency conditions and that $q'\gamma$ is an estimable function of γ. Then if we put

$$\begin{bmatrix} c_1 & c_2 \\ c_3 & -c_4 \end{bmatrix} = \begin{bmatrix} y' & 0 \\ 0 & q' \end{bmatrix} \begin{bmatrix} V & X \\ X' & 0 \end{bmatrix}^- \begin{bmatrix} y & 0 \\ 0 & q \end{bmatrix} \tag{2}$$

we have $c_2 = c_3$ and

(i) The best linear unbiased estimate of $q'\gamma$ is

$$q'\hat{\gamma} = c_2 \quad . \tag{3}$$

(ii) The variance-covariance matrix of the estimate is

$$V(q'\hat{\gamma}) = \sigma^2 c_4 \quad . \tag{4}$$

(iii) An unbiased estimator of σ^2 is

$$\hat{\sigma}^2 = f^{-1} c_1, \qquad f = \text{rank}(V \quad X) - \text{rank}(X). \tag{5}$$

Algebraic expressions for the submatrices of the g-inverse in (2) have been given by Rao and Mitra [1971]. These, however, are not easily adopted for computational purposes. More recently Golub and Styan [1973] have developed numerical computational procedures for the univariate case $V = I$. They also give procedures for the usual F-test and for updating the solution. In this paper we will develop an efficient and numerically stable algorithm for the general model. It will be closely related to methods given by Björck [1968] for the linear least squares problem with linear constraints.

2. BASIC MATRIX DECOMPOSITIONS

In this section we briefly review some well-kown matrix decompositions, which will be used in our algorithm. For a more complete presentation of matrix decompositions with relevance to statistical computations we refer to Golub [1969].

2.1 The Cholesky trapezoidal decomposition

Let V be an $m*m$ real, symmetric and positive semi-definite matrix. Then V can be decomposed

$$P\,V\,P' = \begin{bmatrix} L_1 & 0 \\ L_2 & 0 \end{bmatrix} \begin{bmatrix} L_1' & L_2' \\ 0 & 0 \end{bmatrix} \Big\} s \quad , \quad \text{rank}(V) = s,$$

where P is a permutation matrix, L_1 is a non-singular, lower triangular $s*s$ matrix and L_2 an $(m-s)*s$ matrix. This decomposition can be computed by the sequential Cholesky algorithm: Let $v_{ij}^{(1)} = v_{ij}$, and for $k=1,2,\dots,s$

$$\left.\begin{aligned} l_{kk} &= (v_{kk}^{(k)})^{1/2}, \quad l_{jk} = v_{jk}^{(k)}/l_{kk}, \quad j=k+1,\dots,m, \\ v_{ij}^{(k+1)} &= v_{ij}^{(k)} - v_{ik}^{(k)} l_{jk}/l_{kk}, \quad j=k+1,\dots,m, \quad i=j,\dots,m. \end{aligned}\right\} \quad (6)$$

Note that all reduced matrices $\left\{v_{ij}^{(k)}\right\}$, $k \leq i,j \leq m$, are symmetric and positive semi-definite. Therefore, unless this matrix is identically zero, there is a positive diagonal element, $v_{pp}^{(k)} > 0$, $k \leq p \leq m$. Then, if both rows and columns p and k are interchanged, this element can be taken as pivot in the next step. In practice we take the rank of V to be $s = k$, if for some tolerance ε we have

$$v_{pp}^{(k+1)} \leq \varepsilon, \quad k+1 \leq p \leq m .$$

Finally we note that (6) can also be written

$$L^{-1} P V P'(L^{-1})' = \begin{bmatrix} I & 0 \\ 0 & 0 \end{bmatrix} \Big\} s, \quad L^{-1} = L_s \cdots L_2 L_1, \qquad (7)$$

where L_k is called an elementary elimination matrix, and equals the identity matrix except for the elements in the k:th column on and below the main diagonal, which are

$$l_{kk}^{-1}(1, l_{k+1,k+1}, \dots, l_{m,k})'.$$

2.2 Trapezoidal decomposition of a general matrix

A trapezoidal decomposition can also be computed for a general rectangular $m \times n$ matrix X. If X has rank r, then after r transformations we obtain

$$Q P_1 X P_2 = \begin{bmatrix} R & S \\ 0 & 0 \end{bmatrix} \Big\} r, \quad Q = Q_r \cdots Q_2 Q_1, \qquad (8)$$

where P_1 and P_2 are permutation matrices. Here R is $r \times r$, non-singular and upper triangular, S an $r \times (n-r)$ matrix and the last $(m-r)$ rows of the right hand side matrix are zero. We can again take $Q_k = L_k$, where L_k is an elementary elimination matrix. However it is also possible to take $P_1 = I$ and Q_k to be an orthogonal Householder transformation

$$Q_k = I - 2u_k u_k', \quad u_k' u_k = 1.$$

Then also Q is orthogonal and we call (8) the orthogonal trapezoidal decomposition of X. The numerical stability of this orthogonal decomposition is enhanced if P_2 is chosen so that the diagonal elements of R satisfy

$$|r_{11}| \geq |r_{22}| \geq \dots \geq |r_{rr}| .$$

Then the numerical rank of X is taken to be $r = k$ if

$$|r_{kk}| \geq \varepsilon > |r_{k+1,k+1}| .$$

Computational details can be found in Golub [1965].

A decomposition closely related to (8) can be computed by the modified Gram-Schmidt method [Björck 1967]. If this is applied to X and to an adjoined $m*p$ matrix Y, it produces the decompositions

$$XP_2 = Q'(R\ S), \qquad Y = Q'Z + W \tag{9}$$

where Q' is an $m*r$ matrix with orthogonal columns and the matrix W is orthogonal to Q', $Q'W = 0$.

2.3 The singular value decomposition

Again let X be a real $m*n$ matrix of rank r. Then we can find orthogonal matrices U and V such that

$$U'X\,V = \begin{bmatrix} D & 0 \\ 0 & 0 \end{bmatrix} \}r , \qquad D = \mathrm{diag}(\sigma_1, \sigma_2, \ldots, \sigma_r) \tag{10}$$

The singular values σ_i of X are the positive square roots of the eigenvalues of $X'X$ or $X\,X'$, and we can assume

$$\sigma_1 \geq \sigma_2 \geq \ldots \geq \sigma_r > 0.$$

The main advantage of the singular value decomposition in this context is that the rank of the matrix X can more reliably be estimated from (10) than from the trapezoidal decomposition. However, the singular values must be determined by an iterative algorithm, which is considerably more involved than that for the other decompositions. The singular value decomposition can be computed by the algorithm given by Golub and Reinsch [1970].

3. A UNIFORM ALGORITHM

From theorem 1 it follows that we have to compute the expression $Z'C^-Z$, where Z is an $(m+n)*(s+1)$ matrix,

$$C = \begin{bmatrix} V & X \\ X' & 0 \end{bmatrix} , \qquad Z = \begin{bmatrix} y & 0 \\ 0 & q \end{bmatrix} .$$

The algorithm we propose essentially reduces this expression by a sequence of symmetric transformations, until the generalized inverse can be written down explicitely. We make repeated use of the following simple lemma.

Lemma 1 Let C be a symmetric matrix and S a non-singular transformation.

Then $B = S\,C\,S'$ is also symmetric and if B^- is a symmetric g-inverse of B then $S'B^-S$ is a symmetric g-inverse of C.

Proof: We have $B\,B^-B = B$, or $(S\,C\,S')B^-(S\,CS') = S\,C\,S'$. Since S is non-singular this implies $C(S'B^-S)C = C$.

From this lemma it follows that the expression $Z'C^-Z$ is transformed according to

$$Z'C^-Z = (SZ)'B^-(SZ), \qquad B = S\,C\,S' . \tag{11}$$

The algorithm now proceeds in the following three basic steps:

(i) Transformation of the variance-covariance matrix

Since V is positive semi-definite we use Cholesky's method to transform V. Thus L^{-1} and P_1 are computed so that

$$L^{-1}P_1V\,P_1'(L')^{-1} = \begin{bmatrix} I & 0 \\ 0 & 0 \end{bmatrix} \begin{matrix} \}m_1 \\ \}m_2 \end{matrix}, \quad \text{rank}(V) = m_1 . \tag{12}$$

Now the same transformation must be applied from the left to X and according to (11) also to y,

$$L^{-1}P_1X = \begin{bmatrix} \tilde{X}_1 \\ \tilde{X}_2 \end{bmatrix} \}m_1 , \qquad L^{-1}P_1y = \begin{bmatrix} \tilde{y}_1 \\ \tilde{y}_2 \end{bmatrix} \}m_1 , \tag{13}$$

and after these transformation C and Z will be transformed into

$$C_1 = \left[\begin{array}{cc|c} I & 0 & \tilde{X}_1 \\ 0 & 0 & \tilde{X}_2 \\ \hline \tilde{X}_1' & \tilde{X}_2' & 0 \end{array}\right] , \qquad Z_1 = \left[\begin{array}{c|c} \tilde{y}_1 & 0 \\ \tilde{y}_2 & 0 \\ \hline 0 & q \end{array}\right] .$$

(ii) Transformation of the constraint matrix

The matrix X_2 corresponds to linear constraints imposed on the parameter γ. We now compute the trapezoidal decomposition of the $n{\times}m_2$ matrix X_2',

$$Q_1P_2X_2'P_3 = \begin{bmatrix} R_1 & S_1 \\ 0 & 0 \end{bmatrix} \begin{matrix} \}p \\ \}n\text{-}p \end{matrix} , \quad \text{rank}(X_2') = p . \tag{14}$$

In this step also the following submatrices of C_1 and Z_1 are transformed

$$Q_1P_2\tilde{X}_1' = \begin{bmatrix} W_1' \\ W_2' \end{bmatrix} \}p , \quad Q_1P_2q = \begin{bmatrix} \tilde{q}_1 \\ \tilde{q}_2 \end{bmatrix} , \quad P_3\tilde{y}_2 = \begin{bmatrix} \hat{y}_{21} \\ \hat{y}_{22} \end{bmatrix} . \tag{15}$$

We remark that if $p < m_2$, then certain consistency conditions must be satisfied by the dependent variable y. After this step C_1 and Z_1 have been transformed into

$$C_2 = \left[\begin{array}{cc|c|cc} I & & 0 & W_1 & W_2 \\ \hline & & & R_1' & 0 \\ 0 & & 0 & S_1' & 0 \\ \hline W_1' & R_1 & S_1 & & \\ W_2' & 0 & 0 & 0 & \end{array}\right], \quad Z_2 = \left[\begin{array}{c|c} \tilde{y}_1 & 0 \\ \hline \hat{y}_{21} & 0 \\ \hat{y}_{22} & \\ \hline 0 & \tilde{q}_1 \\ & \tilde{q}_2 \end{array}\right].$$

(iii) Transformation of the reduced design matrix

After step (ii) we have obtained the reduced design matrix W_2 of dimension $m_1 \times (n-p)$. We now compute the orthogonal trapezoidal decomposition of this matrix,

$$Q_2 W_2 P_4 = \begin{bmatrix} R_2 & S_2 \\ 0 & 0 \end{bmatrix} \begin{matrix} \}\, r \\ \}\, m_1 - r \end{matrix}, \quad \text{rank}(W_2) = r. \tag{16}$$

This also induces the following transformations:

$$Q_2 W_1 = \begin{bmatrix} \tilde{W}_{11} \\ \tilde{W}_{12} \end{bmatrix} \}\, r, \quad Q_2 \tilde{y}_1 = \begin{bmatrix} \hat{y}_{11} \\ \hat{y}_{12} \end{bmatrix}, \quad P_4 \tilde{q}_2 = \begin{bmatrix} \hat{q}_{21} \\ \hat{q}_{22} \end{bmatrix}. \tag{17}$$

Note, however, that since $Q_2 Q_2' = I$ the identity matrix in the upper left corner of C_2 is not changed. Since the non-singular transformations applied preserves the rank of the submatrices we now find

$$\text{rank}(V \;\; X) = m_1 + \text{rank}(R_1'), \quad \text{rank}(X) = p + \text{rank}(W_2),$$

and thus

$$f = \text{rank}(V \;\; X) - \text{rank}(X) = m_1 - r. \tag{18}$$

We have now finally reduced C and Z into

$$C_3 = \left[\begin{array}{cc|cc|ccc} I & & 0 & & \tilde{W}_{11} & R_2 & S_2 \\ & & & & \tilde{W}_{12} & 0 & 0 \\ \hline & & & & R_1' & 0 & 0 \\ 0 & & 0 & & S_1' & 0 & 0 \\ \hline \tilde{W}_{11}' & \tilde{W}_{12}' & R_1 & S_1 & & & \\ R_2' & 0 & 0 & 0 & & 0 & \\ S_2' & 0 & 0 & 0 & & & \end{array}\right], \quad Z_3 = \left[\begin{array}{c|c} \hat{y}_{11} & 0 \\ \hat{y}_{12} & \\ \hline \hat{y}_{21} & 0 \\ \hat{y}_{22} & \\ \hline & \tilde{q}_1 \\ 0 & \hat{q}_{21} \\ & \hat{q}_{22} \end{array}\right].$$

We can now easily obtain the estimates in theorem 1, and we collect the result in:

Theorem 2 Let C_3 and Z_3 be computed from (12) - (17) and put

$$v_1 = (R_1')^{-1}\hat{y}_{21}, \qquad \tilde{u}_{11} = (R_2')^{-1}\hat{q}_{21} \,. \tag{19}$$

Then y is consistent if

$$S_1'v_1 = \hat{y}_{22} \,, \tag{20}$$

and $q'\gamma$ is estimable if

$$S_2'\tilde{u}_{11} = \hat{q}_{22} \,. \tag{21}$$

Further, with

$$Q_2(\tilde{y}_1 - W_1v_1) = \begin{bmatrix} w_{11} \\ w_{12} \end{bmatrix} \Big\} r \tag{22}$$

we have

(i) The estimate of $q'\gamma$ is given by

$$q'\hat{\gamma} = \tilde{q}_1'v_1 + \hat{q}_{21}'R_2^{-1}w_{11} \,. \tag{23}$$

(ii) The variance-covariance matrix of $q'\hat{\gamma}$ is

$$V(q'\hat{\gamma}) = \sigma^2\, \tilde{u}_{11}'\tilde{u}_{11} \,. \tag{24}$$

(iii) An unbiased estimator of σ^2 is

$$\hat{\sigma}^2 = (m_1 - r)^{-1}w_{12}'w_{12} \,. \tag{25}$$

Note that if the vector $(\tilde{y}_1 - W_1v_1)$ is transformed as in (22), then it is not necessary to perform the transformations of W_1 and $\tilde{y}_1$ in (17). The proof of this theorem is straightforward. In order to compute $C_3^-Z_3$ we consider the system of equations

$$C_3 \begin{bmatrix} u \\ v \end{bmatrix} = \begin{bmatrix} \hat{y} \\ 0 \end{bmatrix} \,.$$

If the vectors here are partitioned consistenly with C_3, then this system splits up into the following seven subsystems:

(i) $\quad u_{11} + \tilde{W}_{11}v_1 + R_2v_{21} + S_2v_{22} = \hat{y}_{11}$

(ii) $\quad u_{12} + \tilde{W}_{12}v_1 = \hat{y}_{12}$

(iii) $\quad R_1'v_1 = \hat{y}_{21}$

(iv) $\quad S_1'v_1 = \hat{y}_{22}$

(v) $\quad \tilde{W}_{11}'u_{11} + \tilde{W}_{12}'u_{12} + R_1u_{21} + S_1u_{22} = 0$

(vi - vii) $\quad R_2'u_{11} = 0, \quad S_2'u_{11} = 0.$

From (iii), (vi) and (vii) we immediately find

$$u_{11} = 0, \quad v_1 = (R_1')^{-1}\hat{y}_{21}.$$

The equation (iv) expresses the consistency condition (20), and from (ii) we get

$$u_{12} = \hat{y}_{12} - \tilde{W}_{12}v_1. \tag{26}$$

Obviously u_{22} and v_{22} can be chosen arbitrarily, and from (i) and (v) we can solve for

$$v_{21} = R_2^{-1}(\hat{y}_{11} - \tilde{W}_{11}v_1 - S_2v_{22}), \quad u_{21} = R_1^{-1}(-\tilde{W}_{12}'u_{12} - S_1u_{22}).$$

Now, from theorem 1, the estimate for $q'\gamma$ is

$$\hat{q}'v = \tilde{q}_1'v_1 + \hat{q}_{21}'R_2^{-1}(\hat{y}_{11} - \tilde{W}_{11}v_1) - (\hat{q}_{21}'R_2^{-1}S_2 - \hat{q}_{22}')v_{22} \, .$$

This expression is independent of v_{22} if (21) is satisfied, and by (17) and (22) equals (23).

The estimate of $f\sigma^2$ now becomes

$$\hat{y}'u = \hat{y}_{12}'u_{12} - \hat{y}_{21}'R_1^{-1}\tilde{W}_{12}'u_{12} - (\hat{y}_{21}'R_1^{-1}S_1 - \hat{y}_{22}')u_{22} \, .$$

If y is consistent, then by (20) the last term is zero, and if we substitute for $\hat{y}_{12}$ from (26) and note that $u_{12} = w_{12}$ we get (25).

Finally, to get the variance-covariance matrix we have to consider also the system

$$C_3 \begin{bmatrix} \tilde{u} \\ \tilde{v} \end{bmatrix} = \begin{bmatrix} 0 \\ \hat{q} \end{bmatrix}, \quad (\hat{q}_1 = \tilde{q}_1) \, .$$

If we number the subsystems as above, we find from (ii), (iii) and(vi)

$$\tilde{v}_1 = 0, \quad \tilde{u}_{12} = 0, \quad \tilde{u}_{11} = (R_2')^{-1}\hat{q}_{21} \, .$$

Further, (vii) gives the condition for $q'\gamma$ to be estimable, and from (i) we get

$$\tilde{v}_{21} = R_2^{-1}(-\tilde{u}_{11} - S_2\tilde{v}_{22}),$$

where again $\tilde{v}_{22}$ is arbitrary. It now follows that

$$\hat{q}'v = -\hat{q}_{21}'R_2^{-1}\tilde{u}_{11} - (\hat{q}_{21}'R_2^{-1}S_2 - \hat{q}_{22}')\tilde{v}_{22} \, ,$$

which verifies (24).

The algorithm has been described above in a way, which not directly allows the use of the modified Gram-Schmidt's method in step (iii). Since this method in the authors experience is the most accurate to use, we give below formulas based on this decomposition. Instead of (16) we then compute

$$W_2P_4 = Q_2'(R_2 \quad S_2)\,\}\, r\;, \qquad \text{rank}(W_2) = r\;. \tag{16'}$$

The vector $(\tilde{y}_1 - W_1v_1)$ should be adjoined as a right hand side during the orthogonalization, and then we also obtain

$$\tilde{y}_1 - W_1v_1 = Q_2'w_{11} + w_{12}\;, \tag{22'}$$

where $Q_2'w_{12} = 0$. With this definition of R_2, S_2, w_{11} and w_{12} theorem 2 still holds.

4. COMMENTS

It is well-known that if an algorithm for regression analysis is used which forms the matrix $X'X$ of the normal equations, then the solution can be greatly influenced by rounding errors. Our algorithm works directly with X and uses only stable matrix decompositions. It therefore enjoys a high degree of reliability. It is also formulated so that some flexibility remains in the choice of transformations. For example, in step (ii) we can use either elimination transformations or, for a slightly higher price, the same orthogonal transformations as in step (iii). In problems where there are high correlations in X, but not any exact linear dependencies which are easy to find, the determination of rank in step (iii) can be critical. Then one might consider using the singular value decomposition in this step.

Fortran programs for the basic matrix decompositions described in section 2 are available in many scientific subroutine packages. Given these, the implementation of the algorithm should be a simple task.

ACKNOWLEDGEMENT

The problem posed by Rao [1971] was brought to the authors attention by G.H. Golub, whose help is gratefully acknowledged.

REFERENCES

Björck, Å. Solving linear least squares problems by Gram-Schmidt orthogonalization. BIT 7, 1-21, 1967.

Björck, Å. Iterative refinement of linear least squares solutions II. BIT 8, 8-30, 1968.

Golub, G.H. Numerical methods for solving linear least squares problems. Numer. Math. 7, 206-216, 1965.

Golub, G.H. Matrix decompositions and statistical calculations. Statistical Computation, Milton, R.C. and J.A. Nelder (eds.), Academic Press, New York 365-395, 1969.

Golub, G.H., and C. Reinsch. Handbook series linear algebra: Singular value decomposition and least squares solutions. Numer. Math. 14, 403-420, 1970.

Golub, G.H., and G.P.H. Styan. Numerical computations for univariate linear models. J. of Statistical Computation and Simulation, 3, 1973.

Rao, C.R. Unified theory of linear estimation. The Indian J. of Statistics, 33, 371-394, 1971.

Rao, C.R. and S.K. Mitra. Generalized inverse of matrices and its applications. J. Wiley & Sons, New York, 1971.

Inverse Data Analysis

Vidal Cohen, Paris, and **Jacques Obadia**, Jouy-en-Josas

One of the goals of Data Analysis is to represent a data set by using various structures such as : trees, factorial spaces, scales and so forth ... The chosen structure depends on the nature of the collected data as well as on the type of analysis (cluster analysis, factorial analysis, multidimensional scaling etc ...) and a different number of hypotheses about the metric and optimality criterion.

I - INVERSE ANALYSIS OF DATA

1.1 - Stability and compatibility

Virtually, the hypotheses are adapted according to a real situation. In this respect they sometimes appear as too weak or arbitrary and oblige us to explore the stability of the chosen structure. In this case, the type of analysis being defined, we may e.g. vary the metric in order to study empirically the fact of such a modification on the structure. It happens that the measured values suffer of uncertainty ; therefore, it is quite difficult to foresee the effect of a modification, even only a slight one, could have on the representation. We suggest a different point of view : our interest will be the research of data sets compatible with a fixed structure i.e.the research of the charaterization of data sets which have a given strucçture according to a chosen type of analysis.

1.2 - Summary of problems

(i) If by definition a structure of representation F is the result of a particular type of analysis G (eventually depending on a parameter family M such as metrics) of a data set E , what are the sets (or at least some sets) which would have the same image F if they were analysed through the type of analysis G ?

(ii) Let be a data set E which through G has the image F ; how to determine among the data sets having the same image F through G , the one or the ones, that might be the closest (or at least close enough) in terms of a certain metric to the data set E ?

iii) Let be a data set E , a type of data analysis G and a structure F ; how to pick the set of parameters M of G so that E (or a data set which is close to it) has the image F through the application of G ?

As you may note, these problems are in the field of what we have called "Inverse Data Analysis".

This paper will deal only with a particular case namely with the principal component analysis. In addition, we shall examine but some of these problems.

1.3 - Some applicationsof the Inverse Data Analysis

In a fair number of cases, the structures are considered as a sum-up of the analysed data.One may consider that it shows their main features or it can even reveal them. Through the Inverse Data Analysis we can, for instance, examine if two populations upon which the same variables have been applied, can be considered as significatively different, at least from the latent point of view which the structure is meant to reveal. In fact, if we have two data matrices T_1 and T_2 we can associate to T_1 a structure F_1 , then find out a matrix $\hat{T}_2$ having the same structure F_1 and being as close as possible to T_2 . It will be convenient to estimate if the distance between T_2 and $\hat{T}_2$ must be considered as the consequence of perturbations or sampling variability etc ... We can also obtain a structure from a matrix joining T_1 and T_2 and we can go on operating as before. Likely we could use such a method to compare data sets collected at different moments and conduct a multidimensional time series analysis. One could suspect us of having the idea to establish that we can reach any result fixed in advance by manipulating metrics and the uncertainty of measures. But, however, that is not our goal of course ! For it happens that a structure can be associated with a data set before any statistical analysis e.g. variables measured on the sick people. Therefore a clustering must be done according to the short- or long dated issue of the sickness.It is a normal attitude to find out with special caution the type of analysis leading to the structure fixed in advance ; in this respect, our research follows the same way as canonical and discriminant analysis.

II - APPLICATION TO THE PRINCIPAL COMPONENT ANALYSIS

2.1 - Introduction to the problems

Let be $X_1, X_2, \ldots, X_p$ p variables measured on the n individuals of a population (n>p). They have led to the constitution of a data matrix X of n rows and p columns. The k-column of this data matrix gives the n observations of the number k variable. The principal component analysis lead to extract eigen values and eigen vectors of the variance-covariance matrix V :

$$V = R'R \qquad \text{and} \qquad R = \frac{1}{\sqrt{n}}(X-\overline{X})$$

$\overline{X}$ is a matrix with identical n rows and p columns : the k - column being constituted with the average of the variable X_k n times repeated.

The data matrix X is said centered if $\overline{X} = 0$. (The symbol A' is for the transpose of the matrix A).

The factors $f_1, f_2, \ldots f_p$ are the eigen vectors of V with the norm 1 (they are set according to the decreasing order of the eigen values λ_j : $\lambda_1 \geqslant \lambda_2 \geqslant \ldots \geqslant \lambda_p$) ; they thus prove :

$$Vf_j = \lambda_j f_j \qquad j = 1,2,\ldots p \qquad (1)$$

$$f'_j . f_k = \delta_{kj} \qquad k = 1,2,\ldots p \qquad (2)$$

(δ_{kj} =Kronecker symbol).
Let us note : F the p-square matrix

$$F = (f_1, f_2, \ldots, f_p)$$

The j-column of F is constituted by the components of the eigen vector f_j (with regard to a given basis).
Δ the p-diagonal matrix having as diagonal coefficients the eigen values λ_j ; these coefficients λ_j will be all supposed positive ones. The relations (1) and (2) are written as follows

$$VF = F\Delta \qquad (3)$$

$$F'F = I_p \qquad (4)$$

(I_m will indicate the unity m-matrix).

2.1.1 - First problem : constitution of a centred matrix $(X-\overline{X})$ leading to a given factorial structure

We shall call factorial structure and we shall note (F,Δ), any couple of matrices F and Δ such as :

a) F is a p-square matrix such as $F'F = I_p$

b) Δ is p-diagonal matrix ; the diagonal terms of which are positive.

We shall say that *a matrix* X *with* n *rows and* p *columns admits the factorial structure* (F,Δ) *if the principal component analysis of* X *leads to the factors defined by the columns of* F *with the diagonal terms of the diagonal matrix* Δ *as corresponding eigen values.*

(i) Existence of a solution

Let's put down

a) $L = \Delta^{1/2}$

b) $\mathbb{1}$ the vector of R^n whose components equal 1

c) Q a matrix with p rows and n columns such as :

$$QQ' = I_p \qquad (5)$$

$$Q.\mathbb{1} = 0 \qquad (6)$$

d) $R' = FLQ \qquad (7)$

Under these conditions the matrix $V=R'R$ will verify

$$V = F'\Delta F$$

and therefore $Vf_j = \lambda_j f_j (\forall j)$, which proves that $\sqrt{n}R = \sqrt{n}Q'LF'$ suits as the sought matrix $(X-\overline{X})$, the constraint (6) on Q yielding $\sqrt{n}.\overline{R} = 0$.

(ii) Uniqueness of Q with fixed R

The λ_j are supposed to be all positive, L is regular. Any centered matrix admitting the factorial structure (F,Δ) can be written as follow $\sqrt{n}Q'LF'$ with $Q = L^{-1}F'R'$ which verifies the constraints (5) and (6). As a conclusion our problem has generally an infinity of solutions which are all reached.

2.1.2 - Second Problem

A centered matrix $(Y-\overline{Y})$ with n rows and p columns being given, how to find out a centered matrix $(X-\overline{X})$ with n rows and p columns close to the factorial structure (F,Δ) considered above ?

From the first problem, if we have $S = \frac{1}{\sqrt{n}}(Y-\overline{Y})$, this problem can be considered as the research of a matrix Q such as FLQ is close to it in term of a certain metric under the constraints (5) and (6). A quadratic metric between matrices being adopted, we'll try to minimize the sum of the squares of the elements of the matrix $S'-FLQ$ and we'll have to solve the following problem of optimization :

$$(M_1) \begin{cases} \text{Find out } Q \text{ such as} \\ QQ' = I_p \\ Q.\mathbb{1} = 0 \\ \text{and } t = \operatorname{trace}(S'-FLQ)(S-Q'LF') \text{ minimum} \end{cases}$$

$t = \operatorname{trace}(S'S) + \sum_1^p \lambda_k - 2\operatorname{trace}(FLQS)$ and if we have $B=SFL$, the problem (M_1) becomes :

A matrix B being given, what is the maximum of trace (QB) under the constraints $QQ'=I_p$ and $Q.\mathbb{1} = 0$.

(i) Geometrical interpretation : In R^n or more precisely in the plane orthogonal to the vector $\mathbb{1}$, we consider p vectors : $b_1, b_2, \ldots, b_p$ (column-vectors of B). We have to build in this plane p orthonormal vectors $q_1, \ldots, q_p$ (row-vectors of Q) such as :

$$K = \sum_{j=1}^{j=p} q_j . b_j$$

is maximum.

(ii) Algebraic solution - From the previous remark it results that if the vectors b_j are orthogonal by pairs, K is maximum if each q_j is colinear with b_j and with the same direction. Now for any p-square matrix T such as $T'T=I_p$

$$\operatorname{trace}(QB) = \operatorname{trace}(T'QBT)$$

and Q verifies the constraints (5) and (6) if and only if $T'Q$ verify them. We'll choose T such as the columns of BT form an orthogonal system :

$$T'B'BT = \begin{pmatrix} a_1 & & & & \\ & a_2 & & & \\ & & \ddots & & \\ 0 & & & \ddots & 0 \\ & & & & a_p \end{pmatrix} \quad \text{or again } B'BT = T \begin{pmatrix} a_1 & & & & \\ & a_2 & & & \\ & & \ddots & & \\ 0 & & & \ddots & 0 \\ & & & & a_p \end{pmatrix}$$

It results that the p columns of $T: T_1, T_2, \ldots, T_p$ are the eigen vectors of $B'B$ and that $a_1, a_2, \ldots, a_p$ are the associated eigen values (we suppose them non equal to zero). Therefore the maximum of $K = \text{trace } (T'QBT)$ under the constraints $T'QQ'T = I_p$ and $T'Q\mathbb{1} = 0$ (which are equivalent to (5) and (6)) is realized if the rows of $T'Q$ are of norm 1 colinear to the same index column vectors of BT and with the same direction :

$$Q'T = BT\overline{M}^1 \quad \text{putting down} \quad M^2 = \begin{pmatrix} a_1 & & & \\ & a_2 & & 0 \\ 0 & & \ddots & \\ & & & a_p \end{pmatrix}$$

$$\text{Finally} \quad Q = T\overline{M}^1 \, T'B'$$

$$\text{Therefore} \quad \text{Max trace } (QB) = \text{trace } (TMT') = \text{trace } M = \sum_{j=1}^{j=p} \sqrt{a_j}$$

Remarks 1 : At the optimum, the matrix QB is symmetric ,which is interpreted like this:at the maximum of $\sum q_i b_i$ the projection of the vector b_i on q_k is equal to the projection of the vector b_k on q_i for any couple (i,k) $(i \neq k)$

2 : If the factorial structure is the one of the matrix S , then

$$K = \text{trace } S'S + \sum_{j=1}^{j=p} \lambda_j - 2 \text{ trace } FLQS = 2 \sum_{j=1}^{j=p} \lambda_j - 2 \sum_{j=1}^{j=p} \sqrt{a_j}$$

But in this case :

$$B'B = LF'S'SFL = L\Delta L = \Delta^2$$

and necessarily $a_j = \lambda_j^2$ $(\forall j)$; which implies $K=0$.

3 : The relation $Q = T\overline{M}^1 T'B'$ shows that at the optimum the vectors $q_1, \ldots, q_p$ (rows of Q) belong to the sub-space of R^n generated by the vectors $b_1, \ldots, b_p$ (columns of B). This property we can establish directly, gives the opportunity of determinating the q_j by successive rotations from an orthonormal system of this sub-space (cf. Jacobi method).

2.2 - Extension of the previous study

When we make a principal component analysis, we are generally interested in the first axes only, those that are corresponding to the biggest eigen-values of the variance -covariance matrix. Let's have an incomplete factorial structure (F_r, Δ_r) in which $r < p$ and $F_r'F_r = I_r$. The matrix F_r has p rows and r columns whereas Δ_r is a r-diagonal matrix whose diagonal terms $\lambda_j (j=1,2,\ldots,r)$ are supposed to be positive.

2.2.1 - The first problem we considered before lead us to a set of R matrices with n rows and p columns wider than the preceding one sin we can "play" upon the vectors completing the family $\{f_j\}$ constituted with the r column-vectors of F_r to yield an orthonormal basis of R^p, and the associated eigen-values under the condition that they do not overtake $\min_j \lambda_j$.

We'll write, if we adopt the matricial writing by blocks :

$$F = (F_r, F_{p-r})$$

the (p-r) columns of the matrix F_{p-r} are constituted by vectors completing the r vectors $\{f_j\}$ to build an orthonormal basis of R^p. We'll write $L = \Delta^{1/2}$, N^2 will indicate a (p-r) diagonal matrix whose diagonal terms verify $(\forall k)\ \mu_k < \min_j \lambda_j$, and Q any matrix with p rows and n columns verifying

$$QQ' = I_p$$

$$Q.\mathbb{1} = 0$$

Under these conditions, any matrix R such as

$$R' = \begin{pmatrix} F_r , F_{p-r} \end{pmatrix} \begin{pmatrix} L & 0 \\ 0 & HN \end{pmatrix} Q$$

(where H is an arbitrary (p-r) orthogonal matrix)

is a solution of our first modified problem : $\sqrt{n}R$ admits the incomplete factorial structure (F_r, Δ_r).

2.2.2 - The second problem consists in constructing a matrix admitting the incomplete factorial structure (F_r, Δ_r) and to a centred matrix (in ten of a certain metric)

If we adopt the quadratic distance, as we did previously we'll have to choose R such as trace $(S-R)'(S-R)$ is minimum or :

$$2 \text{ trace } SR' - \text{trace } R'R = 2\text{trace }(SR') - \sum_{k=1}^{k=p-n} \mu_k \quad \text{minimum}$$

We propose here some indication on construction of such a matrix R without guarantying its optimality. Let's put down :

$$Q = \begin{pmatrix} Q_r \\ Q_{p-r} \end{pmatrix}$$

where Q_r is a r rows matrix and Q_{p-r} is a (p-r) matrix (both with n columns) such as

$$Q_r Q_r' = I_r \; ; \; Q_{p-r} Q_{p-r} = I_{p-r} \; ; \; Q_r Q_{p-r}' = 0$$

$$Q_r \mathbb{1} = 0 \qquad Q_{p-r} \mathbb{1} = 0$$

Then

$$SR' = (SF_r L \; , \; SF_{p-r} HN) \begin{pmatrix} Q_r \\ Q_{p-r} \end{pmatrix}$$

and trace $(S'R) = \text{trace }(SF_r LQ_r) + \text{trace }(SF_{p-r} HNQ_{p-r})$

$$= \text{trace}(Q_r \; SF_r L) + \text{trace}(NQ_{p-r} \; SF_{p-r} H)$$

We can determine as previously the optimal matrix Q_r under the constraints $Q_r Q_r' = I_r$ and $Q_r \mathbb{1} = 0$. We know that the r row -vectors of Q_r belong to the sub-space of R^n generated by the r vectors of $SF_r L$. If the (p-r) column -vectors of SF_{p-r} were orthogonal to this sub-space we might find out the maximum of trace $(NQ_{p-r} SF_{p-r} H)$ only under the contraints $Q_{p-r} Q_{p-r}' = I_{p-r}$ and $Q_{p-r} \mathbb{1} = 0$. To get closer to this particular situation, we choose a matrice H_1 with (p-r) rows and (p-2r) columns such as

$$GH_1 = 0 \text{ and } H_1 H_1' = I_{p-r} \text{ with } G = F_r' S' SF_{p-r}$$

The column vectors of $SF_{p-r} H_1$ are then orthogonal to those of the matrix $SF_r L$ and also to the row vectors of the Q_r optimal matrix. Writing Q_{p-r} as :

$$Q_{p-r} = \begin{pmatrix} Q_3 \\ Q_4 \end{pmatrix}$$

Q_3 and Q_4 have respectively (p-2r) and r rows (and n columns); then trace $(NQ_{p-r} SF_{p-r} H_1) = \text{trace }(NQ_3 SF_{p-r} H_1) + \text{trace }(NQ_4 SF_{p-r} H_1)$ and we can find out Q_3 verifying only $Q_3 Q_3' = I_{p-2r}$ and $Q_3 \mathbb{1} = 0$ since $Q_r Q_3' = 0$ will be realized from the fact that the rows of Q_3 belong to the sub-space generated by the columns of $SF_{p-r} H_1$. On the other hand, by post multiplying $SF_{p-r} H_1$ by a well chosen orthogonal (p-r) matrix T_1 (its columns will be eigen-vectors of $H_1' F_{p-r}' S'S F_{p-r} H_1$), we obtain a matrix $SF_{p-r} H_1 T_1$ with orthogonal column vectors : the row-vectors of Q_3 could then be chosen colinear to column vectors of $SF_{p-r} H_1 T_1$. But the post multiplication of SF_{p-r} by $H_1 T_1$ has modified the norms of column vectors of SF_{p-r} (but however, this point will be dealt with later concerning the determination of the matrix N). Let be H_2 the matrix which has as columns the r rows of G which were previously orthonormalized.

The matrix $H=(H_1T_1,H_2)$ verifies as required :

$$HH' = I_{p-r} \ ; \ H_2'H_1T_1 = 0 \ ; \ H_2'H_2 = I_r$$

By projecting the column vectors of $SF_{p-r}H_2$ on E_r sub-space of R^n orthogonal to the sub-space generated by the $r + p-2r$ row vectors of Q_r and Q_3 determined above, we will obtain a problem which we have encontered already earlier.

We now have to post- multiply $SF_{p-r}H_2$ by an orthogonal r order matrix T_2 such as the projection on the sub-space E_r of the column vectors of $SF_{p-r}H_2T_2$ will be orthogonal, and the rows vectors of Q_4 could then be chosen colinear to them.

Finally we have to determine the diagonal matrix N the elements $\sqrt{\mu_j}$ of which, have to be such as $\mu_j < \min_k \lambda_k = \lambda$ $(\forall j)$ and maximize

$$2 \sum_{j=1}^{j=p} \sqrt{\mu_j}\, \ell_j - \sum_{j=1}^{j=p} \mu_j \ ; \ \ell_j \text{ is the norm of the}$$

j-column vector of $SF_{p-r}H_1T_1(j=1,\ldots,p-2r)$ or the norm of the projection of the j-column vector of $SF_{p-r}H_2T_2(j=p-2r+1,\ldots,p-r)$.

To reach this aim we now have to note that one of the two following cases is realized:

(i) if $\ell_j^2 \leq \lambda$ the expression $a_j = 2\sqrt{\mu_j}\ell_j - \mu_j$ with $\mu_j \leq \lambda$ is maximum when $\mu_j = \ell_j^2$. The maximum value is $a_j = \ell_j^2$.

(ii) if $\ell_j^2 \geq \lambda$ the expression $a_j = 2\sqrt{\mu_j}\ell_j - \mu_j$ with $\mu_j \leq \lambda$ is maximum when $\mu_j = \lambda$. The maximum value of $a_j = 2\sqrt{\lambda}\ell_j - \lambda$

Remark : We have chosen to build a "suitable" matrix H prior to the matrix N which appears optimal conditionally relative to H .

The optimality of the couple (H,N) is not proved. We have already mentionned that the norms of the column-vectors of SF_{p-r} are altered through our different transformations. However if ℓ_j is the length of the j-column vector of $SF_{p-r}H$, a post-multiplication by an orthogonal matrix yields vectors of length b_j: such as $\Sigma\ell_j^2 = \Sigma b_j^2 = c^2$. One can state that if $(\forall j)$ $\ell_j^2<\lambda$ Max $(\Sigma 2 \sqrt{\mu_j}\ell_j - \mu_j) = c^2$ thus the post-multiplication by H has not altered the maximum relative to N . Otherwise an upper bound of that maximum can be computed which provide an opportunity to test a given structure in the point of view we have outlined previously

We have completed our study by various computer programs and futher applications are in progress.

An Error Estimation Technique for a Chebyshev Method

Salah E. El-Gendi, Benghasi

1. INTRODUCTION

A unified approach has recently been suggested by EL-GENDI [1969, 1970, 1973 and 1974]; to provide an estimate for the Chebyshev solution of various equations. The method, which may be applied to some problems in statistics as well as in other fields is reviewed in the next section.

In the present paper, it is shown that the computed solution satisfies exactly a perturbed equation. A computational procedure which provides the perturbation in the original equation and hence an estimation for the error, is suggested.

The technique is illustrated by numerical examples.

2. CHEBYSHEV METHOD

We assume that the equation or its integrated form is written as

$$y(x) - T\,y(x) = f(x), \qquad -1 \leq x \leq 1 \tag{1}$$

and that the solution may be approximated by

$$y^*(x) = \sum_{r=0}^{N}{}' \; a_r T_r(x) \tag{2}$$

Here $T_r(x) = \cos(r \cos^{-1} x)$ and the summation symbol denotes a sum with the first term halved. For the case $0 \leq x \leq 1$, the solution is obtained in terms of $T_r^*(x)$ where $T_r^*(x) = T_r(2x - 1)$.

The present method reduces the problem to the system of algebraic equations

$$(I - L)\,[y^*] = [f] \tag{3}$$

where I is the unit matrix and the matrix L can be defined as below, for different problems. The vector $[y^*]$ denotes an estimate for the solution at the Chebyshev points

$$x_i = -\cos\left(\frac{i\pi}{N}\right) \quad ; i=0,1,\ldots,N. \qquad (4)$$

The elements of $[f]$ are given by

$$f_i = f(x_i) \quad ; i=0,1,\ldots,N.$$

Now, to define the matrix L we recall the following matrix approximations for $\int_{-1}^{x} y(x)\,dx$ and $y(g(x))$, $-1 \leq x, g(x) \leq 1$ at the points (4);[EL-GENDI, 1969 and 1973]:

$$\left[\int_{-1}^{x} y(x)\,dx\right] = B\,[y] \qquad (5)$$

$$[\,y(g(x))\,] = C_g [\,y\,] \qquad (6)$$

where the matrices B and C_g are of order (N+1) and the elements of B can be computed once and for all for a given N. Relations (5) and (6) will be exact whenever y(x) is a polynomial of degree N or less.

As an illustration of manipulations with approximations (5) and (6), consider the functional-integro-differential equation

$$y''(x) = p(x)\,y(x) + q(x)\,y(g(x)) + \int_{-1}^{x} K(x,s)\,y(s)\,ds + h(x);$$

$$-1 \leq x, g(x) \leq 1$$

with the boundary conditions

$$y(-1) = y_{-1} \quad \text{and } y(1) = y_1$$

We deal with the integrated form of the problem i.e. eq.(1) with

$$Ty = \left(\int_{-1}^{x} - \frac{1+x}{2}\int_{-1}^{1}\right)\int_{-1}^{x}\{p(x)\,y(x) + q(x)\,y(g(x)) + \int_{-1}^{x} K(x,s)y(s)ds\}\,dx\,dx$$

and

$$f(x) = \frac{1-x}{2}\,y_{-1} + \frac{1+x}{2}\,y_1 + \left(\int_{-1}^{x} - \frac{1+x}{2}\int_{-1}^{1}\right)\int_{-1}^{x} h(x)\,dx\,dx.$$

Now, the matrix L in (3) may be defined as follows,

$$L = S\,\{\,(p) + (q)\,C_g + K\,\}$$

where $$S = RB \text{ with } r_{ij} = b_{ij} - \frac{1+x_i}{2}\,b_{Nj}\ ;\ i,j=0,1,\ldots,N.$$

The elements of K are given by $K_{ij} = K(x_i,x_j)\,b_{ij}$; $i,j=0,1,\ldots,N$, and (p) and (q) are diagonal matrices with diagonal elements $p(x_i)$ and $q(x_i)$ respectively where x_i is defined in (4).

We may also need to approximate the integral in f(x) using the matrix S.

Having obtained the solution of (3) one can easily compute the Chebyshev coefficients in (2).

Other types of linear problems may be similarly treated. For non-linear problems, the method can be applied in conjunction with an iterative process, for example, a Gauss-Seidel approach has been used [EL-GENDI, 1970] to treat non-linear ordinary differential equations.

3. ERROR ESTIMATION TECHNIQUE

Having defined system (3) we may need to solve it for two or more values of N, compute the Chebyshev coefficients and thereby decide when to stop by inspection. Though this process may give a valuable check, it does not however give indication of how the error varies with x and it is desirable not only to produce the Chebyshev coefficients in (2) but also some estimate for the error e(x) given by

$$e(x) = y(x) - y^*(x) \tag{7}$$

In the present method, the computed solution $y^*(x)$ satisfies exactly the perturbed equation

$$y^* - T\, y^* = f(x) - E(x) \tag{8}$$

Except for special classes of problems it is difficult to obtain the perturbation E(x) in a closed form. Now, using (7) it is clear that the error e(x) satisfies the equation

$$e(x) - T\, e(x) = E\,(x) \tag{9}$$

We seek an approximation for e(x) in the form

$$e(x) = \sum_{r=0}^{N+p}{}' \, \varepsilon_r \, T_r(x) \tag{10}$$

where the coefficients ε_r can be computed if we know e(x) at the points

$$X_i = -\cos\left(\frac{i\,\pi}{N+p}\right) \quad ; \quad i=0,1,\ldots,N+p \tag{11}$$

Approximations for E(x) at the points (11) may be obtained from

$$[f\,] - (I - L)C_x[\,y^*\,] \tag{12}$$

where L is now of order N + p + 1 and the elements of [f] are given by $f(X_i)$; i = 0,1,..., N+p.

One may use the barycentric form for the interpolation of y(x) that is given at the points (4), [SALZER, 1972], to show that the elements C_{ij} of the matrix C_x satisfy

$$C_{ij} = C_{N+p-i,\; N-j} \quad ; \; i=0,1,\ldots,N+p; \; j=0,1,\ldots,N$$

and this gives some economy in computation.

Having computed the perturbation E(x), a method similar to that of [FOX and PARKER 1968], may be used to estimate e(x) from (9). We may use, for instance;

$$[e] = L[e] + [E]$$

iteratively to get approximations for e(x) at the points (11) and hence the coefficients ε_r in (10).

From the relation between the coefficients in the exact representation of y(x) and the coefficients in (2) we notice that

$$\varepsilon_{N+1} = -\varepsilon_{N-1} \simeq a_{N+1}, \quad \text{[FOX and PARKER, 1968]}$$

Also the points (4) are the roots of $T_{N+1}(x) - T_{N-1}(x) = 0$ and hence the present Chebyshev method is a collocation method based on the zeros of the term $\varepsilon_{N-1} T_{N-1} + \varepsilon_{N+1} T_{N+1}$ in the error expansion (10).

Finally, comparing with the method of OLIVER[1969], the present method does not need any independent information for the neglected Chebyshev coefficients of the required solution.

The extra computation involved in our technique may however be used if we decide to get more accurate solution.

4. NUMERICAL EXAMPLES

Example 1

Our first example is the simple differential equation

$$y' = x(1+x)y, \quad y(1) = 1; \quad -1 \leq x \leq 1$$

The same example has been solved by BOGGS and SMITH,[1971] using different Chebyshev methods.

In table 1 we show the distribution of errors for the case N = 3 and p = 2.

We may notice that the second iterate gives a very good estimate for the error. Here we have taken $[e^{(1)}] = [E]$.

Table 1

x	-1	-.8	-.6	-.4	-.2	0	.2	.4	.6	.8	1
$10^3 e^{(1)}(x)$	-51	-83	-86	-59	-15	29	57	57	30	-3	0
$10^3 e^{(2)}(x)$	-74	-104	-105	-77	-30	18	50	54	30	-2	0
$10^3 e(x)$	-77	-107	-102	-73	-30	13	43	51	32	0	0

Example 2

We consider the boundary-value problem

$$y'' + y \sin x = e^x, \quad y(0) = 1, \; y(1) = 0; \quad 0 \leq x \leq 1$$

In table 2 we show the Chebyshev coefficients of the solution for N = 2 and

$N = 4$ together with the values of ε_r in (10) of the first iterate $e^{(1)} = E$ for the case $N = p = 2$.

Table 2

r	$\acute{a}_r$ (N = 2)	a_r (N = 4)	ε_r of the first iterate (N = p = 2)
0	0.80163	0.80268	0.00121
1	-.5	- .50945	- 935
2	.09918	.09773	- 152
3		945	935
4		93	92

Example 3

For the functional differential equation

$$y'(x) = -y(.8x) - y(x), \qquad y(0)=1;\ 0 \leq x \leq 1$$

the technique produces the perturbation

$$E(x) = \alpha_{N+1}(T^*_{N+1} - T^*_{N-1}) \text{ where } \alpha_{N+1} \simeq a_{N+1};$$

as can be theoretically verified. For the general problem, we do not expect such a simple form. Successive approximations produce the α_{N+1} values

N	2	3	4
α	$-.149 \times 10^{-1}$	$.143 \times 10^{-2}$	$-.103 \times 10^{-3}$

We have here results similar to those obtained by FOX, MAYERS and OCKENDON, [1971] but with different perturbation term. We may also take

$$e(x) = \alpha_{N+1}(T^*_{N+1} - T^*_{N-1})$$

as an estimate for the error.

Example 4

Our final example is the Volterra integral equation,

$$y(x) = x - \int_0^x (x - s)\, y(s)\, ds \qquad 0 \leq x \leq 1$$

If we take $N = 2$ we get for $p \geq 2$

$$E(x) = -0.00213 - 0.00437\,(T^*_3 - T^*_1) + 0.00197\,T^*_2 + 0.00016\,T^*_4$$

as can be theoretically verified.

5 CONCLUSION

We have a Chebyshev method which may be applied to various problems.

With little extra computation we can provide a reasonable error estimate as a function of the independent variable. This extra computation may be used to provide more accurate solution.

REFERENCES

BOGGS, R. and F. SMITH, A note on the integration of ordinary differential equations in Chebyshev series, Comp. J. 14 pp. 270-271; 1971.

EL-GENDI, S. E. , Chebyshev solution of differential, integral and integro-differential equations, Comp. J. 12 pp. 282-287; 1969.

EL-GENDI, S. E. , Chebyshev solution of differential equations, Proc. of Inaugural Conf., Scientific Comp., Cairo Univ. pp. 114-122; 1970.

EL-GENDI, S. E. , Chebyshev Solution of a class of functional-differential equation, (under publication), 1973.

EL-GENDI, S. E. , On a Chebyshev method for functional-integro-differential equations, Proc. of Stat. and scientific comp. conference, Cairo Univ., 1974.

FOX, L ; and I. PARKER, Chebyshev polynomials in numerical analysis , Oxford Univ., 1968.

FOX, L; D. MAYERS; J. OCKENDON and A. TAYLOR; On a functional differential equation, J. Inst. Maths. Applics. 8 pp. 271-307, 1971

OLIVER, J. , An error estimation technique for the solution of ordinary differential equations in Chebyshev series. Comp. J. 12 pp. 57-62, 1969.

SALZER, H. E. , Lagrangian interpolation at the Chebyshev points Xn = cos (J π /n), J = 0(1) n; Some unnoticed advantages; Comp. J. 15 pp. 156-159; 1972.

Construction of a Vector Equivalent to a Given Vector from the Point of View of the Analysis of Principal Components

Yves Escoufier, P. Robert, and J. Cambon, Montpellier

I. INTRODUCTION - All random vectors considered in this paper are assumed to be centered, with elements belonging to the set $L_2(\Omega, \mathcal{A}, \mathcal{P})$ of random variables with finite variances on $(\Omega, \mathcal{A}, \mathcal{P})$.

Let X_1 and X_2 be two such vectors ; let Z_1 and Z_2 be the random vectors of their principal components, respectively. In section III we study the problem of attaching weights to the elements of X_2 in such a way that the vector of principal components Z_2^* of this weighted vector be as closed as possible to Z_1. In section IV we study ways of choosing a subvector X_2^* of X_2 , simultaneously attaching weights to the selected elements , so that the vector Z_2^* of principal components of the resulting weighted X_2^* will best approximate Z_1 , over all choices of X_2^* and all weighting rules.

When conducting a study based on principal components of a vector difficult or costly to observe, one would welcome a theoritically sound method of substituting to it a vector easier to measure and having approximately the same principal components. At this time not much is available. On the one hand, it is known that the simplest non-singular transformation of the original vector may perturb deeply the principal components ; on the other hand, the results to be presented here indicate that the currently proposed techniques ([BEALE et alt 1967], [JOLIFFE 1973]) may suffer from insufficient theoritical foundations.

In section II we recall essential results obtained by ESCOUFIER ([1970] , [1973-a] , [1973-b]) which are the basis of the solutions to the problems investigated in this paper. To comply with printing space regulations, proofs are omitted (see quoted references).

II. METHEMATICAL BASES - Let X be a pxl random vector and $k \leq p$ be the rank of $E(XX')$. It is known that there exists a pxp orthogonal matrix H^* such that

$$H^{*\prime}\, E(XX')\, H^* = \left[\begin{array}{c|c} D & 0 \\ \hline 0 & 0 \end{array}\right]$$

where D is a kxk diagonal matrix with positive diagonal elements. Let H be the pxk submatrix of H^* such that $H'\, E(XX')\, H = D$.

<u>Definition 1</u>. The kx1 random vector $\underset{\sim}{Z} = \underset{\sim}{H}'\underset{\sim}{X}$ is called the vector of principal components of $\underset{\sim}{X}$.

<u>Definition 2</u>. The px1 random vector $\underset{\sim}{X}_1$, with the vector of principal components $\underset{\sim}{Z}_1$, and the gx1 random vector $\underset{\sim}{X}_2$, with the vector of principal components $\underset{\sim}{Z}_2$, are said to be equivalent if there exist a positive integer $k \leqslant \min(p,g)$, a constant α and a kxk matrix $\underset{\sim}{C}$ such that :

i) $\underset{\sim}{C}\underset{\sim}{C}' = \underset{\sim}{C}'\underset{\sim}{C} = \alpha \underset{\sim}{I}_k$, ii) $\underset{\sim}{Z}_1 = \underset{\sim}{C}'\underset{\sim}{Z}_2$.

($\underset{\sim}{I}_k$ is the kxk identity matrix).

It is easily verified that the relation established in Definition 2 is an equivalence relation. Its value, in our context, stems from the following properties [ESCOUFIER, 1973-b] :

<u>Proposition 1</u>. If $\underset{\sim}{X}_1$ and $\underset{\sim}{X}_2$ are equivalent vectors, then

i) for each diagonal element $\delta_i^{(1)}$ of $E(\underset{\sim}{Z}_1\underset{\sim}{Z}'_1)$ there is a diagonal element $\delta_j^{(2)}$ of $E(\underset{\sim}{Z}_2\underset{\sim}{Z}'_2)$ such that $\delta_j^{(2)} = \alpha\delta_i^{(1)}$;

ii) there exist permutation matrices $\underset{\sim}{P}$ and $\underset{\sim}{Q}$ such that

$$\underset{\sim}{Q}\,\underset{\sim}{C}\,\underset{\sim}{P} = \left|\begin{array}{c|c|c|c} \underset{\sim}{A}_1 & 0 & 0 \cdots & 0 \\ \hline 0 & \underset{\sim}{A}_2 & 0 \cdots & 0 \\ \hline \vdots & \ddots & \vdots \cdots & \ddots \\ \hline 0 & 0 & \cdots & \underset{\sim}{A}_r \end{array}\right|$$

where r is the number of distinct diagonal elements of $E(\underset{\sim}{Z}_2\underset{\sim}{Z}'_2)$;

iii) if all the diagonal elements of $E(\underset{\sim}{Z}_2\underset{\sim}{Z}'_2)$ are distinct, there is a permutation σ of the indices 1,2,...,k for which $z_{1i} = \pm\sqrt{\alpha}\, z_{2,\sigma(i)}$.

Proposition 1 should make clear what is to be understood by the "equivalence" of $\underset{\sim}{X}_1$ and $\underset{\sim}{X}_2$.

The next step is to find an easily computable criterion for the equivalence of two vectors. ESCOUFIER [1973-a] associates to a vector $\underset{\sim}{X}$ a linear transformation $U_{\underset{\sim}{X}}$ defined on $L_2(\Omega, \mathcal{a}, \mathcal{P})$ by

$$U_{\underset{\sim}{X}}(Y) = \sum_{i=1}^{p} [E(X_i Y) \cdot X_i] \text{ for all } Y \in L_2(\Omega, \mathcal{a}, \mathcal{P}).$$

The following properties are then established :

a)- the set of operators $U_{\underset{\sim}{X}}$ is a subset of the class of Hilbert-Schmidt operators on L_2 and so is a Hilbert space ;

b)- if $\underset{\sim}{\Sigma} = \begin{bmatrix} \underset{\sim}{\Sigma}_{11} & \underset{\sim}{\Sigma}_{12} \\ \underset{\sim}{\Sigma}_{21} & \underset{\sim}{\Sigma}_{22} \end{bmatrix}$ is the covariance matrix of the vector $\underset{\sim}{X} = \begin{bmatrix} \underset{\sim}{X}_1 \\ \underset{\sim}{X}_2 \end{bmatrix}$, the scalar product for the Hilbert space considered is given by

$$\langle U_{\underset{\sim}{X}_1} , U_{\underset{\sim}{X}_2} \rangle = \text{Tr} \, (\underset{\sim}{\Sigma}_{12} \, \underset{\sim}{\Sigma}_{21}) \; ;$$

c)- if Z_k is a principal component of $\underset{\sim}{X}$ associated with the eigenvalue λ_k, then

$$U_{\underset{\sim}{X}} \, (Z_k) = \lambda_k \, Z_k \, .$$

From the above, the following practical criterion can be deduced [ESCOUFIER, 1973-b] :

<u>Proposition 2</u>. The vectors $\underset{\sim}{X}_1$ and $\underset{\sim}{X}_2$ are equivalent in the sense of Definition 2 if and only if the coefficient

$$RV(\underset{\sim}{X}_1, \underset{\sim}{X}_2) = \text{Tr} \, (\underset{\sim}{\Sigma}_{12} \, \underset{\sim}{\Sigma}_{21}) \Big/ \left[\text{Tr} \, (\underset{\sim}{\Sigma}_{11}^2) . \text{Tr} \, (\underset{\sim}{\Sigma}_{22}^2) \right]^{1/2}$$

is equal to 1.

III. <u>THE FIRST PROBLEM</u> - From the last proposition it can be seen that the first problem can be restated as :

- Given vectors $\underset{\sim}{Y}$ and $\underset{\sim}{X}$ find a diagonal matrix $\underset{\sim}{\Delta}$ so as to maximize RV $(\underset{\sim}{Y} , \underset{\sim}{\Delta}\underset{\sim}{X})$, where

$$RV(\underset{\sim}{Y}, \underset{\sim}{\Delta}\underset{\sim}{X}) = \text{Tr} \left[\underset{\sim}{\Delta} E(\underset{\sim}{X}\underset{\sim}{Y}') E(\underset{\sim}{Y}\underset{\sim}{X}') \underset{\sim}{\Delta}\right] \Big/ \left\{ \text{Tr} \left[E(\underset{\sim}{Y}\underset{\sim}{Y}')\right]^2 . \text{Tr} \left[\underset{\sim}{\Delta} E(\underset{\sim}{X}\underset{\sim}{X}') \underset{\sim}{\Delta}\right]^2 \right\}^{1/2} ;$$

Hence look for $\underset{\sim}{\Delta}$ to maximize

$$\text{Tr} \left[\underset{\sim}{\Delta} E(\underset{\sim}{X}\underset{\sim}{Y}') \, E(\underset{\sim}{Y}\underset{\sim}{X}') \underset{\sim}{\Delta}\right] \Big/ \left\{ \text{Tr} \left[\underset{\sim}{\Delta} E(\underset{\sim}{X}\underset{\sim}{X}') \underset{\sim}{\Delta}\right]^2 \right\}^{1/2} .$$

Denote by m_j the j-th diagonal element of $\underset{\sim}{\Delta}$, by v_{ij} the elements of $E(\underset{\sim}{X}\underset{\sim}{X}')$, by u_{ij} the elements of $E(\underset{\sim}{Y}\underset{\sim}{X}')$ and define the vectors $\underset{\sim}{U}$, $\underset{\sim}{M}$ and the matrix $\underset{\sim}{A}$ by

$\underset{\sim}{U} = (U_j)$ where $U_j = \sum_i u_{ij}$,

$\underset{\sim}{M} = (M_j)$ where $M_j = m_j^2$,

$\underset{\sim}{A} = (A_{ij})$ where $A_{ij} = v_{ij}^2$.

By simply expanding the traces it can be verified that

$$\text{Tr} \left[\underset{\sim}{\Delta} E(\underset{\sim}{X}\underset{\sim}{Y}') \, E(\underset{\sim}{Y}\underset{\sim}{X}') \underset{\sim}{\Delta}\right] = \underset{\sim}{U}'\underset{\sim}{M} \quad ,$$

$$\text{Tr} \left[\underset{\sim}{\Delta} E(\underset{\sim}{X}\underset{\sim}{X}') \underset{\sim}{\Delta}\right]^2 = \underset{\sim}{M}'\underset{\sim}{A}\underset{\sim}{M} \; .$$

Thus we are lead to the quadratic programming problem for $\underset{\sim}{M}$:

$$\text{Problem P1} \quad \begin{cases} \text{Minimize } \underset{\sim}{M}'\underset{\sim}{A}\underset{\sim}{M} \\ \text{Subject to } \underset{\sim}{U}'\underset{\sim}{M} = 1 \text{ and } \underset{\sim}{M} \geqslant 0 . \end{cases}$$

Note that if $E(\underset{\sim}{X}\underset{\sim}{X}')$ is positive semi-definite, so is A [GOWER, 1971]. In such a case, the function $(\underset{\sim}{M}'\underset{\sim}{A}\underset{\sim}{M})$ is convex on the convex set $(\underset{\sim}{U}'\underset{\sim}{M} = 1$ and $\underset{\sim}{M} \geqslant 0)$.

It was found convenient to substitute to Problem P1 on the unknown $\underset{\sim}{M}$ the following equivalent problem on the unknown vector $\underset{\sim}{N}$:

$$\text{Problem P2} \quad \begin{cases} \text{Minimize } \underset{\sim}{N}' \; \underset{\sim}{C} \; \underset{\sim}{N}, \\ \text{Subject to } \underset{\sim}{e}'\underset{\sim}{N} = 1 \text{ and } \underset{\sim}{N} \geqslant 0, \end{cases}$$

where $\underset{\sim}{e}' = (1\ 1 \ldots\ldots 1)$ and where $\underset{\sim}{N}$ and $\underset{\sim}{C}$ are defined by :

$$\underset{\sim}{N} = (N_j) \text{ where } N_j = U_j\, M_j = U_j m_j^2 ,$$

$$\underset{\sim}{C} = (C_{ij}) \text{ where } C_{ij} = \frac{A_{ij}}{U_i U_j} = \frac{v_{ij}^2}{U_i U_j} .$$

The authors have solved Problem P2 for many sets of statistical data. Numerical evidence on these test cases will not be given here but reported on elsewhere [CAMBON, 1974] .

To be noted is the fact that the constraint $\underset{\sim}{e}'\underset{\sim}{N} = 1$ will force most of the currently proposed quadratic programming algorithms into the difficulty of a degenerated case. For this reason the authors have developped a specific algorithm which will now be justified and summarized.

$\underset{\sim}{C}$ being (nxn), denote by I a subset of $\{1,2,\ldots, n\}$ and by J the complement of I. Denote by $\underset{\sim}{C}_I$ the submatrix of $\underset{\sim}{C}$ formed by the C_{ij}'s with $i \in I$, $j \in I$; by $\underset{\sim}{C}_J$ the submatrix of $\underset{\sim}{C}$ formed by the C_{ij}'s with $i \in J$, $j \in I$. From the Kuhn-Tucker conditions for Problem P2, it can be proved that :

<u>Proposition 3</u>. Let $\underset{\sim}{N}_I = \underset{\sim}{C}_I^{-1}\ \underset{\sim}{e}$. If

$$\underset{\sim}{N}_I \geqslant 0 \quad \text{and} \quad \underset{\sim}{C}_J \underset{\sim}{N}_I \geqslant e , \tag{1}$$

then the vector $\underset{\sim}{N} = (N_i)$:

$$N_i = \begin{cases} \dfrac{1}{\underset{\sim}{e}'\underset{\sim}{N}_I}\; N_{Ii} & \text{if } i \in I, \\ 0 & \text{if } i \in J, \end{cases} \tag{2}$$

is an optimal solution of Problem P2, and $N'CN = (e'N_I)^{-1}$.

A technique to find an optimal solution would be to start computing $N_I = C_I^{-1}\, e$ for all subsets of indices I and to retain the first found to satisfy (1). The number of sets I to be considered can be greatly restricted in accordance with the following propositions.

Proposition 4. Let $P = (P_i) = \dfrac{C^{-1}e}{e'C^{-1}e}$, $I^* = \{ i : P_i \geqslant 0 \}$, $J^* = \{ i : P_i < 0 \}$. There is an optimal solution $N = (N_i)$ of Problem P2 such that $N_j = 0$ for at least one index $j \in J^*$.

Proof. It is known that P is an optimal solution of P2 without the positivity constraint. Considerany feasible vector $Q = (Q_i)$ for P2 such taht $Q_j > 0$ for all $j \in J^*$. Let $r \in J^*$ be such taht $Q_r/P_r \geqslant Q_j/P_j$ for all $j \in J^*$, $k = (Q_r/P_r)/(1-Q_r/P_r)^{-1}$, and $Q^k = (1-k)P + kQ$. Since the function F to be minimized is convex and $0 \leqslant k \leqslant 1$, $F(Q^k) \leqslant F(Q)$ But Q^k is feasible for P2 and $Q_r^k = 0$. The conclusion follows.∎

Proposition 5. An optimal solution of Problem P2 can be found as the best of the optimal solutions of the $|J^*|$ (n-1)-dimensional problems on N :

$$\begin{cases} \text{minimize } N' C_{(\mathcal{N} - j)} N , \\ \text{subject to } e'N' = 1 , N \geqslant 0, \end{cases} \tag{3}$$

where $\mathcal{N} = \{1,2,...,n\}$ and where j spans J^*.

Proof. From Proposition 4, it is seen that one can find an optimal solution of P2 by looking at the solutions of each of the $|J^*|$ problems :

$$\begin{cases} \text{minimize } N'CN \\ \text{subject to } e'N = 1 , N \geqslant 0 \\ \text{and the additional constaint } N_j = 0 , \end{cases}$$

for $j \in J^*$. For each j, this last problem is equivalent to (3).∎

Clearly (3) is a problem of type P2 with dimensions reduced by 1. We may therefore repeatedly invoke the above proposition to justify the following algorithm. A queue $\mathcal{Q}$ is built which elements are sets of indices I for which $N_I = C_I^{-1}\, e$ has to be calculated in accordance with Proposition 3. (We recall that in a "queue", elements are added "at the bottom" and removed "from the to the second element moving up at the top).

1 - Set the queue Q to the single element $\{1,2,\ldots,n\}$ and $\alpha = 0$;

2 - If Q is empty, print error diagnostic and stop ; otherwise, select and remove the first element I. Compute $\underset{\sim}{N}_I = \underset{\sim}{C}_I^{-1} \underset{\sim}{e}$;

3 - Let $K = \{k : N_{Ik} < 0\}$. If K is empty, go to Step 5 ; otherwise :

4 - For each $k \in K$, if $(I-\{k\})$ is not a subset of an element of Q, then ádd $(I-\{k\})$ to Q. Go to Step 2 ;

5 - If $\alpha \geqslant \underset{\sim}{e}'\underset{\sim}{N}_I$, go to Step 2 ; otherwise, set $\alpha = \underset{\sim}{e}'\underset{\sim}{N}_I$;

6 - If $\underset{\sim}{C}_j \underset{\sim}{N}_I \geqslant \underset{\sim}{e}$ (see (1)), set N_i as per (2) and stop ; otherwise go to Step 2.

The numerical efficiency of the algorithm will depend greatly on the particular code used to compute $\underset{\sim}{C}_I^{-1} \underset{\sim}{e}$. It will be seen in Section IV that, by using the Choleski decomposition of $\underset{\sim}{C}_I^{-1}$, one can make the test $\underset{\sim}{C}_j \underset{\sim}{N}_I \geqslant \underset{\sim}{e}$ of step 6 an immediate byproduct of step 2.

IV. THE SECOND PROBLEM - Let $\underset{\sim}{Y}$ and $\underset{\sim}{X}$ be two vectors, possibly equal. Assume $\underset{\sim}{X}$ to be n-dimensional and for $k \leqslant n$ denote by $\mathcal{V}_k$ the set of all k-dimensional subvectors of $\underset{\sim}{X}$. If I is a set of k distinct indices, we shall denote by $\underset{\sim}{X}_I$ the subvector of $\underset{\sim}{X}$ retaining those elements of $\underset{\sim}{X}$ with indices in I and by $\hat{\underset{\sim}{Y}}(\underset{\sim}{X}_I)$ the vector $(\underset{\sim}{\Delta}\underset{\sim}{X}_I)$ solution of the first problem studied in Section III. Thus this second problem can be stated as :

- Determine the vector $\underset{\sim}{X}_I \in \mathcal{V}_k$ which maximizes $RV[\underset{\sim}{Y}, \hat{\underset{\sim}{Y}}(\underset{\sim}{X}_I)]$ over $\mathcal{V}_k$.

When n and k are large, it becomes unpractical to try all k-dimensional subvectors of $\underset{\sim}{X}$ and for this reason we propose a sequential algorithm analogous to the one used in step-wise regression.

From the analysis carried in Section III, it is easily seen that for a given set I the vector $\underset{\sim}{X}_I$ is feasible for the present problems if and only if the vector $\underset{\sim}{N}_I = \underset{\sim}{C}_I^{-1} \underset{\sim}{e}$ is non negative and that the coefficient $RV[\underset{\sim}{Y}, \hat{\underset{\sim}{Y}}(\underset{\sim}{X}_I)]^2$ is then proportional to $\alpha_I = \underset{\sim}{e}'\underset{\sim}{C}_I^{-1}\underset{\sim}{e}$. Thus our problem becomes that of maximizing.

$$\alpha_I = \underset{\sim}{e}'\underset{\sim}{C}_I^{-1} \underset{\sim}{e} \quad \text{subject to} \quad \underset{\sim}{C}_I^{-1} \underset{\sim}{e} \geqslant 0,$$

over all subsets I of k distinct indices.

Suppose now that I is a subset of only (k-1) indices which is known to be feasible (i.e. $\underset{\sim}{N}_I = \underset{\sim}{C}_I^{-1} \underset{\sim}{e} \geqslant 0$). The step-wise technique consists in the determination of that index $j \notin I$ such that, with $(I,j) = I \cup \{j\}$, $\underset{\sim}{N}_{(I,j)}$ will be feasible and $\alpha_{(I,j)}$ will be maximized (over all choices of j).

The technicalities will now be described, assuming for simplicity that the set I contains the first (k-1) indices : $I = \{1,2,\dots,k-1\}$.

Consider the matrix

$$\underset{\sim}{C}^{+} = \left[\begin{array}{c|c} \underset{\sim}{C} & \begin{array}{c}1\\1\\\vdots\end{array} \\ \hline 1\,1\,1\cdots & 1 \end{array}\right]$$

where $\underset{\sim}{C}$ is given by (3), and let $\underset{\sim}{C}^{+} = \underset{\sim}{R}\,\underset{\sim}{R}'$ be the Choleski decomposition [FORSYTHE, MOLER, 1967] of $\underset{\sim}{C}^{+}$ (which is positive semi-definite), i.e $\underset{\sim}{R}$ is upper-triangular. We assume that the choice of the elements of the set I has been made sequentially by this same algorithm and that the first (k-1) steps of the Choleski decomposition of $\underset{\sim}{C}^{+}$ have been carried, column-wise, leading to the (n+1)x(k-1) matrix

$$\left[\begin{array}{c} \underset{\sim}{R}_I \\ \hline \underset{\sim}{r}'_{k\bullet} \\ \hline \vdots \\ \hline \underset{\sim}{r}'_{(n+1)\bullet} \end{array}\right] \qquad (4)$$

where $r'_{j\bullet} = (r_{j1}\; r_{j2} \dots r_{j(k-1)})$ for $j \geqslant k$ and

$$\underset{\sim}{C}_I = \underset{\sim}{R}_I\, \underset{\sim}{R}'_I ,$$

$$\underset{\sim}{R}_I\, \underset{\sim}{r}_{(n+1)\bullet} = \underset{\sim}{e} .$$

For $j \notin I$ ($j \neq n+1$), define

$$w_j = c_{jj} - \sum_{i=1}^{k-1} r_{ji}^2$$

$$s_j = 1 - \sum_{i=1}^{k-1} r_{ji}\, r_{(n+1)i} . \qquad (5)$$

($s_j = 1 - \underset{\sim}{r}'_{j\bullet}\, \underset{\sim}{R}_I^{-1}\, \underset{\sim}{e}$). If the index j is to be added to the set I with the feasible vector $\underset{\sim}{X}_{(I,j)}$, we must have the corresponding matrix $\underset{\sim}{C}_{(I,j)}$ non-singular , hence $w_j > 0$. The Choleski decomposition $\underset{\sim}{C}_{(I,j)} = \underset{\sim}{R}_{(I,j)} \underset{\sim}{R}'_{(I,j)}$ would then give :

$$\underset{\sim}{R}^{-1}_{(I,j)} = \begin{bmatrix} \underset{\sim}{R}_I^{-1} & 0 \\ \frac{-1}{\sqrt{w_j}} \underset{\sim}{r}'_{j\bullet} \underset{\sim}{R}_I^{-1} & \frac{1}{\sqrt{w_j}} \end{bmatrix} , \quad \underset{\sim}{R}^{-1}_{(I,j)} \underset{\sim}{e} = \begin{bmatrix} \underset{\sim}{R}_I^{-1} \underset{\sim}{e} \\ s_j / \sqrt{w_j} \end{bmatrix}$$

The coefficient RV $[\underset{\sim}{Y} , \hat{\underset{\sim}{Y}} (\underset{\sim}{X}_{(I,j)})]^2$ will be proportional to

$$\alpha_{(I,j)} = \underset{\sim}{e}' \underset{\sim}{R}'^{-1}_I \underset{\sim}{R}_I^{-1} \underset{\sim}{e} + (s_j^2/w_j) = \alpha_I + (s_j^2/w_j) \qquad (6)$$

The feasibility condition of non-negativity of $\underset{\sim}{N}_{(I,j)} = \underset{\sim}{C}^{-1}_{(I,j)} \underset{\sim}{e}$ must be respected. It can be verified by substitution of the above relations that

$$\underset{\sim}{N}_{(I,j)} = \underset{\sim}{R}'^{-1}_{(I,j)} \underset{\sim}{R}^{-1}_{(I,j)} \underset{\sim}{e} = \begin{bmatrix} \underset{\sim}{N}_I - \frac{s_j}{w_j} \underset{\sim}{Z}_j \\ \frac{s_j}{w_j} \end{bmatrix}$$

where $\underset{\sim}{Z}_j$ is the solution of the triangular system $\underset{\sim}{R}'_I \underset{\sim}{Z}_j = \underset{\sim}{r}_{j\bullet}$.

We can now summarize one basic cycle of the algorithm assuming that k is given, that I is a set of indices with less than k elements and that the corresponding matrix (4) is available, as s_j and w_j for all $j \in I$:

1 - Define $J = \{j : j \notin I, w_j > 0 \text{ and } s_j > 0\}$;

2 - If J is empty, stop. Otherwise select $j_o \in J$ which maximizes (s_j^2/w_j) (see (6)) ;

3 - Solve for $\underset{\sim}{Z}_{j_o}$ the triangular system $\underset{\sim}{R}'_I \underset{\sim}{Z}_{j_o} = \underset{\sim}{r}_{j_o\bullet}$;

4 - If $\underset{\sim}{N}_{(I,j_o)} = \underset{\sim}{N}_I - (s_{j_o}/w_{j_o}) \underset{\sim}{Z}_{j_o}$ is non-negative, then accept j set $I := I \cup \{j_o\}$, $\underset{\sim}{N}_I := \underset{\sim}{N}(I,j_o)$ and go to Step 5. Otherwise set $J := J - \{j_o\}$ and repeat Step 2 ;

5 - Carry one more step of the Choleski decomposition of $\underset{\sim}{C}^+$, computing the column corresponding to column "j_o" of $\underset{\sim}{C}^+$ (to obtain $\underset{\sim}{R}_{(I,j_o)}$) and, for $j \in J$, set ((5)) :

$$s_j := s_j - r_{(n+1)j_o} r_{jj_o} , \quad w_j := w_j - r^2_{jj_o} ;$$

6 - If $k_o < k$, set $k_o := k_o+1$ and repeat a complete cycle from Step 1. Otherwise, stop.

Note that a "stop" a Step 2 implies that the set of indices already chosen connot be enlarged by addition of a single element so as to increase the RV coefficient.

To initiate the algorithm one may choose any one element X_i of $\underset{\sim}{X}$ and set $I = \{i\}$. The square of the RV coefficient is then proportional to $C^{-1}_{i\,i}$; this suggests initiation with that index for which the diagonal element of $\underset{\sim}{C}$ is minimum.

It must be made clear that this algorithm does not necessarily produce the optimal solution. As in the step-wise regression technique, it produces the best solution of order k knowing which solution has been selected to the order (k-1).

Subject to this conditionality restriction upon convergence to the optimal solution, a count of the number of operations for a k-th cycle shows the algorithm to be quite efficient.

We conclude the study of this second problem by pointing out that the first problem (Section III) can be interpreted as a special case of the second : the case where $k = n$. Thus the algorithm of this section could be used in an attempt to solve Problem P2. The following proposition, deduced from the Kuhn-Tucker conditions for Problem P2 and the above analysis, provides a criterion to detect the optimal solution :

Proposition 6. A solution produced by the algorithm of this Section is optimal for Problem P2 of Section III if and only if, at stop,

$$s_j < 0 \quad \text{for all} \quad j \in J$$

or all n variables have been accepted (The proof is omitted).

CONCLUSION. - The theory summarized in Section II has shed new lights on two practical problems of interest to the applied statistician. Arguments on convexity and a factorization of the matrices involved have led to algorithms to solve those problems. A deeper numerical analysis of the problems, particularly the first one, should lead to more efficient algorithms and computer programs.

REFERENCES

BEALE E.M.L., KENDALL M.G. and MANN D.W. (1967) - The Discarding of Variables in Multivariate Analysis - Biometrika, 54, 3 and 4, p. 357-365.

CAMBON J. (1974) - Vecteurs equivalents à un autre au sens des composantes principales : Applications hydroliques - Note du laboratoire d'Hydrologie et d'Aménagement des eaux , USTL-Montpellier n°13-7

ESCOUFIER Y. (1970) - Echantillonage dans une population de variables aléatoires réelles - Publ. Inst. Stat. Univ. Paris - XIX-4 p. 1 à 4

ESCOUFIER Y. (1973-a) - Le traitement des variables vectorielles - Biometrics XXIX, p. 751-760.

ESCOUFIER Y. (1973-b) - Vecteurs aléatoires équivalents du point de vue de l'analyse en composantes principales. Rap. Techn. 7301 , Cen de Recherche en Informatique et Gestion, Av. d'Occitanie , 34000-MONTPELLIER.

FORSYTHE G., MOLER C.B. (1967) - Computer Solution of Linear Algebr Systems - Prentice Hall.

GOWER J.C. (1971) - A General Coefficient of Similarity and Some of its Properties - Biometrics, XXVII, p. 857-874.

JOLIFFE I.T. (1973) - Discarding Variables in a principal Component Analysis I : Artificial Data. Applied Statistics, Vol. 21 n° 2 , p. 160-173.

Numerical Solution of Robust Regression Problems

Peter J. Huber and Rudolf Dutter, Zurich

1. INTRODUCTION.

Traditionally, unknown parameters $\theta = (\theta_1,\dots,\theta_p)$ are fitted to observations $x = (x_1,\dots,x_n)$ by the method of least squares, i. e. by minimizing

$$\sum_i (x_i - f_i(\theta))^2 \qquad (1.1)$$

where $f_i(\theta)$ is the theoretical value of the i-th observation.

Unfortunately, the method is highly sensitive to the presence of occasional gross errors in the observations x_i. We obtain more robust estimates if we replace the square by a less rapidly increasing function ρ and minimize

$$\sum_i \rho(x_i - f_i(\theta)). \qquad (1.2)$$

Somewhat inconveniently, this does not give scale invariant estimates except for $\rho(x) = |x|^\alpha$, and among these, only $\alpha = 1$ leads to estimates which are robust in a technical sense (Hampel (1971)). We therefore prefer to enforce scale invariance by estimating a scale parameter σ simultaneously with θ.

Thus, we propose to minimize an expression of the form

$$Q(\theta,\sigma) = \sum_i \rho\left(\frac{x_i - f_i(\theta)}{\sigma}\right)\sigma + A\sigma, \quad \sigma \geq 0. \qquad (1.3)$$

We shall assume that $\rho \geq 0$ is convex, $\rho(0) = 0$, and satisfies

$$0 < \lim_{|x| \to \infty} \frac{\rho(x)}{|x|} = c \leq \infty. \tag{1.4}$$

If $c < \infty$, Q can be extended by continuity:

$$Q(\theta, 0) = c \sum |x_i - f_i(\theta)| . \tag{1.5}$$

One easily checks that Q is a convex function of (θ, σ) if the f_i's are linear. Thus, unless the minimum $(\hat{\theta}, \hat{\sigma})$ occurs on the boundary $\sigma = 0$ (and then $\hat{\theta}$ is the so-called L_1 - estimate) it can equivalently be characterized by the equations

$$\sum \psi\left(\frac{x_i - f_i(\hat{\theta})}{\hat{\sigma}}\right) \frac{\partial f_i}{\partial \theta_j} = 0, \ j = 1, \ldots, p \tag{1.6}$$

$$\sum \chi\left(\frac{x_i - f_i(\hat{\theta})}{\hat{\sigma}}\right) = A \tag{1.7}$$

with $\psi(x) = \rho'(x)$ and $\chi(x) = x\psi(x) - \rho(x)$. If we want $\hat{\sigma}$ to be asymptotically unbiased for normal errors, we should choose

$$A = (n-p) \frac{\beta}{2} \tag{1.8}$$

where

$$\frac{\beta}{2} = E_{\Phi}(\chi(X)) \tag{1.9}$$

is the expected value of χ for a standard normal argument X.

EXAMPLES.

(i) With $\rho(x) = x^2/2$ we obtain the standard least squares estimates, with $\hat{\theta}$ minimizing (1.1) and $\hat{\sigma}$ satisfying

$$\hat{\sigma}^2 = \frac{1}{n-p} \sum (x_i - f_i(\hat{\theta}))^2.$$

(ii) Let

$$\rho(x) = x^2/2 \qquad \text{for } |x| < c$$
$$= c|x| - c^2/2 \text{ for } |x| \geq c.$$

Then $\chi = 1/2\,\psi^2$, and we obtain the estimates favored by Huber (1964), (1973); these are asymptotically most robust against ε-contamination, with c depending on ε.

The standard algorithm for the minimization of (1.1) can easily be adapted to (1.3); moreover, for linear f_i it can be shown to converge. Its speed of convergence is relatively low (apparently linear), but this algorithm has the great advantage to be very simple.

In the special case (ii), with linear f_i, there are much faster algorithms which typically reach the exact minimum of (1.3) after only 3 to 5 iterations (but occasionally they fail to converge). At the expense of some programming costs and memory space, it is possible to combine the advantages of both and to obtain a guaranteed convergence which is almost always fast.

2. THE "SIMPLE" ALGORITHM.

The algorithm is the same for linear and for non-linear f_i. We need a tolerance level $\varepsilon > 0$ and starting values $\theta^{(o)}$, $\sigma^{(o)}$. If the f_i are linear, we recommend to use the least squares estimates as starting values. The algorithm now can be described as follows:

1. $m := 0$;
2. $\Delta_i := x_i - f_i(\theta^{(m)})$, $i = 1,\ldots,n$;
3. $(\sigma^{(m+1)})^2 := \frac{1}{A}\sum\chi(\frac{\Delta_i}{\sigma^{(m)}})(\sigma^{(m)})^2$;

4. $\Delta_i := \psi\left(\frac{\Delta_i}{\sigma^{(m+1)}}\right)\sigma^{(m+1)}$, $i = 1,\ldots,n$;

5. $c_{ik} := \frac{\partial f_i(\theta^{(m)})}{\partial \theta_k}$;

6. solve $\sum(\Delta_i - \sum_k c_{ik}\hat{\tau}_k)^2 = \min.$

   i.e. solve $C^T C\hat{\tau} = C^T\Delta$

   for $\hat{\tau}$;

7. $\theta^{(m+1)} := \theta^{(m)} + q\hat{\tau}$,

   where $0 < q < 2$ is an arbitrary relaxation factor;

8. if $\|\hat{\tau}\| < \varepsilon$, go to 10.;

9. $m := m+1$, go to 2.;

10. estimate θ by $\theta^{(m+1)}$,

    var (x_i) by $(\sigma^{(m+1)})^2$,

    cov (θ) by $K^2 \frac{1}{n-p}\sum\Delta_i^2$.

Here, K is a correction factor (Huber (1973), p. 812 ff.)

$$K = \frac{1 + \frac{p}{n}\frac{\mathrm{var}\,\psi'}{(\mathrm{ave}\,\psi')^2}}{\mathrm{ave}\,\psi'}$$

where

$$\mathrm{ave}\;\psi' = \frac{1}{n}\sum\psi'\left(\frac{\Delta_i}{\sigma^{(m+1)}}\right) = \mu$$

and

$$\mathrm{var}\;\psi' = \frac{1}{n}\sum\left[\psi'\left(\frac{\Delta_i}{\sigma^{(m+1)}}\right)-\mu\right]^2.$$

In example (i) (ordinary least squares) we have $K = 1$; in example (ii), we obtain

$$K = \frac{1 + \frac{p}{n}\frac{1-\mu}{\mu}}{\mu}.$$

Loosely speaking, we can say that we use the usual least

squares algorithm, except that we "Winsorize" the residuals in steps 3. and 4. (note that in the least squares case, 3. and 4. amount to

$$(\sigma^{(m+1)})^2 := \frac{1}{n-p} \sum \Delta_i^2$$

and

$$\Delta_i := \Delta_i,$$

respectively).

We shall now show under mild assumptions that this algorithm converges. The basic idea is to show that

$$Q(\theta^{(m)}, \sigma^{(m)}) \geq Q(\theta^{(m)}, \sigma^{(m+1)}) \geq Q(\theta^{(m+1)}, \sigma^{(m+1)}),$$

and that the inequalities are strict except if $\sigma^{(m)} = \sigma^{(m+1)}$ or $\theta^{(m)} = \theta^{(m+1)}$, respectively.

<u>LEMMA 2.1.</u> <u>Assume that</u> $\rho(x)/x$ <u>is convex for</u> $x < 0$ <u>and concave for</u> $x > 0$. <u>Then</u>

$$Q(\theta^{(m)}, \sigma^{(m)}) - Q(\theta^{(m)}, \sigma^{(m+1)}) \geq A(\sigma^{(m+1)} - \sigma^{(m)})^2 / \sigma^{(m)}.$$

<u>Proof.</u> Let $\Delta_i = x_i - f_i(\theta^{(m)})$ and define

$$U(\sigma) := \sum \chi\left(\frac{\Delta_i}{\sigma^{(m)}}\right)\left(\frac{(\sigma^{(m)})^2}{\sigma} - \sigma^{(m)}\right) + A(\sigma - \sigma^{(m)}) + Q(\theta^{(m)}, \sigma^{(m)}).$$

At $\sigma = \sigma^{(m)}$, the value and the first derivative with respect to σ of $U(\sigma) - Q(\theta^{(m)}, \sigma)$ are 0. As a function of $z = 1/\sigma$, $U(\sigma) - Q(\theta^{(m)}, \sigma)$ is convex: $f(z) = U(\sigma) - Q(\theta^{(m)}, \sigma) = -\sum \rho(z\Delta_i)/z + a_0 + a_1 z$; hence $U(\sigma) - Q(\theta^{(m)}, \sigma) \geq 0$ for all $\sigma \geq 0$. Note that U reaches its minimum at $\sigma^{(m+1)}$, namely

$$U(\sigma^{(m+1)}) = Q(\theta^{(m)}, \sigma^{(m)}) - A(\sigma^{(m+1)} - \sigma^{(m)})^2 / \sigma^{(m)}.$$

The assertion of the Lemma now follows.

LEMMA 2.2. *Assume* $0 \leq \rho'' \leq 1$ *and that the* f_i *are linear. Keep* $\sigma = \sigma^{(m+1)}$ *fixed. Without loss of generality choose the coordinates in the parameter space such that* $C^T C = I$. *Then*

$$Q(\theta^{(m)},\sigma) - Q(\theta^{(m+1)},\sigma) \geq \frac{q(2-q)}{2\sigma} \sum_j \Big(\sum_i \sigma \psi\Big(\frac{\Delta_i}{\sigma}\Big) c_{ij}\Big)^2$$

$$= \frac{q(2-q)}{2\sigma} \|\hat{\tau}\|^2 = \frac{2-q}{2q\sigma} \|\theta^{(m+1)} - \theta^{(m)}\|^2 .$$

Proof. Put $\Delta_i = x_i - f_i(\theta^{(m)})$, $c_{ij} = \partial f_i / \partial \theta_j$, $\tau = \theta - \theta^{(m)}$, and let

$$W(\tau) = \frac{1}{2\sigma} \sum \Big(\sigma\psi\Big(\frac{\Delta_i}{\sigma}\Big) - \sum_k c_{ik}\tau_k\Big)^2 - \frac{1}{2}\sum \sigma\psi\Big(\frac{\Delta_i}{\sigma}\Big)^2 + Q(\theta^{(m)},\sigma).$$

At $\tau = 0$, the value and the first derivatives with respect to τ of $W(\tau) - Q(\theta^{(m)} + \tau, \sigma)$ are 0. The matrix of second derivatives

$$\frac{\partial^2}{\partial \tau_j \partial \tau_k}\Big[W(\tau) - Q(\theta^{(m)} + \tau, \sigma)\Big] = \sum_i \Big[1 - \psi'\Big(\frac{\Delta_i - \sum c_{i\ell}\tau_\ell}{\sigma}\Big)\Big] c_{ij} c_{ik} \frac{1}{\sigma}$$

is positive semidefinite, hence $W(\tau) - Q(\theta^{(m)} + \tau, \sigma) \geq 0$ for all τ. The minimum of W occurs at $\hat{\tau} = C^T \sigma \psi(\frac{\Delta}{\sigma})$ and has the value

$$W(\hat{\tau}) = Q(\theta^{(m)},\sigma) - \frac{1}{2\sigma}\|\hat{\tau}\|^2 .$$

Since $W(q\hat{\tau}) - Q(\theta^{(m)},\sigma)$ is quadratic in q and vanishes for q = 0 and for q = 2, we obtain more generally

$$W(q\hat{\tau}) - Q(\theta^{(m)},\sigma) = -\frac{q(2-q)}{2\sigma}\|\hat{\tau}\|^2,$$

and the assertion of the Lemma follows.

Assume now that (1.7) cannot be satisfied for arbitrarily small δ (i.e. assume that the maximum number p' of residuals which can be made simultaneously 0 by a suitable choice of θ satisfies

$(n-p')\chi(\pm\infty) > A$; note that $p' \leq p$ with probability 1 if the x_i have densities).

<u>THEOREM.</u> (i) <u>The sequence</u> $(\theta^{(m)}, \sigma^{(m)})$ <u>has at least one accumulation point.</u> (ii) <u>Every accumulation point</u> $(\hat{\theta}, \hat{\sigma})$ <u>which satisfies</u> $\hat{\sigma} > 0$, <u>is a solution of</u> [illegible] (1.7) <u>and minimizes</u> (1.3).

<u>Proof.</u> The existence of an accumulation point follows from the compactness of $\{(\theta,\sigma) \mid \sigma \geq \iota,\ Q(\theta,\sigma) \leq r\}$. Lemmas 2.1 and 2.2 now imply that $(\hat{\theta}, \hat{\sigma})$ satisfies (1.6), and also (1.7) unless $\hat{\sigma} = 0$; but this case can be excluded with the help of the above assumption. That $(\hat{\theta}, \hat{\sigma})$ minimizes (1.3) follows now from the convexity of Q.

<u>CONJECTURE.</u> We conjecture that $(\theta^{(m)}, \sigma^{(m)})$ <u>always</u> converges to some minimum of (1.3), even without the above assumption, and even if the minimum is not unique.

3. SOPHISTICATED ALGORITHMS.

Assume that ρ is as in example (ii) of Section 1, and that the f_i are linear. Then ψ is piecewise linear, and if it would be known to which linear piece of ψ each residual belongs, the solution of (1.6), (1.7) is a problem of elementary algebra.

Thus, once the trial value $(\theta^{(m)}, \sigma^{(m)})$ is sufficiently close to the final value $(\hat{\theta}, \hat{\sigma})$, so that both induce the same classification of the residuals, the exact solution is obtained in one step. If the trial value $(\theta^{(m)}, \sigma^{(m)})$ is too far away, the formal solution $(\tilde{\theta}, \tilde{\sigma})$ of (1.6), (1.7) will lead to a different classification of the residuals, and there is a tendency to overshoot the minimum of Q. It is intuitively obvious that one obtains quick convergence if one selects the next trial value $(\theta^{(m+1)},\sigma^{(m+1)})$ on the segment between $(\theta^{(m)},\sigma^{(m)})$ and $(\tilde{\theta},\tilde{\sigma})$ such that it minimizes the value of Q. However, the literal implementation of this idea seems to be impractical; several approximate versions are being tested, and a detailed technical report is under preparation.

REFERENCES.

HAMPEL, F.R., A general qualitative definition of robustness. Ann. Math. Statist. 42, 1887-1896, 1971.

HUBER, P.J., Robust estimation of a location parameter. Ann. Math. Statist. 35, 73-101, 1964.

HUBER, P.J., Robust regression: asymptotics, conjectures and Monte Carlo. Ann. Statist. 1, 799-821, 1973.

Aspects of Axiomatization of Behavior: Towards an Application of Rasch's Measurement Model to Fuzzy Logic

Oskar Itzinger, Vienna

1. INTRODUCTION

Some 40 years ago, the fundamental importance of the "new" logic - that is, the extension of the "old" Aristotelian logic through calculisation (propositional, functional, relational calculus) - for the social sciences was pointed out in a well-known paper by MORGENSTERN [1936]. In his view, an application of that modern logic to the social sciences causes a sharp cut in the scientific activity, and he stated that "...die zunaechst geschaffene ausserordentliche Einengung der Bewegungsmoeglichkeiten bietet aber andrerseits die Gewaehr, dass man in diesem Bereich wirklich festen Boden unter den Fuessen gewinnt" [verbally cited from MORGENSTERN 1936, p.11]. Here, if we speak of "mathematical logic" or "logic", for short, we mean basically that new logic in its current usage.

Consider now the mathematical apparatus of psychology, social psychology and sociology (let us call these three fields collectively "behavioral sciences"). Without doubt, these areas form a subset of the social sciences in the sense of MORGENSTERN (among the social sciences he ranged also economic disciplines). In view of the paper mentioned above, there is no a priori reason for a renunciation of mathematical logic in the behavioral sciences. But of what kind is the mathematical apparatus of the behavior sciences now? If we disregard methods of mathematical statistics such as distribution analysis, significance tests, sampling theory, and the like, standard procedures are game theory, mathematical learning theory, stimulus sampling theory, and so on. These (mathematical) "theories" share the following features: (a) they list the undefined terms of the theory, and (b) they state the axioms of the theory, using for that purpose these undefined terms, together with a certain body of informal terminology that will presumably be readily understood by the reader. One then proceeds to derive theorems from these axioms, using whatever forms of reasoning seem sound. Therefore, the three "theories" cited above are formulated by a method which is called <u>axiomatic</u>. A look in two of the most important journals of mathematical psychology (Psychometrika and Journal of Mathematical Psychology) shows for example the frequent use of axiomatic methods in psychology: perhaps 70 percent of all papers are related to the previously mentioned procedures. Only a few papers are concerned with behavioral science applications of more abstract structures such as abstract algebra, graph theory, and so on. It seems that MORGENSTERN's assertion "...die neue Logik ist bisher wohl zur Gaenze von den Sozialwissenschaften uebersehen worden" [1936, p. 1] is fully valid today.

A further feature of most of the methods used in the behavioral sciences is the usage of "probability", i.e., some numerical value in the closed interval $[0,1]$. This concept is in favor of axiomatic methods, but in disfavor of <u>formalized theories</u>, that is, theories which are stated within some precisely formulated system of mathematical logic. The main reason for this is that the latter traditionally allow only <u>two</u> values for propositions or formulas of the basic logic, namely $\{true, false\}$ or $\{1,0\}$. But for most

applications to problems of the behavior sciences this is not enough because most data are subject to chance influences.
On the other hand,logic provides tools for the analysis of logical structures of social systems which can not be handled through probability measurements.Because a behavioral scientist is also interested in such logical structures,it would be nice if one could combine probability theory with symbolic logic.But we do not seem to know how to do this [McCARTHY and HAYES 1968].
In this paper we try a first approach to such a combination which hopefully should provide a possibility for the behavioral scientist to apply logic to questions in the social sciences more frequently.

2.FUZZY LOGIC

We start with some language L of the first-order predicate logic which does not include predicate variables.There may or may not be function symbols.Terms,atomic formulas and formulas are defined in the usual way.A first-order theory of L is a collection of formulas which do not contain free variables.Theories which are developed within L are called <u>elementary theories</u>.
In the so-called fuzzy logic,the truth-value of a formula S,instead of assuming two values(0 and 1),can assume <u>any</u> value in the closed interval $[0,1]$ and is used to indicate the "degree of truth" T(S) represented by the formula S.Note that continuum-many truth-values are allowed.We may view fuzzy logic as a special kind of many-valued logic.Properties of many-valued logic are described, for example,in ACKERMAN [1967],ROSSER and TURQUETTE [1952], and RASIOWA [1974].Given a formula S,T(S) can be evaluated as follows [after LEE 1972]:

(1) $T(S) = T(A)$ if $S = A$ and A is a ground atomic formula

(2) $T(S) = 1 - T(R)$ if $S = \neg R$

(3) $T(S) = \min[T(S_1), T(S_2)]$ if $S = S_1 \wedge S_2$

(4) $T(S) = \max[T(S_1), T(S_2)]$ if $S = S_1 \vee S_2$

(5) $T(S) = \inf[T(B(x))/x \in D]$ if $S = \bigwedge xB$ and D is the domain of x

(6) $T(S) = \sup[T(B(x))/x \in D]$ if $S = \bigvee xB$ and D is the domain of x

The reader should note that two-valued logic is a special kind of fuzzy logic;all the rules stated above are applicable in two-valued logic.

At this point,a crucial problem arises:<u>how</u> should a behavioral scientist determine the truth-value $T(S)$ of a formula S which expresses,for example,a certain assertion about behavior?Obviously, the evaluation procedure starts with a <u>given</u> $T(S)$.It should be noted that,basically,the problem is the same in two-valued logic. For formal treatments of logical systems it is certainly sufficient to <u>assume</u> truth-values,that is,logic itself makes no assumptions about the origin of the truth-value of a formula.On the other hand, however,the estimation of the truth-value of a statement might be a serious problem;in fact,it was pointed out by ZADEH [1965,1968] that this has to be subjective.But exactly this is the crucial difficulty with any practicable usage of logic in the behavioral sciences.So,what can someone do to overcome this difficulty?Unfortunately,neither logic nor we can offer a complete solution for that problem but instead we suggest the following first step in that direction.

At a first glance,the fact that logic does not provide clues for the a priori determination of truth-values is a certain disadvantage.On the other hand,this is an advantage,too,because one can use for that purpose any method which seems sound.One such method is the subjective estimation mentioned above;however, whether this method is sound is doubtfully.It seems that one can get better results if the truth-values for formulas can be "measured" in some way.A possibility for such a procedure is the following: first,one defines certain behavioral traits,say,of a social group, and tries to state relevant assertions in the form of formulas of L,i.e.,one constructs a theory.On the other hand,the same assertions should be measureable by some measurement model as used in the behavioral sciences.Measurements have to be in the interval [0,1].These values can be interpreted as truth-values if the used measurement model has some properties which makes such an interpretation plausibly.With that truth-values,the truth-degrees of the given formulas can be evaluated.Obviously,the admissibility of such an interpretation depends very strong on the chosen measurement model.As an appropriate model we suggest the measurement model of RASCH which is outlined in the next section.

3.MEASUREMENT MODEL OF RASCH

We are assuming that the reader is already familiar with the basic notions of latent trait models,for example with latent structure analysis [LAZARSFELD 1959] ,or the logistic test model of BIRNBAUM[1965].Briefly,latent variables are variables which are not observable in a direct manner.In the same context,RASCH[1961,1966] developed models which are based on the concept of "specific objectivity in scientific comparisons".Roughly,a comparison is "specifically objective" if and only if the outcome of the comparison only depends on the objects to be compared and on chance, but on nothing else. To allow for specific objectivity in comparisons the measurements must not depend on some specific sample of items or subjects, but must characterize the objects per se, i.e., the measurements must be "samplefree". RASCH has shown that his models are the only ones that permit samplefree measurements. Samplefreeness is achieved by conditional inference procedures and it can be shown, that RASCH's models are the only ones to permit the required procedures.

Consider now a social system (a group of persons). Social behavior consists of interactive processes. These are said to depend on individual behavior traits and on interindividual relationship-parameters simultaneously. For a determination of the logical structure of such a group, it is necessary to separate individual traits from group traits. Based on RASCH's models, measurement models can be derived which lead to measurements of individual traits which are independent from the social system. On the other hand, interindividual relations can be measured which are independent from individual traits, that is, system parameters can be won. Briefly, the assumptions of such a model are as follows [after SCHEIBLECHNER 1971]:

(1) Latent parameters: it is assumed that each individual is characterized by social behavior traits and each relationship is characterized by a system parameter.

(2) Indicators: events, that depend on latent dimensions and chance influences only, have to be used.

(3) Specific objectivity in comparisons: comparisons between agents (everything that exerts a measureable influence on a random varia-

ble is called an agent) should depend on the agents and chance only.

(4) Probability laws: probabilities of events can be represented by continuous distribution functions. Probability laws will be called "choice characteristic".

(5) Local stochastic independence: events are stochastically independent given the parameters.

(6) Minimal sufficient statistics: the model should provide minimal sufficient statistics for parameters, i.e., the statistics should contain all relevant information about parameters available in the data.

From such assumptions, RASCH derived his probabilistic models and could show that they are the only ones to conform to this set of conditions under the parametric framework of test theory. The most fundamental point is that measurements of interpersonal relations can be separated from individual traits.

Because we are not interested in formal properties of such models, the foregoing sketch is sufficient for our purpose. In the next section it is shown by example how to fit measurements from such a model in fuzzy logic.

4. APPLICATION OF RASCH'S MEASUREMENT MODEL TO FUZZY LOGIC

Recall now the difficulty with the estimation of the truth-value of a statement. The following example should show how this can be relieved.

Given some social system, say a group of persons, assume that we would like to examine the latent variable "playing-strength". For this purpose, we design for example a Hex tournament. Because it can be shown that with Hex a "draw" is impossible and that the first player can always win, there are only two events. As proper parameter, we define "tendency to a win for the first-moving player" and "tendency to a loss for the second-moving player". Clearly, the tournament should consist of several rounds so that a "round-parameter" for each player is necessary. Finally, the system "player v - player u", for all v and u, v≠u, in the group requires a parameter, too. The data of the tournament should fit in some measurement model as described above. So, for each pair (v,u) of players we get - among other measurements - measurements for the probability that player v wins against player u. On the other hand, this last fact can also be expressed as a 2-place relation R(v,u). Therefore, using such relations, we can formulate certain assertions about the playing behavior of the group as a set of formulas of L. Consider the simple formula

$$\bigwedge x \bigvee y \, R(x,y) \tag{1}$$

which states that for each person x there is another person y in the group, such that x wins against y. Clearly, the domain of the relation consists of all group members, and is usually finite. Because the experiment and the formulas are so chosen that they both cover the same relevant facts, the probability measurements can now be interpreted as "truth-values" (the special properties of the used measurement model contribute further to the admissibility of such an interpretation). If they are substituted into the formula, one gets with the use of the evaluation procedure (see section 2) a single value in the interval 0,1 interpreted as the truth-degree of the formula (1). It should be noted that there is no problem with an evaluation if the theory consists of more than one formula because all formulas can be combined into a single formula.

5. DISCUSSION

Assuming that given a set of measurements for some defined variables from an appropriate measurement model, and given a set of formulas of some language L, the combination of both opens new views for the application of logic to behavioral sciences. In common practice, logical theories are accepted iff their truth-degree is equal to one. Now one can fix a certain lower bound $b<1$ for the truth-degree of a theory; the theory is refused iff its truth-degree is less than b. A set of axioms for a theory is a set of sentences with the same consequences as the theory. Clearly, the theory is a set of axioms of itself. Therefore, each acceptable theory is a set of axioms of the considered behavioral traits; furthermore, it is a testable set of conditions for the variables in question. Fuzzy logic allows to accept a theory which is not completely satisfied. Obviously, this is a more natural way for an application of logic to the behavioral sciences than the two-valued logic is. The foregoing ideas suggest that before one can get plausible insights in the logical structure of a social group he should try to measure the interesting facts. Finally, a last possibility should be mentioned. Assume, that one examines several social groups relativeto some variables which remain fixed. Clearly, one has a set of truth-degrees for the corresponding theory. The question arises if the truth-degrees are different in a statistical sense. If it is possible to apply appropriate tests for an answering of that question, and the result is "yes", this would be an indicator for significant differences between logical structures.

REFERENCES

Ackerman,R.Introduction to Many Valued Logics.Dover,New York,1967

Birnbaum,A.Some latent trait models and their use in inferring an an examinee's ability.ETS,Princeton,1965

Lazarsfeld,P.F.Latent structure analysis,in:Psychology:A study of a science,Vol.III.Ed.Koch,S.McGraw-Hill,New York,1959

Lee,R.C.T.Fuzzy Logic and the Resolution Principle.JACM,Vol.19, pp.109-119,1972

McCarthy,J.,P.J.Hayes.Some philosophical problems from the standpoint of artificial intelligence,in:Machine Intelligence,Vol.4. Ed.Meltzer,B.,Michie,D.American Elsevier,New York,1968

Morgenstern,O.Logistik und Sozialwissenschaften.Zeitschr.f.Nationaloekonomie,Vol.8,pp.1-24,1936

Rasch,G.On general laws and the meaning of measurement in psychology,in:Proceedings of the fourth Berkeley symposium on mathematical statistics and probability.University of California Press,1961

Rasch,G.An item analysis which takes individual differences into account.Brit.J.Math.Statist.Psychol.Vol 19,49-57,1966

Rasiowa,H.An Algebraic Approach to Non-Classical Logics.American Elsevier,New York 1974

Rosser,J.B.,A.R.Turquette.Many-valued Logics.American Elsevier, New York,1952

Scheiblechner,H.The seperation of individual- and systeminfluences on behavior in social contexts.Acta Psychologica,Vol.35,442-460, 1971

Zadeh,L.A.Fuzzy sets.Inform.Contr.8,338-353,1965

Zadeh,L.A.Fuzzy algorithms.Inform.Contr.12,94-102,1968

Estimating the Credibility of Results of Statistical Computations when Variables Are Subject to Errors with Non-Zero Sum

Mirek Karasek*,Toronto

1. INTRODUCTION

It has been an established fact that certain fields of mathematics such as algebra and numerical analysis are indispensable tools of present day statistics. Yet, there exists at least one branch of statistics - economic statistics - where a more intensive application of some basic concepts of numerical mathematics, particularly those concerned with the theory of errors, could substantially improve the present state of affairs.

The implicit thesis of our analysis is that in statistics the classical definition of "error" (with its basic zero-sum requirement) does not necessarily cope with problems of real-life statistics of social sciences. Instead, we consider the possibility that the error may not conform to the zero-sum rule, so that our definition of "error" appears to be closer to reality. The former zero-sum-element (also called "noise" in physics) then will be referred to as the "random departure".

Although errors in data-series are normally taken into account in some computational methods, such as the method of "instrumental variables", the mere fact that economic statistics are based almost exclusively on the Central Limit Theorem and the Gauss-Markov Theorem make it impossible to derive inferences in cases of small-sample statistics or systematic errors.

The general problem, in short, is that we are probably dealing most of the time with errors in the variables (data-series, sets) which not only refuse to have zero means (in contrast to the random departures), but the expected values of errors may be quite large and different for each given data-series.

The essence of our exposition is to tie the credibility (see e.g. SHACKLE (1968)) of results of statistical computations with an indicator of, shall we say, trustworthiness of every set of data entering this computation.

---

* The Author is currently with the Econometric Research Branch, Central Statistical Services, Ministry of Treasury, Economics and Intergovernmental Affairs, Government of Ontario. The opinions expressed in this article are those of the author and do not necessarily represent the views of the Ministry.

The outcome of any such computation then should display, apart from the numerical result, the estimated component of error which can be expected to be absorbed in the output of this operation. Again, as we have already mentioned above, this estimated error in result should be denoted by a special numerical characteristic closely attached to the real numerical datum.

The question is how to transplant this idea into a working hypothesis so that the subsequent modelling scheme may offer a plausible explanation of the process of forming the error in result.

We propose to go about it in the following way. In part 2, the necessary prerequisites, basic definitions and notations are revealed. The theory and some of its implications on multiple linear regression models is outlined in part 3. These theoretical conclusions are tested and/or verified by several numerical examples in part 4 while a brief discussion of the results is carried out in the final part 5.

2. PREREQUISITES

2.1 Basic Definitions and Notations

We begin the exposition by defining the elements of our theory. Suppose we are interested in a single value x of an arbitrary measurable phenomenon. It is clear that what we observe in reality is usually not the correct numerical value of the characteristic x^* but some other value, say x, which contains an unknown proportion (or component) of error.

Let us, therefore, introduce and redefine the basic elements of the errors-in-variables models by using the following notation and assumption.

We assume that all variables are subject to errors

$$x_{it} = x^*_{it} + \varepsilon_{it} \quad , \quad t = 1 \,....\, T \quad , \quad i = 1 \,....\, k, \qquad (1)$$

where x^*_{it} is the true value, x_{it} is the observed value and ε_{it} is the error. We maintain, however, that for most of above mentioned problems the assumption $E(\varepsilon_i) \neq 0$ is much more appropriate than the traditional $E(\varepsilon_i) = 0$.

<u>DEFINITIONS</u>:

When $E(\varepsilon_i) = 0$, then ε_{it}'s in eq. (1) are called "random departures" whereas when $E(\varepsilon_i) \neq 0$, then ε_{it}'s will henceforth be called "errors".

Thus, "error"....in this context is the general term for the cumulative (and/or systematic) effect of departures from the average with non-zero sum which in statistics are represented by: errors of observation, errors due to inappropriate definition or classification, sampling errors, etc. (for details, see MORGENSTERN (1963) and for systematic errors see MALINVAUD (1966), pp. 327-328).

2.2 Δx_i as an Indicator of Data Trustworthiness

Assuming

$$E(\varepsilon_i) \neq 0 \quad , \quad i = 1 \,....\, k \quad , \tag{2}$$

two interesting corollaries to eq. (2) immediately emerge:

a) SEGEL (1972) shows that there is a definite advantage in expressing the error magnitude in relative terms.

b) While we are talking about the relative error

$$\delta x_{it} = \frac{\varepsilon_{it}}{x_{it}} \quad , \tag{3}$$

we should consider using its limit value (see BEREZIN-ZIDKOF (1962))

$$\Delta x_i = \frac{\mathrm{Sup}\ |\ \varepsilon_{it}\ |}{|\ x_{it}\ |} \tag{4}$$

instead, because this characteristic has several considerable advantages that become apparent when we use the hypothetical relative error limit Δx_i as an indication of trustworthiness of any given series x_i of basic data. Let us list some of them.

1) Limit values are, generally speaking, much easier to estimate or assess than specific figures. It is quite true that we do not know the value of any given x^* (or x^*_{it}) and hence neither the value of ε_{it}. Despite of this, we should be able to estimate either the expected value of the maximum relative error, where $E(\delta x_i) = \Delta x_i$ (Δx_i then could be assumed to be constant for the entire set of data expressing a certain economic phenomenon), or we could directly attribute one known value of Δx_i, as defined in eq. (5), to any given data-series x_i (or set $\{x_i\}$) and investigate the appropriate computational method for its stability; "stability" thus compares the relative error of the result Δ_R with that of Δx_i. The first alternative gives us

$$\frac{\sum_{t=1}^{T} \varepsilon_{it}}{T\ |x_{it}|} = \Delta x_i \quad , \tag{5}$$

where the assumption (2) holds, while the second alternative, the one with Δx_i expressed as an arbitrary constant, will be discussed in item 3.

2) It is obviously convenient to express the error in percentage terms for it is not only a common denominator but also a well understood phenomenon.

3) Even if we are unable to estimate the mean percentage of the error limit of a given set of data, the battle is still not lost. The main attraction of the second alternative rests with the fact that we, after all, do not need to know any precise level of reliability of each incoming set(or series) of data in terms of Δx_i at all. Instead, we can approximately assess the truthfulness of investigated data-series by any of the verbal-scoring techniques being widely used in the management science. Assume, for instance, that WIENER (1963) is correct claiming that "....economics is a one or two digit science....". Then, one of many possible conversion scales may look like TABLE 1 (adapted from WHITE (1968)).

Δx_i (or Δ_R)	0.10	0.20	0.30	0.40	0.50	0.60
VERBAL INTERPRETATION OF DATA RELIABILITY (OR CREDIBILITY OF RESULTS)	excellent	good	fair	poor	bad	worthless

TABLE 1

3. MODELLING THE CREDIBILITY OF COMPUTATIONAL RESULTS IN STATISTICS

3.1 Elements of the Theory

We have investigated above one way to model an error-prone statistics. It appears that the appropriate numerical scale, employing the characteristic Δx_i , as seen in eq. (4) and (5), should be closely attached (at least in our mind) to the basic quantitative datum with the final result reminding us of the function of a complex variable[1]; especially of its polar form.

Now, since we have just mentioned some points in favor of the bi-characteristic notation

$$(x_i , \Delta x_i) \quad , \quad i = 1 \; .... \; k \quad , \tag{6}$$

for any given series (or set) of data x_i from the field of social sciences, there is obviously one basic question to be asked immediately.

The question is:

What rules can we set or derive for characteristics Δx_i's such that the credibility of the result of any statistical computational method that engages the series (sets) x_i's, i = 1 .... k, could be estimated in terms of Δ_R .

1) In the sense, this idea of bi-characteristic statistics having emerged in our paper, resembles (as far as its utility is concerned) the principle of ϕ-surfaces studied by SHACKLE (1968).

DWYER (1951) gives the following theorems on relative errors.

THEOREMS:

Let $P^* = x^*y^*$, $Q^* = \frac{y^*}{x^*}$, $U^* = x^{*p}$ and $V^* = x^{*1/p}$;

then after some manipulations (with the use of logarithmic differentiation) we have

$$\begin{aligned} |\delta_P| &\leq |\delta x| + |\delta y| , \\ |\delta_Q| &\leq |\delta x| + |\delta y| , \\ |\delta_U| &\leq p\,|\delta x| , \\ |\delta_V| &\leq \frac{1}{p}\,|\delta x| , \end{aligned} \tag{7}$$

where all δ's conform to the definition in eq. (3).

3.2 Corollaries to Formulas (7)

Since it is known (see e.g. CHOW (1966)) that in the model of multiple linear regression

$$y^* = X^* \beta^* + u \quad , \tag{8}$$

the least squares estimate of β^* (when all incoming series of eq. (8) contain, hypothetically that is, only the true values) equals[2)]

$$\beta^* = (X^{*\prime} X^*)^{-1} X^{*\prime} y^* \quad ; \tag{9}$$

the following corollaries to THEOREMS (7) easily stem from eq. (9).

Suppose that we move closer to reality such that we shall deal with errors-in-variables series (6) denoted as $(y , \Delta y)$, $(x_i , \Delta x_i)$ and also suppose that the results of our computations are regression coefficients β_i . Then, from formulas (7), eq. (9) and with $E(\delta y) > 0$, $E(\delta x_i) > 0$, we have

$$|\delta\beta_i| \leq |\delta y| - |\delta x_i| \tag{10}$$

and when we introduce the characteristics Δy , Δx_i into eq. (10), from eq. (4) we have

$$\Delta\beta_i \sim |\Delta y - \Delta x_i| \; , \quad \Delta\beta_i \equiv \Delta_R \; . \tag{11}$$

Similarly, THEOREMS (7), eq. (4) and eq. (9) yield

$$\Delta\beta_i \sim (\Delta y + \Delta x_i) \tag{12}$$

2) We are aware of the existence of, at least, half a dozen other estimators which are perhaps more suitable for the case of errors in variables, but it is quite clear that the above formulas (7) could be applied to whatever formula we face; the OLS estimator was just chosen to be an example because of its general popularity.

when either $E(\delta y) < 0$ and $E(\delta x_i) > 0$ or $E(\delta y) > 0$ and $E(\delta x_i) < 0$.

Finally, when $E(\delta y) < 0$ and $E(\delta x_i) < 0$, from formulas (7), eq. (9), eq. (10) and eq. (4) we have

$$\Delta\beta_i \sim | \Delta x_i - \Delta y | \quad . \tag{13}$$

4. NUMERICAL EXPERIMENTS

We have tested formulas (11) - (13) on two numerical multiple regression examples adapted from MALINVAUD (1966, p. 17) and HALD (1952). These examples yield the "true" data-series $y^{1*}, x_1^{1*}, x_2^{1*}, x_3^{1*}, x_4^{1*}, y^{2*}, x_1^{2*}, x_2^{2*}, x_3^{2*}, x_4^{2*}$ in the APPENDIX.

To obtain the "observed" series for the first alternative (see eq. (5)), we put $E(\delta y^1) = 0.15$, $E(\delta x_1^1) = 0.15$, $E(\delta x_2^1) = 0.30$, $E(\delta x_3^1) = 0.05$, $E(\delta x_4^1) = 0.25$, $E(\delta y^2) = 0.25$, $E(\delta x_1^2) = 0.20$, $E(\delta x_2^2) = 0.10$, $E(\delta x_3^2) = 0.30$ and $E(\delta x_4^2) = 0.15$ thus deriving the series $y^{11}, x_1^{11}, x_2^{11}, x_3^{11}, x_4^{11}, y^{12}, x_1^{12}, x_2^{12}, x_3^{12}, x_4^{12}, y^2, x_1^2, x_2^2, x_3^2$, and x_4^2 to be seen in the APPENDIX.

The second alternative (i.e. Δx_i = arbitrary constant) is again derived from the "true" series, only instead of randomly distributed relative errors with non-zero means, as in the first case, we have used the above values $E(\delta x)$, $E(\delta y)$ as constants Δy , Δx_i holding for each element of the appropriate "true" series. Thus, we have obtained the "observed" series $\overline{y}^1$, $\overline{x}_1^1, \overline{x}_2^1, \overline{x}_3^1, \overline{x}_4^1, \overline{y}^2, \overline{x}_1^2, \overline{x}_2^2, \overline{x}_3^2, \overline{x}_4^2$ in the APPENDIX.

To understand the notation of TABLE 2, where the results of all thirteen final runs[3] k, k = 1 .... 13, are displayed, let us take a look at one of them, say the one denoted by 3 in the first column. The symbol

$$\{\overline{y}^1, x_1^{1*}, \overline{x}_2^1, x_3^{1*}, \overline{x}_4^1; y^{1*}, \overline{x}_1^1, x_2^{1*}, \overline{x}_3^1, x_4^{1*}\} \tag{14}$$

shows that series $\overline{y}^1, x_1^{1*}, \overline{x}_2^1, x_3^{1*}, \overline{x}_4^1$ (with their index numbers and notation referring to the APPENDIX) have been used as the "true" data-set for the computation of "true" correlation coefficients denoted by β^*_{i3} . The other five elements of eq. (14), i.e. $y^{1*}, \overline{x}_1^1, x_2^{1*}, \overline{x}_3^1, x_4^{1*}$, then specify the series used as the "observed" ones thus yielding the "observed" coefficients β_{i3} .

3) It is also worth mentioning that each single trial run k consists of a numerical solution of 2 x 15 single and multiple regressions, where all possible combinations of 2 x 5 variables (both "true" and "observed" series) are investigated.

k	$\{y^{j*}, x_1^{j*}, x_2^{j*}, x_3^{j*}, x_4^{j*}; y^j, x_1^j, x_2^j, x_3^j, x_4^j\}$	$\Delta_{\beta_{1k}}^T$	$\Delta_{\beta_{2k}}^T$	$\Delta_{\beta_{3k}}^T$	$\Delta_{\beta_{4k}}^T$	$E(\Delta_{\beta_{1k}})$	$E(\Delta_{\beta_{2k}})$	$E(\Delta_{\beta_{3k}})$	$E(\Delta_{\beta_{4k}})$
1	$\{y^{1*}, x_1^{1*}, x_2^{1*}, x_3^{1*}, x_4^{1*}; \bar{y}^1, \bar{x}_1^1, \bar{x}_2^1, \bar{x}_3^1, \bar{x}_4^1\}$	0	0.15	0.10	0.05	0.01	0.20	0.11	0.05
2	$\{y^{2*}, x_1^{2*}, x_2^{2*}, x_3^{2*}, x_4^{2*}; \bar{y}^2, \bar{x}_1^2, \bar{x}_2^2, \bar{x}_3^2, \bar{x}_4^2\}$	0.05	0.15	0.05	0.10	0.08	0.13	0.06	0.12
3	$\{\bar{y}^1, x_1^{1*}, \bar{x}_2^1, x_3^{1*}, \bar{x}_4^1; y^{1*}, \bar{x}_1^1, x_2^{1*}, \bar{x}_3^1, x_4^{1*}\}$	0.30	0.15	0.20	0.10	0.33	0.19	0.24	0.07
4	$\{\bar{y}^2, x_1^{2*}, \bar{x}_2^2, x_3^{2*}, \bar{x}_4^2; y^{2*}, \bar{x}_1^2, x_2^{2*}, \bar{x}_3^2, x_4^{2*}\}$	0.45	0.15	0.55	0.10	0.61+	0.20	0.65	0.14
5	$\{y^{1*}, x_1^{1*}, x_2^{1*}, x_3^{1*}, x_4^{1*}; y^{12}, x_1^{12}, x_2^{12}, x_3^{12}, x_4^{12}\}$	0.00	0.15	0.10	0.05	0.02	0.10	0.14	0.05
6	$\{y^{2*}, x_1^{2*}, x_2^{2*}, x_3^{2*}, x_4^{2*}; y^2, \bar{x}_1^2, x_2^2, \bar{x}_3^2, \bar{x}_4^2\}$	0.05	0.15	0.05	0.10	0.10	0.17	0.10+	0.14
7	$\{y^{2*}, x_1^{2*}, x_2^{2*}, x_3^{2*}, x_4^{2*}; y^2, \bar{x}_1^2, \bar{x}_2^2, \bar{x}_3^2, \bar{x}_4^2\}$	0.05	0.15	0.05	0.10	0.08	0.19	0.13+	0.11
8	$\{y^{1*}, x_1^{1*}, x_2^{1*}, x_3^{1*}, x_4^{1*}; y^{11}, x_1^{11}, x_2^{12}, x_3^{11}, x_4^{12}\}$	0.00	0.15	0.10	0.05	0.01	0.14	0.10	0.02
9	$\{y^{12}, x_1^{12}, x_2^{1*}, x_3^{1*}, x_4^{1*}; y^{1*}, x_1^{1*}, x_2^{12}, x_3^{12}, x_4^{12}\}$	0.00	0.45	0.10	0.40	0.03	0.37	0.24+	0.44
10	$\{y^2, x_1^{2*}, x_2^2, x_3^{2*}, x_4^2; y^{2*}, x_1^2, x_2^{2*}, x_3^2, x_4^{2*}\}$	0.45	0.15	0.55	0.10	0.38	0.22+	0.61	0.24+
11	$\{y^{11}, x_1^{1*}, x_2^{11}, x_3^{1*}, x_4^{11}; y^{1*}, x_1^{11}, x_2^{1*}, x_3^{11}, x_4^{1*}\}$	0.30	0.15	0.20	0.10	0.37	0.14	0.25	0.36+
12	$\{y^{1*}, x_1^{12}, x_2^{1*}, x_3^{12}, x_4^{1*}; y^{12}, x_1^{1*}, x_2^{12}, x_3^{1*}, x_4^{12}\}$	0.30	0.15	0.20	0.10	0.27	0.21+	0.18	0.11
13	$\{y^{2*}, x_1^2, x_2^{2*}, x_3^2, x_4^{2*}; y^2, x_1^{2*}, x_2^2, x_3^{2*}, x_4^2\}$	0.45	0.15	0.55	0.10	0.39	0.18	0.58	0.22+

+ non-significant at 0.05 level

TABLE 2

Now, taking the corresponding β^{*j}_{i3} and β^{j}_{i3}, where j denotes the data base (of the APPENDIX) used for this particular set (run) of multiple regressions and 3 then denotes the 3rd run, i.e. k = 3, we finally arrive (with the help of formulas (3), (4)) at a set of characteristics $\Delta\beta_{i3}$. The next step is the computation of expected values $E(\Delta\beta_{i3})$,

$$E(\Delta\beta_{i3}) = \frac{1}{n_{i3}} \sum_{n_{i3}} \Delta\beta_{i3} \quad , \quad n_{ik} \leq 15, \quad k = 1 \; .... \; 13 \quad . \tag{15}$$

The results of eq. (15)- under the assumption of their normal distributions with $N(\mu_{ik}, \sigma^2_{ik}/n_{ik})$, $\mu_{ik} \equiv \Delta\beta^T_{ik}$ $(\equiv \Delta^T_R)$, i.e. the theoretical results' credibility computed from formulas (11), (12) and (13) - are then investigated for their significance at 0.05 level. The $E(\Delta\beta_{ik})$'s found nonsignificant at this level are carrying the + sign.

Since in our case $E(\delta y^1) = -0.15$, $E(\delta x^1_1) = 0.15$, $E(\delta x^1_2) = -0.30$, $E(\delta x^1_3) = 0.05$ and $E(\delta x^1_4) = -0.25$, we can compute these theoretical characteristics $\Delta\beta^T_{ik}$'s (in the second part of TABLE 2) in the following way.

From eq. (12) we have $\Delta\beta^T_{13}$ (as (0.15 + 0.15)) ~ 0.30; $\Delta\beta^T_{23}$ is then according to eq. (13), as (|0.30 - 0.15|), approximately equal to 0.15; $\Delta\beta^T_{33}$ follows from eq. (12), (0.15 + 0.05), as approximately equal to 0.20 and $\Delta\beta^T_{43}$ is, from eq. (13) as (|0.25 - 0.15|), approximately equal to 0.10.

The entries of the last section of TABLE 2, following from computed characteristics $\Delta\beta_{i3}$ and eq. (15), are means of the real results' credibilities.

5. CONCLUSION

The objective of this paper is to focus attention on the following points:

(i) Assuming that the series (or sets) of the basic statistical data subscribe to the above non-zero-sum-of-errors hypothesis, we maintain (and the evidence in TABLE 2 supports it) that the credibility of numerical results stemming from the use of an arbitrary computational formula of statistics could be estimated under a certain condition.

(ii) This necessary and sufficient condition calls for the evaluation of all basic data-series, involved in computation, as far as their trustworthiness (or reliability) is concerned.

(iii) To describe and measure this phenomenon, i.e. the credibility of results and/or the reliability (or trustworthiness) of basic data, the characteristic Δ_R and/or Δx_i has been introduced. These characteristics define the computational output and/or the incoming data-series in terms of the limit of their relative errors so that constants, widely accepted and understood, such as percentages, could be used here.

(iv) The most important feature of these characteristics Δ_R and Δx_i , however, lies in their ability to be readily converted into a verbally interpreted scale (see TABLE 1) and thus make the basic prerequisite (ii) - and by so doing the whole estimation procedure, such as the one in § 3.2 - possible.

APPENDIX

$y^{1}*$:	12.6	13.1	15.1	15.1	14.9	16.1	17.9	21.0	22.3	21.9	21.0	23.0
x_1^1*:	117.0	126.3	134.4	137.5	141.7	149.4	158.4	166.5	177.1	179.8	183.8	190.0
x_2^1*:	3.1	3.6	2.3	2.3	0.9	2.1	1.5	3.8	3.6	4.1	1.9	2.1
x_3^1*:	84.5	89.7	96.2	99.1	103.2	107.5	114.1	120.4	126.8	127.2	128.7	130.0
x_4^1*:	9.9	10.0	10.2	10.3	8.9	9.5	10.2	10.3	10.4	9.0	8.8	10.1
$y^{2}*$:	79	74	104	88	96	109	103	73	93	116	84	113
x_1^2*	7	1	11	11	7	11	3	1	2	21	1	11
x_2^2*	26	29	56	31	52	55	71	31	54	47	40	66
x_3^2*:	6	15	8	8	6	9	17	22	18	4	23	9
x_4^2*:	60	52	20	47	33	22	6	44	22	26	34	12
$\bar{y}^1$:	14.8	15.4	17.8	17.8	17.5	18.9	21.1	24.7	26.2	25.8	24.7	27.0
$\bar{x}_1^1$:	137.6	148.5	158.1	161.7	166.6	175.7	186.3	195.8	208.3	211.4	216.1	233.4
$\bar{x}_2^1$:	4.4	5.1	3.3	3.3	1.3	3.0	2.1	5.4	5.1	5.9	2.7	3.0
$\bar{x}_3^1$:	89.0	94.5	101.3	104.4	108.7	113.2	120.1	126.8	133.5	134.6	135.5	136.9
$\bar{x}_4^1$:	12.4	12.5	12.8	12.9	11.1	11.9	12.8	12.9	13.0	11.3	11.0	12.6
$\bar{y}^2$:	105	99	139	117	128	145	137	97	124	155	112	151
$\bar{x}_1^2$:	9	1	14	14	9	14	4	1	3	26	1	14
$\bar{x}_2^2$:	29	32	62	34	58	61	79	34	60	52	44	73
$\bar{x}_3^2$:	9	21	11	11	9	13	24	31	26	6	33	13
$\bar{x}_4^2$:	71	61	24	55	39	26	7	52	26	31	40	14
y^{11}:	14.4	16.2	17.0	17.5	17.7	19.5	21.0	24.8	26.2	26.1	25.0	26.3
x_1^{11}:	135.0	149.5	159.0	161.0	168.0	175.0	187.8	195.0	207.0	212.0	218.0	222.1
x_2^{11}:	4.2	5.0	3.6	3.0	1.3	3.2	2.0	5.5	5.6	5.8	2.0	3.2
x_3^{11}:	88.0	94.0	100.0	107.0	106.0	113.2	120.0	128.0	134.2	136.0	135.5	136.6
x_4^{11}:	12.4	13.0	13.0	13.0	10.5	12.7	12.8	11.5	12.8	12.0	11.0	12.5
y^{12}:	14.6	15.5	18.0	17.8	17.7	19.0	21.3	24.5	26.5	25.5	24.6	26.7
x_1^{12}:	137.0	148.0	158.1	162.2	167.3	176.0	186.3	195.0	208.0	211.5	216.8	223.3
x_2^{12}:	4.2	5.0	3.5	3.3	1.5	3.0	2.0	5.0	5.2	5.7	3.0	3.2
x_3^{12}:	89.0	94.0	101.3	104.4	109.0	113.5	120.1	126.0	133.8	134.0	135.7	137.7
x_4^{12}:	12.4	12.6	12.7	12.9	11.2	12.0	12.8	12.7	13.0	11.2	11.1	12.6

y^2 :	104	100	135	115	130	145	140	100	122	151	113	154
x_1^2 :	8	2	13	14	8	14	5	2	3	24	2	15
x_2^2 :	28	28	64	34	60	61	74	38	56	58	40	77
x_3^2 :	10	19	11	8	10	14	26	25	30	8	34	12
x_4^2 :	70	59	26	50	38	25	13	50	28	30	38	19

REFERENCES

BEREZIN, I. S., N. P. ZIDKOF, Numerical Methods, GIFML, Moscow, 1962, (I), 43.

CHOW, G., A Theorem on Least Squares and Vector Correlation in Multivariate Linear Regression, JASA, 61, 413-414.

DWYER, P., Linear Computations, J. Wiley & Sons, Inc., 1951, p. 26.

HALD, A., Statistical Theory and Engineering Applications, J. Wiley & Sons, New York, 1952, 647-650.

MALINVAUD, E., Statistical Methods of Econometrics, Rand McNally, Chicago, 1966.

MORGENSTERN, O., On the Accuracy of Economic Observation, Princeton, 1963, § 2.

SEGEL, L. A., Simplification and Scaling, SIAM Rev., 14, 547-571.

SHACKLE, G. L. S., Uncertainty in Economics (and Other Reflections), Cambridge University Press, Cambridge, 1968, § 2.

WIENER, N., see MORGENSTERN, p. 116, footnote 14.

A Dynamic Model for Measuring Individual and Situative Influences on Social Behavior

Wilhelm F. Kempf, Kiel

1. INTRODUCTION

A basic concept in stochastic test theory is the *item characteristic curve* $f_i(\xi_v)$. Let ξ_v be an individual parameter representing e.g. the ability of an individual v in an achievement test or the aggressiveness of the individual in an aggressivity test. Then we introduce a random variable a_{vi} such that $a_{vi} = 1$ if the individual v gives a positive ("correct", "aggressive") response to item number i and $a_{vi} = 0$ if the individual's response is a negative ("incorrect", "non-aggressive") one. $f_i(\xi_v)$ is the probability that the individual v will respond positively to item i, and under the assumption of *local stochastic independency*

$$(1.1) \quad p\{a_{v1}, a_{v2}, \ldots, a_{vg}\} = \prod_{i=1}^{g} p\{a_{vi}\} \quad \text{for all } g = 1, \ldots, k$$

the probability distribution of an individual's responses to k items is defined uniquely by

$$(1.2) \quad p\{(a_{vi})\} = \prod_{i=1}^{k} f_i(\xi_v)^{a_{vi}} \cdot (1 - f_i(\xi_v))^{1-a_{vi}}$$

in which $(a_{vi}) = (a_{v1}, a_{v2}, \ldots, a_{vk})$ denotes the *response vector* of the individual.

As regards the structural form of $f_i(\xi_v)$ as a function of the latent variable ξ, the literature contains several suggestions. For a number of reasons which are clearly emphasized in the papers by RASCH (1960, 1961) and ANDERSEN (1973a, 1973b), for instance, the logistic test model

$$(1.3) \qquad f_i(\xi_v) = \frac{\xi_v}{\xi_v + \sigma_i} \quad ,$$

in which σ_i is the difficulty of the item, is the most attractive statistical model. According to a well known theorem by NEYMAN & SCOTT (1948) the traditional methods of parameter estimation fail if the number of parameters to be estimated does not tend towards a fixed numeral while the number of observations grows to infinity. From eq. (1.2), however, it follows, that each $\underline{S}$ that is added to the sample will cause the introduction of an additional parameter. The only models in which consistent estimators exist in such a situation are those in which the individual parameters can be separated from the structural parameters pertaining to the items by use of *conditional inference* methods (cf. ANDERSEN, 1973a) and as ANDERSEN (1972b) has shown, within the framework of stochastic test theory, these are the models suggested by RASCH (1960, 1961) only.

In many psychological applications of stochastic test theory, however, this framework turnes out to be too narrow. Many psychological concepts such as *learning* or *catharsis* conflict with the assumption of local stochastic independency and KEMPF (1974), therefore, has suggested an extension of stochastic test theory to what may be called *dynamic test theory*.

2. THE MODEL

The basic conception of dynamic test theory is to replace the assumption of local stochastic independency by the concept of *local serial dependency*

$$(2.1) \qquad p\{(a_{vi})\} = \prod_{i=1}^{k} p\{a_{vi} | s_{vi}\}$$

in which $(a_{vi}) = (a_{v1}, a_{v2}, \ldots, a_{vk})$ denotes the vector of the individual's responses to k items and in which s_{vi} stands for the partial response vector $(a_{v1}, a_{v2}, \ldots, a_{vi-1})$. The item characteristic curves $f_i(\xi_v)$ are replaced by *conditional item characteristic curves*

$$(2.2)\quad f_{i.s_{vi}}(\xi_v) = p\{a_{vi} = 1 \mid (a_{v1}, a_{v2}, \ldots, a_{vi-1}) = s_{vi}\}$$

so that

$$(2.3)\quad p\{a_{vi} \mid s_{vi}\} = f_{i.s_{vi}}(\xi_v)^{a_{vi}} \cdot (1 - f_{i.s_{vi}}(\xi_v))^{1-a_{vi}}$$

On the basis of this formalism, KEMPF (1974) has proposed a dynamic test model in which the conditional item characteristic curves are assumed to depend on the *number of positive responses* to the preceeding items

$$(2.4)\quad r_{vi} = \begin{cases} 0 & \text{for } i = 1 \\ \sum_{j=1}^{i-1} a_{vj} & \text{for } i = 2, 3, \ldots, k \end{cases} ,$$

but *not* to depend on *which* of the preceding items were answered positively:

$$(2.5)\quad f_{i \cdot s_{vi}}(\xi_v) = f_{i \cdot r_{vi}}(\xi_v)$$

for all partial response vectors s_{vi} which are compatible with the partial score r_{vi} .

The functions $f_{i \cdot r_{vi}}(\xi_v)$ are defined by

$$(2.6)\quad f_{i \cdot r_{vi}}(\xi_v) = \frac{\xi_v + \psi_{r_{vi}}}{\xi_v + \sigma_i}$$

in which $\psi_{r_{vi}} \leq \sigma_i$ for all $r_{vi} = 0, \ldots, i-1$ and $i = 1, \ldots, k$.

Before we proceed to a discussion of the model, we note that the *Rasch*-model is a special case of the dynamic model: (2.6) reduces to (1.3) if $\psi_0 = \psi_1 = \ldots = \psi_{k-1} = 0$. This suggests that ξ_v ist a generalized ability parameter and that σ_i is an item difficulty parameter. Fig. 2.1 shows that the probability (2.6) is a monotone *increasing* function of the individual parameter ξ_v and a monotone *decreasing* of the item parameter σ_i. (2.6) tends to 0 for $\sigma_i \to \infty$ and it tends to 1 for $\xi_v \to \infty$ as well as for $\sigma_i \to \psi_r$. If $\xi_v \to 0$, finally, (2.6) tends to ψ_r/σ_i.

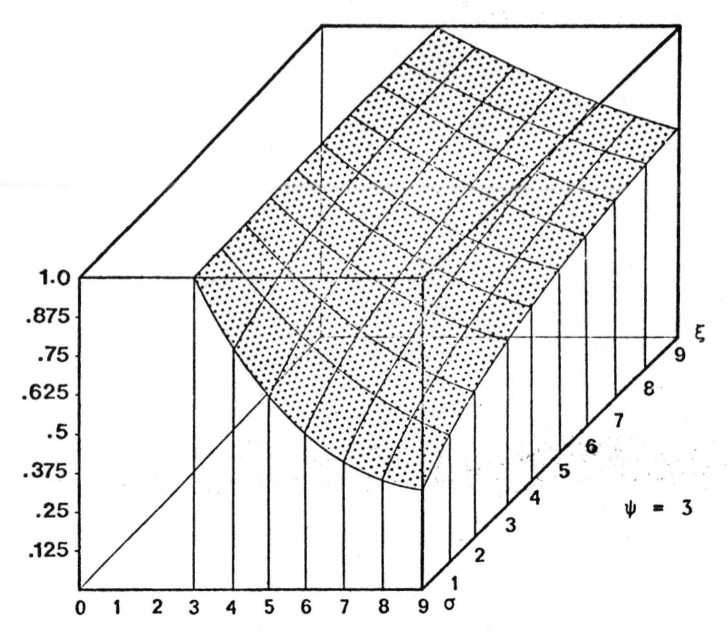

Fig. 2.1.: The conditional item characteristic curve (2.6) as a function of the latent trait variable ξ and of the item difficulty σ; two different values of ψ.

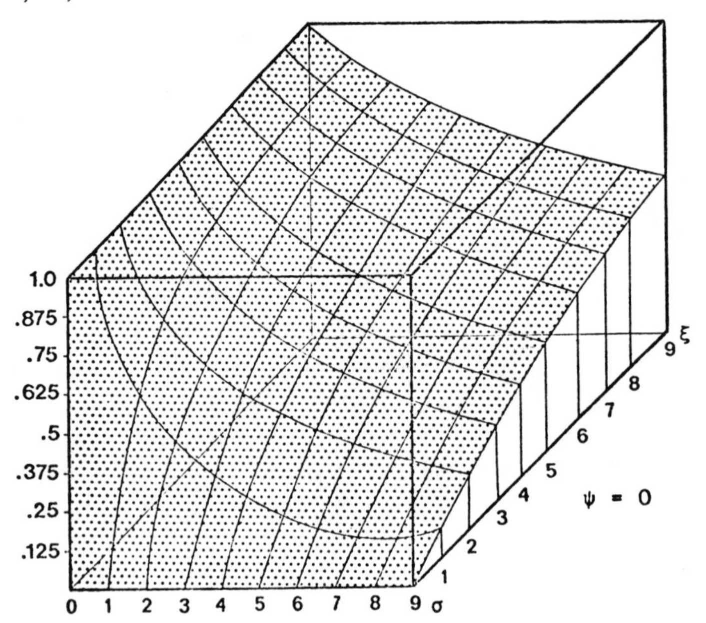

Regardless of an individual's initial ability ξ_v, after r positive responses to the preceeding items, the individual's probability of sucess on item i will not be less than ψ_r/σ_i.

ψ_r will be referred to as a transfer parameter. It describes how an individual's prcbability of success is effected by his prior responses. (2.6) is a linear increasing function of ψ_r. It tends to 1 if $\psi_r \to \sigma_i$ and to $\xi_v/(\xi_v + \sigma_i)$ if $\psi_r \to 0$. The transfer ψ_r is a *learning effect*, if it is an *increasing* function of r (*positive transfer);* it is a *reactive inhibition* (such as catharsis, for instance), if it is a *decreasing* function of r (*negative transfer),* and it is a *fluctuation* that can be explained by concurricung positive and negative transfer, if it is a *non-monotone* function of r.

As for the simple model with $\psi_0 = \psi_1 = \ldots = \psi_{k-1} = 0$, the model is slightly overparametrized by $\xi_v,\ldots,\xi_n$, $\sigma_1,\ldots,\sigma_k$ and $\psi_0,\ldots,\psi_{k-1}$ The parameters of the model are not determined uniquely by eq. (2.6) and the model will still hold when the parameters are multiplied by a positive constant c_1 and when an arbitrary constant c_2 is added to the individual parameters ξ_v, and, at the same time, is subtracted from the item- and transfer-parameters σ_i and ψ_r. The parameters, therefore, are measured on *interval scales* only and we may look for a proper standardization such as

(2.7) $\quad \mathrm{MIN}(\psi_r) = 0 \qquad$ for $r = 0,\ldots,k-1$

and

(2.8) $\quad \prod_{i=1}^{k} \sigma_i = 1 \quad ,$

by which we introduce the same parametrization conditions that are usual in the *Rasch*-model.

As KEMPF (1974) has shown, the dynamic test model (2.6) also has the same mathematical properties as the *Rasch*-model. The test scores $a_{vo} = \Sigma_{i=1}^{k} a_{vi}$ are minimal sufficient statistics for the latent trait parameters ξ_v and the item difficulties σ_i can be estimated by use of the CML-method.

3. SUFFICIENT STATISTICS

The test score a_{vo} is a sufficient statistic for estimating the individual parameter ξ_v *iff* the conditional likelihood of the individual's response vector (a_{vi})

$$(3.1) \qquad p\{(a_{vi}) \mid a_{vo}\} = \frac{p\{(a_{vi})\}}{p\{a_{vo}\}}$$

does not depend on the individual parameter ξ_v.
Reformulating (2.6) as

$$(3.2) \qquad p\{a_{vi} \mid r_{vi}\} = \frac{(\xi_v + \psi_{r_{vi}})^{a_{vi}} (\sigma_i - \psi_{r_{vi}})^{1-a_{vi}}}{\xi_v + \sigma_i}$$

and inserting (3.2) into (2.1) yields the unconditional likelihood of the response vector

$$(3.3) \qquad p\{(a_{vi})\} = \frac{\prod_{i=1}^{k} (\xi_v + \psi_{r_{vi}})^{a_{vi}} (\sigma_i - \psi_{r_{vi}})^{1-a_{vi}}}{\prod_{i=1}^{k} (\xi_v + \sigma_i)} .$$

$$= \frac{\prod_{r=0}^{a_{vo}-1} (\xi_v + \psi_r) \cdot \prod_{i=1}^{k} (\sigma_i - \psi_{r_{vi}})^{1-a_{vi}}}{\prod_{i=1}^{k} (\xi_v + \sigma_i)}$$

The likelihood of the individual's test score a_{vo} is obtained from (3.3) by summation of the probabilities $p\{(a^*_{vi})\}$ of all possible response vectors (a^*_{vi}) which are compatible with the score so that $\Sigma_{i=1}^{k} a^*_{vi} = a_{vo}$

$$(3.4) \qquad p\{a_{vo}\} = \sum_{(a^*_{vi}) \mid a_{vo}} p\{(a^*_{vi})\}$$

$$= \frac{\prod_{r=0}^{a_{vo}-1} (\xi_v + \psi_r) \cdot \sum_{(a^*_{vi}) \mid a_{vo}} \prod_{i=1}^{k} (\sigma_i - \psi_{r^*_{vi}})^{1-a^*_{vi}}}{\prod_{i=1}^{k} (\xi_v + \sigma_i)}$$

r^*_{vi} is defined by $r^*_{vi} = \Sigma_{j=1}^{i-1} a^*_{vi}$ for $i = 2,3,\ldots,k$ and $r^*_{vi} = 0$ for $i=1$.

The conditional likelihood of the response vector, finally, is obtained from inserting the equations (3.3) and (3.4) into formula (3.1):

$$p\{(a_{vi})|a_{vo}\} = \frac{\prod_{i=1}^{k} (\sigma_i - \psi_{r_{vi}})^{1-a_{vi}}}{\sum_{(a^*_{vi})|a_{vo}} \prod_{i=1}^{k} (\sigma_i - \psi_{r^*_{vi}})^{1-a^*_{vi}}} \tag{3.5}$$

(3.5) is dependent on the item- and transfer-parameters σ_i and ψ_r only and does not depend on the individual parameter ξ_v. Consequently, a_{vo} is a sufficient estimator for ξ_v and any extra information about *which* of the items were answered positively is useless as a source of inference about ξ_v. However, it can be used for inferring the item- and transfer-parameters independently from the individual parameters if $0 < a_{vo} < k$. [1)]

4. CML-ESTIMATORS

Let us consider the responses of n individuals with $0 < a_{vo} < k$. Then, the individual's responses can be arranged in a $n \times k$ response matrix $((a_{vi}))$ and the individual's scores can be arranged in a n-dimensional score vector (a_{vo}). The conditional likelihood of the response matrix follows from inserting (3.5) into

$$p\{((a_{vi}))|(a_{vo})\} = \prod_{v=1}^{n} p\{(a_{vi})|a_{vo}\} \quad , \tag{4.1}$$

which yields

$$p\{((a_{vi}))|(a_{vo})\} = \frac{\prod_{i=1}^{k} \prod_{r=0}^{i-1} (\sigma_i - \psi_r)^{n_{ri}}}{\prod_{v=1}^{n} \sum_{(a^*_{vi})|a_{vo}} \prod_{i=1}^{k} (\sigma_i - \psi_{r^*_{vi}})^{1-a^*_{vi}}} = L \tag{4.2}$$

---

1) If $a_{vo} = 0$ or $a_{vo} = k$, then $p\{(a_{vi})|a_{vo}\} = 1$ and does not provide any information about the parameters to be estimated.

in which n_{ri} is the number of individuals who responded negatively to item i after $r_{vi} = r$ positive responses to the preceding items $j = 1,2,\dots,i-1$.

Now let N_{k-s} be the number of individuals who gave a total of s negative responses to the k items so that $a_{vo} = k-s$, and let

$$\delta_m(k-s) = \begin{cases} 1 & \text{for } m = 0 \\ \sum_{j_1=0}^{k-s} \sum_{j_2=j_1}^{k-s} \cdots \sum_{j_m=j_{m-1}}^{k-s} \prod_{t=1}^{m} \psi_{j_t} & \text{for } m = 1,2,\dots,s \,. \end{cases}$$

Then, according to KEMPF & HAMPAPA (1974), the denominator in (4.2) can be written as

$$(4.3) \qquad \prod_{v=1}^{n} \sum_{(a^*_{vi})|a_{vo}} \prod_{i=1}^{k} (\sigma_i - \psi_{r^*_{vi}})^{1-a^*_{vi}} =$$

$$= \prod_{s=1}^{k-1} \left\{ \sum_{m=0}^{s} \delta_m(k-s) \cdot \gamma_{s-m}(k) \cdot (-1)^m \right\}^{N_{k-s}}$$

in which $\gamma_{s-m}(k)$ denotes the elementary symmetric function of order s-m of the parameters $\sigma_1,\dots,\sigma_k$. Inserting (4.3) into (4.2) and differentiating

$$(4.4) \qquad \ln(L) =$$

$$= \sum_{i=1}^{k} \sum_{r=0}^{i-1} n_{ri} \cdot \ln(\sigma_i - \psi_r) - \sum_{s=1}^{k-1} N_{k-s} \cdot \ln\left(\sum_{m=0}^{s} \delta_m(k-s) \cdot \gamma_{s-m}(k) \cdot (-1)^m\right)$$

with respect to the item-parameters $\sigma_\alpha, \alpha = 1,\dots,k$ and to the transfer-parameters ψ_β, $\beta = 0,\dots,k-1$, finally, leads to the necessary estimation equations $\partial \ln(L)/\partial\sigma_\alpha = 0$ for $\alpha = 1,\dots,k$ and $\partial \ln(L)/\partial\psi_\beta = 0$ for $\beta = 0,\dots,k-1$. The resulting CML-estimators are

$$(4.5) \qquad \sum_{r=0}^{\alpha-1} \frac{n_{r\alpha}}{\sigma_\alpha - \psi_r} =$$

$$= \sum_{s=1}^{k-1} N_{k-s} \cdot \frac{\sum_{m=0}^{s-1} \delta_m(k-s) \cdot \gamma^{(\alpha)}_{s-m-1}(k) \cdot (-1)^m}{\sum_{m=0}^{s} \delta_m(k-s) \cdot \gamma_{s-m}(k) \cdot (-1)^m}$$

for estimating the item-difficulties σ_α and

$$(4.6) \quad \sum_{i=\beta+1}^{k} \frac{n_{\beta i}}{\sigma_i - \psi_\beta} = \sum_{s=1}^{k-1} N_{k-s} \cdot \frac{\sum_{m=1}^{s} \gamma_{s-m}(k) \cdot \left\{ \sum_{j=0}^{m-1} \psi_\beta^j \cdot \delta_{m-1-j}(k-s) \right\} \cdot (-1)^m}{\sum_{m=0}^{s} \delta_m(k-s) \cdot \gamma_{s-m}(k) \cdot (-1)^m}$$

for estimating the transfer parameters ψ_β .

$\gamma_{s-m-1}^{(\alpha)}(k)$ denotes the elementary symmetric function of order s-m-1 of the parameters $\sigma_1, \ldots, \sigma_{\alpha-1}, \sigma_{\alpha+1}, \ldots, \sigma_k$.

A *FORTRAN*-program for the numerical solution of the estimation equations has been written by KEMPF & MACH (1974).

REFERENCES

ANDERSEN, E.B.: Conditional inference and models for measuring. Copenhagen: Mentalhygiejnisk Forlag, 1973a.

ANDERSEN, E.B.: Conditional inference for multiple-choice questionnaires. Br.J.math.statist.Psychol. 1973b, 26, 31-44.

KEMPF, W.F.: Dynamische Modelle zur Messung sozialer Verhaltensdispositionen. In: KEMPF, W.F. (Hrsg.): Probabilistische Modelle in der Sozialpsychologie. Bern: Huber, 1974.

KEMPF, W.F. & P. HAMPAPA : The numerical solution of a set of conditional estimation equations arising in a dynamic test model. Br.J.math.statist.Psychol. 1974 (forthcoming).

KEMPF, W.F. & G. MACH : Unveröffentliches Computerprogramm zur CML-Parameterschätzung in einem dynamischen Testmodell. Kiel:IPN (Inst.f.d.Päd.d.Naturwiss.), 1974.

NEYMAN, J. & E.L. SCOTT : Consistent estimates based on partially consistent observations. Econometrica 1948, 16, 1-32.

RASCH, G.: Probabilistic models for some intelligence and attainment tests. Copenhagen: Nielsen & Lydiche, 1960.

RASCH, G.: On general laws and the meaning of measurement in psychology. In: Proceedings of the fourth Berkeley symposium on mathematical statistics and probability. Vol IV. Berkeley: Univ.of Calif.Press, 1961, 321-334.

On the Benzecri's Method for Computing Eigenvectors by Stochastic Approximation (The Case of Binary Data)

Ludovic Lebart, Paris

1. INTRODUCTION

This paper deals with a method for computing eigenvectors and eigenvalues of a covariance matrix (or, more generally of a moment matrix) directly from the data array. This iterative technique will give us the dominant and subdominant eigenvalues (and their associated eigenvectors) after a few readings of the data array row by row.

The data array may be stored in an auxiliary memory (tape or disk) for the algorithm does not need it in the central core. The method which we discuss is referred to as "direct reading method" : this technique is most useful in the treatment of very large arrays (up to 1 000 variates), or in the treatment of large arrays on small size computers.

If the moment matrix can be put into the central core, classical methods will be in general more interesting (e.g. Householder and QR), except if the data array has the "binary disjonctive form", a structure of data which occurs very frequently in survey's data analysis. Although the results of section 2 and 3 hold for any kind of data, we shall recommend the use of the algorithm when the data array have the form mentionned above. The example given in the section 4 relates only to this case. In this section, we shall briefly expose the statistical problem which leads to our computational method, and give some definition and notationnal conventions.

1.1. The statistical problem

We are mainly interested in arrays filled with responses to "politomous" questions. We shall denote by S a set of individuals (or observations, or households in an economic survey), by Q a set of questions. The numbers of elements of the sets S and Q are denoted respectively card S and card Q. A question $q \in Q$ consist of a set J_q of card J_q possible responses. If the answer to a question q must be one (and only one) response out of the card J_q possible responses, the binary arrays Z_q (Z_q has card S rows and card J_q columns) of the responses given by the card S individuals to the question q is said to have the binary disjonctive form ; an element (i,j) of the array Z_q has the value 1 if the individual i has given the response j, the value 0 in any other case.

We shall denote by J the set of all possible responses of the questionnaire Q.

$J = \{\cup J_q \mid q \in Q\}$, $\text{card } J = \Sigma\{\text{card } J_q \mid q \in Q\}$.

The binary disjonctive array associated with the questionnaire Q is the array Z obtained by placing the arrays Z_q side by side :

$Z = (Z_1, Z_2, \ldots Z_{\text{card } Q})$

The array Z is a (card S $\times$ card J) matrix. In one row of Z, card Q elements have the value 1, the others elements having the value 0.

If we denote by Z^T the transpose of the matrix Z, the symmetric matrix $B = Z^T Z$ is often referred to as a "BURT Contingency Table" associated with the matrix Z. The partition of the matrix Z in card Q blocks $Z_1, Z_2, \ldots Z_{\text{card } Q}$ imply a partition of the matrix B in $(\text{card } Q)^2$ blocks. Because of the disjonctive form of the data, the card Q diagonal blocks of B are diagonal matrices ; these matrices are of the form $Z_k^T Z_k$. The card Q (card Q - 1) non diagonal blocks of B are the classical contingency tables, associated with each pair of questions : the block (k,k') is the contingency table $Z_k^T Z_{k'}$, giving a cross tabulation of the answers to the questions k and k' ; this block has a size(card $J_k \times$ card $J_{k'}$). Let D be the diagonal(card J $\times$ card J)matrix the diagonal elements of which are identical to those of the matrix B defined above. The multiple correspondence analysis | BENZECRI (1972)|, |LEBART, TABARD (1973) | , a technique of factor analysis which generalizes the correspondence analysis of contingency table | BENZECRI (1973) |, will enable us to grasp the structure of the data having the binary disjonctive form (all these methods are of course closely related to principal component analysis). The major computing problem occurring during a multiple correspondance analysis lies in the calculations of eigenvectors and eigenvalues of the matrix $D^{-1}B = A$. This matrix A may have a dissuasive size : if the questionnaire has 100 questions, each of them having about 10 possible responses, A will be a (1 000 $\times$ 1 000)matrix. These sizes of data are very often met in economic or sociological surveys. The advantage of the direct reading method which we discuss in the next sections is that the amount of calculation involved in the process depends only on the number card Q of questions, and not on the number card J of all possible responses.

It is easy to see the principle of the reduction of the number of operations : as it has already been said, a row Z_k^T of the matrix Z has only card Q non-zero elements. The information brought by such a row may be condensed in a vector with card Q components, each component beeing the address of a component of Z_k^T having the value 1.

Therefore, the scalar product of one row of Z by any vector of the vector space $R^{\text{card } J}$ requires only card Q operation (instead of card J). This property holds only if the metric is diagonal ; precisely, the process we shall study uses only scalar products of rows of the matrix Z by vectors in a diagonal metric.

In the previous example, the analysis of the array of answers to a questionnaire with 100 questions does not,with regard to the amount of calculation,depend on the number of all possible responses (1 000) : in this case, the algorithm using stochastic approximation is of course the cheapest one, and the easiest to implement on medium size computers.

1.2. Definitions and notations

An empirical moment matrix B, computed from a set S of observations, can always take the form $B = \Sigma\{ B(s) \mid s \in S \}$, where the matrix B(s), which has a rank one, is the contribution of the observation s to B.

For the matrix B defined above, $B = Z^T Z = \Sigma\{ z_s^T z_s \mid s \in S \}$

This decomposition is extensively used in computer programs when the data matrix Z does not hold in the central core, and therefore must be read row by row (or blocks by blocks).

We shall use this decomposition for the matrix $A = D^{-1}B$ of the previous section. We obtain :

$$A = D^{-1}Z^T Z = \Sigma\{ D^{-1} z_s^T z_s \mid s \in S \}$$

Let us define A(s) by the equality $\quad A(s) = (\text{card } S)\, D^{-1} z_s^T z_s$

Then we have : $A = (1/\text{card } S) \Sigma \{ A(s) \mid s \in S \}$

This equality point out the connection with the classical problems of stochastic approximation : the matrix A may be considered as the expectation of random matrices A(s). BENZECRI (1969) has demonstrated the convergence of a stochastic approximation algorithm under general assumptions : the elements A(s) were supposed to constitue a sequence (not necessarily finite) of random elements of a normed algebra of finite dimension. A beeing the expectation of the A(s).

Under weak assumptions concerning the law of the random elements A(s), the iterative scheme defined as follows :

$$\begin{cases} u_1 = \text{any vector of } R^{\text{card } J} \\ u_2 = u_1 + A(1)\, u_1 \\ \vdots \\ u_{s+1} = u_s + A(s)\, u_s \\ \vdots \end{cases}$$

gives us, when s tends of infinity, a vector u_s proportionnal to the eigenvector of the matrix A associated with the largest eigenvalue of A (if this largest eigenvalue is unique).

In the next sections, we shall demonstrate again this convergence under assumptions required by practical use and evaluations of the algorithm.

Let E be a normed vector space. (E.g. E will be the vector space $R^{\text{card } J}$, with a quadratic norm associated with the diagonal matrix D). The norm of x is denoted by $\|x\|$. It is usual to provide the vector space L(E,E) of the linear mappings of E in itself with a norm (sometimes called the subordinate matrix norm) defined by :

for any $M \in L(E,E)$, $\|M\| = \sup_{x \neq 0} \frac{\|Mx\|}{\|x\|} = \sup_{\|x\|=1} \|M x\|$

The space L(E,E) has a structure of non commutative normed algebra, for the multiplication of linear operators is compatible with the norm in the following sense :

$$\| M_1 \cdot M_2 \| \leqslant \| M_1 \| \cdot \| M_2 \| \ ;$$

We shall be concerned with a finite sequence $\{A(s) \mid s = 1,2,\ldots, \text{card } S\}$ (the index s will denote indifferently an element of the set S or an integer $\leqslant$ card S) of elements of a such non-commutative normed algebra. A will be the empirical mean of the elements A(s) :

$$A = (\Sigma\{A(s) \mid s \in S\}) / \text{card } S$$

The execution of the algorithm will imply a certain number of reading of the sequence $\{A(s) \mid s=1,\ldots \text{card } S\}$. We call "iteration k" the kth reading of the sequence. The algorithm involves also a positive function h(s,k) ; h(s,k) is a decreasing function of s, and a strictly decreasing function of k. h(s,k) will have in general the following form :

$$h(s,k) = \frac{c}{(s + (k-1) \text{ card } S)^{\gamma}}$$

We shall see that the coefficient c must be greater than 1, and that $\frac{1}{2} < \gamma \leqslant 1$. The algorithm works also successfully if h(s,k) is constant during each iteration , namely : h(s,k) = h(k).

The algorithm we shall discuss consist in building, from any vector u_o of $R^{\text{card } J}$, a sequence of vector u_i satisfying :

$$u_i = u_{i-1} + h(s,k) A(s) u_{i-1}$$

The index i is connected with the indexes s and k by : i = s + (k − 1) card S. For each value of k (i.e. for each iteration), the index s varies from 1 to card S. In fact, the iteration k consist in premultiplying the vector u issued from the previous iteration by the operator :

$$H(k) = \prod_{s=1}^{s=\text{card } S} (I + h(s,k) A(s)).$$

We shall mention the background theory of the algorithm in section 2. We discuss the implementation of the algorithm in section 3, and give an example in section 4.

2. SOME THEORETICAL RESULTS

We shall begin by recalling a lemma [BENZECRI (1969)] which enable us to compare products of elements of a non commutative algebra $\mathcal{A}$:

Lemma 0 : let V_k and v_k be two finite sequences of elements belonging to $\mathcal{A}$, indexed by k = 1,2,...K.

We denote : $\Pi(V_k) = V_K V_{K-1} \cdots V_1$ and $\Pi(V_k + v_k) = (V_K + v_K) \cdots (V_1 + v_1)$

If, for each $k \leqslant K$, $\| V_k \| \geqslant 1$ and if $\sum \|v_k\| \leqslant 1$

Then : $\left\| \Pi(V_k) - \Pi(V_k + v_k) \right\| \leqslant \Pi \left\| V_k \right\| \ . \ 2\Sigma \left\| v_k \right\|$ (1)

We shall not recall the proof of this lemma.

We shall begin by discussing the case of a function h(s,k) which does not depend on s (h(s,k) = h(k)). Then we shall study the case of a more general function h(s,k). At last, we examine the consequences of reading the sequence A(s) in differen direction (i.e. the sequence A(s) is read from 1 to card S, then from card S to 1, and so on). This reading "there and back" will strongly improve the efficiency of the algorithm.

2.1. The case of a function h(s,k) = h(k)

Lemma 1 : if the elements A(s) satisfy the conditions $\|A(s)\| \leqslant a$ for any $s \in S$, if $\|A\| \leqslant 1$, and if the scalar m satisfies $m \geqslant a.\text{card } S$, then

$$\left\| \prod_{s=1}^{s=\text{card } S} \left(I + \frac{A(s)}{m}\right) - \exp\left\{\frac{\text{card } S}{m} \ . \ A\right\} \right\| \leqslant (a^2 + 1) \left(\frac{\text{card } S}{m}\right)^2 \qquad (2)$$

($\exp\{X\}$ denote the sum of the formal serie : $I + \frac{X}{1!} + \frac{X^2}{2!} + \ldots + \frac{X^n}{n!} + \ldots$)

Proof : the left part of the inequality is less than :

$$\Pi\left(1 + \frac{a}{m}\right)^{\text{card } S} - 1 - \frac{a \text{ card } S}{m} + \exp\left\{\frac{\text{card } S}{m}\right\} - 1 - \frac{\text{card } S}{m}$$

Now, we have also : $\Pi\left(1 + \frac{a}{m}\right)^{\text{card } S} \leqslant \exp\left\{\frac{a \text{ card } S}{m}\right\} \leqslant 1 + \frac{a \text{ card } S}{m} + \left(\frac{a \text{ card } S}{m}\right)^2$

By comparing these two expressions, we deduce the right side of the inequality (2).

We can now give the proof of the convergence of the algorithm, when the function h(s,k) does not depend on s : h(s,k) = h(k). (h(k) is a constant within the iteration k). In order to use the results of the lemma 0, we define V_K and v_k by :

$$V_k = \exp\left\{\text{card } S.h(k) \ . \ A\right\}$$

$$V_k + v_k = H(k) = \prod_{s=1}^{\text{card } S} (I + h(k) \ A(s))$$

The lemma 1, where 1/m = h(k), gives us an upper bound for $\|v_k\|$:

$$\|v_k\| \leqslant (a^2 + 1) \left| h(k) \ . \ \text{card } S \right|^2$$

The inequality (1) of the lemma 0 can now be written : (the number k_1 beeing chosen so as to assume $\Sigma \|v_k\| \leqslant 1$. This will be always possible, for the serie h(k) will be chosen so as to make the serie $\Sigma \|v_k\|$ converge).

$$\left\| \prod_{k=k_1}^{k=k_2} h(k) - \exp\left\{\sum_{k_1}^{k_2} h(k) \ . \ \text{card } S \ . \ A\right\} \right\|$$

$$\leqslant 2(a^2 + 1) \ (\text{card } S)^2 \ . \ \exp\left\{\sum_{k_1}^{k_2} h(k) \ \text{card } S\right\} \ . \ \sum_{k_1}^{k_2} h^2(k) \qquad (3)$$

From this inequality follows the theorem 1 :

Theorem 1 : let u_i be the sequence defined at the end of the section 1. Under the assumptions of the lemma 1, and if the largest eigenvalue of the matrix A is unique, the vector u_i tends when i tends to infinity, to the eigenvector (unnormalized) of A associated with the largest eigenvalue of the matrix A, if the serie h(k) diverges, and if the serie $h^2(k)$ converges.

If these two last conditions are filled, it is possible to find an integer K having the following property :

For any numbers k_1 and k_2 greater than K, the quantity $\sum_{k_1}^{k_2} h^2(k)$ is arbitrarily small, (this quantity is on the right side of the inequality (3)) and the quantity $\sum_{k_1}^{k_2} h(k)$ is arbitrarily large. Therefore, the operator $\exp\{\sum_{k_1}^{k_2} h(k).\text{card } S.A\}$ after beeing normed, tends to a mapping of $R^{\text{card } J}$ on to the subspace of the first eigenvector, and tends simultaneously to the operator $\prod_{k_1}^{k_2} H(k) = \prod_{k_1}^{k_2} \prod_{s=1}^{s=\text{card } S} (I + h(k)\ A(s))$ which is the operator associated with the algorithm.

The simplest example of serie h(k) filling the conditions of theorem 1 is the harmonic serie : $h(k) = 1/k$.

Another serie, suggested by J.P. BENZECRI, enable us to give an explicit and exact form to the inequality (3) : let us define $h(k) = (1/\text{card } S).h'(k)$, where $h'(k)$ is the serie : $1+(1/2+1/2)+(1/4+1/4+1/4+1/4)+(1/8+...+1/8) + ...$ From the rank $k_1 = 2^{n-1}$ to the rank $k_1' = 2^n - 1$, the general term of this serie is $(1/2)^{n-1}$. If we define $k_1 = 2^{n-1}$ and $k_2 = 2^{p-1}$, with $p \geqslant n$, thenwe have $\sum_{k_1}^{k_2} h'(k) = p - n$ (for the sum of the termes included in one pair of parenthesis is 1).

The serie $h'^2(k)$ is a geometrical progression the ratio of which is 1/2. From all this, we get the explicit inequality :

$$\left\| \prod_{k_1}^{k_2} H(k) - \exp\left\{ \text{Log}_2 \frac{k_2}{k_1} . A \right\} \right\| \leqslant \frac{2}{k_1} (a^2 + 1) . \exp\left\{ \text{Log}_2 \frac{k_2}{k_1} \right\}$$

2.2. The function h(s,k) depends on s :

When the function h(s,k) is not constant within an iteration k, the upper bound given by the lemma 1 must be modified. We shall study the case called to mind in the section 1.2 : $h(s,k) = \dfrac{c}{(s + (k - 1)\ \text{card } S)^\gamma}$

Let $m = (k-1).\text{card } S$ and $T(m,\gamma,c) = \Sigma\left\{\dfrac{c}{(m + s)^\gamma} \;\middle|\; s \in S\right\}$

Lemma 2 : if the elements A(s) satisfy the conditions $\|A(s)\| \leqslant 1$, and if $\|A\| \leqslant 1$; if the scalàr m satisfies $m^\gamma \geqslant c.a.\ \text{card } S$; then :

$$\left\| \prod_{s=1}^{s=\text{card } S} \left(I + \frac{c\ A(s)}{(m+s)^\gamma}\right) - \exp\{T(m,\gamma,c) . A\} \right\|$$
$$\leqslant c^2(a^2 + 1)\left(\frac{\text{card } S}{m^\gamma}\right)^2 + \frac{2\ a\gamma\ c\ (\text{card } S)^2}{m(m + \text{card } S)^\gamma}$$

We will not give the proof of the lemma 2 for it is just a matter of calculations, (similar to the demonstration of the lemma 1). The main consequence of the lemma 2 is that the conditions of convergence of the algorithm are not changed : the exponent γ must be $\leqslant 1$, in order to assure the divergence of the serie $\Sigma\{T(m,\gamma,c) \mid k \geqslant K\}$ and $>$ 1/2 in order to assure the convergence of the series of the majorants.

2.3. The effect of the reading "there and back" :

To read the sequence A(s) in different directions during two successive iterations improves the efficiency of the algorithm. Intuitively, we may say this kind of reading remedies the non-commutativity of the algebra ;

Lemma 3 : The elements of the finite sequence $A(s)$ satisfying always the conditions $\|A(s)\| \leqslant 1$. If $\|A\| \leqslant 1$, and if m is a scalar $\geqslant 2$ a card S ; then :

$$\left\| \prod_{s=1}^{s=\text{card } S} \left(I + \frac{A(s)}{m}\right) \prod_{s=\text{card } S}^{s=1} \left(I + \frac{A(s)}{m}\right) - \exp\left\{\frac{2 \text{ card } S}{m} \cdot A\right\} \right\| \leqslant$$

$$\frac{a^2 \text{ card } S}{m^2} + \frac{a^3+1}{3} \left(\frac{2 \text{ card } S}{m}\right)^3 \qquad (4)$$

Proof : in the left side of the inequality (4), the terms of degree 0 and 1 vanishes. The difference between the terms of 2nd degree is : $\Sigma\{A^2(s)/m^2 \mid s \in S\}$ The norm of this quantity is less than : a^2 card S/m^2.
The terms of degree 2 ar less than $(f(a) + f(1))$, the function f beeing defined by :

$$f(a) = \exp\left\{\frac{2 \text{ a card } S}{m}\right\} - 1 - \frac{2 \text{ a card } S}{m} - \frac{(2 \text{ a card } S)^2}{2 m^2}$$

Now, $f(a)$ is less than $\frac{a^3}{3} \left(\frac{2 \text{ card } S}{m}\right)^3$.Hence, the relation (4) holds.

The bound of the inequality (4) is lower than that obtained previously.

Example : if, within the kth iteration, m = 2 k card S, and if a = 1 the right side of the relation (4) can be written : $(1/(4\ k^2 \text{ card } S) + 1/3\ k^3)$. The majoration given by the lemma 1 would be, in the same conditions : $2/k^2$. (We must remember that card S, number of individuals, is generally a large number).

3. IMPLEMENTATION OF THE ALGORITHM

We shall now note u_{k-1} the state of the vector u_i at the issue of the iteration k-1, and we shall assume that the norm of u_{k-1} is 1. As we pointed out above, the kth iteration consist in premultiplying u_{k-1} by the operator $H(k) = \prod_{s=1}^{s=\text{card } S} (I + H(k,s)\ A(s))$ The calculation of $H(k)u_{k-1}$ requires the same number of operations than the calculation of Au_{k-1}, this last calculation beeing an iteration of the classical iterated power algorithm adapted to the direct reading of data : as a matter of fact, calculating $H(k)u_{k-1}$ requires card S assignments like $\{u^s := u^{s-1} + h(s,k)A(s)u^{s-1}\}$, whereas calculating Au_{k-1} requires card S assignments like $\{u_k := u_k + (1/\text{card } S)A(s)\ u_{k-1}\}$.

The algorithm we are discussing will be worthy if (u beeing the vector we are trying to find) : cosine(u, $H(k)\ u_{k-1}$) > cosine (u, Au_{k-1}). Let us suppose that H(k) and A have the same eigenvectors (the pratical situation will not be so favourable). Let M be the matrix defining the metric of $R^{\text{card } J}$, A and H(k) beeing M-symmetrics, and u beeing normed. $(M(u,u)^* = M(u_{k-1}, u_{k-1}) = 1)$. The above inequality may be written.

$$\frac{M(u, H(k)u_{k-1})}{M(H(k)u_{k-1}, H(k)u_{k-1})} > \frac{M(u, Au_{k-1})}{M(Au_{k-1}, Au_{k-1})}$$

The numerators of the left and right side are respectively :

$$\lambda_H{}^1 \cdot M(u,u_{k-1}) \text{ and } \lambda_A{}^1 \cdot M(u,u_{k-1})$$

λ_H^1 and λ_A^1 beeing the largest eigenvalues of H(k) and A.

The denominators are respectively : $(\Sigma\ \lambda i_H^2 \cdot h_i^2)^{\frac{1}{2}}$ and $(\Sigma \lambda i_A^2 \cdot h_i^2)^{\frac{1}{2}}$
(h_i is the ith component of u_{k-1} in the basis of the common eigenvectors of the two matrices).

These expressions lead to the inequality, where $w_i^2 = \left(\frac{\lambda_H{}^i}{\lambda_H{}^1} + \frac{\lambda_A{}^i}{\lambda_A{}^1}\right) h_i^2$

*M(u,u) denotes the quadratic form associated to M.

$\sum_{i>1}\left(\frac{\lambda_H^i}{\lambda_H^1} - \frac{\lambda_A^i}{\lambda_A^1}\right) w_i^2 < 0$. Now : $\lambda_H^i = e^{\alpha(k).\lambda_A^i}$. Supposing $\lambda_A^1 = 1$, and putting $\lambda_i = \lambda_A^i$: $\sum_{i>1}\left(e^{\alpha(k)(\lambda_i - 1)} - \lambda_i\right) w_i^2 < 0$

This inequality shows us that we will have to replace H(k) by A as soon as the coefficient $\alpha(k)$ is less than 1. When $h(s,k) = c/(s + (k-1) \text{ card } S)^{\gamma}$ an eigenvalue of H(k) can be written $e^{\alpha(K)\lambda}$, λ beeing the corresponding eigenvalue of the matrix A.
With $\alpha(k) = \Sigma\left\{\frac{c}{(s + (k-1) \text{ card } S)^{\gamma}} \middle| s \in S\right\}$
Let us study briefly three examples : (assuming always $\lambda_A^1 = \| A \| = 1$)

i) Suppose c = 1, card S = 10^3, $\gamma = 1$. (Hence h(s,k) is the harmonic serie). For the first iteration $\alpha \simeq \text{Log } 10^3 + \ldots \gg 1$. For the second iteration $\alpha \simeq \text{Log } 2 < 1$. It is not worth using the algorithm in this case.

ii) More generally, if $\gamma = 1$ and if $k > 1$, $\alpha(k) \approx c \text{ Log } (\frac{k}{k-1}) < \frac{c}{k-1}$. The algorithm will be interesting until the iteration k only if the coefficient c is greater than k-1. We would be tempted to use a function h(s,k) with a very large coefficient c, but the bounds given by the previous lemmas depends on c, (the coefficient m depends on c), and the algorithm may fail because of a too large difference between the eigenvectors of H(k) and those of exp{A}.

iii) If $\gamma < 1$, we have :

$$\alpha(k) \approx \frac{c}{1-\gamma} (\text{card } S)^{1-\gamma} \left[k^{1-\gamma} - (k-1)^{1-\gamma} \right] \leqslant \frac{(\text{card } S)^{1-\gamma}}{k^{\gamma}} c$$

We must also mention that, as it is the case for the iterated power algorithm, we can obtain directly several eigenvectors ; it is then necessary, for a matter of the accuracy, to orthonormalize regurarly the set of vectors involved in the process. (E.g. after each reading of 100 elements A(s)). If we start, for example, with 4 vectors, we obtain finally a 4-dimensionnal basis containing the first four eigenvectors which will in the end be obtained by the diagonalization of a 4 × 4 matrix.

4. NUMERICAL EXAMPLES

The array of data we are going to study will have the following characteristics : (we use the notation of the section 1).

Card Q = 5 (5 questions)
Card J = 43 (43 possible responses)
Card S = 1 000 (1 000 individuals).

(1 000 households are characterized by 5 variables : occupation of the father, (8 possible responses : card $J_1 = 8$), occupation of the mother, if any, (card $J_2 = 12$), occupation of the father (card $J_3 = 9$), occupation of the father — of the mother (card $J_4 = 9$), activity of the mother (card $J_5 = 5$). The moment matrix issued from these data beeing a 43 × 43 matrix, the calculation was likely to be performed by a classical method. We shall however use this set of data to put the algorithm to the test.

The sequence of the first 5 exact eigenvalues (up to 7 figures) is :

0.453716

0.411627

0.298538

0.282402

0.268375

We shall describe the results of each iteration by the cosines of the angles between the exact eigenvectors and their estimations.

The function $h(s,k) = c/(s + (k-1) \text{ card } S)^{\gamma}$ depends on the two free parameters c and γ . We have pointed out that, when $\gamma = 1$, we must have $c > K$, K beeing the number of iterations required.

i) In the first step, we shall compare the efficiency of "simple reading" (reading of the sequence A(s) always in the same direction) and of "there and back" reading (cf. section 2), for the first two eigenvectors, and for 8 iterations, with $\gamma = 1$ and c = 10.

Table of cosines

Iteration	Simple reading		"There and back" reading	
	f1	f2	f1	f2
1	0.992	0.991	0.992	0.991
2	0.998	0.997	0.9995	0.9998
3	0.9990	0.998	0.9998	0.9998
4	0.9993	0.9992	0.99994	0.99996
5	0.9996	0.9994	0.99996	0.99996
6	0.9997	0.9996	0.99998	0.99998
7	0.9998	0.9997	0.99998	0.99998
8	0.9998	0.9998	0.999990	0.999992

The numerical results are satisfying. We notice that the fact of reading the sequence in the reverse order during the second iteration leads us to the level of accuracy of the 5th simple reading. After three reading, the second beeing in the reverse order, the level of accuracy is the same as with 8 simple readings. The estimations of eigenvalues, after 8 readings, are 0.453712 and 0.411624 (all the figures but the last are exact ; cost for 8 reading = 15 sec. on C.D.C. 6600).

ii) Let us compare the efficiency of this algorithm with the iterated power method :

Table of cosines

Iteration	f1	f2
1	0.59	0.42
2	0.84	0.72
3	0.95	0.90
4	0.990	0.960
5	0.997	0.990
6	0.9990	0.995

We notice that 5 iterations are necessary to reach the level of accuracy obtained by the stochastic approximation method at the end of the first iteration.

iii) Numerous experience for different values of the parameters c and γ , and in the case of a function h(s,k) = h(k), constant within an iteration, have been worked out.

A definitive strategy, concerning the form of the function h(s,k), must be settled. However, the first results seem to be quite promising.

REFERENCES

BENZECRI J.P. Approximation stochastique dans une algèbre normée non commutative Bull. Soc. Math. France. 97. pp. 225-241 (1969).

BENZECRI J.P. Sur l'analyse du tableau binaire associé à une correspondance multiple. (Multigra. paper of the Labo. of math. stat. Faculty of Science, University Paris VI) (1972).

BENZECRI J.P. Analyse des données. Tome II : "Analyse des correspondances". Dunod Ed. Paris. (1973).

FENELON J.P. Deux contributions à une programmathèque d'analyse de données. Thesis, 3° cycle, University Paris VI. (1973).

LEBART L. FENELON J.P. Statistique et informatique appliquées. Dunod Ed. Paris. (2nd ed : 1973).

LEBART L. TABARD N. Recherches sur la description automatique des données socio-économiques. Multigra. Report for the C.O.R.D.E.S. 30, rue Las Cases, Paris 7. (1973).

A Study of the Relationship between Qualitative Variables

A. Leclerc, Le Vésinet

1. The relationship between two qualitative variables

Two methods for the study of the relationship between two qualitative variables are presented in this chapter. Canonical analysis is the first approach wich is discussed. The results obtained correspond to those found by correspondence analysis, which is developped by J.P. BENZECRI (1973). Certain particularly interesting properties which can be generalized are finally discussed.

1.1 Canonical Analysis

Two qualitative variables are considered :

A - indexed 1................ i max

B - indexed 1................ j max

I and J denote the sets of possible events

$$I = \{1 \ldots\ldots\ldots i \text{ max}\}$$

$$J = \{1 \ldots\ldots\ldots j \text{ max}\}$$

and f_i is the frequency associated with event i belonging to A

f_j is the frequency associated with event j belonging to B

f_{ij} is the joint frequency associated with event i for A and event j for B.

(as a general rule, the index i will refer to A and j to B).

f_i, f_j and f_{ij} are calculated from a continqency table the lines of which correspond to the events of A and the columns to the events of B. The total number of outcomes is n. The study of the relationship between A and B will be restricted almost entirely to a description.

A and B define the identifying fonctions :

g_1................. g_{imax}

h_1................. h_{jmax}

Where, for a given subject k belonging to the population

$gi(k) = 1$ if and only if $A(k) = i$

$hj(k) = 1$ if and only if $B(k) = j$

Canonical analysis can be performed on the fonctions gi and hj ; a set of triplets is obtained:

$$(r_q , \varphi_q^I , \varphi_q^J)$$

Where φ_q^I is a linear combination of the gi,

φ_q^J is a linear combination of the hj,

and r_q is the empirical correlation between these two fonctions.

If φ_q^i denotes the coefficient associated with gi and φ_q^j that associated with hj, then for a given subject k :

$$\varphi_q^I (k) = \sum_{i=1}^{imax} \varphi_q^i \, g_i (k)$$

$$\varphi_q^J (k) = \sum_{j=1}^{jmax} \varphi_q^j \, h_j (k)$$

and so,

$$\varphi_q^I (k) = \varphi_q^i \quad \text{if and only if } A(k) = i$$

$$\varphi_q^J (k) = \varphi_q^j \quad \text{if and only if } B(k) = j$$

<u>Properties of φ_q^I and φ_q^J :</u>

These fonctions have a zero mean value.

$$\sum_{i \in I} f_i \varphi_q^i = 0 \quad ; \quad \sum_{j \in J} f_j \varphi_q^j = 0 \qquad (1)$$

and unit variance :

$$\sum_{i \in I} f_i {\varphi_q^i}^2 = 1 \quad ; \quad \sum_{j \in J} f_j {\varphi_q^j}^2 = 1 \qquad (2)$$

Pairs of fonctions φ_q^I and $\varphi_{q'}^I$, φ_q^J and $\varphi_{q'}^J$ are uncorrelated :

$$\sum_{i \in I} f_i \varphi_q^i \varphi_{q'}^i = 0 \quad \text{and} \quad \sum_{j \in J} f_j \varphi_q^j \varphi_{q'}^j = 0 \qquad (3)$$

for all such pairs (q, q') where $q \neq q'$

Under the above conditions, the correlation between φ_q^I and φ_q^J is maximum :

$$\sum_{i \in I} \sum_{j \in J} b_{ij} \varphi_q^i \varphi_q^j = r_q \tag{4}$$

The index q corresponding to pair (φ_q^I, φ_q^J) indicates the rank of the pair considered as being one of all possible pairs arranged in descending order of correlation.

1.2 Matrix notation and factor determination

The classical results of canonical analysis are :

φ_q^J is the eigenvector of $D_J^{-1} F' D_I^{-1} F$

$$D_J^{-1} F' D_I^{-1} F \varphi_q^J = r_q^2 \varphi_q^J \tag{5}$$

Similarly :

$$D_I^{-1} F D_J^{-1} F' \varphi_q^I = r_q^2 \varphi_q^I$$

Where :

D_J is the diagonal matrix (f_1...... fj...... f j max)

D_I is the diagonal matrix (f_1...... fi...... f i max)

F is the i max x j max matrix (fij)

φ_q^J is the column vector of the φ_q^j

φ_q^I is the column vector of the φ_q^i

The functions φ_q^I and φ_q^J are such that :

$$F\varphi_q^J = r_q D_I \varphi_q^I \quad \text{and} \quad F'\varphi_q^I = r_q D_J \varphi_q^J \tag{6}$$

The triplets (r_q, φ_q^I, φ_q^J) are called factors. In extension, the couple (φ_q^I, φ_q^J) could be termed a factor, as could a single element of this pair. The number of factors is not equal to the rank of matrix $D_J^{-1} F' D_I^{-1} F$ as one eigenvalue of this matrix corresponds to an eigenvector which cannot fulfill conditions (1) and (2) :

The unit vector is an eigenvector of this matrix and corresponds to an eigenvalue of 1. This root will be called the "trivial eigenvalue" or "constant eigenvalue" and has the first rank when the roots are arranged in decreasing order. The set of indices corresponding to the other factors, in decreasing order of correlation is called Q :

$$Q = \{2, \ldots\ldots\ldots\ldots\ldots\ldots q\ max\}$$

q max being bounded by min (i max, j max).

Remark : The matrix $D_J^{-1}F'D_I^{-1}F$ is non-symmetric, but matrix $D_J^{-1/2}F'D_I^{-1}F D_J^{-1/2}$ [4] where $D_J^{-1/2}$ is the diagonal matrix ($1/\sqrt{f_j}$) is symmetric, has the same eigenvalues as $D_J^{-1}F'D_I^{-1}F$ and its eigenvectors u_q^J are related to φ_q^J by the equation :

$$\varphi_q^J = D_J^{-1/2} u_q^J \tag{7}$$

The trivial factor φ_1^J also has a corresponding trivial vector u_1^J

It may be convenient to calculate first the vectors u_q^J; these vectors may also be found to be useful for further calculations.

1.3 Application of correspondence analysis

The factors φ_q^I and φ_q^J can be obtained using a different approach : principal component analysis can be applied to represent the lines and columns of the contingency table. A few modifications will be necessary due to the nature of the table as it is preferable to obtain the same results whether the method is applied to the A x B table or the B x A table. In effect this is done by correspondence analysis a general description of which is given in (BENZECRI, 1973) or (LEBART, 1973) and may be summed up as follows :

The i rows of the table can be represented as points in a j max-dimensional space, each point being attributed a mass fi and having as co-ordinate on the j axis the expression $f_{ij}/f_i\sqrt{f_j}$. A classical Euclidian metric is applied to the space.

Principal component analysis is applied to the cluster of points in this space and the factors are the functions which apply to the vectors having as co-ordinates not the $f_{ij}/f_i\sqrt{f_j}$, but the f_{ij}/f_i (relative frequency of j for a fixed i). These factors in fact turn out to be the φ_q^j functions defined earlier and corresponding to the eigenvalues $\lambda_q = r_q^2$

The analysis of the j columns of the table, represented by points having the co-ordinates $f_{ij}/f_j\sqrt{f_i}$ and of mass fj in an i max-dimensional space corresponds to factors which are the φ_q^I fonctions, the eigenvalues being the λ_q

The method immediately implies a number of properties. Among these we note that "distributional equivalence" applies i.e. if two rows of the table are proportional, then the results of the analysis will remain unchanged if they are replaced by their sum.

A simultaneous graphical representation of the line-points and of the column-points is possible : the line-points can be represented in the subspace of $\mathbb{R}^{j\,max}$ generated by the principal axes of inertia.

Similarly, the column-points can be represented in "the best" subspace of $\mathbb{R}^{imax}$. It is usual to superimpose the representations in the corresponding planes. This surperposition illustrates the good symmetrical qualities of the line and column representations :

the point corresponding to line i will be the weighted centre of gravity of the points corresponding to the columns j displaced by a small amount.

the point representing column j will also be the "pseudo-centre of gravity" of the points representing the lines.

Despite the results common to both, the two approaches of correspondence analysis and canonical analysis are different.

The φ_q^i coefficient of the canonical analysis directly represents outcome i of the A variable, or row i.

In correspondence analysis φ_q^J is considered to be the function corresponding to the f_{ij}/f_i vector and to characterise the "profile" of row i. For example, the graphical representation of i will be given by the co-ordinate of i on the factorial axis q = $\varphi_q^J(i) = \sum_{j \in J} \varphi_q^j \, f_{ij}/f_i$

From équation (6) this is equal to $r_q \varphi_q^i$: the φ_q^i of the canonical analysis together with coefficient r_q (similarly the coordinate of j on the q factorial axis will be $r_q \varphi_q^j$).

1.4 Decomposition of inertia

From I.2 we have :

trace $(D_J^{-1/2} F' D_I^{-1} F D_J^{-1/2}) = 1 + \sum_{q \in Q} r_q^2$

(1 is the first eigenvalue)

and so :

$$\sum_{q \in Q} r_q^2 = \sum_i \sum_j f_{ij}^2 / f_i f_j - 1$$

or

$$n \sum_{q \in Q} r_q^2 = \sum_i \sum_j \frac{(n_{ij} - n f_i f_j)^2}{n f_i f_j} \qquad (8)$$

with n = number of observation

nij = number corresponding to element (i,j) of the contingency table.

The right-hand expression is a measure of the correlation between A and B as expressed in the contingency table. This quantity, under the hypothesis of independance, follows a chi-square distribution with (imax - 1) (jmax - 1) degrees of freedom.

This quantity will be called the "total inertia of the table" as it is, according to correspondence analysis, n times the total inertia of one or other of the scatter of points obtained.

Formula (8) can be summed up in the following way :
the measurement of the degree of non-independance between the lines and the columns of a contingency table classically calculated by a chi-square, can be decomposed into the sum of positive terms each corresponding to a pair of functions (φ_q^I, φ_q^J) .

1.5 Reconstruction of the initial table

It is not very difficult to see that formula (6) will give the result :

$$f_{ij} = f_i f_j \left[1 + \sum_{q \in Q} r_q \varphi_q^i \varphi_q^j \right] \qquad (9)$$

Formula (8) showed that each factor contributed to the description of the initial table in that each contributed to the chi-square.

Formula (9) shows that each factor contributes to the reconstruction of each individual element (i,j). The constant factor initialises the fij value to the fifj value which corresponds to the hypothesis of independance.

Formulae (8) and (9) both imply that it is meaningful to speak of the approximation obtained by a subset Q' of the non-trivial factors.

- Approximation to the total inertia

$$n \sum_{q \in Q'} r_q^2$$

- Approximation to nij corresponding to element (i,j)

$$n f_i f_j \left[1 + \sum_{q \in Q'} r_q \varphi_q^i \varphi_q^j \right] \qquad (10)$$

There is a big difference between these two approwimations :

- each factor contributes positively to the total inertia. In other words, the addition of a factor reduces the distance between the initial value for the inertia and the real value.

- however, there is no reason why the $r_q \varphi_q^i \varphi_q^j$ quantities should be positive. Consequently it is possible to increase the distance between nij and its approximation by the addition of a factor.

1.6 A property of the reconstruction by a subset of the factors

It is easy to see that the reconstruction of the initial table using formula (10) will leave the marginal totals unaltered.

Let $n_{ij}^{Q'}$ be an approximation to nij by a non-trivial subset Q' of the factors. Then

$$\sum_{i \in I} n_{ij}^{Q'} = \sum_{i \in I} n_{ij} \quad \text{and} \quad \sum_{j \in J} n_{ij}^{Q'} = \sum_{j \in J} n_{ij}$$

The grand total of the table will hence remain unaltered.

This property demonstrates that the reconstruction of the table by a subset Q' of the factors will provide a table of the same type in that it will have the same marginals and grand total.

2. The relationship between two groups of qualitative variables

2.1 Introduction

L qualitive variables A_1 .... A_L and M qualitative variables B_1 ......B_M. are to be considered. The relationship between the two variables A_l and B_m is of interest.

A first method for the study of this relation would be :

- to transform the L variables A_l into one single qualitative variable, the events being the cartesian products of the events of the individual variables.
- to transform the M variables B_m in the same way.
- to apply the results of chapter 1 to these two new variables.

This approach may be useful in certain cases. In practice it is not applicable when the cartesian products correspond to a large number of events in comparison to the size of the sample available.

Each pair of variables $A_\ell \times B_m$ can be studied individually, but this method is obviously laborious and clumsy. A global method is needed which will provide the "common" factors obtainded by the study of the individual tables. The following proposition was forwarded with this aim in view, and in practice has turned out to be a fruitful means of analysing data of this type (Davidson, 1973).

The table $(A_1 \cup \ldots \cup A_L) \times (B_1 \cup \ldots \cup B_M)$ is formed by the union of all the $A_\ell \times B_m$ contingency tables, each calculated on the same set of subjects.

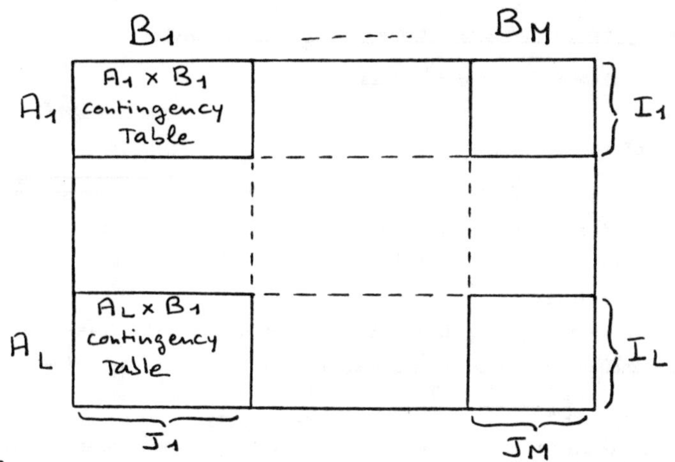

where

I_l = the set of rows of the table corresponding to the events of A_l

J_m = the set of columns corresponding to the events of Bm

$$I = \{I_1 \cup \ldots\ldots\ldots\ldots\ldots\ldots \cup I_L\}$$

$$J = \{J_1 \cup \ldots\ldots\ldots\ldots\ldots\ldots \cup J_M\}$$

A row of the table is denoted by i, a column by j. Hence each row i belongs to a set I_l, and to only one, and each column j to one and only one set Jm.

nij, the number in element (i,j), is the number of subjects corresponding to a row i (one event of one of the A_ℓ variables) and to a row j (one event of one of the Bm).

The "united contingency table" defined in this way is treated by the method of chapter I. Factors (r_q, φ_q^I, φ_q^J) are obtained which it is not easy to interpret. φ_q^I and φ_q^J can no longer be considered as being functions of the observations and r_q no longer represents a correlation. Only the $\varphi_q^{I\ell}$ or φ_q^{Jm} subsets of the factors corresponding to the I_ℓ or J_m subset can be regarded as being factors of the observations. In this chapter a few properties of these "factors" will be given to justify the method and to aid the interpretation of results from experiment or survey. The proofs are all very simple but are not given here, they may be found in. (LECLERC, 1973).

2.2 Some characteristics of the table and properties of the factors

The sum of the elements of the table is

$$N = n.L.M. \tag{11}$$

where n is the number of the sample.

Associated with each pair (i,j) we have

- frequencies calculated for the individual contingency table $I_l \times J_m$ to which (i,j) belongs, and written :

$$f'ij \quad ; \quad f'i \quad ; f'j$$

- frequencies calculated in the global or "united contingency table". These are calculated exactly as though the table were a contingency table, and are given by :

$$f_{ij} = n_{ij}/N \qquad f_i = \sum_{j\in J} f_{ij} \qquad f_j = \sum_{i\in I} f_{ij}$$

It is easy to establish the following relationships between these two types of frequencies :

$$f'ij = L.M.\ fij \tag{12}$$

$$f'i = L.\ fi \tag{13}$$

$$f'j = M.\ fj \tag{14}$$

if, as has been assumed, the separate contingency tables each correspond to the same population.

All the properties of the factors as defined by (1), (2), (3), (4), (5), (6), (8) and (9) remain here with fi, fj, fij the frequencies on the global table and with n being replaced by N, its grand total.

When the φ_q^{Il} and φ_q^{Jm} functions are considered, factors corresponding to the subsets I_l or J_m, one might suppose that these functions which are factors over the observations might at least have properties (1), (2) or (3) :

zero mean, unit variance and zero correlation between φ_q^{Il} and $\varphi_{q'}^{Il}$ or between φ_q^{Jm} and $\varphi_{q'}^{Jm}$ for $q \neq q'$

In fact these conditions are not verified with the exception of property (1) : <u>the φ_q^{Il} or φ_q^{Jm} factors of subsets I_l or J_m, considered as functions of the observations, have a zero mean</u> :

$$\sum_{i\in Il} f'_i \varphi_q^i = 0 \quad ; \qquad \sum_{j\in Jm} f'_j \varphi_q^j = 0 \tag{15}$$

Condition (2) provides one property of the variances :

$$\frac{1}{L}\sum_{l=1}^{L} \text{var}\, \varphi_q^{Il} = 1 \quad ; \qquad \frac{1}{M}\sum_{m=1}^{M} \text{var}\, \varphi_q^{Jm} = 1 \tag{16}$$

2.3 Interpretation of the pseudo-correlation rq

2.3 a - rq as a linear combination of correlations

It is meaningful to speak of the empirical correlation between the φ_q^{Il} and φ_q^{Jm} functions defined above :

$$r(\varphi_q^{Il}, \varphi_q^{Jm})$$

Then the following result follows :

$$r_q = \sum_{l=1}^{L} \sum_{m=1}^{M} \frac{\sqrt{\text{var}\,\varphi_q^{Il} \;\; \text{var}\,\varphi_q^{Jm}}}{L.M} \; r(\varphi_q^{Il}, \varphi_q^{Jm}) \qquad (17)$$

The first non-trivial factor -in this case the second factor- will give rise to :

- a series of functions $\varphi_2^{I_1} \ldots\ldots \varphi_2^{I_L}$; each with a zero mean value and the variances being related by (16)
- a series of functions $\varphi_2^{J_1} \ldots\ldots \varphi_2^{J_M}$ also with zero means and obeying the same variance relation
- these two series are such that (17) is a maximum

2.3 b - Significance of the sum of the rq values

Relation (8) can be written as :

$$N \sum_{q \in Q} r_q^2 = \sum_{i \in I} \sum_{j \in J} \frac{(n_{ij} - N f_i f_j)^2}{N f_i f_j}$$

It is easy to see that the right hand expression, which we can again call the total inertia :

- is the sum of L.M. terms each of which measures the degree of departure from the row-column independance hypothesis for a contingency table - i.e. the sum of the L.M. chi-square values for the separate contingency tables.
- under the independance hypothesis for the Al and Bm variables, follows a chi-square distribution with (card I - L) (card J - M) degrees of freedom.

This implies that the pseudo-correlations rq, represent a quantity which is a precise measure on the global table.

In practice the total inertia of the table is compared to its approximate estimation calculated from a subset Q' of the factors : $N \sum_{q \in Q} r_q^2$

2.4 Factors corresponding to the subsets I x Jm or Il x J

The subset I_1 x J of the global table may be considered. It consists of the union of the M contingency tables corresponding to the relationships between variable Al and all the Bm variables. We now consider whether the factors of this subset provide any further information about the relationships expressed in this sub-table.

The following relationship can be established :

$$\sum_{i \in I\ell} \sum_{j \in J} \frac{(n_{ij} - N f_i f_j)^2}{N f_i f_j} = \frac{N}{L} \sum_{q \in Q} r_q^2 \operatorname{var} \varphi_q^{I\ell} \qquad (18)$$

where the left-hand expression follows, under the independance hypothesis between A1 and each of the B variables, a chi-square distribution with (card I_1 - 1) (card J - M) degrees of freedom. It provides a measure of the degree of departure from the independance hypothesis.

Equation (18) expresses this index as a sum of positive terms each dependant on a single factor. As this index can also be interpreted as representing inertia, it is meaningful to speak of the percentage of inertia accounted for by a subset Q' of the factors. These percentage values may be of assistance for the interpretation of the factors.

2.5 Reconstruction of the Contingency Tables

Formula (9) is still applicable :

$$fij = fifj \left[1 + \sum_{q \in Q} r_q \varphi_q^i \varphi_q^j \right]$$

Application of this formula to a subset Q' of the factors will provide an approximation $n_{ij}^{Q'}$ to the nij:

$$n_{ij}^{Q'} = N f_i f_j \left[1 + \sum_{q \in Q'} r_q \varphi_q^i \varphi_q^j \right]$$

The following property can be shown to exist : each approximation to the original table using a subset Q' of the factors will preserve unchanged the marginal values of each separate contingency table.

Hence

$$\sum_{i \in I\ell} n_{ij}^{Q'} = \sum_{i \in I\ell} n_{ij}$$

and

$$\sum_{j \in Jm} n_{ij}^{Q'} = \sum_{j \in Jm} n_{ij} \qquad (19)$$

It follows that the total each separate table will also remain unchanged.

The relationships (19) imply that each separate contingency table $I_1 \times Jm$ will be approximated by the subset of factors chosen, the grand and marginal totals remaining unchanged. The only difference will be in the relationship between the I_1 and Jm as expressed by the approximating tables.

In practice one could :

- measure the degree of departure from the original table obtained by the approximation used (by calculating the chi-square distance using the reconstituted table).

- base the interpretation given to a single factor on the tables which this factor, when used alone, is able best to reconstitute.

- isolate groups of contingency tables which are described by the same subsets of the factors.

REFERENCES

BENZECRI J.P.
L'analyse des données II (Analyse des correspondances)
Dunod 1973

DAVIDSON F., CHOQUET M. , DEPAGNE M.
Les lycéens devant la drogue et les autres produits psychotropes
I.N.S.E.R.M. 1973

LEBART L., FENELON J.P.
Statistique et informatique appliquées
Dunod 1973

LECLERC A.
Etude de certains types de tableaux par l'analyse des correspondances. Application à une enquête de Santé Publique.
Thèse 3e cycle - Université Paris VI 1973

SRIKANTAN K.S.
Canonical association between nominal measurements.
JASA vol 65 N° 329 P 284-292 - 1970

Statistical and Numerical Aspects of Methods for Identification of Distributed-Lag-Models

Hans-J. Lenz, Berlin

1. PRELIMINARIES

The existence of distributed lags between variables seems to be an inherent feature of dynamic situations at all if, of course, the interval of observation is not too large. Let us give two examples. The total reaction of an industry is spread over some period of time during which the demanded quantities are produced. A firm has to wait until the outgoing invoices are cashed. Evidently, these delays and distributed lags exist because of economic, technological, institutional and psychological reasons.

No doubt, distributed lag models have become a convenient and workable tool in economic analysis. Unfortunately, it seems rather difficult to discriminate between alternative models when trying to apply them to real world datas, because even the incorporated estimation and numerical methods are too different. Recently, MORRISON[197o] and SMITH[1973] have published some comparative studies, but for very special cases only, i.e. geometric lag structures.

"To take more stock" this paper is devoted to a larger class of models, i.e. to rational lag structures with additive ARMA disturbances in the discrete single input-single output case with no feedback. We try to present the models, the estimation methods and the numerical techniques on an uniform basis. We restrict ourselves, of course, to "data-guided" specification and estimation (identification). In these cases the lag structure is not known a priori but must be found out by an appropriate data-analysis.

2. THE GENERAL MODEL STRUCTURE

First of all, we consider the following general parametric class of models as discussed extensively by HANNAN[197o], BOHLIN[1970], BOX and JENKINS[1970] and DHRYMES[1971]. The equation system is given on the next page and shown diagrammatically in FIG.1.

plant equation: $A(L)x_t=B(L)u_{t-\tau}$ (1)

error equation: $D(L)e_t=C(L)\varepsilon_t$ (2)

observation eq.: $y_t=x_t+e_t$ (3)

where A,B,C,D are polynomials in L of the form,

$$\left.\begin{array}{l} A(L)=1+a_1L+\dots+a_nL^n \\ B(L)=b_o+b_1L+\dots+b_{m-1}L^{m-1} \end{array}\right\} m\le n$$

$$\left.\begin{array}{l} C(L)=1+c_1L+\dots+c_rL^r \\ D(L)=1+d_1L+\dots+d_sL^s \end{array}\right\} r\le s \qquad (4)$$

L is the well known backward shift operator and u_t is a deterministic digital input. x_t, y_t and e_t are, respectively, the non-observable error-free output, the observed output and the error term, all at the tth ampling instant. ε_t is independent (or uncorrelated) of the input u_t and specified as usual, i.e.

$$E\{\varepsilon_t\}=0 \qquad E\{\varepsilon_t^2\}=\sigma_\varepsilon^2 \qquad E\{\varepsilon_t,\varepsilon_k\}=0 \ (k\neq j) \qquad (5)$$

Finally, τ is the (integer) number of pure time delay.

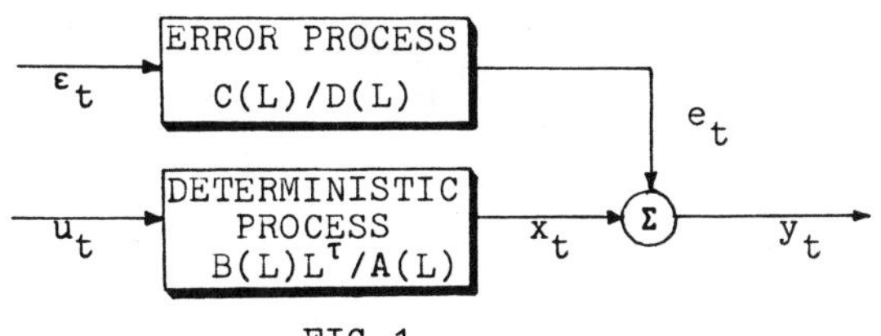

FIG.1
model structure

For obvious reasons it is desirable to place stability restrictions on the polynomials, i.e. D, A should have roots outside the unit circle. However, to introduce "stochastic non-stationarities" into the process one can relax this requirement and allow D to have roots on the boundary. BOX and JENKINS refer to this as an ARIMA error model. The advantage of distributed-lag models consisting of polynomial fractions is its build in flexibility as it can represent quite a large number of encountered representations as listed in table 1 below, the evidence of the model's dynamic characteristics and its "parsimony" which is, however, rather obvious from an approximation theory point of view.

Table 1

Model Structures

No.	Structure	Name	Used Polynomials A	B	C	D
I	$y_t = \frac{1}{A} u_t + \frac{1}{A} \varepsilon_t$	linear regression	•			
II	$y_t = B\, u_t + \varepsilon_t$	Moving-average		•		
III	$y_t = \frac{b}{1-aL} u_t + \varepsilon_t$	KOYCK	•	•		
IV	$y_t = \frac{b}{(1-aL)^r} u_t + \varepsilon_t$	SOLOW	•	•		
V	$y_t = \frac{B}{A} u_t + \varepsilon_t$	JØRGENSON uncorrelated error	•	•		
VI	$y_t = \frac{B}{A} u_t + \frac{1}{AD} \varepsilon_t$	JOHNSTON CLARKE HASTINGS special AR error	•	•		•
VII	$y_t = \frac{B}{A} u_t + \frac{C}{A} \varepsilon_t$	ÅSTRÖM	•	•	•	
VIII	$y_t = \frac{B}{A} u_t + \frac{C}{D\,A} \varepsilon_t$	rational lag structure with special ARMA disturbances	•	•	•	•
IX	$y_t = \frac{B}{A} u_t + \frac{C}{D} \varepsilon_t$	rational lag structure with ARMA disturbances	•	•	•	•

3. ALTERNATIVES OF IDENTIFICATION

Before going into any details it seems worth while to survey the main four approaches to identification as listed in table 2 below. Authors are referred to only in cases when it is felt that they did studies of outstanding character in the respective areas.

Table 2

Approaches to identification

kind of system loop		analysis domain		analysis line		mathematics	
open	closed	time	frequency	on	off	analytic	numeric
	AKAIKE [1967] PRIESTLEY [1971]		AKAIKE [1964] HANNAN [1970]				BOX& JENKINS [1970] DHRYMES [1971]
•		•		••	•••	•••	••

CODE: • majority of studies in engineering and econometrics
•• " " " in engineering
••• " " " in econometrics

4.PARAMETER ESTIMATION METHODS

The parameter estimation methods will now be dealt with concentrating ,of course,to the main areas of research in the above table, i.e.open-loop systems,estimation in the time-domain(parametric estimation) combined with both kinds of implementation.To show evidently the high degree of freedom in identification even of very simple models we start with geometric lag structures,cf.No.III in table 1 above,although it would appear that some of the following results are of historical interest only.

Table 3

Identification of KOYCK-type models

Technique	Estimator	asympt.properties		computation	
		consist.	effic.	effort	feature
OLS 1	$\begin{pmatrix}\hat b\\ \hat a\end{pmatrix}=\begin{pmatrix}\Sigma u_t^2 & \Sigma u_t y_{t-1}\\ \Sigma x_t y_{t-1} & \Sigma y_{t-1}^2\end{pmatrix}^{-1}\begin{pmatrix}\Sigma x_t y_t\\ \Sigma y_t y_{t-1}\end{pmatrix}$	-	- 5 > 1	low	off-line analytic
KOYCK-KLEIN 2	$\begin{pmatrix}\hat b\\ \hat a\end{pmatrix}=\begin{pmatrix}\Sigma u_t^2 & \Sigma u_t y_{t-1}\\ \Sigma x_t y_{t-1} & \Sigma y_{t-1}^2\end{pmatrix}^{-1}\begin{pmatrix}\Sigma x_t y_t\\ \Sigma y_t y_{t-1}+\frac{a}{1+a^2}\Sigma e_t^2\end{pmatrix}$	+	- 5 > 2	low	off-line analytic
LIVIATAN (IV) 3	$\begin{pmatrix}\hat b\\ \hat a\end{pmatrix}=\begin{pmatrix}\Sigma u_t^2 & \Sigma u_t y_{t-1}\\ \Sigma u_t u_{t-1} & \Sigma y_{t-1} u_{t-1}\end{pmatrix}^{-1}\begin{pmatrix}\Sigma y_t u_t\\ \Sigma y_t u_{t-1}\end{pmatrix}$ $E(e_t u_{t-1})=0$	+	- 5 > 3	low	off-line analytic
HANNAN (asympt. GLS or ML) 4	$\begin{pmatrix}\hat b\\ \hat a\end{pmatrix}\cong\left(\begin{pmatrix}u^T\\ \tilde y_-^T\end{pmatrix}\tilde V^{-1}(u,y_-)\right)^{-1}\begin{pmatrix}u^T\\ \tilde y_-^T\end{pmatrix}\tilde V^{-1}y$ $\tilde V=E\{\tilde e\tilde e^T\}$; $\tilde e_t=(1-\tilde b-\tilde a L)y_t$ $\tilde y_t=\Sigma_{i=0}^{\infty}(\tilde a L)^i y_t$; plim $\tilde b=b$; plim $\tilde a=a$	+	+	high	off-line analytic
DHRYMES (ML) 5	$\mathcal{L}(\hat a)=\max_{\lvert a\rvert\in(0,1)}\mathcal{L}(a)=-\frac{T}{2}(\ln 2\pi+1)-\frac{T}{2}\ln\hat\sigma^2(a)$ $\hat b=\hat b(\hat a)$	+	+	medium	off-line analytic &numeric
STEIGLITZ &Mc BRIDE (appr.ML) 6	$\dot y_0=\dot u_0=\ddot u_0=0$ $\hat a^{(0)}=\hat a^{OLS}$ $\dot y_t=\frac{1}{1-\hat a^{(k)}}y_t$; $\dot x_t=\frac{1}{1-\hat a^{(k)}L}u_t$; $\ddot x_t=\frac{1}{(1-\hat a L)^2}u_t$ $\dot y_t=\hat a^{(k)}\dot y_{t-1}+y_t$; $\dot u_t=\hat a^{(k)}\dot u_{t-1}+u_t$; $\ddot u_t=\hat a^{(k)}\ddot u_{t-1}+\dot u_t$ $\begin{pmatrix}\hat b\\ \hat a\end{pmatrix}^{(k+1)}=\begin{pmatrix}\Sigma \dot u_t^2 & \Sigma \dot y_{t-1}\dot u_t\\ \Sigma \dot u_t\ddot u_{t-1} & \Sigma \dot y_{t-1}\ddot u_{t-1}\end{pmatrix}^{-1}\begin{pmatrix}\Sigma \dot y_t\dot x_t\\ \Sigma \dot y_t\dot x_{t-1}\end{pmatrix}$	+	+ 5 ⩾ 6	medium	off-line analytic iterative

To simpify notation we make use now of matrix representation. Let

$$\underline{y}^T=(y_n, y_{n+1},\ldots,y_{n+N}) \quad \text{observed output-vector} \tag{6}$$

$$\Psi=\begin{pmatrix}-y_{n-1} & \cdots & -y_0 & \vdots & u_n & \cdots & u_0\\ \vdots & & & \vdots & \vdots & & \\ -y_{N+n-1} & \cdots & -y_N & \vdots & u_{N+n} & \cdots & u_N\end{pmatrix} \quad \text{design-matrix} \tag{7}$$

$$\underline{\theta}^T=(a_1,a_2,\ldots,a_n \;\vdots\; b_0,b_1,\ldots b_n) \quad \text{(unknown) parameter-vector} \tag{8}$$

Models of type V can now be written as

$$\underline{y}=\Psi\underline{\theta}+\underline{\varepsilon} \tag{9}$$

Note that the vector of residuals $\underline{\varepsilon}$ is defined in EYKHOFF's sense.

4.1 OLS-estimators

The OLS-estimator is, of course, given by

$$\underline{\hat{\theta}} = (\Psi^T\Psi)^{-1}\Psi^T \underline{y} \tag{10}$$

where $\Psi^T\Psi = (\boxed{\diagdown})$ is a symmetry, so storage reducing matrix. The OLS-properties of models from V are well known, i.e.

$$E(\underline{\hat{\theta}}) = \underline{\theta} \quad \text{and} \quad \underset{N\to\infty}{\text{plim}}\ \underline{\hat{\theta}} = \underline{\theta} \tag{11}$$

and variances $\hat{\sigma}^2_{\underline{\hat{\theta}}} \geq \mathcal{J}^{-1}(\underline{\theta})$ under the normal hypothesis, where $\mathcal{J}^{-1}(\underline{\theta})$ is the information matrix derived from the RAO-CRAMÉR-inequality. The main <u>numerical</u> disadvantage of this OLS approach lies in its "off-line" implementation. For each new pair of observation the matrix $\Psi^T\Psi$ must be recalculated and inverted. To avoid this recursive estimation seems appropriate.

4.2 RLS-estimators (recursive OLS)

Let

$$k = n+N,\ \underline{\Psi}_k^T = (-y_{k-1}, \ldots, y_{k-n} \,\vdots\, u_k, \ldots, u_{k-n}) \ \text{and}\ P_k = (\underline{\Psi}_k^T \underline{\Psi}_k)^{-1} \tag{12}$$

We can write the estimator $\underline{\hat{\theta}}$ at the kth step

$$\underline{\hat{\theta}}_k = P_k \underline{\Psi}_k^T \underline{y}_k \tag{13}$$

After some troublesome algebraic manupulations making use of HO's [1962] matrix inversion lemma, it follows as a recursive OLS-version (RLS):

$$\underbrace{\underline{\hat{\theta}}_{k+1}}_{\text{new estimate}} = \underbrace{\underline{\hat{\theta}}_k}_{\text{old}} + \underbrace{P_{k+1}\underline{\Psi}_{k+1}}_{\text{corr. factor}}(\underbrace{y_{k+1}}_{\text{new observat.}} - \underbrace{\underline{\Psi}_{k+1}^T\underline{\hat{\theta}}_k}_{\text{predicted value}}) \tag{14}$$

and

$$P_{k+1} = P_k[I - \underline{\Psi}_{k+1}\underline{\Psi}_{k+1}^T P_k(\underline{\Psi}_{k+1}^T P_k \underline{\Psi}_{k+1} + 1)^{-1}] \tag{15}$$

Notice that because (.) is a scalar no matrix inversion is more necessary. Argueing heuristically the convergence of the RLS-estimators follow from $\underset{N\to\infty}{\text{plim}}\,\underline{\hat{\theta}} = \underline{\theta}$ in the pure OLS-case, of course, modulo starting values. To start the recursion up one can make use of a priori knowledge in a similiar manner as BAYES' followers will do it in regard to complete distributions, off-line pilot studies or even simply $\underline{\hat{\theta}}_0 = \underline{0}$; the arguments follow the same line as to the variance-covariance-matrix P, except for the last case. Using a diagonal matrix with "very great" elements on the main diagonal a compromise is necessary between speed of convergence and stability of the estimator trajectories in the parameter space.

4.3 SA-estimators(stochastic approximation)

An alternative form of (14) is

$$\underline{\hat{\theta}}_{k+1}=\underline{\hat{\theta}}_k-P_{k+1}(\underline{\Psi}_{k+1}\underline{\Psi}_{k+1}^T\underline{\hat{\theta}}_k - \underline{\Psi}_{k+1}y_{k+1}) \tag{16}$$

This is an interesting equation since it will be noted that the term in the brackets (.) is proportional to the gradient of the instantaneous OLS loss-function at the (k+1)th instant,i.e

$$S_{k+1}=(\underline{\Psi}_{k+1}^T\hat{\underline{\theta}}_k-y_{k+1})^2 \quad \text{or} \quad \frac{1}{2}\frac{\partial S}{\partial \underline{\hat{\theta}}_k}=\underline{\Psi}_{k+1}\underline{\Psi}_{k+1}^T\hat{\underline{\theta}}_k-\underline{\Psi}_{k+1}y_{k+1} \tag{17}$$

HO[1962] pointed out that RLS is,in fact, a special stochastic approximation algorithm controlled by the weighting matrix P_k. Generally, the SA-algorithm in a KIEFER-WOLFOWITZ-version takes the form

$$\hat{\underline{\theta}}_{k+1}=\hat{\underline{\theta}}_k-\gamma_k\frac{\partial}{\partial \underline{\theta}}E\{e_k^2\}=\hat{\underline{\theta}}_k+2\gamma_k\underline{\Psi}_k(y_k-\underline{\Psi}_k^T\underline{\theta}_k) \tag{18}$$

for modells of the type $y_k=\underline{\Psi}_k^T\underline{\theta}+\varepsilon_k$ .A necessary condition for convergence of (18) is given by the DVORETZKY-theorem

$$\sum_{k=1}^{\infty}\gamma_k=\infty \qquad \sum_{k=1}^{\infty}\gamma_k^2<\infty \tag{19}$$

Evidently, the SA-algorithm is indeed of a very simple construction.Addition and multiplication besides some storing of the new incoming parameters is only necessary without any matrix inversion.Unfortunately, it behaves ill in regard to biasedness when the conditions of V (cf.table 1) are not fulfilled.

4.4 GLS-estimators(general-least-squares)

Let $\quad A(L)y_k=B(L)u_k+e_k \quad$ and $\quad D(L)e_k=\varepsilon_k$ (20)

Substitution results in the AR-error-model

$$A(L)D(L)y_k=B(L)D(L)u_k+\varepsilon_k \tag{21}$$

and (21) turns out to be nothing but a prefiltered V-version.

Defining $\quad \dot{y}_k=D(L)y_k \quad$ and $\quad \dot{u}_k=D(L)u_k$

gives $\quad A(L)\dot{y}_k=B(L)\dot{u}_k+\varepsilon_k$ (22)

being a structure like (9).An iterative procedure starts with OLS-estimates and residuals

$$\hat{\underline{\theta}}=(\Psi^T\Psi)^{-1}\Psi^T\underline{y} \qquad \hat{e}_k=\hat{A}(L)y_k-\hat{B}(L)u_k \tag{23}$$

giving first estimates of the coefficients of the D-polynomial

$$\hat{\underline{d}}=(E^TE)^{-1}E^T\hat{\underline{e}} \tag{24}$$

where E is the estimated (suitable defined)matrix of residuals. With D(L) so estimated one starts again with prefiltering etc. It is clear from the above reasoning that this procedure is extremely storage and CPU-time consuming(cp.table 5 below).

4.5 RGLS-estimators(recursive-general-least-squares)

RGLS is the two-stage version of simple RLS in full analogy to the GLS-case. During recursion the algorithm starts with

(cf.(14)and (15)) (25)

turns to (26)

before starting again with new observations.
Although it is easy to implement RGLS, this procedure is pretty CPU-time and storage consuming. However, the need for storage may be reduced by a proper storage overlay-technique.

4.6 ML-estimators(maximum-likelihood)

Perhaps the best behaving identification procedure up to the IX - class of models is the ML-method as suggested by BOX,JENKINS [1970],BOHLIN[1970] and others. As both methods are fully described elsewhere it will suffice to outline the underlying philosophy of both off-line techniques only.
Because $S(\underline{\theta})=S(\underline{a},\underline{b},\underline{c},\underline{d})=\Sigma\varepsilon_k^2$, where ε_k is defined by $\frac{C}{D}\varepsilon_k=y_k-\frac{B}{A}u_{k-\tau}$ is highly non-linear in the parameter vector $\underline{\theta}$, both methods minimize S iteratively by means of TAYLOR-expansion. BOHLIN approximates S up to the second order and makes use of the NEWTON-RAPHSON-algorithm(cp.table4). Interactive steering of the computer-guided program is incorporated as is ,actually, the same in the scheme of BOX&JENKINS. Thus it is hoped to combat difficulties as non-uniqueness,unfavourable surfaces,singularities etc. Finally, a heuristic search procedure determines the appropriate polynomial-degrees. Besides of prewithening by ARMA-models an(eventually)already digitally filtered input u_k and output y_k in order to get rough ideas of the orders of the polynomials and starting values from the cross-correlation-function $\hat{\rho}_{\dot{u}\dot{y}}$ BOX&JENKINS use a modified MARQUARDT-algorithm (cf.table 4,D7),i.e. first-order TAYLOR series expansion of S combined with a steepest-descent-technique,for an (approximate) ML parameter estimation. No doubt, residual analysis and several statistical checks are applied to improve the fitting and the model's characteristics of dynamics. Computationally the method of BOX&JENKINS is leading because, by means of "backprediction" of the state variables, they solve the problem of starting values in an optimal manner. Statistically, their approach has the

advantage, that the estimators ($\underline{\hat{a}},\underline{\hat{b}}$) are asymptotically uncorrelated with the estimators ($\underline{\hat{c}},\underline{\hat{d}}$) of the disturbance model.

4.7 IV/AML-estimators(instrumental variables/appr.maxim.likelih.)

"Off-line" IV-estimation of $\underline{\theta}$ in $\underline{y}=\Psi\underline{\theta}+\underline{e}$ is done by a suitable transformation of the state variables with the aid of the matrix W^T, consisting of well defined instrumental variables, i.e.

$$E(W^T\underline{e})=\underline{0} \quad \text{and } E(W^T\Psi) \text{ non singular} \tag{27}$$

It follows immediately $\quad W^T\underline{y}=W^T\Psi\underline{\theta} + W^T\underline{e}$ (28)

or $\quad \underline{\hat{\theta}}=(W^T\Psi)^{-1}W^T\underline{y}$ (29)

Quite clear, there is some degree from freedom in the choice of the instrumental variables.

YOUNG's[1970] fully "on-line"approach ,using the state variables ($\underline{x},\underline{u}$) as instruments , seems to the best known compromise between "on-line" realisation and a minimal norm of the covariance matrix of $\underline{\theta}$.Using

$$\underline{\hat{\Psi}}_k^T=(-\hat{x}_{k-1},-\hat{x}_{k-2},\dots,-\hat{x}_{k-n};u_k,\dots,u_{k-n}) \tag{30}$$

as vector of instrumental variables generated by the "auxiliary model" $\hat{A}(L)\hat{x}_k=\hat{B}(L)u_k$ the IV-estimator $\underline{\hat{\alpha}}_k^T=(\hat{a}_1,\dots,\hat{a}_n;\hat{b}_o,\dots,\hat{b}_n)$ of($\underline{a},\underline{b}$) at the (k+1)th instant is then given by(cp.(16),(17))

$$\underline{\hat{\alpha}}_{k+1}=\underline{\hat{\alpha}}_k+\hat{P}_k\underline{\hat{\Psi}}_{k+1}(\underline{\Psi}_{k+1}^T\hat{P}_k\underline{\hat{\Psi}}_{k+1}+1)^{-1}(y_{k+1}-\underline{\Psi}_{k+1}^T\underline{\hat{\alpha}}_k) \tag{31}$$

$$\hat{P}_{k+1}=\hat{P}_k[I-\underline{\hat{\Psi}}_{k+1}\underline{\Psi}_{k+1}^T\hat{P}_k(\underline{\Psi}_{k+1}^T\hat{P}_k\underline{\hat{\Psi}}_{k+1}+1)^{-1}] \tag{32}$$

The parameter vector $\eta=(\underline{c},\underline{d})$ of the noise model(2) is estimated from $\hat{e}_k=y_k-\hat{x}_k$ by an recursive approximate ML-procedure,

$$\hat{\eta}_{k+1}=\hat{\eta}_k+P_k\underline{\hat{v}}_{k+1}[\underline{\hat{v}}_k^T P_{k-1}\underline{\hat{v}}+1]^{-1}(\hat{e}_{k+1}\underline{\hat{v}}_{k+1}^T\hat{\eta}_k) \tag{33}$$

$$P_{k+1}=P_k[I-\underline{\hat{v}}_{k+1}\underline{\hat{v}}_{k+1}^T P_k(\underline{\hat{v}}_{k+1}^T P_k\hat{v}_{k+1}+1)^{-1}] \tag{34}$$

to obtain an estimate $\hat{\eta}_k$ of η occurring in the noise model

$$e_k=\underline{v}_k^T\eta+\varepsilon_k \tag{35}$$

where $\quad \underline{v}_k^T=(-e_{k-1},\dots,-e_{k-s};\varepsilon_{k-1},\dots,\varepsilon_{k-r})$ (36)

and $\underline{\hat{v}}_k^T$ defined analogously.The components $\hat{e}_k$ are generated by

$$\hat{\varepsilon}_k=\hat{e}_k-\underline{\hat{v}}_k^T\hat{\eta}_k \tag{37}$$

An analytic proof of consistency of the IV/AML estimator could only be given in largely hypothetical situations(YOUNG[1970]).From real and simulated data,however,it is felt that YOUNG's procedure is rapidly convergent in operation above as a rule of thumb 1oo pairs of data under normal operating conditions.As all recursive algorithm IV/AML is designed in the spirit of the KALMAN-filter-theory and derives ,moreover, profit from the incorporated ML-procedure.Unnecessary to note that various precautions must be build in to ensure stability and a nice convergence of the parameter estimates.

5. PROGRAMMING DETAILS

To make this study not too tedious the most important statistical and numerical aspects of the above mentioned identification methods are carried together in this section. No comment more will be given for it is felt that the results are evidently enough and are along well known lines .By the way, because of rare space the references at the end of this paper are cut down sharply.The author hopes that the quoted studies of DHRYMES,YOUNG and GÖHRING makes it possible to find ,nevertheless,the cited sources.

Table 4 Numerical techniques

code		numerical techniques	name	solution	programming effort	effectivity
A	1	matrix inversion	GAUSS	$\Psi^T\Psi\underline{\hat{\theta}} = \Psi^T\underline{y}$	low	+
	2		GAUSS-BANACHIEWICZ	$\underline{\hat{\theta}} = (\Psi^T\Psi)^{-1}\Psi^T\underline{y}$		
	3		GAUSS-JORDAN			
	4		CHOLESKY			
	5	recursion	recursive algorithm	$\underline{\hat{\theta}}_{K+1} = \underline{\hat{\theta}}_K + \underline{K}_{K+1}(y_{K+1} - M^T_{K+1}\hat{\theta}_K)$	low	+
B	1	orthogonal transformation (triangulat	GIVENS	$H_K\underline{\hat{\theta}}_K = \varepsilon_K \xrightarrow{T} H^{\blacktriangle}_K\underline{\hat{\theta}}_K = \varepsilon^{\blacktriangle}_K$	medium	+
	2		HOUSHOLDER			
C	1	search techniques	Univariate	$\underline{\hat{\theta}}_{K+1} = \underline{\theta}_K + \beta\Delta\underline{\hat{\theta}}_K$	low	
	2		POWELL			
	3		SOUTHWELLrelaxation			
	4		ROSENBROCK			
	5		FIBONACCI			
D	1	gradient techniques	steepest decent	$\underline{\hat{\theta}}_{K+1} = \underline{\hat{\theta}}_K - \beta_K R\frac{\partial S}{\partial\theta}\Big\vert_{\hat{\theta}_K}$	medium/ high	
	2		Acceleration			
	3		PARTAN			
	4		NEWTON-RAPHSON			±
	5		FLETCHER-REEVES			
	6		POWELL			
	7		MARQUARDT-LEVENBERG			+
	8		SHANNO-SMITH			++
E		stochastic search procedures	Monte Carlo	$S_{K+1} = S(\underline{\hat{\theta}}_K \pm \Delta\underline{\hat{\theta}}_K)$	low	-

FIG.2

Convergence studies of a_1 for 2-degree polynomials and alternative identification methods

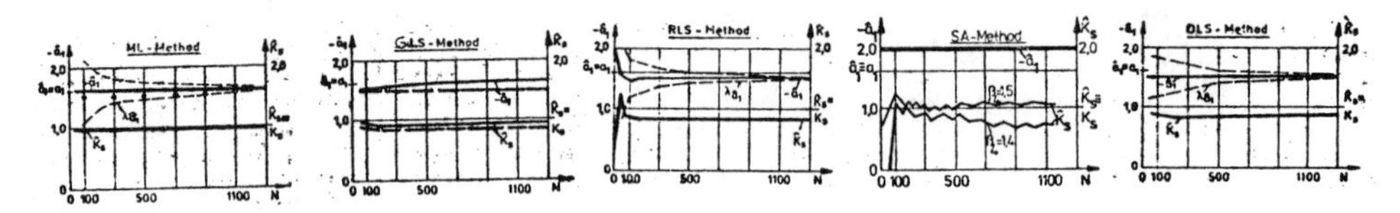

Table 5

Alternative Identification methods

No.	Name	CODE	STRUCTURE	NUMER. TECHNIQUE	IMPLEMENTATION off	IMPLEMENTATION on	APRIORI KNOWL.	CPU 0-100	MEMORY 0-100	CONVERGENCE	PROGR.-EFFORT
1	ordinary least squares	OLS	V	A1,A2,A3	•			4	59	±	ex.low
2	recursive least squares	RLS	V	A4,A5,B		•	starting values	7	37	±	ex.low
3	general least squares	GLS	VI	A1,A2,A3	•		error-structure	30	100	+	high
4	recursive general least squares	RGLS	VI	A4,A5,B		•	starting values	20	65	+	medium/high
5	instrumental variables	IV	V-IX	A1,A2,A5	•	•	instrum. variables	15	63	+	low/medium
6	stochastic approximation	SA	I,II, VII	A4,A5,D	•	•	starting values	5	30	-	ex.low
7	approx. maximum likelihood	ML	IX	D	•	•	normal hypothesis starting values step size	12	60	+	ex.high
8	instrumental variables-appr.maxim. likelihood	IV-ML	IX	A5,D		•	cf. 7	25	85	+	ex.high

FIG.3

Studies of CPU-time dependent on poly.-degrees

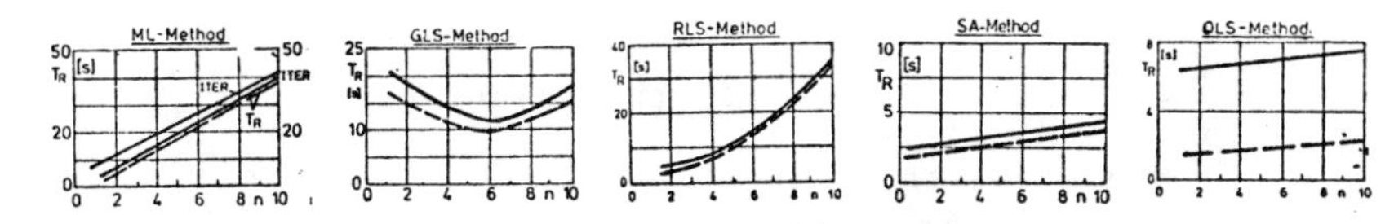

ACKNOWLEDGEMENT

The author has to thank deeply P.YOUNG(Cambridge) and B.GÖHRING (Stuttgart) for fruitful discussions on this topic and making available quite a lot of preprints used here to complete this study.

REFERENCES

Dhrymes,P. : Distributed Lags Problem of Estimation and Formulation, Holden-Day, San Francisco, 1971

Young,P.et al.: A Recursive Approach to Time -Series Analysis,Dep. of Eng.,Univ.of Cambridge,1971

Göhring,B. : Erprobung statistischer Parameterschätzmethoden und Strukturprüfverfahren zur experimentellen Identifikation von Regelsystemen,Diss. TU Stuttgart, 1973

Correspondence Factorial Analysis

M. F. Maignan, Zurich

1 INTRODUCTION

We expose hereafter some informational aspects of the correspondence factorial analysis. It was developped since the mid-sixties by Professor J.P. Benzecri and his research team in mathematical statistics.

The correspondence factorial analysis performs a multivariate analysis of the mutual information between the rows and the columns of a given matrix, say $P=\{p_{ij}\}$. The correspondence analysis extracts the eigenvalues and eigenvectors of the covariance matrix C of R'R, with $R=(p_{ij} - p_{i.}p_{.j})/\sqrt{p_{i.}p_{.j}}$, where $p_{i.}=\Sigma_j p_{ij}$, and $p_{.j}=\Sigma_i p_{ij}$. $i=1,..n$ $j=1,..,p$

Instead of considering this transformation $P \longrightarrow R$ as a transformation on the variables p_{ij}, we use this expression for the definition of the so-called χ^2-distance.

Between two points i,i' of R^p, the classical distance would be:

$$d^2(i,i')=\Sigma_{j=1}^{p}\left(p_{ij} - p_{i'j}\right)^2$$

By using:

$$d^{2*}(i,i')=\Sigma_j^p \frac{1}{p_{.j}}\left[\frac{p_{ij}}{p_{i.}} - \frac{p_{i'j}}{p_{i'.}}\right]^2$$

, we have the advantage that d^* is in fact the χ^2-distance, and that there are close relationships between χ^2 and informational quantities.

In the following developments, we use generally the formalism of Benzecri [1] .

2 INFORMATION

We shall proceed stepwise, by exposing the different definitions of information quantity and related elements at different levels of mathematical rigorness.

2.1 Hartley definition

Let Ω be a finite set of n elements.

We have for wellknown definition of the information quantity given by the knowledge of an element:

$$I = \log_2 n \quad (1)$$

2.2 Shannon Formula

Let Ω be a finite set of n elements,and $\{A_i\}$ a partition of Ω, and $P(A_i)=P_i = n_i/n$,probability that A_i occurs,n_i being the number of elements in A_i.

By wishing additivity,if we note $I_\Omega(\omega)$ the information brought by the realization of a basic event ω in Ω, $I_{Ai}(\omega)$ the information brought by the same ω in A_i,and $I_\Omega(A_i)$ the information brought in Ω by A_i,we write:

$$I_\Omega(\omega) = I_\Omega(A_i) + I_{A_i}(\omega) \Longrightarrow I_\Omega(A_i) = -\log_2 \frac{n_i}{n} = -\log_2 P_i \quad (2)$$

We consider the entropy as the "mean information":

$$H = -\Sigma_{i=1}^{n} p_i \log_2 p_i \quad (3)$$

2.3 Information in Hypothesis testing

Let us consider a probability space $(X, S, P=\{p_i\})$, $\{f_i(x)\}$ a set of density functions $f_i(x)=f(x|H_i)$,and two hypothesis

$H_0 = \{p_{o1},\ldots,p_{oc}\}$, $H_1 = \{p_{11},\ldots,p_{1c}\}$,where p_{ij} is the probability of a character i among c possible characters in a basis set.

We have $f_o(x) = \{p_{oj}\}$ $f_1(x) = \{p_{1j}\}$

The mean information quantity brought per observation when testing H_o against H_1 (resp H_1 against H_o) is:

$$I(H_o:H_1) = -\Sigma_{j=1}^{c} p_{oj} \log \frac{p_{1j}}{p_{oj}} \quad (4) \quad (\text{resp } I(H_1:H_o) = -\Sigma_{j=1}^{c} p_{1j} \log \frac{p_{oj}}{p_{1j}})$$

We define as divergence between H_o and H_1 the sum of both informations:

$$D(H_o,H_1) = I(H_o:H_1) + I(H_1:H_o) = n\, \Sigma_{j=1}^{c} (p_{1j} - p_{oj}) \log \frac{p_{1j}}{p_{oj}} \quad (4)$$

In the case of Null-against hypothese testing,we have for n_i samples

Null hypothesis H_o $p_{ij}=p_{i.}p_{.j}$

Against hypoth. H_1 $p_{ij} \neq p_{i.}p_{.j}$

$$\text{Divergence}: \Sigma_{i=1}^{n} n_i \Sigma_{j=1}^{c} (p_{ij} - p_{i.}p_{.j}) \log \frac{p_{ij}}{p_{i.}p_{.j}} \quad (5)$$

2.4 Mutual Information

With the same notations as above,we consider two partitions $A = \{A_i\}$ and $B = \{B_j\}$ of Ω .We have $I_\Omega(A \text{ and } B) \leqslant I_\Omega(A) + I_\Omega(B)$,with equality if $\{A_i\}$ and $\{B_j\}$ are undependent.

By definition of the mutual information between A and B:

$$I_m(A,B) = I_\Omega(A) + I_\Omega(B) - I_\Omega(A \cap B) \text{ ,i.e } I_m(A,B) = p_{ij} \log_2 \frac{p_{ij}}{p_{i.}p_{.j}} \quad (6)$$

By approximation,when $p_{ij} \sim p_{i.}p_{.j}$,we have

$$I_m(A,B) = 1/\text{Log } 2 \left\{\Sigma_i \Sigma_j (p_{ij} - p_{i.}p_{.j})^2 / p_{i.}p_{.j}\right\} \quad (7)$$

2.5 Information under Bayesian Inference

Let us consider the likelihood curve $l(\theta|y)$ for a parameter θ and data y. We say that the likelihood curve is data translated if $l(\theta|y)$ is expressible in the form: $l(\theta|y) = g[\phi(\theta) - f(y)]$.
A particular case is when $l(\theta|y)$ is approximately normal; the logarithm of the likelihood is then approximately quadratic so that:

$$L(\theta|y) = \log l(\theta|y) = \log \prod_{i=1}^{n} p(y_i|\theta)$$

$$= L(\hat{\theta}|y) - n/2\ (\theta - \hat{\theta})^2\ (- 1/n\ \partial^2 L/\partial\theta^2)_{\hat{\theta}} \quad (8)$$

where $\hat{\theta}$ is the maximum of likelihood estimate.
Let us consider that $I(\hat{\theta}) = (- 1/n\ \partial^2 L/\partial\theta^2)_{\hat{\theta}}$ is a function of $\hat{\theta}$ only. If $\phi\ (\theta)$ is a one-to-one transformation:

$$I(\phi) = (- 1/n\ \partial^2 L/\partial\phi^2)_{\hat{\theta}} = (- 1/n\ \partial^2 L/\partial\theta^2)_{\hat{\theta}}\ (d\theta/d\phi)^2 = J(\theta)(\frac{d\theta}{d\phi})^2$$

We note $J(\hat{\phi}) = (- 1/n\ \partial^2 L/\partial\phi^2)$, and chose $\left|\frac{d\theta}{d\phi}\right|_{\hat{\theta}} = J^{-1/2}(\hat{\theta})$ (9)
The likelihood is now data translated in words of ϕ.

In the general case when $(- 1/n\ \partial^2 L/\partial\theta^2)_{\hat{\theta}}$ is a function of y, we have for each given θ:

$$- 1/n\ \partial^2 L/\partial\theta^2 = - 1/n \sum_{i=1}^{n} \partial^2 \log p(y_i|\theta)/\partial\theta^2$$

For $n \to \infty$, the average converges in probability to $E\left[-\frac{\partial^2 \log p(y|\theta)}{\partial\theta^2}\right]$ if E exists.
The expression $I_F(\theta) = E\ (-\partial^2 L/\partial\theta^2)$ is the Fisher's information measure.

2.6 INFORMATION MATRIX (FISHER)

Let us consider x, point of RP, with density $f(x,\theta), \theta \in \Theta \subset R^k$, so that $E_\theta = \{x \in R^p\ ,\ f(x,\theta) > 0\}$ is undependent of θ.
Under restriction of regularity of $\{\partial \text{Log}\ f(x,\theta)/\partial\theta_j\}$, j=1,k, the log. covariance matrix of these points, is defined by

$$I_{Fij} = \text{Cov}\ \{\partial \text{Log}\ f(x,\theta)/\partial\theta_i, \partial f(x,\theta)/\partial\theta_j)$$
$$= -E\left\{\partial^2 \text{Log}\ f(x,\theta)/\partial\theta_i\partial\theta_j\right\} \quad \text{"Information Matrix" (10)}$$

We define further:

$h_1(\theta),\ldots,h_s(\theta)$	s functions of θ
$T_1,\ldots,T_s$	s estimators without bias of $h_1,\ldots,h_s$
V	Matrix (s,s) of covariances of T_r, T_s
I_n	information matrix of the n-sample
D	matrix $D_{ij} = \partial h_i(\theta)/\partial\theta_j$ i=1,s; j=1,h

Under evident regularity conditions, the matrix M

$M = V - D\ I_n^{-1}\ {}^tD$ is positive definite (Cramer Rao inequality)

In the case of one estimator T only: $\text{Var}\ T \geq [\partial E(T)/\partial\theta]^2/I_n(\theta)$ (11)
(analog to Heisenberg relation in quantum physics)

3 KARHUNEN LOEVE EXPANSION IN PATTERN RECOGNITION

3.1 K.L. Expansion

Let us consider a set of n dimensional centralized random vectors $\{x\}$,with p_i the probability of occurence of pattern class ω_i ,i=1,p. We intend to express x_i into a finite expansion $x_i = \Sigma_{j=1}^{n} v_{ji} u_j$ (12)
under the conditions: $u_j \, {}^{t}\bar{u}_i = \delta_j^i$ ($^{t-}$ complex conjugate)

$$\Sigma_{i=1}^{p} p_i \; E(v_{ji}\bar{v}_{ki}) = \varrho_j^2 \, \delta_k^j$$

It is easily shown that the u_j are the eigenvectors of the covariance matrix of $\{x_{ij}\}$,and that the ϱ_j^2 are the associated eigenvalues. Let us write some details of the computational aspect.

3.2 Adjustment of a vectorsubspace in R^n

We intend to adjust a set of p points $\{r_{.j}\}$,j=1,p in R^n,by a straight line V,with unit vector U,in fact with the method of least squares.

We write: $S^2 = \Sigma_{j=1}^{p} \{U' \, r_{.j}\}^2 = \Sigma_{j=1}^{p} \{U' r_{.j} \, r'_{.j} \, U\}$ (13)

We use Lagrange multiplicators for expressing the extremum conditions: $U'RR'U + \lambda(U'U - 1) = \max$,with $U'U - 1 = 0$.
We obtain then $2 R'RU - 2\lambda U = 0$, $U'R'RU = \lambda U'U = \lambda$
For representing the x of R^n in a space of dimension p or less,we will then,at successive levels of approximation,consider the eigen vectors spaces,ordered by descending order of the eigenvalues.

3.3 Properties of the K.L expansion.

It is shown without any difficulty,that the K.L expansion
-minimizes the mean square error committed by taking only the first terms of the expansion series,
-minimizes the entropy function defined over the variances of the random coefficients in the expansion. [4] pp 56,60
These properties lead to an optimal procedure for feature selection and ordering.
Especially when the informational divergence is taken as criterion for the selection of features,it can be shown that the maximal divergence between two pattern classes which can be achieved by varying the number of classes is then m/(m-1)×the total divergence of the set:

$$d^2 \leqslant \frac{m \, J(\Omega)}{m - 1}$$

In the case of discriminant analysis,this K.L expansion under consideration of the informational divergence as selection criterion leads to distance considerations of the form:

$D = (C - B\,M)/\,B'B^{\frac{1}{2}}$ (matrix notations),where B and C are elements of the family of linear discriminant functions chosen,and M is built by the means of the variables.
We will consider hereafter the so-called " χ^2 distance",which shows close relationships to information quantities,in particular to the divergence,and it will be proceed to a K.L. expansion.

4 CORRESPONDENCE ANALYSIS UNDER COMPUTATIONAL ASPECTS

4.1 Input Matrix

Let be I,J finite sets of n,resp.p elements, f a function $I\times J \xrightarrow{f} R^{+o}$

Let us consider the marginal laws $f_I = \{f_{i.}\} = \{\Sigma_j f_{ij}\}$ and $f_J = \{f_{.j}\}$ (also noted f^I, f^J)

Two probabilistic transitions are defined:

$$f_I^J = \{f_i^j \mid i\in I, j\in J\} \qquad f_i^j = p\,(i|j) = f_{ij}/f_{.j}$$

$$f_J^I = \{f_j^i \mid i\in I, j\in J\} \qquad f_j^i = p\,(j|i) = f_{ij}/f_{i.} \qquad (14)$$

4.2 Objectives

We wish to define a transformation χ of the input matrix $\{f_{ij}\}$ into $\{f^*_{ij}\}$ so that, by effectuing the K.L expansion of the $\{f^*_{ij}\}$, we get the following properties:

symetry of sets I and J during the analysis, equivalence of the analysis of IxJ, IxI, JxJ, "distributional" equivalence (stability), representation of the observations (I) and of the variables (J) in the same euclidean space, for graphical output.

We define for this purpose, under evident restrictions of strict positiveness, the transformation χ: $f_{ij} \longrightarrow (f_{ij} - f_{i.}f_{.j})/\sqrt{f_{i.}f_{.j}}$

Let us show the meaning of this transformation.

4.3 χ^{2*} Metric

We can write the following statements:

for f:
$$f_{ij} \xrightarrow{\quad\chi\quad} f^*_{ij} = \frac{f_{ij} - f_{i.}f_{.j}}{\sqrt{f_{i.}f_{.j}}} \qquad (15)$$

for covariance (scalar product) of $f_{i.}$ and $f_{i'.}$

$$d^2_{ii'} = f_{ij}f_{i'j} \xrightarrow{\quad\chi^{2*}\quad} d^{2*}_{ii'} = \Sigma f^*_{ij}\, f^*_{i'j} \quad (\Sigma \text{ on } j)$$

$$d^2_{jj'} = f_{ij}f_{ij'} \longrightarrow d^{2*}_{jj'} = \Sigma f^*_{ij}\, f^*_{ij'} \quad (\Sigma \text{ on } i)$$

The expression of $d^{2*}_{ii'}$ is then:
$$\Sigma_j \frac{f_{ij} - f_{i.}f_{.j}}{\sqrt{f_{i.}f_{.j}}} \cdot \frac{f_{i'j} - f_{i'.}f_{.j}}{\sqrt{f_{i'.}f_{.j}}} \qquad (16)$$

4.4 Properties of the correspondence analysis

The K.L expansion of $\{f^*_{ij}\}$ shows:

- distribution equivalence: 2 variables, "columns", or 2 observations, "rows", can be replaced by their sum $I_0 = I_1 + I_2$, for analysing a matrix $(n-1)\times p$ or $n\times(p-1)$

$$\frac{p_{i1j}}{p_{i1.}} = \frac{p_{i2j}}{p_{i2.}} \Longrightarrow \frac{p_{ioj}}{p_{io..}} = \frac{p_{i1j}+p_{i2j}}{p_{i1.}+p_{i2.}} \qquad (17)$$

- $\Sigma_{i,i'} d^2_{ii'}$, $\Sigma_{j,j'} d^2_{jj'}$ are the approximations of the mutual information between I and J (!) (re formula 7)

-the expansions of $d^{2*}_{II}, d^{2*}_{JJ}, f^{*}_{IJ}, {}^{t}f^{*}_{IJ}$ deliver the same results,i.e in words of applied statistician,the analysis of matrices IxI,JxJ,IxJ lead to same eigenvectors and proportional eigenvalues.
-When $p_{ij}, p_{i.}, p_{.j}$ are normally distributed,it is obvious that d^{2*} is a χ^2,that is a gamma Γ (n-1/2,1/2).

4.4 Correspondence analysis formulas

We will write the principal formulas of correspondence analysis. The demonstration are to be found at paragraph 6 ,in [1] and in [3]. Of course,instead of working in R^I and R^J,as computers do,exact demonstrations are to be done with elements of Hilbert spaces $H_2^X(P_X)$

Analysis of I	Analysis of J	
φ^I eigenvector of:	φ^J eigenvector of:	
$\varphi^I \circ f_I^J \circ f_J^I = \lambda(\varphi)\,\varphi^I$	$\varphi^J \circ f_J^I \circ f_I^J = \lambda(\varphi)\,\varphi^J$	(18)
$\varphi^I \circ f_I^J = \sqrt{\lambda(\varphi)}\;\varphi^J$	$\varphi^J \circ f_J^I = \sqrt{\lambda(\varphi)}\;\varphi^I$	(19)
$\Sigma_i \varphi^i f_{ij}/f_{.j} = \sqrt{\lambda(\varphi)}\;\varphi^j$	$\Sigma_j \varphi^j f_{ij}/f_{i.} = \sqrt{\lambda(\varphi)}\;\varphi^i$	(20)
$\Sigma_i \varphi^i \varphi^{i'} f_{i.} = \delta_{i'}^{i}$	$\Sigma_j \varphi^j \varphi^{j'} f_{.j} = \delta_{j'}^{j}$	(21)

We have the so-called "reconstitution formula" of the input datas:

$$f_{ij} = f_{i.}\, f_{.j} \cdot \sum_{\lambda} \left\{ \sqrt{\lambda(\varphi)}\; \varphi^i\, \varphi^j \right\} \quad (22)$$

φ^I,(φ^J) are eigenvectors of S^I (S_J^J) : $s_{i'}^{i} = \Sigma_j f_{ij} f_{i'j}/f_{.j}\sqrt{f_{i.}f_{i'.}}$ (23)

I is represented in an euclidean space,with χ^2 distance,and the factors,proportional to the eigenvectors,can be represented on the same graph

5 χ^2 DISTANCE IN PARAMETRIC STATISTICS

We distinguish R^I vectorspace of functions on I,and R_I V.S. of measures.

5.1 χ^2 distance between probability laws

We will only emphasize that the methodic of χ^2 test consists in the calculation of a χ^2 distance between population law r_I and sample $q_I = p_I$,if we define for 3 laws p_I, q_I, r_I ,the χ^2 distance relative to center p_I,between laws q_I and r_I by: (24)

$$d^{\chi 2}_{pI}(r_I, q_I) = \|r_I - q_I\|^2_{pI} = \langle (r_I - q_I), (r_I - q_I) \rangle_{pI} = \Sigma_i (r_i - q_i)^2/p_i \quad i \in I$$

5.2 Information Matrix in a variety of probability laws

Let be I a finite set,P_I space of probability laws on I,L differentiable subvariety of dimension r.L has a Riemann's structure induced by the χ^2 metric.We suppose L can be defined by a system of parametric equations (allways the case locally).

We note $\theta = \{\theta^1,..,\theta^r\} \in \Theta$ open $\subset R^r, p_I^\theta = \{p_i^\theta | i \in I\} \in L \subset P_I \subset R_I$
p_i^θ is a derivable function of variables θ.
We note in a point p_i^θ, for parameter θ_0:

$$g_{kk'} = \sum \left\{ (p_i^{\theta_0})^{-1} \, (\partial p_i^\theta / \partial \theta^k) \, (\partial p_i^\theta / \partial \theta^{k'}) \,\middle|\, i \in I \right\} \tag{25}$$

We have for the Riemann's structure $ds^2 = g_{kk'} \, d\theta^k \, d\theta^{k'}$ (26)
We can say that $g_{kk'}$ is the scalar product, for the χ^2 metric of center p_i^θ of the 2 vectors $\partial . / \partial \theta^k$, $\partial . / \partial \theta^k$. We hereby find again the information matrix of Fisher (27)

5.3 χ^2-metric and maximum of likelihood
Let be E euclidean space of dimension r, with a basis $\{e_k\}$.
Let be $\langle , \rangle$ the scalar product and $\{g_{kk'}\} = \langle e_k, e_{k'} \rangle$, $\{g^{kk'}\}$ its inverse
We can write, for the norm of a vector u : $\|u\|^2 = \sum g^{kk'} u_k u_{k'}$ (28)
Let consider R_I again, and its tangent space T_0, with basis e_k

$$e_k = (\partial p_i^\theta / \partial \theta^k) \tag{29}$$

We wish to test if the frequence law f_I of a sample (n) is the law p_I^θ: we project orthogonaly (in the meaning of the χ^2 metric) f_I on the tangential variety T_0 in p_I^0. We name f_I^0 this projection.
We have, since $(f_I - p_I^0)$ is orthogonal to T_0:

$\langle f_I - p_I^0 , e_k \rangle = \langle f_I^0 - p_I^0 , e_k \rangle$, and easy to show $\langle p_I^\theta, e_k \rangle = 0$

We can then write:

$$\|f_I^0 - p_I^0\|^2 = \sum \left\{ g^{kk'} \langle f_I^0 - p_I^0, e_k \rangle . \langle f_I^0 - p_I^0, e_{k'} \rangle \,\middle|\, k,k' \in]r] \right\} \tag{30}$$
$$= \sum \left\{ g^{kk'} \langle f_I , e_k \rangle . \langle f_I , e_{k'} \rangle \,\middle|\, k,k' \in]r] \right\}$$

with

$$\langle f_I, e_k \rangle = \sum \left\{ f_i (\partial p_i^\theta / \partial \theta^k / p_i^0 \,\middle|\, i \in I \right\} \tag{31}$$

We consider the case when $f_I^0 = p_I^{\theta_0} \Longrightarrow \langle f_I , e_k \rangle = 0$

f_I is the frequence law of sample $\{i(s_\cdot) | s \in]n]\}$
The probability of this sample is:

$$\Pi \left\{ p_{i(s)}^0 \,\middle|\, s \in]n] \right\} = \text{Prob} \left\{ i(s) \,\middle|\, s \in]n] \right\}$$

and we have for the logarithmic derivative:

$$\partial \text{ Log Prob } \{i(s)\} \;/\; \partial \theta^k = \sum \left\{ (\partial p_{i(s)}^\theta / \partial \theta^k) / p_{i(s)}^0 \,\middle|\, s \in]n] \right\} \tag{32}$$
$$= \sum \quad n_i \; \partial p_i^\theta / \partial \theta^k) / p_i^0$$
$$= \quad n \; \langle f_I , e_k \rangle$$

Then $f_I^0 = p_I^0$ means that the probability of the sample has a zero derivate for $\theta = \theta_0$. If this is an extrema, we arrive at the estimation of the maximum of likelihood.

6 χ^2 DISTANCE IN CORRESPONDENCE ANALYSIS

6.1 Notations

We consider the notations of paragraph 4.1:
P(I) (resp. P(J)) is the simplex of probability laws in I(resp J)

We note $N(J) = \left\{(f_I^j, f^j), j\in J\right\} \subset P(I) \subset R_I$

$$N(I) = \left\{(f_J^i, f^i), i\in I\right\} \subset P(J) \subset R_J \qquad (33)$$

We define the quadratic form, positive definite: $m^{ii'} = \delta^i_{i'}/f_{i.}$ (34)
(Kronecker's delta)
We note $n_{ii'}$ the inverse of $m^{ii'}$ $\quad n_{ii'} = \delta^i_{i'}.f_{i.}$ (35)

6.2 Inertia

The quadratic form of inertia moment against origin for the set N(J) is given by:

$$\sigma_{II} = (f_I^J \times f_I^J)\, f_{.J} \quad (36) \quad \text{i.e. } \sigma_{ii'} = \sum_j f_i^j f_{i'}^j . f_{.j} = \Sigma_j (f_{ij} f_{ij}/f_{.j})$$

If we consider the inertia moment against the gravity center $f_{I.}$ of set N(J), we apply the Huyghen's theorem: (37)

$$\tau_{II} = \sigma_{II} - f_I o f_I \;,\; \tau_{ii'} = \sigma_{ii'} - f_{i.} f_{i'.} = \Sigma_j (f_i^j - f_{i.})(f_{i'}^j - f_{i'.}) f_{.j}$$

6.3 Definition of a factor

A function φ^I defined on I is a factor relative to eigenvalue λ if φ^I is an eigenvector of $m \, o \, \tau$ so that:

$$\Sigma_{i'i''} m^{ii''} \tau_{i''i'} \varphi^{i'} = \lambda \varphi^i \qquad \text{i.e. } \Sigma_{i'j} \varphi^{i'} f_{i'}^j f_j^i - \Sigma_{i'} \varphi^{i'} f_{i'.} = \lambda \varphi^i$$

$$\text{or } \varphi^I o f_I^J o f_J^I - (\varphi^I o f_I) \delta^I = \lambda \varphi^I \qquad (38)$$

We have $m \, o \, \sigma = f_I^J \, o \, f_J^I$

It is easy to prove that the factors form an orthogonal basis of R^I.
The expression of a vector X_I of R_I in this basis is the following:

$$X_I = \Sigma \left\{ (\varphi^I \, o \, X_I)\, (\delta_I^I . \varphi^I) \, o \, f_I \quad \forall \varphi^I \right\} \quad (39)$$

6.4 Comparison of the factors of N(I) and N(J)

φ^I factor: $\varphi^I \, o \, (f_I^J \, o \, f_J^I) = \lambda \varphi^I$, $(\varphi^I \, o \, f_I^J) \, o \, (f_J^I \, o \, f_I^J) = \lambda \, (\varphi^I \, o f_I^J)$
i.e. $\varphi^I \, o \, f_I^J$ factor of N(I) (40)

If we consider the norms in correspondence, we easily show that, for norm n_{JJ}, the scalar product of vectors of R^J, transformed of φ^I and ψ^I by f_I^J is λ times the scalar product of vectors φ^I and ψ^I of R^I; we have

$$\varphi^J = (1/\sqrt{\lambda(\varphi)})\, \varphi^I \, o \, f_I^J \quad \text{, so called transition formula} \qquad (41)$$

It can be shown that $0 \leqslant \lambda \leqslant 1$.

6.5 Expression of the input datas in the factor basis (Reconstitution formula)

$f_{Ij} = \Sigma(\varphi^I \circ f_{Ij})(\delta_I^I \times \varphi^I) \circ f_I \quad \forall \varphi^I$ with $\varphi^I \circ f_{Ij} = f_{j.}(\varphi^I . f_I^j)$

then $f_{ij} = f_i f_j \Sigma\{(\sqrt{\lambda(\varphi)}\ \varphi^i \varphi^j\}$ (42)

6.6 Representation of both I and J, analysis of N(I,J)

It proceeds of typical tensorial calculous that the analysis of N(K), K=I ∪ J delivers the same results as the analysis of N(I) or N(J), [1], tome 2, pp 159 - 164 .

7 CONTINUOUS CORRESPONDENCE ANALYSIS [5]

A generalized correspondence analysis can be studied rigorously in the continuous case, with I and J as Hilbert spaces. The extraction of eigenvectors of covariance matrix is changed into a spectral analysis of an integral operator. This operator can be decomposed as $U = T \cdot {}^tT$. Applying expanded Schmidt theorem, we arrive at the same reconstitution formula: $T(x,y) = T_1(x) T_2(y) \Sigma_i \sqrt{\lambda_i}\ \psi_i(x)\ \varphi_i(y)$ (43), T_1 and T_2 density measures.

8 COMPUTER PROGRAMS

During the past years, a lot of computer programs for the correspondence analysis and related topics have been developped by the team of Professor J.P. Benzecri, at the Institute of Statistics of the University of Paris (ISUP).
The program for correspondence analysis listed in [1] is broadly employed, and is characterized by its quickness (Golub Reinsch singular values decomposition, etc..). This main program is complete by programs resulting of recent advances in applications of the correspondence analysis: weightening of submatrices in case of analysis of heterogeneous datas, graphical outputs, calculous of inerti ellipsoids for subsets, correspondence analysis for very large matrices, regression on factors, and so on.
On the other hand, the consideration of the χ^2 distance in other domains of multivariate analysis has promoted the development of other program packages at the I.S.U.P : automatic classification, etc. Computer programs for automatic classification are presented in [1], tome 1 .
The properties of correspondence analysis allow to consider different kinds of input matrices of positive numbers: frequence arrays, contigence arrays, measure arrays, logic description arrays, intensit arrays, multiple arrays.
The domains where the applications of the correspondence analysis has given remarkable results are as large as the whole domain of multivariate analysis:
sociology, ethnology, climatology, marketing, industrial production datas, planning of experiments and so on.

9 GRAPHICAL OUTPUT. example

We observe hereafter a typical graphical output.It concerns an aluminium electrolyse plant.The variables involved are:production , intensity ,ausb Faraday efficiency ,span tension,anet net anode consumption,ener and enek specific energy consumption.The observations are monthly values of these variables during nearly three year: A discrimination between the months with or without energy shortage is obvious. Abundant numeric outputs help the interpretation of the results: distance between each variable and the origin,each observation and the origin,correlations of variables and observations with each factor,part of inertia extracted by each factor,and so on.

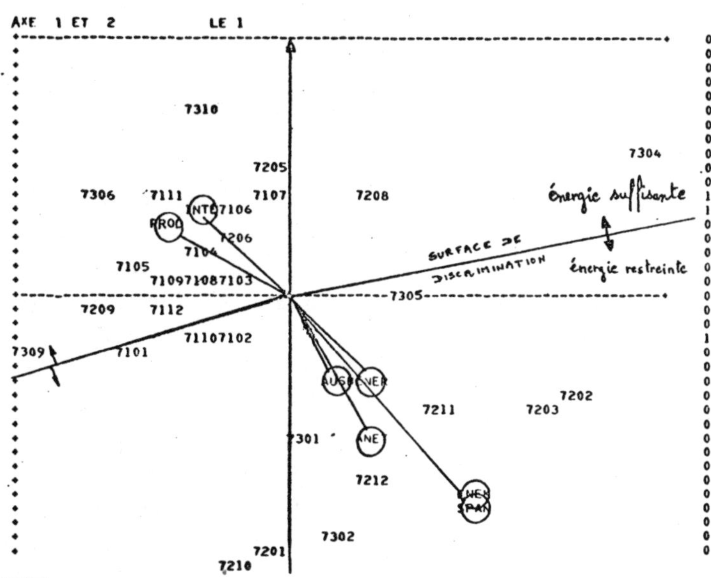

10 REFERENCES

[1] BENZECRI J.P. L'Analyse des données Tome 1 La Taxinomie, Tome 2 L'Analyse des correspondances.Dunod,Paris (1973)
[2] BOX,TIAO Bayesian inference in statistical analysis.Addison Wesley,Reading Mass. (1973)
[3] LEBART,FENELON Statistique et Informatique appliquées.Dunod, Paris (1971)
[4] MENDEL J.M.,FU K.S. Adaptive,Learning,and Pattern Recognition Systems.Academic Press N.Y. (1970)
[5] NAOURI J.C. Analyse factorielle des correspondances continue, in Pub. de l'Institut de Stat. Univ. Paris (1970) Vol XIX F 1
[6] RAO C.R. Linear statistical inference and its applications. Wiley,N.Y. (1965)
[7] ZACKS S. The theory of statistical inference.Wiley,N.Y. (1970)

A Statistical Method for Error Reduction in an Integration Process Involving Experimental Data

T. J. Stahlie, Amsterdam

SUMMARY

The numerical integration of a non-linear function $f(a_1(t), a_2(t),\ldots a_K(t))$ with mutually independent variables a_k is considered for the case that the initial conditionsfor the integration process and time series of a_k are determined by experiment.

In order to minimize the error in the result of the integration process a technique is developed based on the stochastic characteristics of the measured variables. This technique is applied to the problem of determining aircraft trajectories from in-flight recordings of acceleration components and attitude of the aircraft (Euler angles). Starting from given initial conditions for position and speed of the aircraft in addition to a (measured) end condition a minimum-error estimate of the trajectory is derived.

Special attention is paid to the following aspects of the technique: a comparison to simple non-stochastic error reduction techniques, the effect of errors in the additional ("updating") information on the resulting accuracies and the validation of the technique.

INTRODUCTION

In evaluating the reliability and accuracy of the end result of numerical computations an error analysis is nearly always inevitable. Such an analysis generally comprises an investigation of the different types of error (truncation error, measuring error, round-off error, etc.) which may be involved in the particular problem and subsequently a stochastic or deterministic quantification of the error sources. With the results obtained it can be checked whether the accuracy requirement of the problem considered has been satisfied. If the error in the outcome of the computations can not be accepted either a different calculation process or an error reduction technique should be applied.

A typical example of a numerical computation which leads to unacceptable errors in the results is the integration of a sampled function f(t) with given initial conditions. Because the sampled values of f(t) are obtained by experiment they are affected with errors; a process of open integration, applied on these data, will therefore lead to an error cumulation as a function of the independent variable t. It is clear that in such cases nearly always a need for error reduction exists.

This paper deals with a typical problem from the mentioned class of integration processes, namely the integration of a non-linear function $f(a_1(t), \ldots, a_K(t))$ of mutually independent variables a_k. For this problem a statistical method for error reduction is presented, which uses the specific characteristics of this integration process. After a discussion of the fundamentals of the technique an aerospace application is described concerning the determination of aircraft trajectories from in-flight recordings of aircraft acceleration and attitude. Some results of a flight-test program set up to obtain realistic data for the application of the technique and its validation will be shown. Finally, attention is paid to some practical aspects involved.

2 THE STATISTICAL METHOD FOR ERROR REDUCTION

2.1 Error analysis of the underlying problem

As has been mentioned earlier the numerical integration of a non-linear function $f(a_1(t),\ldots,a_K(t))$ over a varying time interval $[t_o,t_i]$ $(i=1,\ldots N)$ is considered. The variables a_k have been measured at discrete time intervals while the initial conditions for the integration process are also available.

In order to analyse the influence of the different errors with which the variables a_k are affected they have to be characterized (e.g. stochastic or deterministic, systematic or random). According to the "Gaussian law of errors" [Topping, 1963] the errors involved are considered as normally distributed stochastic variables; consequently only the characteristic parameters of the distribution have to be estimated.

The resulting error in f_i (being the value of the function f at time t_i) follows from relation:
the

$$f_i + \Delta f_i = f(a_1(t_i) + \Delta a_1, \ldots, a_K(t_i) + \Delta a_K) \tag{1}$$

An expansion of the function f in a Taylor series (omitting second and higher order terms) results into:

$$\Delta f_i = \left(\frac{\partial f}{\partial a_1}\right)_{t_i} \Delta a_1 + \ldots + \left(\frac{\partial f}{\partial a_K}\right)_{t_i} \Delta a_K \tag{2}$$

The probability distributions of Δa_k being known the distribution of Δf_i can be derived making use of the independency of the variables a_k. The propagation of the errors Δf_i (if necessary in combination with other error sources, such as truncation error, roundoff error) in the integration process can now be analyzed.

2.2 Characteristics of the error propagation

Three important characteristics of the error propagation in the problem considered can be mentioned:

- In equation (2) the partial derivatives are time dependent because of the non-linearity of the function f. This implies that the local errors Δf_i are a function of the sampled values of the variables a_k and the error in the end result of the integration can not be determined in advance (see also section 4).
- The error in the end result of the integration over the interval $[t_o, t_i]$ can be divided in two components, namely the error in the end result over the interval $[t_o, t_{i-1}]$ and the error over the last time interval Δt. This implies that the errors at time t_i and t_{i-1} are strongly correlated.
- The standard deviation of the error distribution in the end result at time t_i is according to the above greater than at time t_{i-1}; thus the error in the end result is a continuously increasing function of the integration interval.

The last mentioned characteristic underlines the necessity to apply an error reduction technique for the class of problems considered.

2.3 Fundamentals of the error reduction technique.

Having described the errors in terms of probability the above mentioned characteristic of correlation can be expressed by the coefficient ρ_{ij} $(i \neq j)$, indicating the correlation between the error at time t_i and at time t_j. This coefficient can be calculated from the given "local" error distributions. With the help of this correlation coefficient the joint normal distribution of the resulting error at time t_i and time t_j can be determined.

Based on the available information of the errors involved (e.g. correlation, joint distribution) a powerful method for error reduction can now be applied. The fundamentals of this method will be formulated below:

If at the end of the integration interval $[t_o, t_j]$ additional information can be obtained concerning the actual value of the local error, the conditional error distribution of the end result of the integration over the time intervals $[t_o, t_i]$ $i=o,\ldots,j-1$ can be derived. The mean value of this conditional distribution is the "most probable value of the error at time t_i based on the information of time t_j" and can be used for correcting the calculated integration results over $[t_o, t_i]$ which leads to "updated" results. The standard deviation of this conditional distribution indicates the accuracy of the updated results at time t_i.

In the following section mathematical and statistical details are described; it will be shown that because of the characteristics of the class of problems considered (both an increasing inaccuracy of the end result and a stronger correlation between successive errors in the end results in time) the error reduction is in general very powerful.

3 APPLICATION OF THE ERROR REDUCTION TECHNIQUE

The earlier defined problem (section 2.1) arises when an aircraft trajectory has to be determined from in-flight recordings of acceleration components ($a_1(t)$, $a_2(t)$ and $a_3(t)$) and attitude of the aircraft (the "Euler" angles: $a_4(t)$, $a_5(t)$ and $a_6(t)$). With given initial conditions for position and speed of the aircraft at time t_o the aircraft position x_i at time t_i $(i=o,\ldots N)$ can be calculated by solving the differential equation:

$$\frac{d^2x}{dt^2} = f\ (a_1(t), \ldots a_K(t)) \tag{3}$$

with

$$x(t_o) = x_o \text{ and } \left(\frac{dx}{dt}\right)_{t_o} = v_o$$

From the error analysis of the process of measuring, recording and processing of the experimental data the following error sources have been derived:

- Δx_o: the error in the (photographically determined) position x_o
- Δv_o: the error in the recorded speed v_o
- Δa_k: the overall error in the sampled values of the acceleration components and Euler angles.

Characterizing all these errors by appropriate probability distributions, the distribution of the error Δf_i can be derived according to eq. (2); further the error distribution $N(o, \sigma_{\Delta x_i})$ of the integration result over the interval $[t_o, t_i]$ (the aircraft position x_i) can be determined for the numerical integration procedure applied. For a detailed description of the error analysis and the calculation of the error distributions of the aircraft positions is referred to [Stahlie, 1972,a].

An actual example of the standard deviation $\sigma_{\Delta x_i}$, being a measure for the accuracy of the calculated aircraft position at time t_i, is shown in figure 1; the result presented is based on the trapezoidal rule as the integration procedure. Because the maximum allowable standard deviation of the error in the calculated aircraft trajectory for the application considered had to be less than 30 m the necessity of an error reduction technique was clear. In order to obtain a minimum-error estimate of the aircraft trajectory the statistical error reduction technique as described in section 2.3 was applied.

According to the fundamentals of this technique an accurate "update" position has to be established at time t_N being the end of the maximum integration interval of interest. The difference between the update position (derived from aerial photography) and the integrated, error-affected, position of the aircraft at time t_N provides information on the actual error at time t_N. Having determined the values of the correlation coefficients $\rho_{i,N}$ (between the error at time t_N and at time t_i $(i=o,\ldots,N)$), the conditional normal distribution of the error at time t_i can be calculated from the known error at time t_N denoted as Δx_N. This distribution has the following parameters:

$$\mu_{\Delta x_i | \Delta x_N} = \rho_{i,N} \cdot \frac{\sigma_{\Delta x_i}}{\sigma_{\Delta x_N}} \cdot \Delta x_N \qquad (4a)$$

$$\sigma_{\Delta x_i | \Delta x_N} = \sigma_{\Delta x_i} \sqrt{1 - \rho_{i,N}^2} \qquad (4b)$$

The original aircraft trajectory can now be corrected with the "most probable error $\mu_{\Delta x_i | \Delta x_N}$ based on the updating information Δx_N" resulting in an updated aircraft trajectory with local error distributions, according to:

$$N\ (o,\ \sigma_{\Delta x_i} \sqrt{1 - \rho_{i,N}^2}) \qquad (5)$$

It can be seen from eq. (5) that a strong correlation ($\rho_{i,N} \cong 1$) between the errors at time t_i and at time t_N provides a considerable error reduction. In figure 2 an example is shown of the correlation coefficient $\rho_{i,N}$ which is characteristic for the integration process considered, while figure 3 shows the values of $\sigma_{\Delta x_i} \sqrt{1-\rho_{i,N}^2}$ being the standard deviation of the local errors after error reduction (both figures 2 and 3 are based on the same data as figure 1).

De data presented in figure 3 are calculated under the assumption that the error in the updating information is zero. The effect of the latter error is dealt with in section 5.

The figures 4 and 5 show two typical examples of aircraft trajectories for the calculation of which the updating technique was applied. Besides the "integrated" trajectory and the "updated" one the figures also show the aircraft trajectory as measured with a high-accuracy radar altimeter. The good agreement between the "updated" and "radar" trajectories clearly illustrates the merits of the technique.

4 A COMPARISON WITH DETERMINISTIC ERROR REDUCTION TECHNIQUES

Error reduction techniques are often based on a deterministic estimate of a relation between error and time. The error propagation in time is then assumed to obey simple relations, such as a linear or quadratic one.

As a consequence of the typical character of the function f in the class of problems considered (see section 2.2) the error propagation is unpredictable and by consequence a simple relation between error and time can not be established. Moreover non-stochastic error reduction techniques do not provide information on the accuracy of the updated trajectory in contrary to the stochastic approach as described in this paper (see eq. (5)).

5 INFLUENCE OF THE ERROR IN THE "UPDATING" INFORMATION

As has been outlined in section 3 the "updating" information - i.e. the actual aircraft position - has to be established in order to determine at time t_N the error Δx_N. In the underlying application the aircraft position at time t_N is determined photographically. However, this "updating" information is also affected with an error; its influence on the error reduction will be discussed below.

It is assumed that the "updating" error can be characterized by a Gaussian distribution $N(o,\sigma_\varepsilon)$. According to the fundamentals of the error reduction technique the conditional probability distribution of the position error at time t_i can also be calculated if the error in the updating information at time t_N is taken into account. In that case the conditional distribution is a Gaussian one with parameters (compare eq. (4)):

$$\mu_{\Delta x_i|\Delta x_N} = \rho_{i,N} \frac{\sigma_{\Delta x_i} \sigma_{\Delta x_N}}{\sigma^2_{\Delta x_N} + \sigma^2_\varepsilon} \cdot \Delta x_N \tag{6a}$$

$$\sigma_{\Delta x_i|\Delta x_N} = \sigma_{\Delta x_i} \sqrt{\frac{\sigma^2_{\Delta x_N}(1-\rho^2_{i,N}) + \sigma^2_\varepsilon}{\sigma^2_{\Delta x_N} + \sigma^2_\varepsilon}} \tag{6b}$$

where σ_ε is the standard deviation of the error in the updating information. As has been mentioned in section 3 the mean value (6a) represents the most probable correction to be applied to the originally calculated aircraft trajectory and the standard deviation (6b) represents the resulting inaccuracy of the updated trajectory.

In order to illustrate the gain in accuracy after the updating compared to the original result the following equation is used:

$$\left\{\frac{\sigma_{\Delta x_i} - \sigma_{\Delta x_i|\Delta x_N}}{\sigma_{\Delta x_i}}\right\} \times 100 = \left\{1 - \sqrt{\frac{\sigma^2_{\Delta x_N}(1-\rho^2_{i,N}) + \sigma^2_\varepsilon}{\sigma^2_{\Delta x_N} + \sigma^2_\varepsilon}}\right\} \times 100 \tag{7}$$

Relation (7) is illustrated in figure 6 for different values of the correlation coefficient $\rho_{i,N}$. It should be noted that a strong correlation ($\rho_{i,N} \cong 1$) in combination with relative accurate updating information ($\sigma_{\varepsilon} \ll \sigma_{\Delta x_N}$) results in a considerable gain in accuracy at time t_i. From figure 2, which illustrates a typical relation between the correlation coefficient and the time, it can be concluded that for the class of problems considered a considerable gain in accuracy can be obtained.

6 CONCLUSION

It has been shown in the paper that by making use of the typical characteristics of a numerical computation process involving experimental data a useful technique can be developed to reduce the effect of the inaccuracy of the data. For this purpose a stochastic approach has been chosen. The application of this technique permitted the calculation of high-accuracy estimates of aircraft trajectories using data from a simple "strapdown" inertial measurement system [Stahlie, 1972, b].

7 REFERENCES

Topping, J. — Errors of observation and their treatment. Chapman and Hall, London, 1963.

Stahlie, T.J. — An updating procedure for correcting measured positions of an aircraft trajectory. National Aerospace Laboratory; NLR Memorandum VG-72-040 Dec. 1972(a).

Stahlie, T.J., Veldhuyzen, R.P. — The measurement of aircraft trajectories during manoeuvres of short duration with a simple strapdown inertial system. National Aerospace Laboratory; NLR Memorandum VG-72-031 Nov. 1972(b).

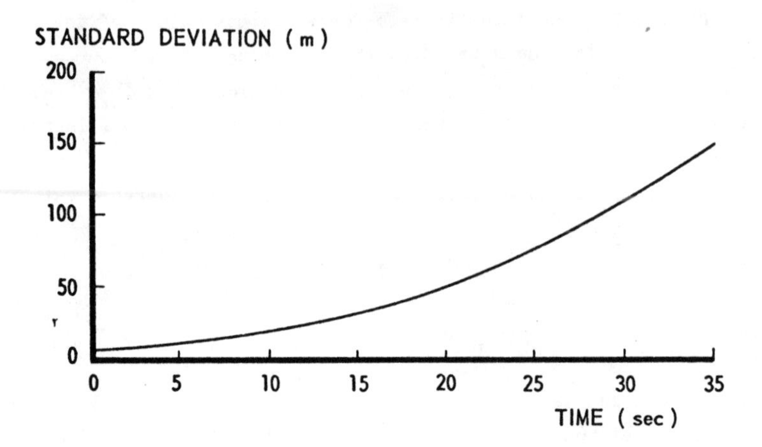

Fig. 1: Example of the standard deviation of the error in a calculated aircraft trajectory before the application of the error reduction technique.

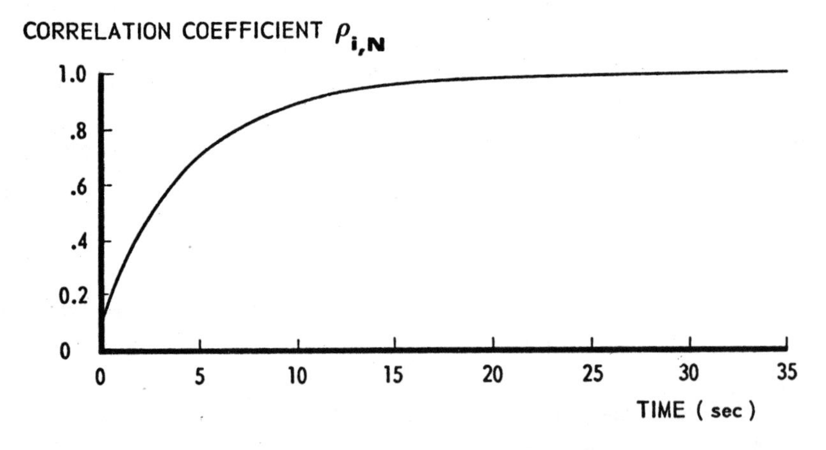

Fig. 2: Example of the correlation coefficient $\rho_{i,N}$ between the error at time t_i and at time t_N = 35 sec.

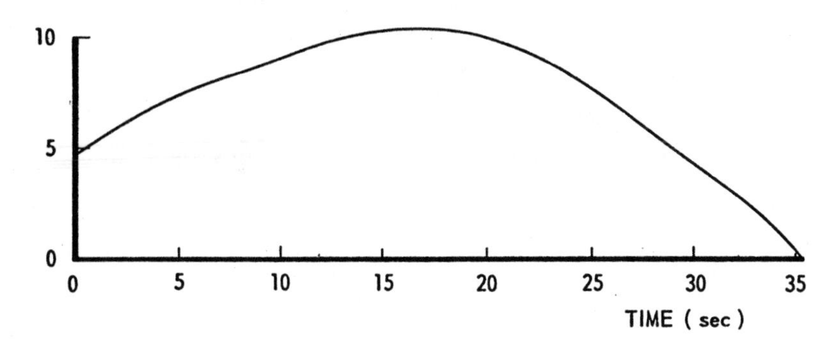

Fig. 3: Example of the standard deviation of the error in a calculated aircraft trajectory after the application of the error reduction technique (t_N = 35 sec)

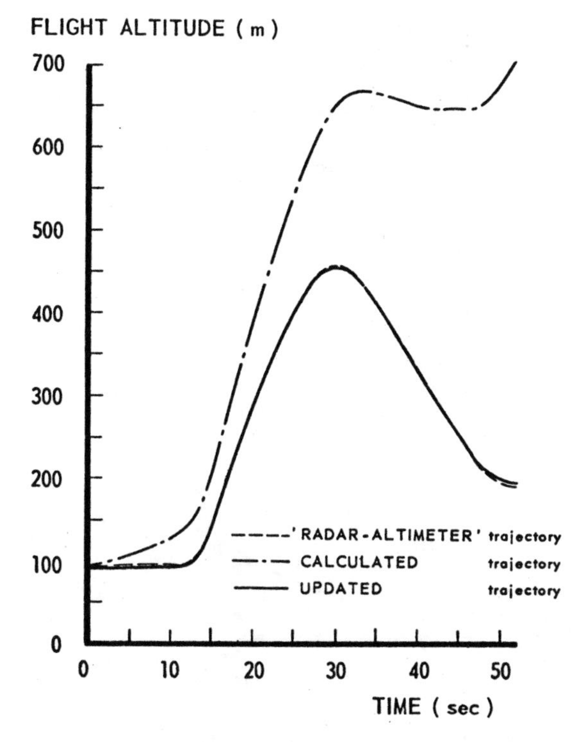

Fig. 4: Example of an aircraft trajectory, obtained after application of the statistical error reduction technique (pitch-up manoeuvre)

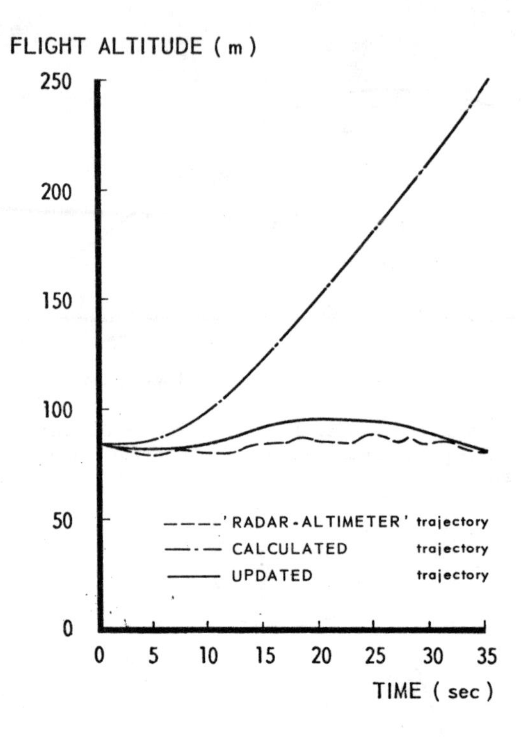

Fig. 5: Example of an aircraft trajectory, obtained after application of the statistical error reduction technique (level flight)

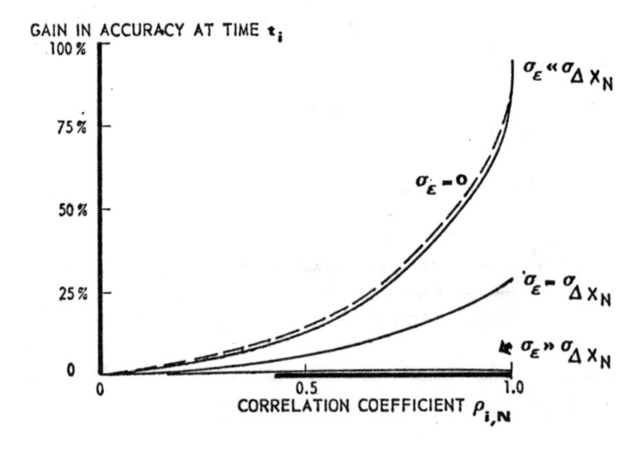

Fig. 6: The gain in accuracy at time t_i resulting from the updating as a function of the correlation coefficient for different "updating" accuracies

Estimating the Parameters of the Factor Analysis Model without the Usual Constraints of Positive Definitness

O. P. van Driel, H. J. Prins, and **G. W. Veltkamp,** Eindhoven

1. INTRODUCTION

The model for factor analysis described by LAWLEY and MAXWELL [1971, p. 6] is

$$\underline{x} = \Lambda \underline{f} + \underline{e}, \qquad (1)$$

where $\underline{x}$ is a vector of p components (test scores), $\underline{f}$ is a vector of $k (k < p)$ components (common factor scores), $\underline{e}$ is a vector of p components (specific part of the test scores) and Λ is a $p \times k$ matrix. The elements of Λ are called factor loadings.

It is assumed that $\underline{f}$ and $\underline{e}$ are normally distributed with $E(\underline{f}\underline{f}') = I$, $E(\underline{e}\underline{f}') = 0$ and $E(\underline{e}\underline{e}') = \Psi$, a diagonal matrix. Consequently $\underline{x}$ is normally distributed with mean zero and covariance matrix Σ defined as (LAWLEY and MAXWELL [1971, p. 6])

$$\Sigma = E(\underline{x}\underline{x}') = \Lambda\Lambda' + \Psi. \qquad (2)$$

This relation may be equivalently expressed as

$$\left.\begin{array}{l} \Sigma = \Pi + \Psi, \\ \text{where} \\ \quad \Pi \text{ is symmetric, positive semidefinite} \\ \quad\quad \text{and of rank} \leqslant k, \\ \quad \Psi \text{ is diagonal and positive definite,} \\ \quad \Sigma \text{ is symmetric and positive definite.} \end{array}\right\} \qquad (3)$$

The problem in factor analysis is to find from a given sample covariance matrix S an estimate $\hat{\Sigma}$, satisfying conditions (3), for the population covariance matrix, Σ_0. Henceforth the population parameters will be denoted by Σ_0, Π_0, Ψ_0, ..., where as Σ, Π, Ψ, ..., are mathematical variables. The estimate $\hat{\Sigma}$ is usually found by minimizing some "pseudo distance" of S and Σ, where Σ is constrained to the set specified by conditions (3). This may give rise to minima situated on the boundary of this set (for instance some diagonal elements of $\hat{\Psi}$ being zero), called improper solutions. Since an improper solution generally does not correspond to a stationary point of the "distance", statistical interpretation is difficult. To circumvent these difficulties, we propose that (at least until the interpretation stage) no constraints of positive (semi-) definiteness should be laid on the parameters Π and Ψ. Consequently we pose as our model

$$\left.\begin{array}{l} \Sigma = \Pi + \Psi, \\ \text{where} \\ \quad \Pi \text{ is symmetric with rank} \leq k, \\ \quad \Psi \text{ is diagonal}, \\ \quad \Sigma \text{ is symmetric and positive definite*)}. \end{array}\right\} \tag{4}$$

In the sequel we shall call matrices Σ that satisfy conditions (4) *admissible*. In the set of admissible Σs we distinguish the subset of *interpretable* Σs that satisfy conditions (3). With the generalization from model (3) to model (4) we now can, using confidence intervals, accept or reject the hypothesis that Σ_o is interpretable, even if the estimate $\hat{\Sigma}$ does not satisfy conditions (3). With this strategy we hope to find better interpretations for cases which give rise to improper solutions with the classical approach.

Section 2 will show in some detail how to obtain an admissible maximum likelihood estimate $\hat{\Sigma}$ for Σ_o. In section 3 the consequences for the interpretation are discussed. Section 4 deals with some examples formerly giving rise to improper solutions. Section 5 contains some concluding remarks.

2. MAXIMUM LIKELIHOOD ESTIMATES

In order to determine how well an admissible $\hat{\Sigma}$ fits a given matrix S a "distance function" must be chosen. From maximum likelihood theory it follows (LAWLEY and MAXWELL [1971, pp. 26 & 35]) that $\hat{\Sigma}$ has to minimize the function

$$F(\Sigma) = \operatorname{tr}(S\Sigma^{-1}) - p - \log \det(S\Sigma^{-1}). \tag{5}$$

It can be shown, using a compactness argument, that for every positive definite S this function has a proper minimum within the set of admissible Σs.

Similar to the classical approach the first step in the construction of a minimizing $\hat{\Sigma}$ is to minimize F as a function of Π for fixed Ψ. It follows from Courant's minimax theorems that for a given Ψ there exist admissible Σs if and only if Ψ has not more than k non-positive diagonal elements. These Ψs will be called *admissible*. The existence of a minimum of F for a fixed admissible Ψ can be shown.

Suppose F is minimal with respect to variations of Π in some "point" $\Sigma = \Pi + \Psi$, where Π has exactly rank k. Then Π and all sufficiently nearby Πs of rank $\leq k$ may be written as

$$\Pi = \Lambda J \Lambda', \tag{6}$$

where J is a $k \times k$ diagonal matrix with elements $+1$ or -1 and Λ is a $p \times k$ matrix. By choosing the symbol Λ we do not suggest that, in case J is not the unit matrix, Λ has an interpretation

---

*) The constraint of positive definiteness of Σ is retained since we wish to use the "pseudo distance" following from the maximum likelihood theory. For other possible pseudo distances (for instance, generalized least squares (JÖRESKOG and GOLDBERGER [1972])) this would not be necessary.

in terms of the model (1). Consequently, F has to be minimal as a function of Λ, hence $\partial F/\partial\Lambda = 0$. Exactly as in the classical case (LAWLEY and MAXWELL [1971, p. 27]) this stationarity condition is equivalent to

$$(\Sigma^{-1}S\Sigma^{-1} - \Sigma^{-1})\Lambda = 0 \tag{7}$$

The consequences of eq. (7) can be formulated in terms of the selection of one out of several eigenvalue problems (LAWLEY and MAXWELL [1971, p. 28], DERFLINGER [1968], JÖRESKOG and GOLDBERGER [1972]). Since S certainly is positive definite, whereas Ψ may be nondefinite, we prefer the general eigenvalue problem:

$$\Psi Z = SZ\Gamma, \quad Z'SZ = I, \tag{8}$$

where Z is the suitably normalized $p \times p$ matrix of generalized eigenvectors and Γ is the diagonal matrix of eigenvalues. As will be shown in appendix I every matrix Π satisfying eq. (7) is determined by a certain choice of k elements of Γ and the corresponding columns of Z, i.e. there exists an ordering of the eigenvalues and eigenvectors such that

$$\Pi = SZ\left(\begin{array}{c|c} I_1 - \Gamma_1 & \\ \hline & o \end{array}\right)Z'S, \tag{9}$$

where I_1 is the $k \times k$ unit matrix and Γ_1 is the $k \times k$ upper part of the partitioning

$$\Gamma = \left(\begin{array}{c|c} \Gamma_1 & \\ \hline & \Gamma_2 \end{array}\right) \cdot \tag{10}$$

From eqs. (8) and (9) it follows that

$$\Psi = SZ\left(\begin{array}{c|c} \Gamma_1 & \\ \hline & \Gamma_2 \end{array}\right)Z'S \tag{11}$$

and

$$\Sigma = SZ\left(\begin{array}{c|c} I_1 & \\ \hline & \Gamma_2 \end{array}\right)Z'S. \tag{12}$$

Substitution of Σ from eq. (12) in the function F gives

$$F(\Sigma) = \sum_{j=k+1}^{p} \phi(\gamma_j) \tag{13}$$

where $\phi(\gamma) = \gamma^{-1} - 1 + \log\gamma$ (since Σ is assumed to be admissible, $\gamma_j > o$ for $j > k$).

The restriction that the minimizing Π has exactly rank k can now be disposed of. If Π would have rank $k' < k$, eq. (13) would hold with k replaced by k'. Since $\phi(\gamma) > o$ for $\gamma \neq 1$, it follows that this Π can not minimize F unless $\gamma_j = 1$ for $j > k'$. Even then, however, the representation (9) is valid.

It follows from the results obtained so far that if all orderings of the eigenvalues and eigenvectors for which $\gamma_{k+1}, \ldots, \gamma_p$ are positive are considered, eqs. (9) and (12) represent all admissible matrices Π and Σ for which F is stationary with respect to variations of Π. Such orderings exist for every admissible Ψ, for it follows from eq. (8) that $\Gamma = Z'\Psi Z$, hence Sylvester's theorem (see for instance MARCUS and MINC [1964, p. 83]) asserts that Γ has the same number of positive diagonal elements as Ψ.

We now have to choose among all admissible Σs making F stationary that one for which F is minimal. Clearly this minimum is attained if the eigenvalues of eq. (8) are ordered in such a way that Γ_2 is positive definite and the right-hand side of eq. (13) is minimal. This ordering is characterized by the following conditions

$$\text{if } j > k \text{ then } \gamma_j > o, \qquad \text{if } i \geq k \text{ and } \gamma_i > o \text{ then } \phi(\gamma_i) \geq \phi(\gamma_j) \text{ for all } j > k. \quad (14)$$

This ordering shall be called *optimal*.

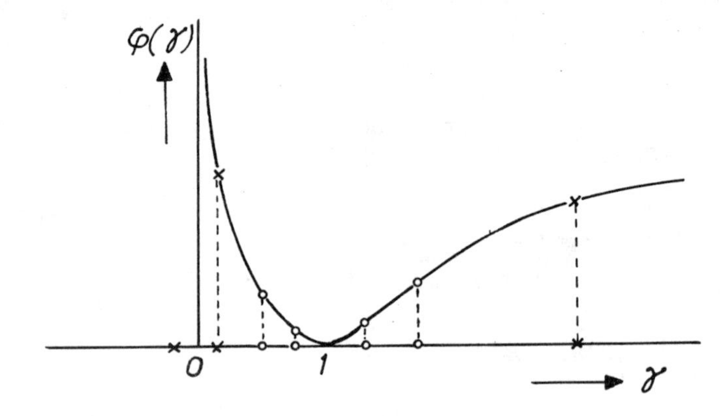

Fig. 1. Optimal ordering of eigenvalues as induced by minimisation of the right-hand side of eq. (13) $(p = 7,\ k = 3,\ m = 2)$

× eigenvalues $\gamma_1, \ldots, \gamma_k$

o eigenvalues $\gamma_{k+1}, \ldots, \gamma_p$

An optimal ordering can easily be found using the following result, illustrated by fig. 1. Let $\bar{\gamma}_1 \leq \bar{\gamma}_2 \leq \ldots \leq \bar{\gamma}_p$ be the eigenvalues of eq. (8) in increasing order. Then (if Ψ is admissible) there exists an m $(o \leq m \leq k)$ such that the set $\{\bar{\gamma}_{m+1}, \ldots, \bar{\gamma}_{m+p-k}\}$ coincides with the set $\{\gamma_{k+1}, \ldots, \gamma_p\}$ belonging to the optimal ordering. It should be remarked that the only difference between the classical approach and ours is that in the former $m = k$ (since Π has to be semidefinite), whereas we have to choose m such that the right-hand side of eq. (15) is minimal. The minimal value of F for a given admissible Ψ will be called $f(\Psi)$, hence

$$f(\Psi) = \sum_{j=m+1}^{m+p-k} \phi(\bar{\gamma}_j). \tag{15}$$

The second step in the construction of a minimizing $\hat{\Sigma}$ is, as in the classical approach, to minimize f over the set of admissible Ψs. Here it should be born in mind that both the eigenvalues $\bar{\gamma}_j$ and the value of m in eq. (15) depend on Ψ. This implies that f, although continuous, will not be differentiable everywhere in Ψ-space. It can be seen, however, that for every local minimum of f there exists a neighbourhood where m is constant and, consequently, f depends analytically on Ψ.

In our computerprogramme the technique to minimize $f(\Psi)$ is the Newton-Raphson method with line minimization. For this purpose we obtained numerically stable formulae for the first and second derivatives of f with respect to the elements of Ψ. As symmetric eigenvalue problems are easier to handle the value of f is calculated from the eigenvalues of the matrix $L^{-1}\Psi(L^{-1})'$, which are the same as the eigenvalues of eq. (8). LL' is the Choleski-decomposition of S.

3. CONSEQUENCES FOR THE INTERPRETATION

In section 2 maximum likelihood estimates $\hat{\Pi}$ and $\hat{\Psi}$ have been obtained which need not be positive definite and correspond to a minimum of $F(\Sigma)$. One of the necessary conditions for the asymptotic properties of maximum likelihood estimates to hold, is that this minimum is a stationary point, which in our model is true (in contrast to the classical approach where improper minima may exist). Assuming that these asymptotic properties hold it is possible

- to test whether Σ_0 is interpretable, i.e. to test whether Π_0 and Ψ_0 are positive definite,
- to give asymptotic confidence intervals for the elements of Π_0 and Ψ_0.

It seems that testing the interpretability of Σ_0 is most conveniently carried out by the use of $\hat{\Gamma}$, which is the estimate of the true Γ_0 corresponding to the true values Ψ_0 and Σ_0:

- testing for positive definiteness of Ψ_0 is equivalent with testing for positive definiteness of Γ_0,
- testing for positive semidefiniteness of Π_0 is equivalent with testing for positive semidefiniteness of the true value of $I_1 - \Gamma_1$.

If one of these tests indicates that Π_0 and/or Ψ_0 is non-definite the analysis of the data must be stopped as no further interpretation in terms of the original linear model (eq. (1)) is possible. This may, for instance, be caused by non-linear relations between factor scores and test scores.

If the model test does not reject the hypothesis that Σ_0 is interpretable, interest is centred on the values of the elements of Π_0 and Ψ_0 and consequently on the confidence intervals for these parameters. The matrix containing the asymptotic sampling variances and covariances of the maximum likelihood estimates of the elements of Π_0 and Ψ_0 is the inverse of the expected value of the "information matrix":

$$\frac{1}{2} n \begin{pmatrix} \frac{\partial^2 F}{\partial \Pi^2} & \frac{\partial^2 F}{\partial \Pi \partial \Psi} \\ \frac{\partial^2 F}{\partial \Psi \partial \Pi} & \frac{\partial^2 F}{\partial \Psi^2} \end{pmatrix}, \tag{16}$$

where n is the sample-size and $\partial/\partial\Pi$ indicates partial differentiation with respect to the independent parameters in Π. If only the confidence interval for Ψ_0 is needed it is not necessary to invert the whole matrix (16), as it can be shown that

$$\mathrm{cov}(\hat{\psi}_i, \hat{\psi}_j) = \frac{2}{n}\left(\frac{\partial^2 f}{\partial \Psi^2}\right)^{-1}_{ij} . \tag{17}$$

From the confidence intervals only those parts are used which correspond to interpretable Σs.

The first part of the model test as mentioned above may be alternatively carried out using the confidence intervals of Ψ_0. However, the elements of $\hat{\Psi}$ can not be used for testing the positive definiteness of Π_0 since, although $\psi_i > s_{ii}$ for some i certainly causes non-definiteness of Π, the opposite need not be true so we still need the covariance matrix of the elements of $\hat{\Gamma}$ for a complete model test.

4. IMPROPER SOLUTIONS

In factor analysis a solution is called improper if at least one component of $\hat{\Psi}$ is zero. Many of these improper solutions are found in the literature (TUMURA, FUKUTOMI and ASOO [1968], MATTSSON, OLSSON and ROSEN [1966]). We examined many cases that caused improper solutions when the classical approach was used. Our experience is that the following causes may give rise to improper solutions (in this section Σ and S are supposed to be scaled to correlation matrices, which transformation does not alter F):

- the true value of one (or more) of the $\hat{\psi}_i$ is so close to 0 or to 1 that the estimate tends to slip out of the interval [0,1],
- the variance of one (or more) of the $\hat{\psi}_i$ is large.

The following three sets of data illustrate the two kinds of causes separately:

1. IC production 11/08/72 (Table 1),
2. Random 13/09/72 - 165 B (Table 2),
3. Maxwell 1961 (LAWLEY and MAXWELL [1971, p. 44]).

Data 1. One of the purposes of our work on factor analysis is investigation of the processing of electrical components. Analyzing one of these sets of data (Table 1), using the classical

1								
.564	1							
-.506	-.720	1						
-.001	-.050	.072	1					
-.579	-.864	.824	.046	1				
.734	.440	-.455	.107	-.464	1			
.008	.009	-.062	.356	-.056	.132	1		
-.137	.121	-.040	.191	-.116	-.151	.151	1	
.119	.100	-.130	.796	-.181	.228	.421	.083	1

Table 1. Correlation matrix of IC production 11/08/72 (data 1; p = 9; n = 251).

1							
.520	1						
.519	.520	1					
.510	.486	.483	1				
.500	.473	.477	.456	1			
.531	.504	.491	.468	.476	1		
.506	.479	.468	.469	.498	.466	1	
.047	.063	.042	.041	.026	-.009	.038	1

Table 2. Correlation matrix of Random 13/09/72 - 165 B (data 2; p = 8; n = 800).

approach, for $k = 3$ an improper solution with $\hat{\psi}_5 = 0$ was found (column 2 in Table 3). The result of our approach was a proper minimum with $\hat{\psi}_5 = -.011$, and its standard deviation $\sigma_5 = .027$ (column 3 and 4 in Table 3).

Data 2. Artificial data are obtained by simulating drawings of $\underline{f}$ and $\underline{e}$ in eq. (1) for given Λ_0 and Ψ_0. We constructed data that yielded improper solutions when the classical approach was used. For one of these sets of data (Table 2) the classical approach for $k = 2$ gave an improper solution with $\hat{\psi}_8 = 0$ (column 6 in Table 3) and our approach resulted in a solution with $\hat{\psi}_8 = 1.030$, $\sigma_8 = .086$ (column 7 and 8 in Table 3), whereas the true value is $\psi_8 = .999$ (column 5 in Table 3). It may be concluded that if some ψ_i are close to 1 the usual approach may lead to a wrong conclusion.

To understand in detail the difference in this respect between the two models the shape of the function $f(\Psi)$ for data 2 is examined. As can be seen from Table 3 columns 6 and 7 it is sensible to vary ψ_8 only. Figure 2 shows that if m equals k (as is a constraint in the classical approach) the function $f(\Psi)$ has no proper minimum. Only if other values of m are admitted can a true minimum be obtained.

Data 3. This set of data is data 2 from the book of LAWLEY and MAXWELL [1971]. For $k = 4$ they found one improper solution with $\hat{\psi}_8 = 0$. Using several starts for the classical approach we found this and two other improper solutions (columns 2, 3 and 4 in Table 4). Using all 3 improper solutions as initial estimate for our approach only one minimum was found (column 5 in Table 4), which however corresponds to an uninterpretable Σ. As σ_8 is large the standard deviation of the factor loadings of variate 8 (the 8th row of Λ in eq. (1)) will be large as well, i.e. the relation between variate 8 and the other variates is indeterminable. In such cases the best thing to do seems to be to omit variate 8. The result is an interpretable solution (column 7 in Table 4). To our surprise we found, however, that omitting variates 6 or 9, respectively, also resulted in interpretable solutions (columns 9 and 11, respectively, in Table 4). It is interesting to note that the well determined elements of $\hat{\Psi}$ nearly coincide in these three cases.

i	data 1			data 2			
	$\hat{\Psi}_I$	$\hat{\Psi}_{II}$	σ_{II}	Ψ_o	$\hat{\Psi}_I$	$\hat{\Psi}_{II}$	σ_{II}
(1)	(2)	(3)	(4)	(5)	(6)	(7)	(8)
1	.261	.259	.083	.5	.450	.459	.032
2	.246	.254	.029	.5	.493	.502	.035
3	.314	.320	.033	.5	.503	.501	.030
4	.243	.242	.068	.5	.537	.535	.031
5	0	-.011	.027	.5	.536	.535	.031
6	.230	.233	.105	.5	.508	.588	.166
7	.805	.805	.073	.5	.533	.530	.031
8	.876	.877	.082	.999	0	1.030	.086
9	.112	.113	.074				

Table 3. Results for data 1 (columns 2, 3 and 4) and data 2 (columns 6, 7 and 8). Column 5 contains the true value Ψ_o for data 2. An index I indicates a result obtained with the classical approach. An index II indicates a result using our approach.

i	data 3					data 3, one variable omitted					
	$\hat{\Psi}_I$	$\hat{\Psi}_I$	$\hat{\Psi}_I$	$\hat{\Psi}_{II}$	σ_{II}	$\hat{\Psi}_{II}$	σ_{II}	$\hat{\Psi}_{II}$	σ_{II}	$\hat{\Psi}_{II}$	σ_{II}
(1)	(2)	(3)	(4)	(5)	(6)	(7)	(8)	(9)	(10)	(11)	(12)
1	.385	.373	.376	.399	.04	.397	.04	.377	.04	.378	.04
2	.623	.611	.602	.628	.06	.630	.06	.612	.06	.605	.06
3	.301	.307	.310	.299	.04	.298	.04	.307	.04	.310	.04
4	.638	.644	.645	.645	.04	.644	.04	.644	.04	.645	.04
5	.347	.406	.403	.329	.13	.324	.13	.402	.07	.401	.07
6	.778	0	.774	.813	.05	.802	.04			.773	.04
7	.286	.313	.304	.275	.04	.275	.05	.312	.04	.298	.04
8	0	.655	.665	2.179	2.49			.674	.04	.696	.04
9	.690	.675	0	.740	.05	.725	.04	.675	.04		
10	.600	.616	.599	.610	.04	.609	.05	.616	.04	.605	.04

Table 4. Results for data 3. Columns 2, 3 and 4 contain three improper solutions obtained by the classical approach. Column 5 contains the solution obtained by our approach. Column 7, 9 and 11 contain solutions obtained by omitting variates 8, 6 and 9, respectively. Columns 6,8,10 and 12 contain σ_is of the foregoing $\hat{\Psi}_i$s.

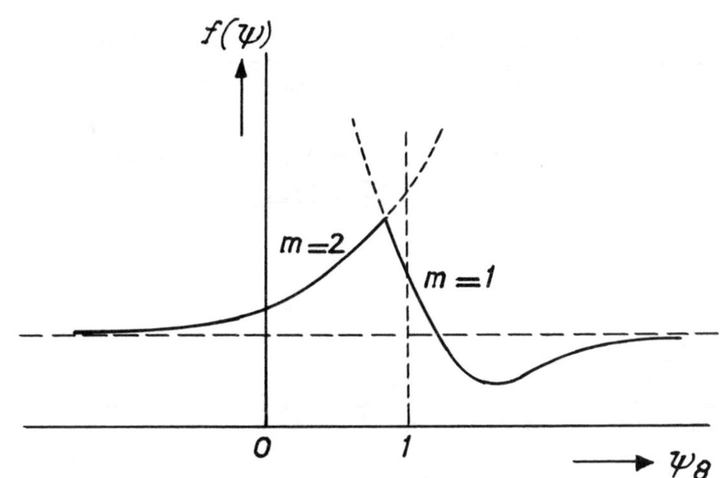

Fig. 2. $f(\Psi)$, data 2, if ψ_8 is changed.

for $m = 1$ $f(\Psi) = \sum_{j=2}^{7} \phi(\bar{\gamma}_j)$; for $m = 2$ $f(\Psi) = \sum_{j=3}^{8} \phi(\bar{\gamma}_j)$

5. FINAL REMARKS

In the foregoing sections we posed and illustrated several arguments for dropping the constraints of positive definiteness for the parameters Π and Ψ. Here we want to summarize the reasons for this decision:

- most of the sets of data that we analyzed with the classical approach yielded one or more improper solutions (for instance for data 3 we found for $k = 5$ five different improper solutions),
- the asymptotic properties of the maximum likelihood estimates hold only if the minimum of the function F is a stationary point, which for an improper solution generally is not true,
- a minimization problem without constraints is generally easier to handle than the corresponding problem with constraints.

Our approach of generalizing the model has the following advantages:

- it now becomes possible to test whether Σ_0 is interpretable, that is whether the model of factor analysis is in accordance with the given n and S,
- the χ^2-test, used for determining k, can now be relied on, as the solution $\hat{\Sigma}$ corresponds to a stationary point of the function F,
- improper solutions can now be better interpreted (immediately or omitting an appropriate variate).

A drawback of our method may seem to be that the maximum likelihood estimates $\hat{\Pi}$ and $\hat{\Psi}$ may not be interpretable in terms of the original model. By using confidence intervals this situation may be remedied by only considering that part of the simultaneous confidence interval for which Π and Ψ are positive semidefinite.

Apart from the contents of this paper the investigation yielded other points of interest, for instance

- the existence of more than one proper minimum (even in the interval [0,1]),
- an indication of the closeness of approximation of χ^2,
- the construction of an algorithm ensuring global convergence of the Newton-Raphson method,
- several formulae for the initial estimates of Ψ,
- results concerning the behaviour of the function f.

APPENDIX I.

Equation (7) is equivalent with

$$\Sigma S^{-1}\Lambda = \Lambda. \tag{18}$$

Let $\Lambda = SV$ where V is $p \times k$ with rank k. Then, using eqs. (4) and (6), eq. (18) turns into

$$\Psi V = SV(I_1 - JV'SV), \tag{19}$$

where I_1 is the $k \times k$ unit matrix. It follows from this equation that the columns of V span an invariant subspace of the mapping represented by the matrix $S^{-1}\Psi$. Hence there exists an ordering of the eigenvalues and eigenvectors of the general eigenvalueproblem (8) such that

$$V = Z_1 W \tag{20}$$

Where Z_1 consists of the first k columns of Z and W is a regular $k \times k$ matrix. Using that $\Psi Z_1 = SZ_1\Gamma_1$ and $Z_1'SZ_1 = I_1$, substitution of eq. (20) in eq. (19) yields

$$SZ_1\Gamma_1 W = SZ_1 W(I_1 - JW'W) = SZ_1(I_1 - WJW')W.$$

Or, since S and W are regular and Z_1 has full column rank,

$$\Gamma_1 = I_1 - WJW'.$$

Consequently,

$$\Pi = \Lambda J\Lambda' = SZ_1WJW'Z_1'S = SZ_1(I_1 - \Gamma_1)Z_1'S,$$

which may also be written as

$$\Pi = SZ\left(\begin{array}{c|c} I_1 - \Gamma_1 & \\ \hline & O \end{array}\right)Z'S. \tag{21}$$

Since eq. (8) implies $Z'\Psi Z = \Gamma$, the Σ corresponding to Π is

$$\Sigma = SZ\left(\begin{array}{c|c} I_1 & \\ \hline & \Gamma_2 \end{array}\right)Z'S.$$

Conversely, it follows by straight algebra that every matrix of form (21) satisfies the stationarity condition (7).

REFERENCES

DERFLINGER, G., Neue Iterationsverfahren in der Faktorenanalyse, Biometr. Z., 10, 58-75, 1968.

JÖRESKOG, K.G. and A.S. GOLDBERGER, Factor analysis by generalized least squares, Psychometri., 37, 243-260, 1972.

LAWLEY, D.N. and A.E. MAXWELL, Factor analysis as a statistical method, Butterworth, London, 1971.

MARCUS, M. and H. MINC, A survey of matrix theory and matrix inequalities, Prindle, Weber & Schmidt, Boston, Massachusetts, 1964.

MATTSSON, Å., U. OLSSON and M. ROSÉN, The maximum likelihood method in factor analysis with special consideration to the problem of improper solutions, Research Report, Institute of Statistics, University of Uppsala, 1966.

TUMURA, Y., K. FUKUTOMI and Y. ASOO, On the unique convergence of iterative procedures in factor analysis, TRU Math., 4, 52-59, 1968.

SUBJECT GROUP D

Simulation and Stochastic Processes

The Assignment of Students to School Branches of a Differentiated School System
– A Study of Some Strategies by Means of Simulation –

Fritz K. Bedall and Helmuth Zimmermann, Munich

1. INTRODUCTION

A constantly re-occurring situation in a society is the following: By means of a criterion, which can be established with sufficient reliability, certain life careers are suggested and/or assigned to members of such a society.

A typical setting fo such a decision-making procedure is the social subsystem "School". A stream of students encounters a branching point. School achievement is determined by a special examination or on the basis of the annual school progress report. Assignment of the students to certain school tracks is effected on the basis of their school achievement. The present simulation study is concerned with this process of assignment and related planning considerations.

In order to better facilitate the later description of the path to a solution, this study is limited to a particular point within the German school system. Without further development it is, however, possible to carry these procedures over to other similar cases both inside and outside the school system.

The location of interest within the German school system is the Orientation Step (Orientierungsstufe), which begins with the fifth school year and ends with the sixth. During this period a decision is reached as to whether a student will in the future attend the Intermediate School (Hauptschule), the Science or Modern Secondary School (Realschule) or the Classically-Oriented Secondary School (Gymnasium). The Orientation Step has not been generally introduced in the Federal Republic of Germany.

There exists a series of suggestions for the realization of the goals of the Orientation Step, that is, a maximally achievement-based classification of the students. We have tried to operationalize these suggestions for a Monte-Carlo-Study and have designated the corresponding strategies of educational planning with the term "planning factors".

It should be noted here that the number of school tracks that are available at the assignment point may also be examined as a planning factor. As the stream of students encounters an assignment point, the number of branches into which the main stream of students will be divided is surely of decisive importance. An awareness about the relationships can occasion a maximal differentiation, or - to the contrary - simply a two-branch division, of the stream of students at an assignment point.

The modification of the number of school branches is, however, a problematic matter in a real school system, even if it could be clearly shown that the actual organizational form is disadvantageous.

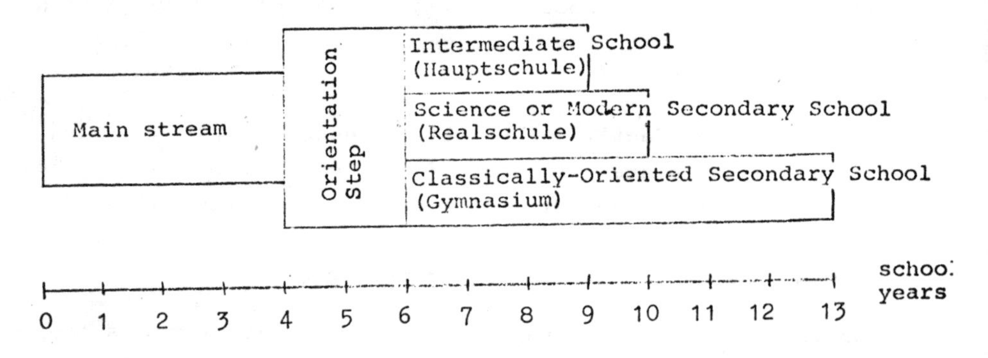

Illustration 1: The case studied in this work: A stream of students is split into three branches at an assignment point (Orientation Step) in accordance with assessed achievement.

For this reason we have studied this planning factor only in a preliminary investigation; we were able to ascertain its strong influence. For this report, however, we have assumed a fixed division of the students into three groups, as specified in the Orientation Step (Illustration 1).

2. SOURCES OF RANDOM VARIATION

Let us stress once more the main points of interest in this study:

- A stream of students encounters a point in the school system where a decision must be reached as to which path every individual student should take;
- School achievement serves as the basis for this decision.

The weak point of the above is the measurement of school achievement; generally no standardized testing methods are employed. The "testing methods" actually used have inferior psychometric characteristics, and their adequacy as prognosticators of future school performance has not been verified. But even if the most carefully devised testing procedures are utilized, measurement errors must be expected. In the field of the empirical social sciences this goes without saying.

2.1 The Reliability of a Test

Reliability is the extent to which test scores are reproduceable. In the present simulation, the reliability of a test was defined by repeated measurement. The more the results of further measurement are similar to the results of the original measurement, the higher is the reliability of the test. Any reliability below 0.50 is considered to be unacceptable. Such tests show different results in measuring the same objects at different times, even though no actual change of the measured object has occurred. It is just the measuring instrument which is so unprecise that undependable test results develop. In the social sciences a reliability coefficient of 0.90 is considered to be a high value. It guarantees to a certain extent that the object has been measured with sufficient precision. The maximum possible value of a reliability coefficient is 1.0 .

In this simulation study reliabilities within the range from 0.50 to 1.0 have been tolerated.

2.2 The Validity of a Test

A predictive validity coefficient is the characteristic value of a test which indicates how well the test measures what it is supposed to measure. The correlation between the test results and the criterion to be predicted by the test is considered to be the predictive validity coefficient. Theoretically the validity may be maximally as large as the root of the reliability of the test - that is, it may be larger than it. However, in practice anyone is happy who can observe validities of 0.40 to 0.60.

In this study we will encounter validities within the scale of one half of the reliability to the root of the reliability. This is a reasonable limitation, even though not a mandatory one.

3. PLANNING FACTORS

The State, who is the planner and supporter of the school system, is in the position to consider a number of planning factors in order that an adequate education in the appropriate school track may be achieved for as many students as possible. The challenge is to reach decisions concerning the school track to be selected as nearly as possible despite the aspect of faultiness of the achievement testing device.

3.1 Achievement Measurement Using One or Several Scales

We wish to differentiate here between achievement test results which in the end consist only of one figure, and those which produce several values. In the first instance one dimension of achievement is postulated; in the second instance, several. The assignment procedure of the school system is fundamentally different. If only one dimension is measured, boundary points will exist on this scale at certain clearly defined places which will permit a classification of the achievement test values (Illustration 2a). This is different in the case of several dimensions - as a simplification we postulate as many achievement branches as school branches, and thus have achievement tests which are specific to the school branches - in this case a measurement takes place in every dimension and the student will be finally directed to that school path to which he brings the best prerequisites (Illustration 2b). The dimensions should be independent.

(a) General Achievement Test

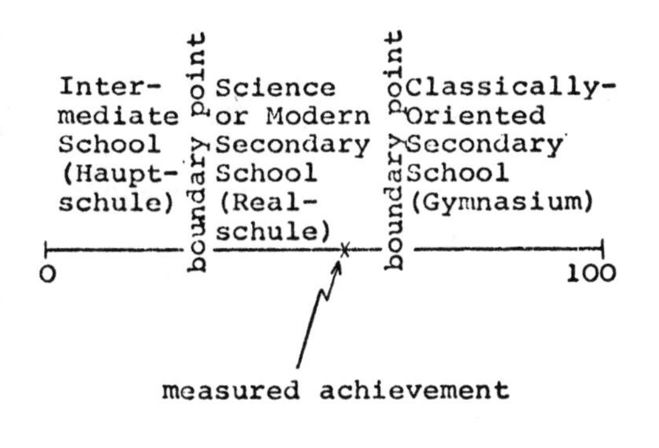

(b) Achievement Tests Specific to School Branches

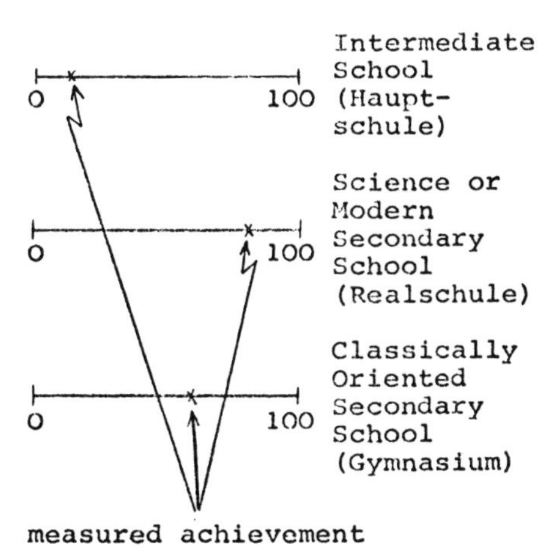

Illustration 2: Single- or multidimensional achievement measurement.

For our simulation study we assume that the scale from Illustratic 2a is identical with the scale for the Gymnasium in Illustration 2b. There is evidence in the German school system which indicates that this assumption is realistic.

3.2 Variation of Duration of the Observation Period Prior to the Final Assignment of a Student

Occasionally a brief decision procedure is felt to be too harsh, because, after all, the school track and thus also the life path of a human being is thereby determined, and this most often at a really early point in time. Efforts are being made to offer to the student more than one opportunity to demonstrate the kind and level of his achievement. This would mean that at any given time the student would attend the school branch to which he is assignec on the basis of the test; should, however, a later test suggest the correction of the previous decision, the student would take the new path. A prerequisite for the effectiveness of the planning factor just described is the permeability of the school branches during the time of observation.

The conditions examined by us extend from a solitary achievement assessment to up to five achievement measurements as the basis for the assignment decision.

This situation is in keeping with the contemplations in the Germa school system within the frame of the Orientation Step: A transition period of two years with a total of five achievement measurements at six-month intervals is under discussion.

3.3 Assignment of Students Through a Single Examination or by Consideration of Their Achievement History

The possibility of carrying out repeated achievement measurements suggests the notion of considering the results of earlier testing sessions in reaching a decision; thereby the achievement history of a student would be recognized. Otherwise, the assignment decision would be based on the test results of the last achievement evaluation.

It is a trivial statement that repeated achievement examinations are sensible only where all test results contribute to the decision reached. It is, however, fitting to keep in mind the extent of the consequences if this trivial relationship is not taken into account in computations.

The planning factor "consideration of achievement history" can have an influence only if more than one achievement measurement is made.

3.4 School Capacity: Massing of Students in the Intermediate School (Hauptschule) or in the Science or Modern Secondary School (Realschule)

Capacity realities and/or quotas depend essentially upon the ideological obligations of the school politicians. In this study we consider two conceptions of the capacities of a school system. One view sees in the Intermediate School the type of school to which the main body of students should go (Illustration 3a); the other view holds that it should be preferable to send a majority to the Science or Modern Secondary School (Illustration 3b).

3.5 School Capacity: Variation of the Share the Standard School has in Relationship to the Total Number of Students

While planning factor 3.4 governs the nature of the distribution capacities within the school system, planning factor 3.5 determines the degree to which planning factor 3.4 is carried through in the school system. We limit our report to three different shares of the standard school with respect to the total number of students available. These are the proportions 2/5, 3/5 and 4/5. The remaining number of students are distributed in equal shares between the other two school branches.

Intermediate School (Hauptschule)	Science or Modern Secondary School (Realschule)	Classically-Oriented Secondary School (Gymnasium)

(a) Intermediate School as standard school

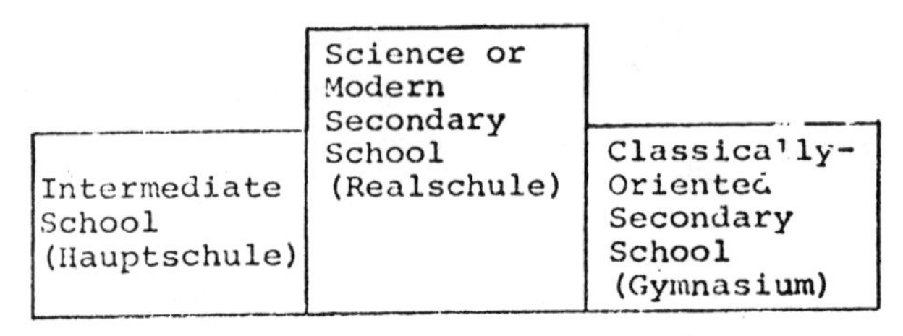

(b) Science or Modern Secondary School as standard school

Illustration 3: Massing of students in the Intermediate School or in the Science or Modern Secondary School

4. QUALITY CRITERIA FOR THE PLANNING FACTORS

It is time now to describe the variables involved in our study concerning the influences of the planning factors. For this purpose we make a distinction between the criteria which deal with the classification of the students and those which deal with the achievement of the classified students.

4.1 Classification of the Students

The criterion group specially considers planning factor 3.2, the observation period until the student is finally assigned to a school branch.

It is convenient to use three percentage figures to estimate the quality of a classification result of an assignment system: The portion (a) of students who are correctly classified at all opportunities, the portion (b) of students properly classified on the first and last occasions and the portion (c) of students who are properly classified on the last occasion.

In addition to the complements two linear combinations of these three groups are of special interest. b - a indicates the group of students whose reclassification was useless, and c - b indicates the group whose reclassification was warranted.

4.2 Achievement of the Students or Achievement Level of the School Branches

From the great number of possible quality criteria in this sector we select three portion values: For every school branch a determination is made as to the portion of students with good achievement for the entire simulation period. In the case of unidimensional measurement of achievement the quality of achievement for all three school branches is evaluated from the view of the Classically-Oriented Secondary School (Gymnasium).

5. SIMULATION

A simulation of a school system is possible under the conditions described above. The simulation process in detail is as follows. A student is created with an achievement level that is typical for him, uni- or multidimensional, depending upon the planning (see section 3.1). Achievement is $N(0,1^2)$-, that is, normally distributed with mean 0 and variance 1^2. To the original achievement values of every student an $N(0,1-r_{tt})$-distributed random value is added. This is the error variance, which takes into account the fact that no test can yield measurements that are one hundred percent accurate. r_{tt} represents the reliability of a test. The test point values thus determined are $N(0,2-r_{tt})$-distributed and decide the school track of a student.

The student will be further examined in regard to his achievement which he demonstrates in the school branch to which he has been assigned on the basis of his test results. The achievement values in question may be estimated by means of the validity of the test applied. Another $N(0,1-r_{xy}^2)$-distributed random value is added to the measured achievement values. This error variance allows for the fact that almost no test is capable of predicting with one hundred percent certainty a future behavior. r_{xy} is the validity of a test in regard to a particular behavior, i.e., the correlation between the test result and the school achievement in a certain future period. The expected achievement is $N(0,3-r_{tt}-r_{xy}^2)$-distributed.

In accordance with the usual grading system used in the Federal Republic of Germany the calculated scores can be divided into six intervals so that it will be possible to estimate the success of a student in the school track to which he is assigned. We prefer here a classification whereby the ratings 1 and 2 are consolidated as "good", 3 and 4 as "medium" and 5 and 6 as "bad". Grades are assigned in such a manner that in the case of $N(0,3-r_{tt}-r_{xy}^2)$, 25 percent of all ratings are "good", 25 percent are "bad" and fifty percent are "medium".

However, the various planning factors necessitate other distributions so that their influence may be quantitatively estimated.

If a student is scheduled for an additional examination date (see section 3.2) he once again traverses the algorithm. Of interest this time, however, is whether his achievement history (see section 3.3) is a relevant factor in the evaluation of his new examination results. If yes, the Spearman-Brown prophecy formula is applied. The test validity for the determination of the expected school achievement is handled similarly (Guilford, 1954).

In assigning the student, the capacity quotas are considered both in regard to nature (see section 3.4) and degree (see section 3.5).

6. COMPUTATIONAL PROCEDURE AND RESULTS

The data generated by simulation would ordinarily be analysed by a crossed, fivefactorial, orthogonal multivariate analysis of variance. Instead, however, we used a rather special computational method (Johnson and Koch, 1971). The goal was to describe most elegantly and simply the necessarily existing interactions among the factors. For this purpose we determined more or less arbitrarily a particular sequence of factors and set up a corresponding hierarchic design matrix for the planning factors so arranged. Further computation was made univariately, i.e., separately for each dependent variable (see section 4). Computations were made with Finn's MULTIVARIANCE Version V (1972).

The results appear in Tables 1 and 2.

Table 1 shows that only three planning factors have significant influence upon the classification of students. These are the planning factors

- 3.2 variation of observation period prior to the final assignment of a student (1, 2, 3, 4 or 5 achievement assessments)
- 3.3 assignment of students by a single examination or by consideration of achievement history
  and
- 3.5 school capacity: variation of the share of the standard school with respect to the total number of students (2/5, 3/5 or 4/5).

As expected, planning factor 3.2 played the deciding role; the number of correctly classified students rapidly diminishes with each achievement assessment. The same applies to the number of students who are correctly classified both at the beginning and

Table 1

Effects of three factors in a quasi-hierarchic design for three quality criteria for the classification of students. Only statistically significant influences of the planning factors are reported. All values in the body of the table are percentages times 10 (e.g., 681 = 68.1 %).

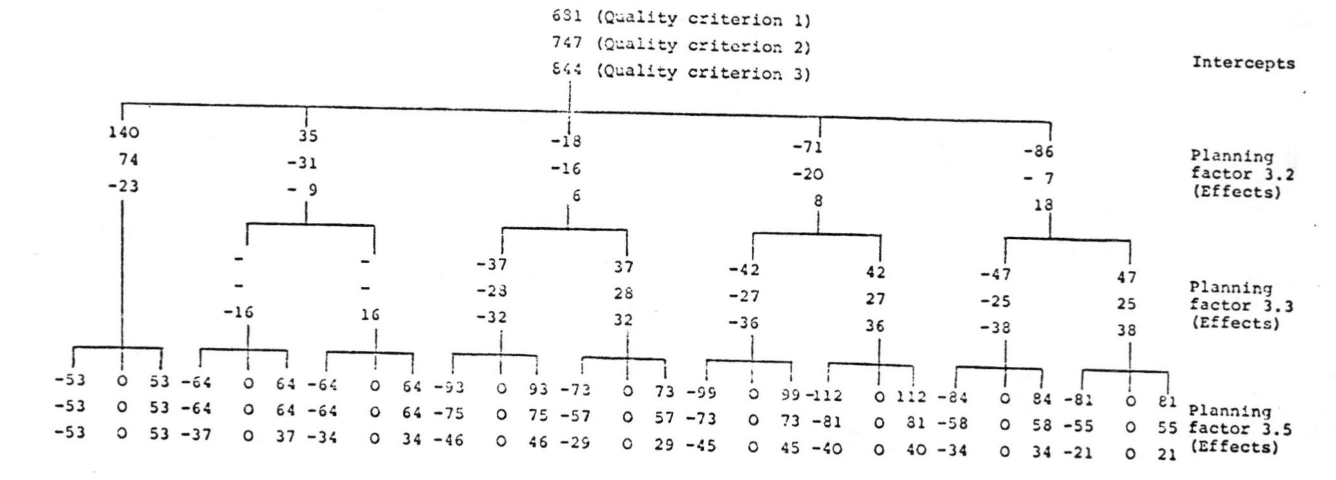

at the end of the observation period. On the other hand the number of students who at least in the end are found to be correctly classified grows only slowly with each achievement assessment. A maximum increase of 4.1 percent of the students finally classified stands in contrast to 22.6 percent of the students who in the same period must be reclassified, even though they might have been initially correctly classified had only one single measurement taken place. If only a single measurement is used - a sort of entrance examination - the percentage of correctly classified students is 68.1 + 14.0 = 74.7 + 7.4 = 84.4 - 2.3 = 82.1, quite high.

Planning factor 3.3 shows a certain effect if several achievement measurements are made. Its consideration may result in classification results that are up to 9.4 % better.

Capacity realities also have effects on the quality criteria regarding the classification of students. The general rule is: The more distinctly a school branch becomes the standard school, the better will be the classification success of the school system. This is perhaps obvious, yet it is advantageous to know the order of magnitude of this influence when it is compared with that of other factors.

The planning factors 3.1 and 3.4 are not of any demonstrable relevance in regard to quality criteria for a school system as far as the classification of students is concerned (see section 4.1).

The roles of the planning factors examined change considerably with regard to the quality criteria concerning achievement level of schools (see section 4.2).

Table 2 makes it clear that the influences of only three of the five planning factors are important; these are the planning factors

Table 2

Effects of three factors in a quasi-hierarchic design for three quality criteria in regard to the achievement level of schools. Only statistically significant influences of the planning factors are reported. All values in the body of the table are percentages times 10 (e.g., 263 = 26.3 %).

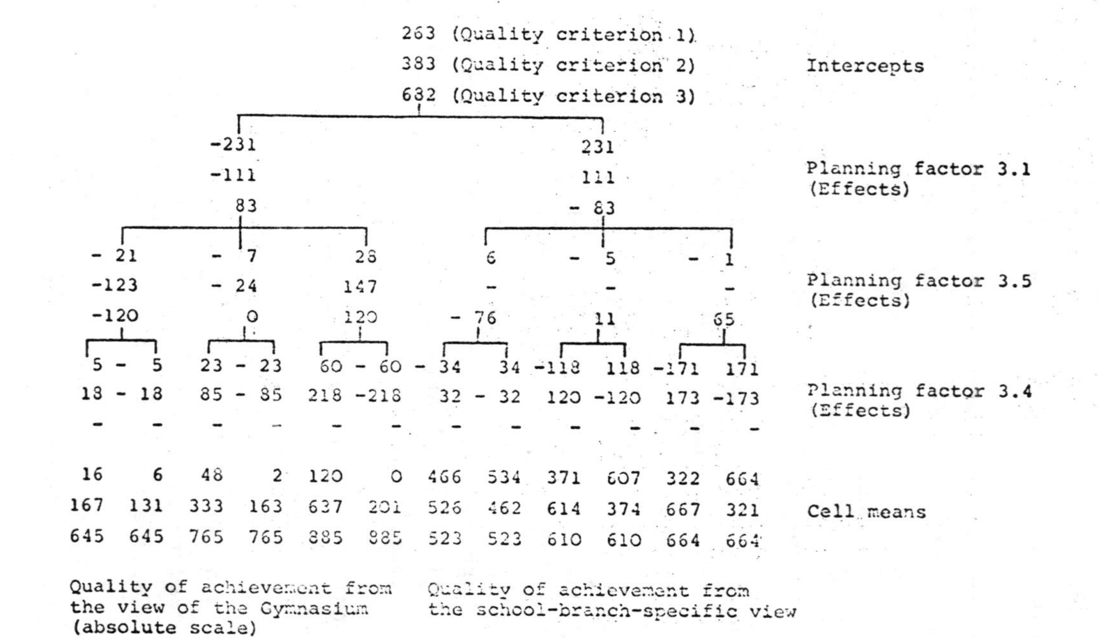

Intercepts: 263 (Quality criterion 1), 383 (Quality criterion 2), 632 (Quality criterion 3)

	Col 1	Col 2	Col 3	Col 4	Col 5	Col 6	Col 7	Col 8	Col 9	Col 10	Col 11	Col 12
Planning factor 3.1 (Effects)	-231						231					
	-111						111					
	83						- 83					
Planning factor 3.5 (Effects)	- 21		- 7		28		6		- 5		- 1	
	-123		- 24		147		-		-		-	
	-120		0		120		- 76		11		65	
Planning factor 3.4 (Effects)	5	- 5	23	- 23	60	- 60	- 34	34	-118	118	-171	171
	18	- 18	85	- 85	218	-218	32	- 32	120	-120	173	-173
	-	-	-	-	-	-	-	-	-	-	-	-
Cell means	16	6	48	2	120	0	466	534	371	607	322	664
	167	131	333	163	637	201	526	462	614	374	667	321
	645	645	765	765	885	885	523	523	610	610	664	664

Columns 1–6: Quality of achievement from the view of the Gymnasium (absolute scale)

Columns 7–12: Quality of achievement from the school-branch-specific view

- 3.1 achievement measurement with one or with several scales;
- 3.5 school capacity: variation of the share of the standard school with respect to the total number of students (2/5, 3/5 or 4/5)

  and
- 3.4 school capacity: massing of students in the Intermediate School or in the Science or Modern Secondary School.

If the achievement of the students in the various school branches in a school system were estimated on an absolute scale, the results using unidimensional achievement measurement (planning factor 3.1) would be catastrophic for the Intermediate School and the Science or Modern Secondary School. However, it is actually the case that there appears in every school branch a specific standard level, relative to which good achievement can be judged. It is difficult to define such standard levels for the simulated apportionments; we were not able to find unequivocal transformations for the purpose of a realistic comparison.

For this reason we plan, in a later study, to distribute the students of a school system in school branches in accordance with an average grade from three school-branch-specific achievement tests. Because the achievement level of every student in his respective school branch would then be known, there would also exist a realistic value for the portion of the students with good achievement in each school branch. Because of time constraints we were not able to incorporate this approach into this report.

In the following we forgo an interpretation of the results with respect to unidimensional achievement measurement.

The situation is completely otherwise if achievement is determined on a school-branch-specific, and therefore multidimensional, basis. Tending to mass in all school branches are students with achievement good for that school branch. With multidimensional achievement assessment (planning factor 3.1) one can, as a rule, expect far more than 25 % of the students to have good grades.

Both planning factors 3.5 and 3.4, which involve capacity assumptions in computation, can have a modifying effect. The condition of being a standard school (planning factor 3.4) always has a disadvantageous effect on the level of a school; this effect is demonstrable as a trend, too (planning factor 3.5).

The achievement level of the Classically-Oriented Secondary School (Gymnasium) is not affected by planning factor 3.4.

The planning factors 3.2 (prolongation of the observation period) and 3.3 (consideration of the achievement history) play no demonstrable role for the expected achievement of the students (see section 4.2).

7. SUMMARY

An assignment mechanism of a school system was simulated. Students of a certain school age were assigned to three school branches with the help of achievement tests. The measurement of the students' achievement, and thus also their assignment, was done under the influence of five planning factors. The effects of the planning factors upon the school tracks of the students and the expected achievement level of the schools were to be determined along with an estimation of their orders of magnitude.

The following planning factors were used:

- uni- or multidimensional achievement measurement
- shorter or longer observation period
- consideration or non-consideration of achievement history
- capacity expectations with respect to nature
  and
- capacity expectations with respect to degree.

In regard to the school tracks of the students during the observation period, the influence of the length of the observation period (number of achievement measurements) was confirmed. A small gain in the number of students who had been properly classified in the end was contrasted by a fivefold number of students who had to be re-classified. The administrative waste of time and money appears to be considerable.

A consideration of the achievement history of a student during the observation period can successfully counteract the negative influence of the time factor.

Capacity expectations can decisively modify the relationships; the more a school branch develops into a standard school, the greater will be the number of correct classification decisions.

The nature of the achievement measurement and the knowledge of which school branch is functioning as standard school are not relevant in connection with the school tracks of students. No pertinent influence could be identified.

The achievement levels of the schools depend to an essential degree upon the nature of achievement measurement; multidimensional achievement measurement is to be preferred.

In the case of multidimensional achievement measurement it proved to be a disadvantage to have one distinct standard school within the school system; the achievement level of the standard school becomes unavoidably lowered. The degree of the level lowering is directly related to the degree to which the standard school is realized.

The length of the observation period prior to the final classification of the student, with or without consideration of his achievement history, has no demonstrable influence upon the achievement level of a school branch.

8. ACKNOWLEDGEMENT

We thank the management of the Leibniz Computing Center of the Bavarian Academy of Sciences in Munich for the computing time they generously made available to us.

9 REFERENCES

Finn, Jeremy D.
MULTIVARIANCE
Univariate and Multivariate Analysis of Variance, Covariance, and Regression. A FORTRAN IV Program. Version V.
Department of Educational Psychology, State University of New York at Buffalo
National Educational Resources, Inc.
Ann Arbor 1972

Guilford, Joy P.
Psychometric Methods
McGraw-Hill
New York 1954

Johnson, William D.; Koch, Gary G.
A note on the weighted least squares analysis of the Ries-Smith contingency table data.
Technometrics, Vol. 13, 438 - 477, 1971

Some Applications of Simulation in an Iron- and Steel-Works Laboratory

P. Booster, Ijmuiden

SOME APPLICATIONS OF SIMULATION IN AN IRON- AND STEEL WORKS LABORATORY

The practical application of statistical methods in research- and control of an iron- and steel works laboratory can be greatly advanced by using appropriate simulation techniques, even though perhaps complicated direct mathematical techniques of approach may be possible. Sometimes the time necessary for the development of a more fundamental solution is not available. In such cases, a simulation procedure, provided that in some way its reliability can be surveyed, is desirable.
In this contribution, three different topics will be considered:

1. An acceptance-test procedure for high quality steel for reinforcing of concrete;
2. Correlation of two variables which ara both submitted to random errors;
3. The statistical power of two outlier-tests.

1. ACCEPTANCE TESTING OF STEEL BARS FOR REINFORCING OF CONCRETE

The acceptance requirements for the mechanical properties resulting from the tensile test: yield strength, tensile strength and allongation, as they have been prescribed by certain surveying authorities [1] can be summarized following flowchart fig. 1.

fig. 1 .

The operating characteristic curve of the first step, requiring :

$$\bar{x}_6 \geqslant G + 2 s_6 \qquad (1)$$

can be calculated as an application of the non-central t- distribution, but the <u>conditional</u> addition of a retest and the requirement that :

$$\bar{x}_{12} \geqslant G + 1.8 s_{12} \qquad (2)$$

for all trials together, makes straightforward computation of the O.C. curve impossible - or at least extremely complicated.
Moreover, in case of rejection after the second step, it was permitted to remove

the maximum value, in order to prevent the material from being rejected due to an "extremely good' result, causing however a more or less excessive value for the standard error s_{12}. It was allowed to repeat this remarkable truncation-procedure six times before a definite decision about rejection or acceptance of the material had to be taken.

It is quite obvious that only a simulation procedure enables the estimation of the OC-curve of this acceptance test.

Previous to the programming of the simulation, a large amount of available trials, usually executed with 3 - 5 repetitions, provided fair estimates of the experimental variances σ^2 for the yield- and tensile strength and for the allongation. Standardisation of the requirements (1) and (2) to:

$$\frac{\bar{x}_6 - \mu}{\sigma} \geq \frac{G - \mu}{\sigma} + 2 \frac{s_6}{\sigma} \qquad (3)$$

$$\frac{\bar{x}_{12} - \mu}{\sigma} \geq \frac{G - \mu}{\sigma} + 1{,}8 \frac{s_{12}}{\sigma} \qquad (4)$$

simplified the simulation process to the generation of 6 random values from a normal N(0,1) distribution, eventually followed by a second and a third group of 6 random values, representing the retests.

Fig. 2 shows the results.

Fig. 2

The points represent the estimates of the probabilities of acceptance, obtained from 500 simulations each.

As to the simple sample system, the theoretical OC-curve has been calculated; the diagram shows a very close agreement between theory and simulation, indicating a sufficient reliability of the simulation process.

Obviously, an eventual second retest will contribute hardly anything to the power of the testing procedure. Furthermore, it is remarkable that the removal of 'extreme good results' from an initial total of 12 tests will have practically the same effect on the power of the testing procedure as the admittance of a third series of 6 tests.

Table 1 shows that in the region of 'doubtful qualities' there are rare cases leading indeed to acceptance after removal of the maximum admitted number of six 'extremely good results' .

Table 1

2. CORRELATION OF TWO VARIABLES, WHICH ARE BOTH SUBMITTED TO RANDOM ERRORS

When two variables, X and Y , which are correlated with one another, are both submitted to random errors, but when it known that elimination of these errors will result in a pure functional relationship, it can be shown that in general the correlation coefficient ρ becomes equal to 1 when the variances of the dinominator are diminished by the variances of the random errors:

$$\rho^2 = \frac{(cov. XY)^2}{\sigma_x^2 \cdot \sigma_y^2} \qquad 1 = \frac{(cov. XY)^2}{(\sigma_x^2 - \sigma_{ox}^2)(\sigma_y^2 - \sigma_{oy}^2)} \tag{5}$$

If at least the ratio:

$$q = \frac{\sigma_{oy}^2}{\sigma_{ox}^2} \tag{6}$$

is known, estimates of these variance components can be found by solving a quadratic equation for σ_{ox}^2 . Let this estimate be equal to u .
From a given collection of n empirical pairs of values (X, Y), this estimate u can be solved as the relevant root of the equation:

$$\frac{\{\sum_i (x_i - \bar{x})(y_i - \bar{y})\}^2/(n-1)^2}{(s_x^2 - u)(s_y^2 - qu)} = \frac{\{\sum_i x_i' y_i'\}^2/(n-1)^2}{(s_x^2 - u)(s_y^2 - qu)} = 1 \tag{7}$$

The problem te be solved is, which function of u can be considered as an unbiased estimate of σ_{ox}^2 and what can be said about its probability function.
Without loss of generality, it is supposed that the linear relationship between X and Y can be presented as: $Y = X$, in other words, the empirical points (X_i, Y_i) scatter about a 45°-line.
The simulation procedure consists of a triple use of a library-subroutine generating random normal numbers A , ε_1 and ε_2 giving the values:

$$\begin{aligned} X_i &= A_i + \varepsilon_{1i} \\ Y_i &= A_i + \varepsilon_{2i} \end{aligned} \qquad = 1, . . . , n \tag{8}$$

The random normal numbers are chosen from normal distributions with the following parameters:

A_i from $N(0, \sqrt{p})$
ε_{1i} from $N(0, 1)$
ε_{2i} from $N(0, \sqrt{q})$

Table 2 shows a survey of the choice of the numerical values of the parameters.

- - - - - - - - - - -
Table 2
- - - - - - - - - - -

Each run consisted of 250 simulations, producing the following results with known expected values:

a) the conventional correlationcoefficient:

$$r^2 = \frac{\{\sum_i x_i' y_i'/(n-1)\}^2}{s_x^2 \cdot s_y^2} \quad ; \quad \rho^2 = \frac{Cov.XY}{(\sigma_A^2 + \sigma_{\varepsilon_1}^2)(\sigma_A^2 + \sigma_{\varepsilon_2}^2)} = \frac{p^2}{(p+1)(p+q)} \tag{9}$$

b) the regressioncoefficient:

$$b = \frac{\sum x_i' y_i'/(n-1)}{s_x^2 - u} = \frac{s_y^2 - qu}{\sum x_i' y_i'/(n-1)} \quad ; \quad \beta = \frac{Cov\,XY}{\sigma_A^2} = \frac{\sigma_A^2}{Cov\,XY} = 1 \tag{10}$$

c) the error-variance of X :

$$s_{0x}^2 = \varphi(u) \tag{11}$$

which function has to meet the requirement that its expected value $\sigma_{0x}^2 = \sigma_{\varepsilon_1}^2 = 1$.

d) the variance ratios:

$$c_1 = s_x^2/(p+1) \quad ; \quad c_2 = s_y^2/(p+q) \tag{12}$$

with expected values both equal to 1 and probabilitydistributions corresponding to $\chi^2/(n-1)$.

e) the Fisher-tranformation of the correlationcoefficient:

$$z = \frac{1}{2} \ln\left(\frac{1+r}{1-r}\right) \tag{13}$$

approximately normally distributed with parameters

$$N\left[\frac{1}{2} \ln\left(\frac{1+\rho}{1-\rho}\right) + \frac{\rho}{2(n+1)} \quad ; \quad \frac{1}{\sqrt{(n-3)}}\right] \tag{14}$$

The frequency-distributions of c_1 , c_2 and z are meaningful to verify wether the simulation process by which A , ε_1 and ε_2 have been generated, was sufficiently random.

Finally, after the execution of each run of 250 simulations, the correlation-matrix of the five variables r , b , u , c_1 and c_2 was calculated. It turned out that the principal variable u was not correlated with any of the variables b , c_1 or c_2 .

Witin the context of this paper, we restrict our considerations to the simulated values of the error-variance, presented by u .

The mean values of u differ systematically from the expected value: $\sigma_{0x}^2 = 1$, as table 3 shows for all simulated parametercombinations of table 2.

- - - - - - - - - - -

Table 3

- - - - - - - - - - -

This systematic difference can be practically eliminated by the removal of one degree of freedom:

$$s_{0x}^2 \backsim \frac{(n-1)\,u}{(n-2)} \tag{15}$$

For all simulations with $n = 10$, fig. 3 shows the sumfrequencies, obviously in excellent agreement with the theoretical $F(n-2;\infty)$-distribution, from which it follows that the estimation method of the residual error is meaningful.

- - - - - - - - - - - -
fig 3 and fig 4
- - - - - - - - - - - -

Fig. 4 shows the result of an application: a series of 30 steel samples was analysed on their oxygen content by two laboratories. Both laboratories applied the same analytical method, but one of them executed the determinations in twofold sothat in eq. (7), $q = 2$. The residual error is presented by an ellipse with axes equal to + and - 2 s_{0x} and $\sqrt{2}.s_{0x}$, respectively. The tangents parallel to the regressionline indicate the scatter of the diagram.
It is our experience that this method of regression -calculation is rather sensitive for outliers among the points-scatter. Graphical survey by means of a scatter-diagram is necessary.

3. THE STATISTICAL POWER OF TWO OUTLIER-TESTS
The detection of outliers within small series of empirical measurements without the use of any additional information of the population-dispersion can be performed by several statistical tests, of which the well-known test of DIXON [2] requires practically no computational work, and the test of DOORNBOS [3] using Student's t , can be easily programmed.
Practical experience with either of these tests shows however that the elimination of an extreme value, after being detected as an 'outlier' , often results in estimates of the dispersion which from a physical point of view seem te be very unlikely.
Statistically speaking, an outlier has to be considered as an element belonging to a different population than the other elements. This implies directly that the extreme value of a series of measurements does not necessarily belong to that 'strange' population: any other value could belong to it as well, particularly if the probability-distributions of the two populations overlap one another. So the question arises wether a test regarding the extreme values which include the given series of measurements can give sufficient information about the existence of a strange population.

By means of simulation, the following strategy has been developed:
assume that two normal populations with equal variaces are in the game, but that their population-means differ by a certain value δ .

Draw $(n-1)$ random values from the distribution $N(\mu;\sigma)$ and one random element from the distribution $N(\mu+\delta;\sigma)$. Simulate this procedure a great number of times and derive thereafter the following probabilities:
a) the chance that the 'strange element' is really an extreme value;
b) the chance that it will be detected as an outlier;
c) the chance that any other element will be detected as an outlier .

Without any loss of generality, it was assumed that the population-variance σ^2 is equal to 1. The mean μ has been arbitrarily chosen as $\mu = 5$, thus avoiding the existence of negative values within the samples.
The change that the 'strange element' is really an extreme value can be calcula directly without simulation by numerical solution of the integral:

$$P\left[X_{EX} \geq X_{max}\right] = n\int_{-\infty}^{+\infty}\left(F(x)\right)^{n-1} \cdot \left(1-F(x-\delta)\right) \cdot \varphi(x)\,dx \qquad (16)$$

This theoretical result can be used to check the validity of the simulation-process. Table 4 shows that there exists a satisfactory relationship between the results of theory and simulation.

The two remaining questions are only soluble by means of a simulation-process, using again random drawings from a normal population with known parameters. The following input-values were chosen:

σ = 1
n = 5 8 12 16
δ = 1,0 1,5 2,0 2,5 3,0 4,0 5,0
α = 0,05 0,10 (significance level of the test)

For one case, viz $n = 8$; $\alpha = 0{,}05$, table 5 shows the results of 7 series of 500 simulations for the test of DOORNBOS.
Not before $\delta \geq 4{,}0$ ($\geq 4\sigma$), the 'strange element' will be identical to 'extreme value' .
The detected outliers fall apart in three groups:
EX.U = the strange element is an outlier
MAX = the maximum value) of the 'normal elements' is an outlier
MIN = the minimum value)

Table 5

The identification of MAX or MIN as outliers is an error of the first kind : a correct null-hypothesis has been rejected.
If the test fails to identify EX as an outlier, an error of the second kind will occur: a false null-hypothesis has not been rejected:

$$\beta = \frac{500 - EX.U}{500} \quad \text{or} \quad 1-\beta = \frac{EX.U}{500} \qquad (17)$$

The total probability of identifying an outlier is:

$$\alpha + (1-\beta) = \frac{EX.U + MAX + MIN}{500} \qquad (18)$$

which, for $\delta = 0$, becomes equal to α and for extreme values of δ tends to $(1-\beta)$.
Thus the probability-values $(\alpha + (1-\beta))$ are the bases of the power-curves of these tests.
Fig 5, a,b,c and d show the final results. For very small samples ($n \leqslant 5$), an outlier-test based on nothing more than the information of the sample itself has hardly any meaning.
It is remarkable that the tests of DIXON and DOORNBOS have practically identical discriminative power curves.

4. REFERENCES

[1] Bulletin officiel du Ministère de l'Equipement et du Logement, France. Fascicule spécial no.68/3 quater; circulaire no.12 du 8.février, 1968.

[2] DIXON,W.J. and F.J. MASSEY (1957): Introduction to Statistical Analysis, 2nd Ed., p.275/276; McGraw Hill New York / Kōgakusha, Tokyo.

[3] DOORNBOS,R and H.J. PRINS (1958): Indagationes Mathematicae 20, p.38 - 46.

==================

Table 1: Number of accepted tests from 500 simulations.

Procedure: test + retest + removal of extreme values

$\frac{G-\mu}{\sigma}$	Accepted after:								Total	
	first test ($n = 6$)	second test ($n = 12$)	Removal of extreme values						Accepted tests	Probability of acceptance
			1	2	3	4	5	6		
0	0	0	0	0	0	0	0	0	0	0
– 0,8	35	6	3	3	3	1	2	2	55	11,0
– 1,6	173	70	20	13	8	4	6	1	295	59,0
– 2,4	384	85	8	4	0	0	0	1	482	96,4
– 3,2	483	17	–	–	–	–	–	–	500	100
– 4,0	498	2	–	–	–	–	–	–	500	100

Table 2: Parameters for the regression-problem

n →	10	15	20	30	50
p	4	4	4	4	4
q	1 2	1 2	1 2	1 2	1 2
p	6				
q	1 2				
p	10				
q	1 2				
p	16				
q	4				

Table 3: Mean values of u and s_{0x}^2

n	p	q	u	$s_{0x}^2 = \frac{n-1}{n-2} \cdot u$
10	4	1 2	0,9065 0,8950	1,020 1,007
15	4	1 2	0,9324 0,9280	1,004 0,9994
20	4	1 2	0,9361 0,9312	0,9881 0,9829
30	4	1 2	0,9524 0,9540	0,9864 0,9880
50	4	1 2	0,9682 0,9677	0,9884 0,9878
10	6	1 2	0,9113 0,9030	1,025 1,016
10	10	1 2	0,9147 0,9082	1,029 1,022
10	16	4	0,9037	1,017

Table 4: Probability that "strange" element is an extreme value

Difference δ	Number of EX per 1000 simulations							
	$n=5$		$n=8$		$n=12$		$n=16$	
	Sim	Int	Sim	Int	Sim	Int	Sim	Int
1,0	491	449	387	361	304	294	247	251
1,5	643	614	557	530	486	459	395	410
2,0	795	758	718	691	647	630	614	585
2,5	884	867	837	822	781	777	757	743
3,0	935	935	920	910	904	883	867	861
4,0	990	990	987	985	982	979	969	974
5,0	999	999	1000	998	998	998	999	997

Sim = results of simulation process
Int = results of integration:

$$P\left[EX \geq X_{max}\right] = n \int_{-\infty}^{+\infty} \left(F(x)\right)^{n-1} \cdot \left(1 - F(x-\delta)\right) \cdot \varphi(x)\, dx$$

$F(x)$ = prob.function } of N (5,1)
$\varphi(x)$ = prob.density }

Table 5: Results of simulation process

$\alpha = 0,05$
$n = 8$

DOORNBOS-test

Difference δ	Per 500 simulations: EX	EX.U	MAX	MIN	$1-\beta$	$\alpha+(1-\beta)$
1,0	186	16	21	27	0,032	0,130
1,5	278	43	7	17	0,086	0,134
2,0	364	70	4	14	0,140	0,176
2,5	411	97		5	0,194	0,204
3,0	460	175		7	0,350	0,364
4,0	497	306		1	0,612	0,614
5,0	500	415			0,830	0,830

$$1-\beta = \frac{EX.U}{500}$$

$$\alpha+(1-\beta) = \frac{EX.U + MAX + MIN}{500}$$

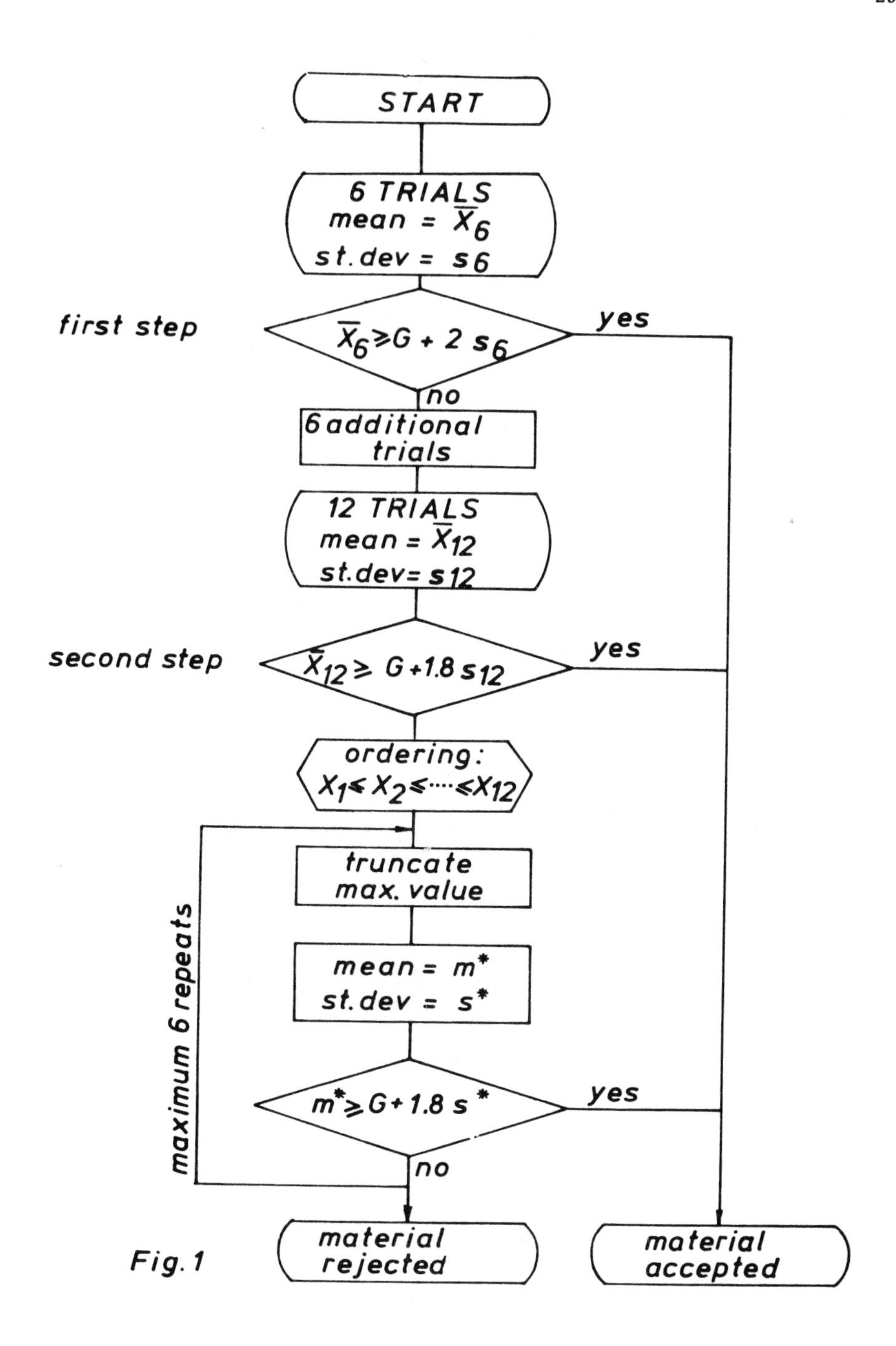
START
6 TRIALS
mean = $\bar{X}_6$
st.dev = s6
first step
$\bar{X}_6 \geq G + 2 s_6$
yes
no
6 additional trials
12 TRIALS
mean = $\bar{X}_{12}$
st.dev= s12
second step
$\bar{X}_{12} \geq G + 1.8 s_{12}$
yes
ordering:
$X_1 \leq X_2 \leq \cdots \leq X_{12}$
truncate max. value
mean = m^*
st.dev = s^*
maximum 6 repeats
$m^* \geq G + 1.8 s^*$
yes
no
material rejected
material accepted

Fig. 1

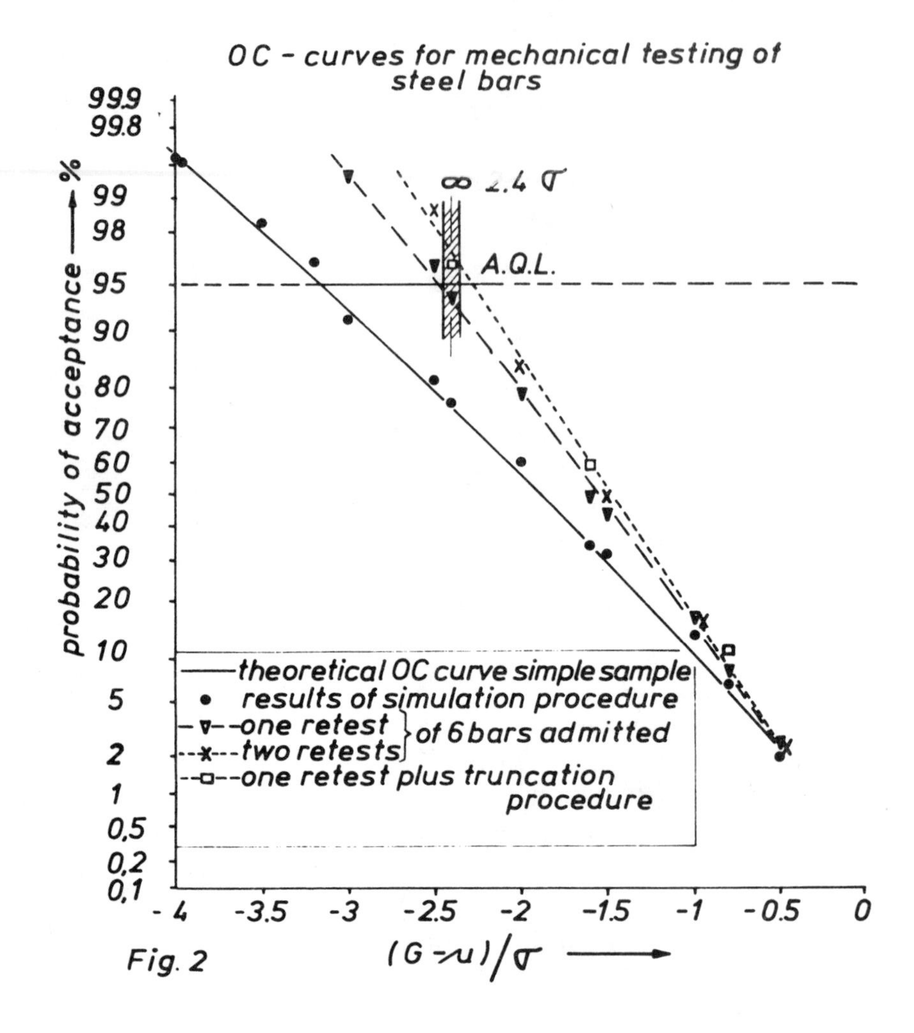
OC - curves for mechanical testing of steel bars
probability of acceptance ⟶ %
99.9
99.8
99
98
95
90
80
70
60
50
40
30
20
10
5
2
1
0,5
0,2
0,1
∞ 2.4 σ
A.Q.L.
theoretical OC curve simple sample
results of simulation procedure
one retest
two retests
of 6 bars admitted
one retest plus truncation procedure
- 4
-3.5
-3
-2.5
-2
-1.5
-1
-0.5
0
(G - μ)/σ ⟶

Fig. 2

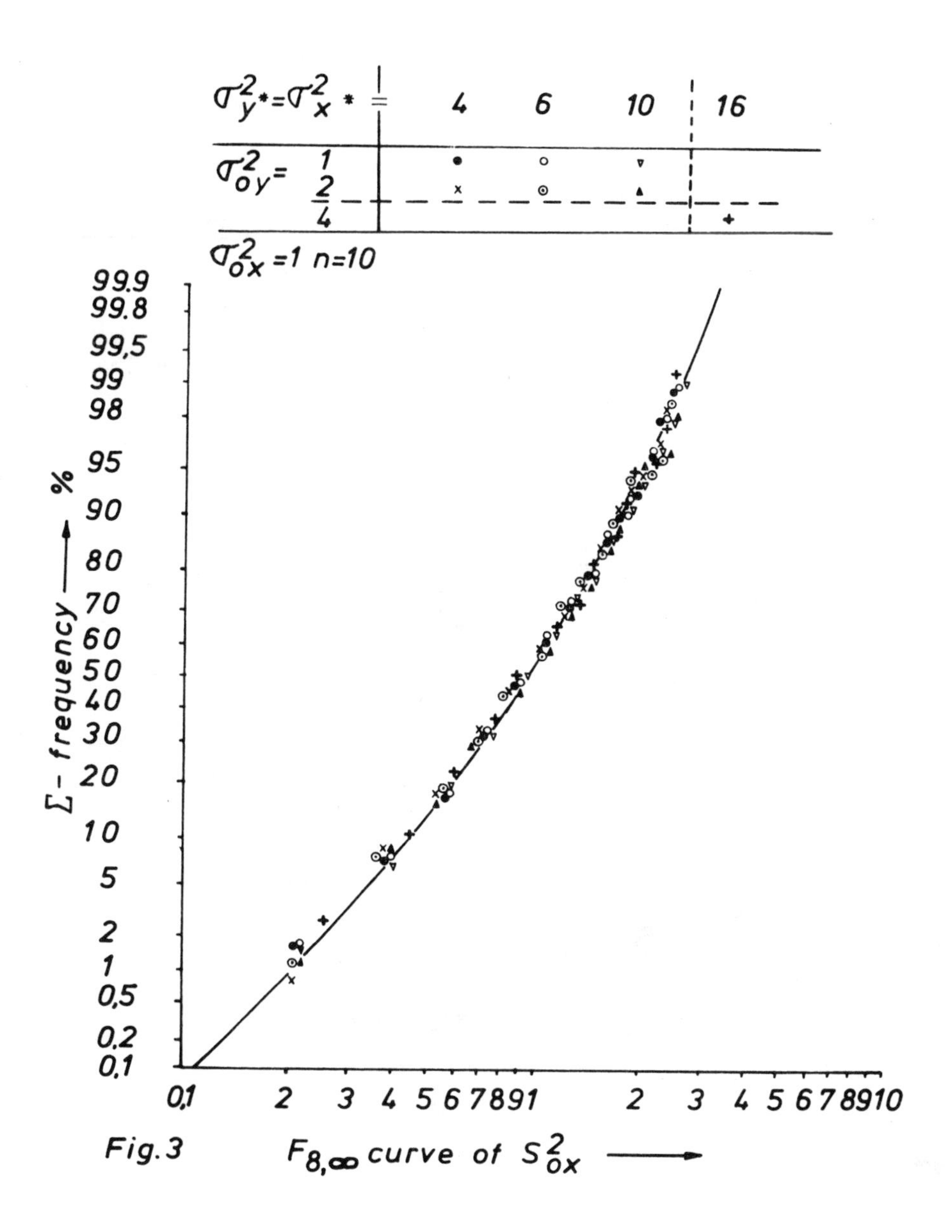

Fig. 3 $F_{8,\infty}$ curve of S^2_{ox} →

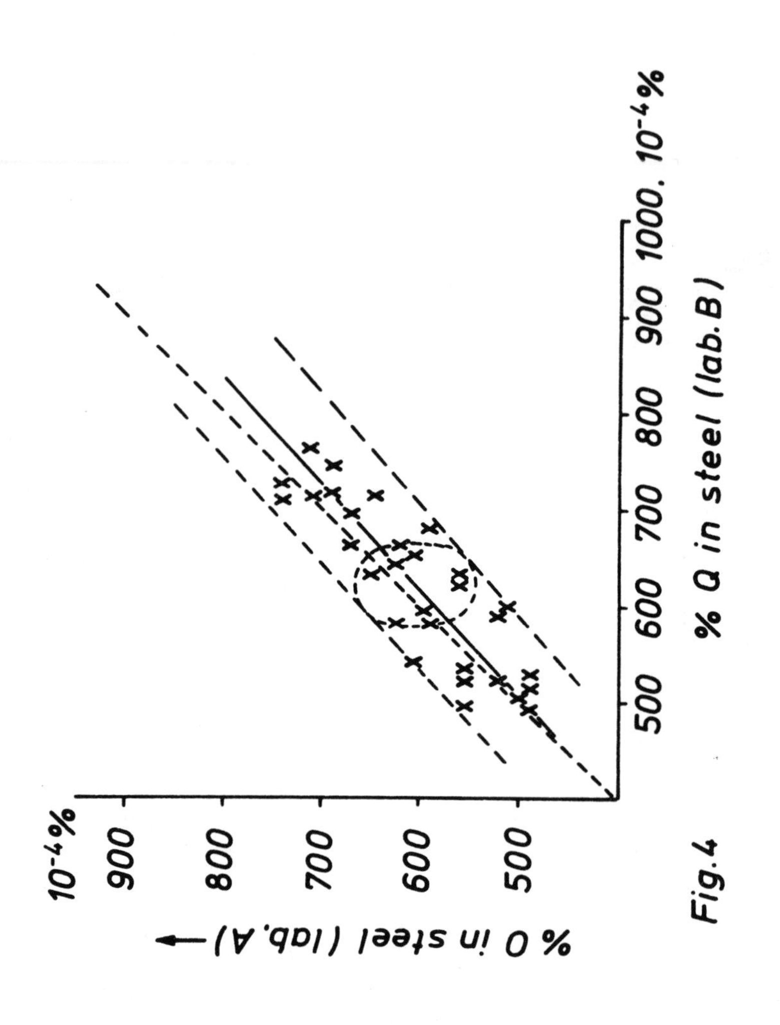

Fig.4

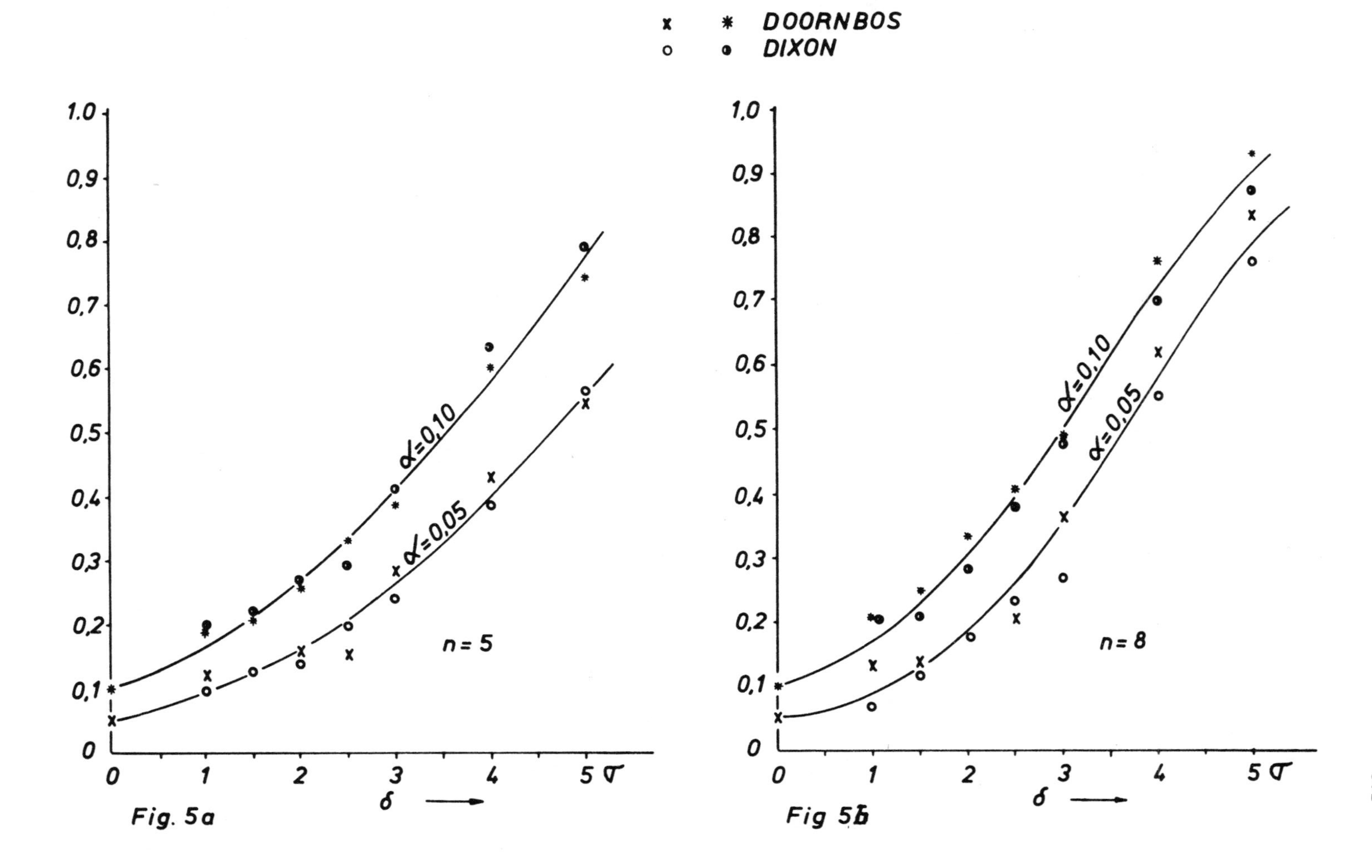

Power curves of outlier-tests:
x * DOORNBOS
o ● DIXON
α=0,10
α=0,05
n=5
n=8
δ
5σ

Fig. 5a

Fig 5b

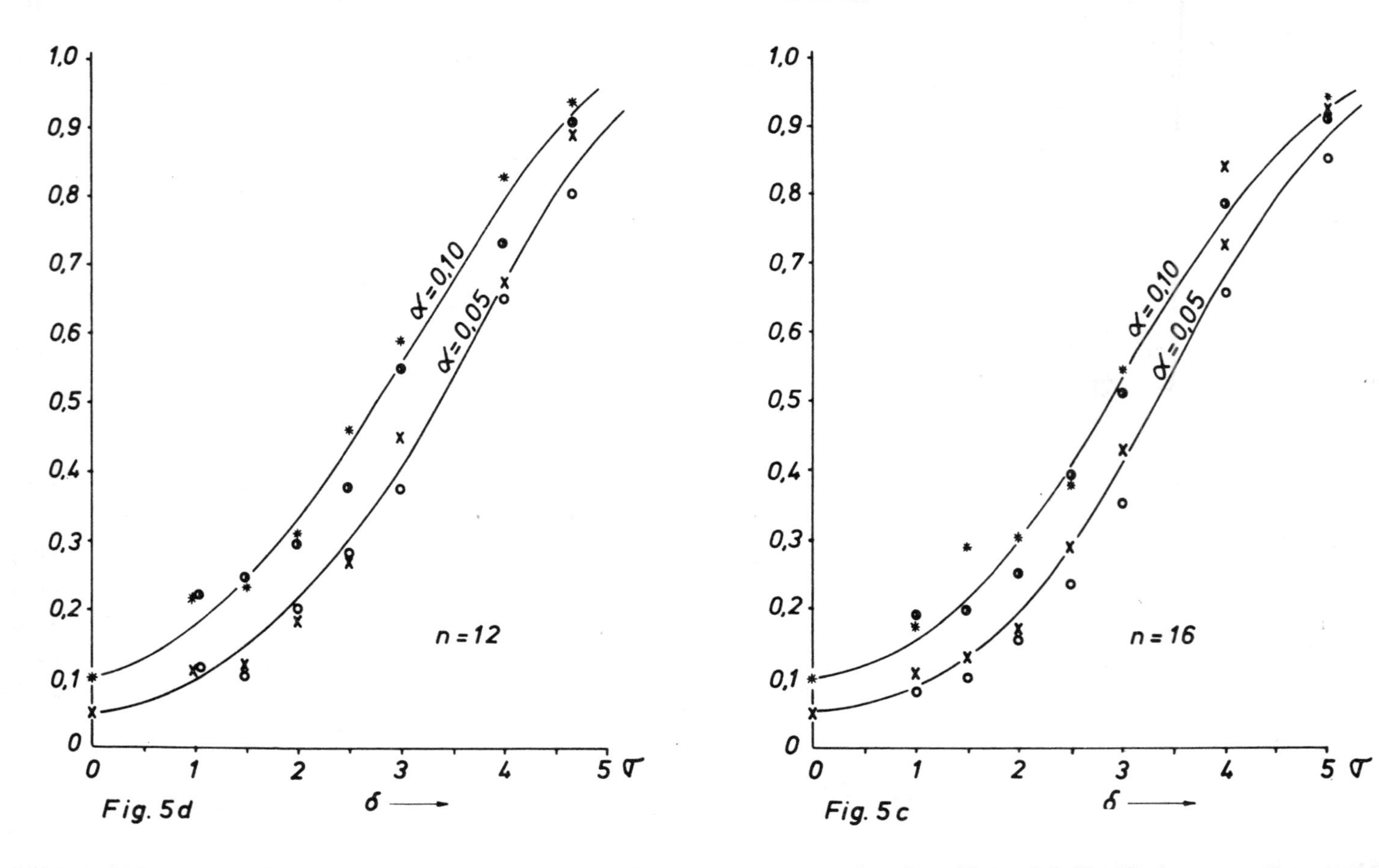
Power curves of outlier-tests
× * DOORNBOS
○ ● DIXON
1,0
0,9
0,8
0,7
0,6
0,5
0,4
0,3
0,2
0,1
0
α = 0,10
α = 0,05
n = 12
n = 16
0
1
2
3
4
5 σ
δ ⟶

Fig. 5d

Fig. 5c

Digital Simulation of Vehicle Vibration

Hermann Bruns, Munich

I. GENERAL INTRODUCTION

Develloping a passenger car, it is of great interest to get informations in vibration behaviour of the system before assembling a prototype of the car. This fact will give a great gain in time and it is possible to reduce test driving on the road. If the model is good enough, there will be the possibility to optimize the most important parameters of the system, that are, for example, spring, damper, wheel-base etc. There is nearly no restriction in desribing very complicated vibration models, but the borders are given by the problem of understanding the results and the program for the computer must have a running time, which may be acceptable, even if a lot of parameters have to be changed. Digital computers have succeeded over the analog computers, having no restrictions in amplitude and no problems in accuracy. Last not least very comfortable simulation languages give the possibility for describing vibration problems very easily. [5]

II. VEHICLE VIBRATION MODELS

To describe the vibration behaviour of the car in one total model is not possible. But for special problems you get good correlation between measurement and calculation. So it is the problem, to find the right model for the given question. Fig. 1 shows a diagram for describing vehicle dynamics, starting with a simple model with one degree of freedom up to three-dimensional systems with several degrees of free-

dom.
A typical model for a passenger car shows the next figure, Fig. 2. This model gives the possibility to describe the influence of driving speed, which may be simulated by a time lag between front and rear axle excitation, the signal itself being the same. The model can be described by differential equations, and assumming small pitching angulars and a linear system, there are no problems in solving the equations.
Fig. 3 shows the response of the system for a step function input, given by the time histories of the main output values.
A more complicated model is given in the next picture, Fig. 4. [3]

. PROBLEMS ON VIBRATION SIMULATION

A linear model often does not give the right answer in vibration analysis. This is the advantage of the theoretical method, to try different parameters in the system without doing any test work.
Several components of the suspension system are nonlinear and cannot be adequately described by differential equations. For example, the damper characteristic of the typical shock absorber shown in Fig. 5 is always different in directions of compression and extension. Therefore the damping force is not always proportional to the relative velocity between car body and the wheel. These characteristic can be directly simulated by digitizing the functional characteristics of the element. A comparison between a linear and nonlinear damper characteristic can be seen from the system answer to a step function, Fig. 6a and 6b and there is quite a difference in the time histories.

IV. SIMULATION OF EXCITATION FUNCTIONS

Especially in the case of nonlinearities in the system, the simulation of input, that is the excitation function, becomes an important problem. In this case, the system answer depends on the amplitude of the input function. Therefore the real system inputs have to be simulated as well as possible. Therefore it is necessary to describe the road surface by the equations of random processes, that is spectral density function and probability distribution, because the road surface is given by a random function. (For first system-investigations a step function or sinusoidal excitation may be sufficient).
Measurements have shown [1,2], that the spectral density of road undulation can be generally described by the equation

$$A\ (f) = C \cdot f^{-w} \qquad (1)$$

with f (frequency), C (constant) describing a good or bad road and w (exponent) giving the relation between amplitude and frequency. As a vehicle passes over the road surface, it will experience mechanical vibrations which depend on the wavelength of the road according to the following formula:

$$f = v / \lambda \qquad (2)$$

v vehicle speed, λ wavelength.
Concerning the distribution, in most cases a Gaussion distribution can be assumed, the crest factor may be in the range from 3 to 4. If the distribution of a random process is Gaussion, the standard deviation σ is equal to the RMS-value of the time history and the square of which is equal to the area of the spectral density function G (f).

$$\text{mean square} = X_{eff}^{2} = \int_{0}^{f} G\ (f)\ df \qquad (3)$$

For simulating a random process a digital noise or random software generator may be used. In most cases you can choose between a normal or an unique distribution. Fig. 7 shows the results from such a software-normal-distribution generator, for two different ensembles, each one having total of 100 points. For to get a random process, the output from the software generator must be sequential, with fixed time intervals. This time interval may be given by the step-width for the simulation run, Fig. 8. By using the Fast-Fourier-Transformation Algorithm [4] on this time series one gets the power spectral density, Fig. 9. Th characteristic is called a "bandlimited white noise" spectrum and can be described in one range by $G(f) \sim f$ and in the second range by an equation like $G(f) \sim f^{-w}$. From this spectrum it is possible to simulate power spect with different exponents w, due to the power spectra of road surfaces.

The last figure, Fig. 10 shows the time histories from the two-dimensional model with random excitation for several different outputs. Comparisons with measurements on the road give sufficient coincidence of the results and justify this method for vehicle vibration analysis and computation.

REFERENCES

1. BRAUN, H. "Untersuchungen über Fahrbahnunebenheiten." Deutsche Kraftfahrtforschung und Straßenverkehrstechnik, Heft 186, VDI-Verlag Düsseldorf, 1966.

2. BRUNS, H. Zur Problematik der Bewertung von Straßenunebenheiten unter Berücksichtigung des Einflusses auf das Schwingungsverhalten von Straßenfahrzeugen. Straße und Autobahn 21 (1970) Heft 8, Seite 297/303.

3. FABIAN, L. Zufallsschwingungen und ihre Behandlung. Springer-Verlag, Berlin-Heidelberg-New-York, 1973.

4. KREMER, H. Praktische Berechnung des Spektrums mit der schnellen Fourier-Transformation. elektronische datenverarbeitung (1969) Heft 6, Seite 281/284

5. WYMAN, D.G. "DSL, an IBM 1800 Program for Simulation of Continuous Process Dynamics featuring User-Interaction and on-line Graphics." IBM Program Nr. 1800-15.1.001

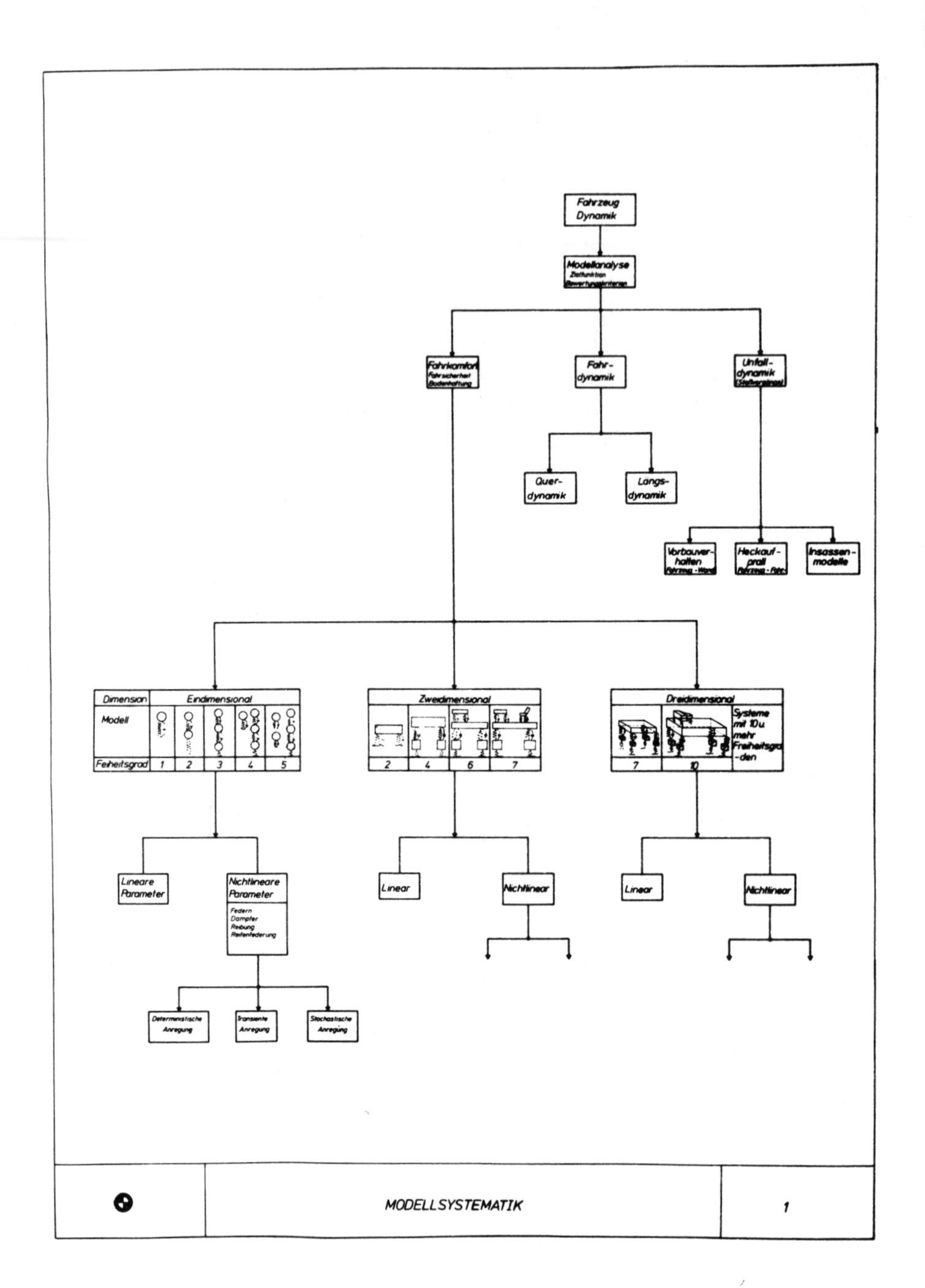

FIG. 1 Classification for vehicle vibration models.

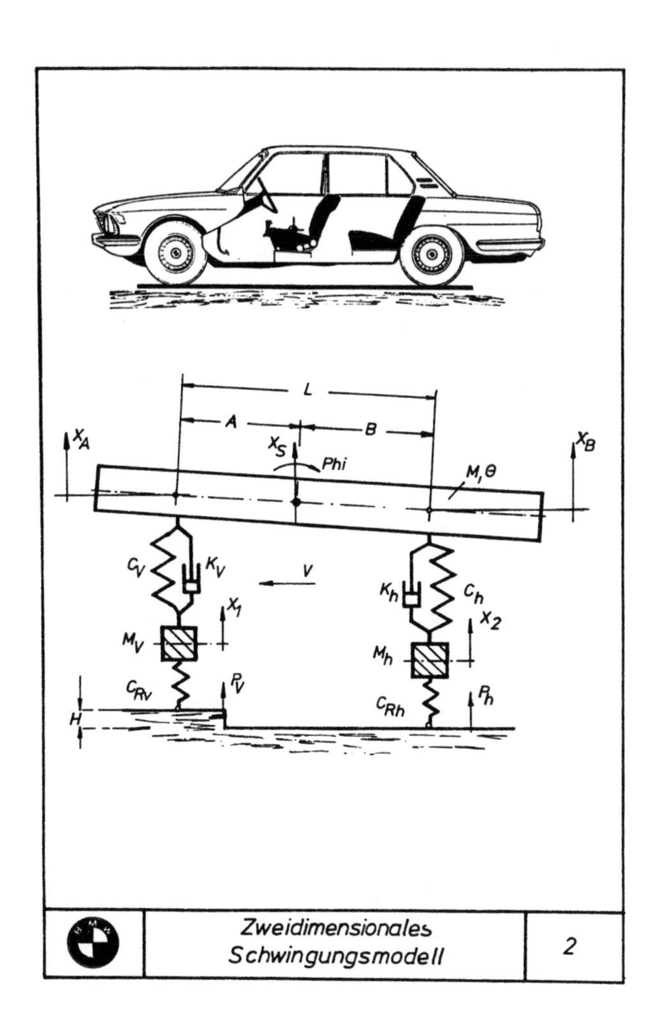

FIG. 2 Two-dimensional model with four degrees of freedom, simulating vibration behaviour of a passenger car.

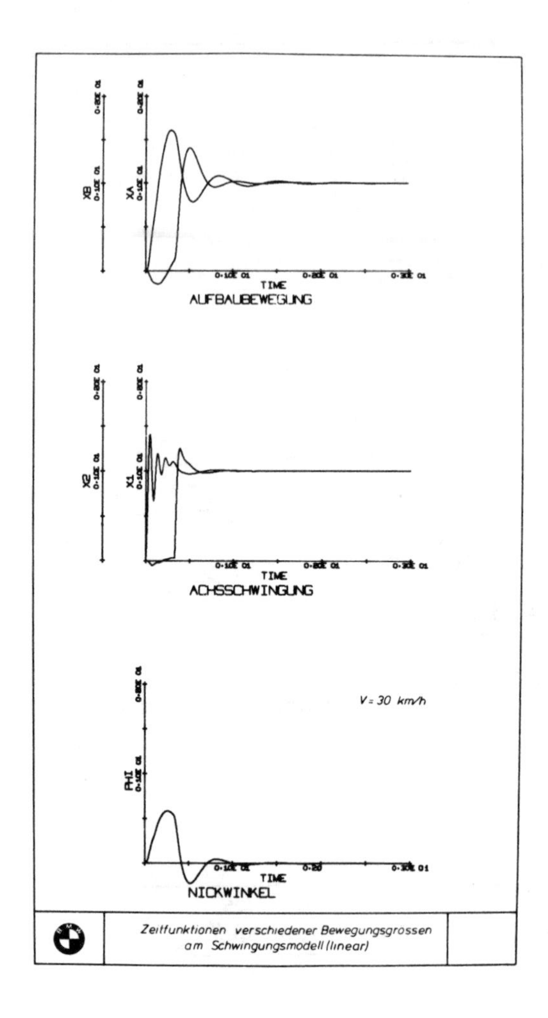

FIG. 3 Response of the system from FIG. 2 to a step function input on front and rear axle.

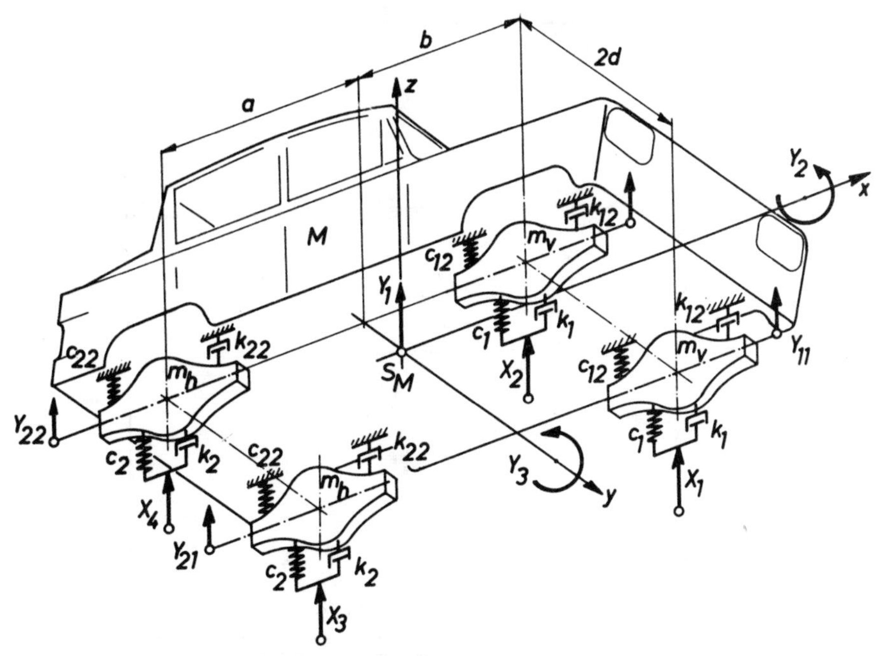

FIG. 4 Three-dimensional model simulating a passenger car.

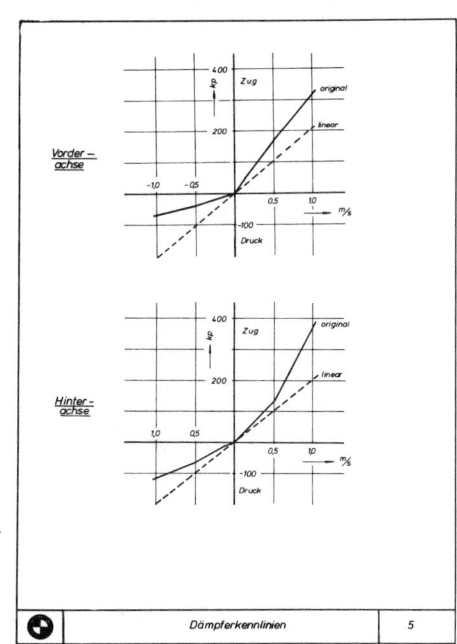

FIG. 5

Damper characteristics of typical shock absorbers from a passenger car.

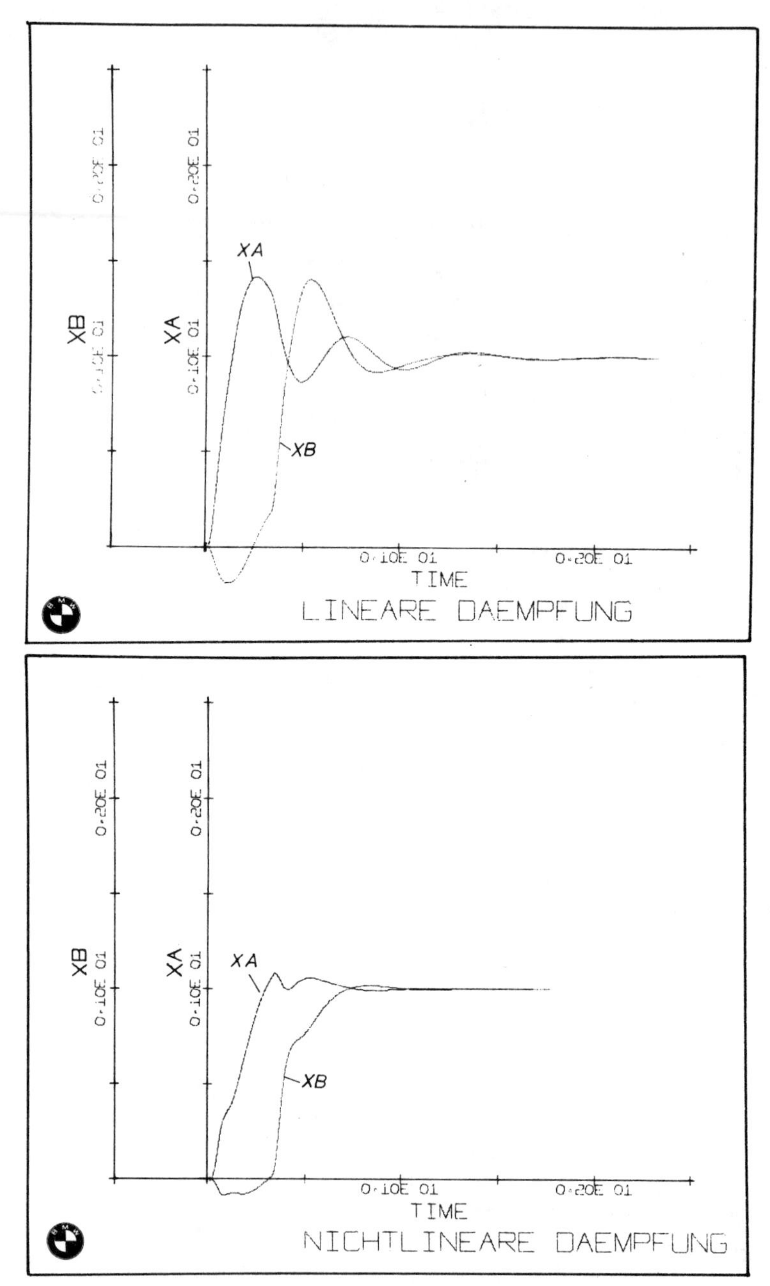

FIG. 6 System output (FIG. 2) for a step function input for linear (a) and nonlinear (b) system.

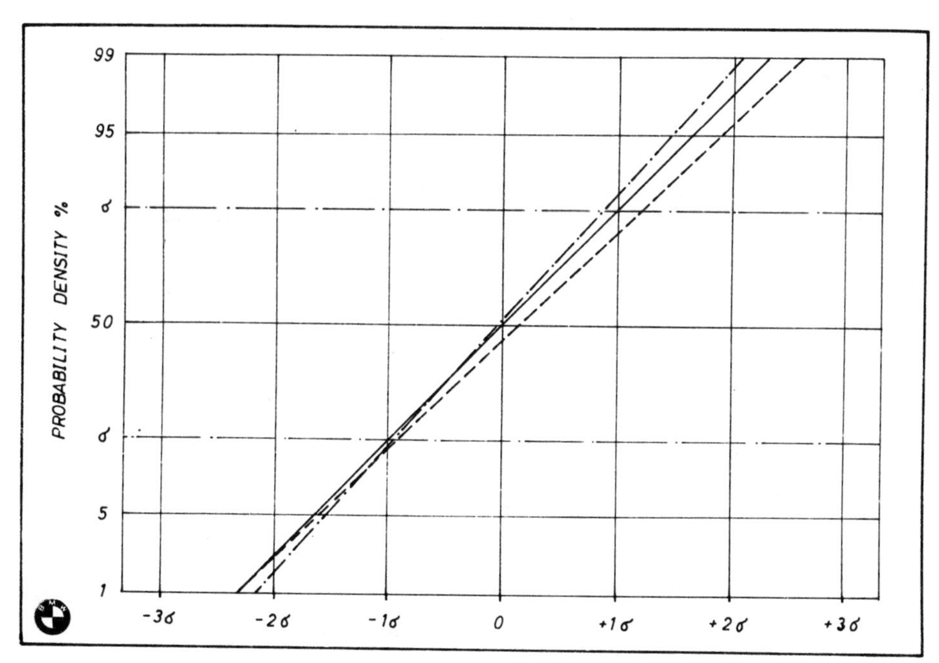

FIG. 7 Probability distribution for two ensembles of 100 points from a software random generator.

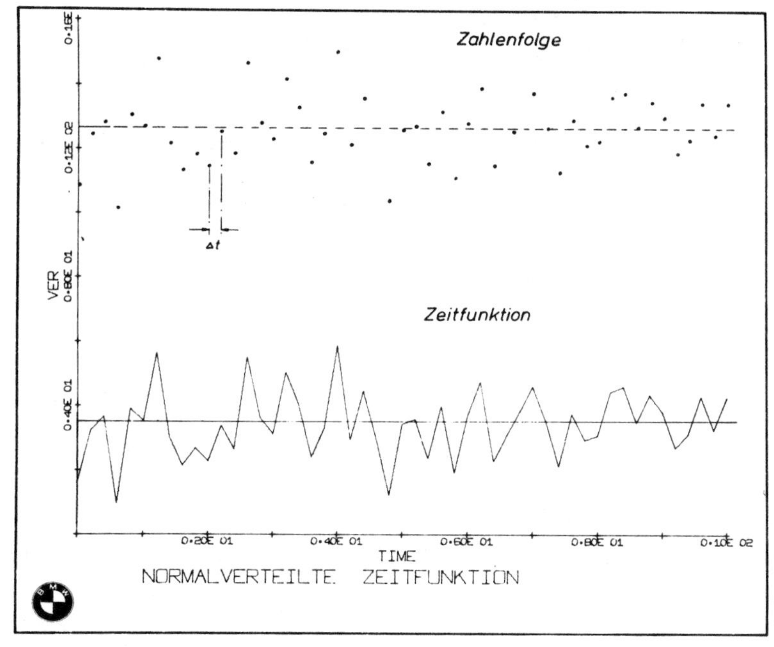

FIG. 8 Simulating a random process from a random generator. The upper row shows the single values, beneath the resulting time history.

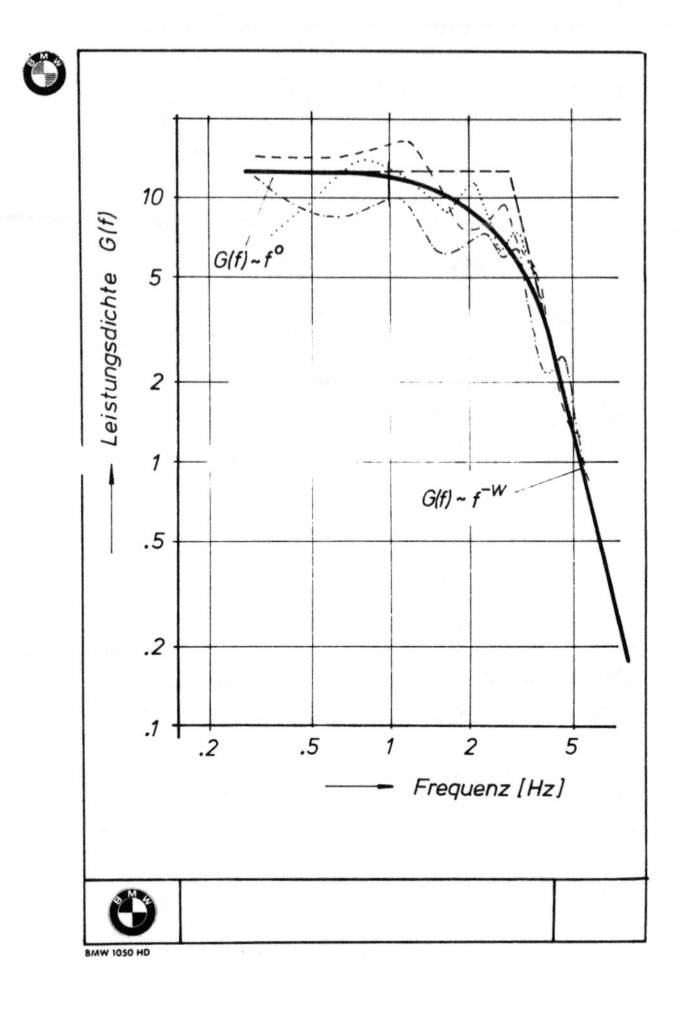

FIG. 9 Bandlimited white noise, power-spectrum from the sample record from FIG. 8

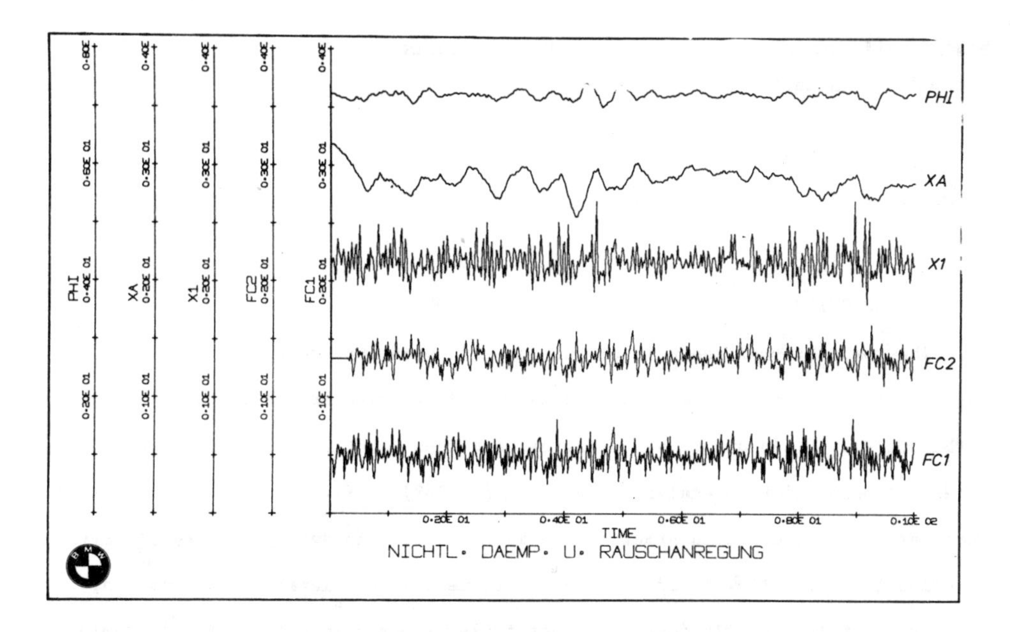

FIG. 10 Some time histories, representing several outputs of the nonlinear passenger car model from FIG. 2. FC 1 and FC 2 random excitation at front and rear axle, X 1 vertical motion of the front axle, XA vertical motion of center of gravity and PHI representing the pitch angular.

Statistical Computation in the Theory of Turbulence

Alexandre J. Chorin, Berkeley

1. INTRODUCTION

The computational analysis of turbulent flow encounters two problems, which appear difficult because they are unusual: (i) particular realizations of the turbulent flow field are to all practical purposes unobtainable by the usual numerical means (for an analysis, see CHORIN [1969]); (ii) such realizations are worthless anyway, since in practice one is interested in average values of certain functionals of the flow field (such as drag, mean square level of fluctuation, etc.) and not in specific details of particular flows. This situation is reminiscent of the situation in kinetic theory, where detailed information about all particles which make up a system is neither obtainable nor interesting. It should be apparent, however, that these two problems, when considered together, constitute an opportunity rather than an obstacle; it is well known [KAC, 1959] that under a wide variety of conditions the evaluation of mean values of a random variable affords an easier task than the evaluation of specific values which that variable may assume.

We thus abandon completely the effort of computing particular flows, and address ourselves directly to the problem of computing means. We present two methods for solving this problem. The first is a Monte-Carlo method which provides a random flow field whose means approximate the means of the solution of the Navier-Stokes equations. The second method computes these means directly; it is closely related to the previous one, and derives from an approximation method for Wiener integrals [CHORIN, 1973b].

2. THE RANDOM VORTEX METHOD

Consider for simplicity's sake a two-dimensional flow in a domain $\mathcal{D}$, with boundary $\partial\mathcal{D}$. The Navier-Stokes equations can be written in the form

$$\partial_t \xi + (\underline{u}\cdot\underline{\nabla})\xi = \frac{1}{R}\Delta\xi \tag{1a}$$

$$\Delta\psi = -\xi \tag{1b}$$

$$u = -\partial_y\psi \quad , \quad v = \partial_x\psi \tag{1c}$$

where $\underline{u} = (u,v)$ is the velocity, ψ is the stream function, ξ is the vorticity, and R is the Reynolds number. The boundary conditions are

$$\underline{u} = 0 \quad \text{on} \quad \partial \tag{2}$$

The main hurdle in the conventional solution of equations (1) stems from the fact that $1/R$ is small in problems of practical interest, and thus the diffusion term cannot be distinguished from the truncation error. The use of a Monte-Carlo algorithm will overcome this problem.

We take for granted the following fact: the vorticity field at large R can be approximated by a sum

$$\xi(x,y) = \xi(\underline{r}) = \sum_j k_j \xi_c(\underline{r} - \underline{r}_j) \tag{3}$$

where the k_j are constants, and the $\underline{r}_j$ are positions of the centers of vortex blobs of the universal form ξ_o. If a length typical of the inertial range is large compared to a diffusion length (for an explanation of the terms, see, e.g., CHORIN [1974]), we may write

$$\xi_o = -\Delta^{-1}\psi_o \quad , \tag{4}$$

where

$$\psi_o = \begin{cases} \frac{1}{2\pi} \log r & r \geq \sigma \\ \frac{r}{\sigma} + \text{constant}, & r < \sigma \quad ; \quad r \equiv |\underline{r}| \quad . \end{cases}$$

If $R^{-1} = 0$, one can show that k_j = constant. If $R^{-1} = 0$ only the normal component of the boundary condition (2) can be imposed; each vortex will move in the velocity field induced by all the others and modified by the boundary, and we have

$$\underline{r}_j^{n+1} \cong \underline{r}_j^{n} + k\, \underline{u}_o^{n} \quad , \tag{5a}$$

where k is a time step, and $\underline{u}_o$ is given by

$$\underline{u}_o^{n} = \underline{u}_\xi^{n} + \underline{u}_p^{n} \quad , \tag{5b}$$

$$\underline{u}_\xi^{n} = (u_\xi^{n}, v_\xi^{n}) \quad , \tag{5c}$$

$$u_\xi^{n} = \frac{1}{2\pi} \sum_{j}{}_1 \frac{y_j^{n} - y}{\rho_j^2} k_j + \frac{1}{2\pi} \sum_{i}{}_2 \frac{y_i^{n} - y}{\sigma \rho_i} k_i \quad , \tag{5d}$$

$$v_\xi = -\frac{1}{2\pi} \sum_{j}{}_1 \frac{x_j^{n} - x}{\rho_j^2} k_j - \frac{1}{2\pi} \sum_{i}{}_2 \frac{x_i^{n} - x}{\sigma \rho_i} k_i \quad , \quad \rho_j = |\underline{r} - \underline{r}_j| \tag{5e}$$

where $\sum_1$ is a sum over $\underline{r}_j$ such that $|\rho_j| > \sigma$ and $\sum_2$ is a sum over $\underline{r}_j$ such that $|\rho_j| \leq \sigma$. $\underline{u}_p$ is a potential flow designed to satisfy the boundary condition on $\partial\mathcal{D}$. Let $R^{-1} \neq 0$. A diffusion is added to the equations above, and it can be taken into account by means of a random walk. Let η_1, η_2 be Gaussian random variables with mean 0 and variance $2k/R$. Replace (5a) by

$$\underline{r}_j^{n+1} = \underline{r}_j^{n} + k\, \underline{u}_o^{n} + \underline{\eta} \quad , \quad \underline{\eta} = (\eta_1, \eta_2) \quad . \tag{6}$$

The resulting vorticities distribution satisfies equation (1) with a statistical error $O(R^{-1/2})$. At a boundary, the tangential condition $\underline{u} \cdot \underline{s} = 0$ must now be satisfied, where $\underline{s}$ is a unit vector tangent to $\partial\mathcal{D}$. This can be done by creating a boundary layer with a total vorticity $\underline{u}\cdot\underline{s}$ per next length of the boundary, distributing it among a discrete set of vortices of the form above, and allowing these vortices to move according to the laws (6).

Thus, the Monte Carlo feature of the computational method is used to handle the singular perturbation which arises when R^{-1} is very small. Numerical results have been displayed, e.g., in CHORIN [1973a]. An analysis is given in MARSDEN [1974].

3. COARSE GRAINING

The method above is particularly attractive in problems where the vorticity is confined to small regions (for example, problems of flow past aerofoils). In other problems, the number of vortices needed may become large, and the evaluation of their interactions prohibitively expensive. One may wish for a method which can provide means of the solution more directly.

Divide the region $\mathcal{D}$ of the flow into cells ω_i, such that the number of vortices per cell is not small. The total vorticity, ξ_i, per cell may be assumed to be a gaussian random variable, with ξ_i, ξ_j independent when $i \neq j$. Equations for the evolution of the means $\overline{\xi}_i$ and variances $\overline{\xi'^2} - \overline{\xi}^2$ can be written.

Rather than write out in full the rather complex equations involved, we shall explain the principles which underlie them. Equation (1a) can be written in the form

$$\partial_t \xi + \mathrm{div}(\underline{u}\xi) = \frac{1}{R}\Delta\xi \tag{7}$$

we first replace the equation by a discrete approximation, e.g.,

$$\xi^{n+1}_{i,j+1/2} - \xi^{n}_{i,j+1/2} = \frac{k}{h}[(u^n\xi^n)_{j+1} - (u^n\xi^n)_j] + \frac{k}{R}\Delta_h\xi^n \tag{8}$$

where k is a time step, h is a spatial increment, Δ_h is a discrete laplacian, and $\xi^n_{i,j+1/2} \equiv \xi(ih,(j+1/2(h,nk)$, etc. We now apply our main assumption to this equation, i.e., rather than try to find k,h small enough so that this algebraic system of equations may be a valid approximation to equation (1a), we apply the statistical assumption to this equation and obtain an equation which approximates equation (1a) in the mean only. The discretization is necessary because the passage to the limit $h = 0$ is not possible in the main assumption. (If it were possible, one could deduce that $\xi,\underline{u}$ are gaussian fields, which is certainly false, see CHORIN [1974b].) To average equation (8), we have to evaluate expressions such as

$$\overline{(v^n\xi^n)_{j+1}} \quad .$$

However,

$$\overline{(v\xi)_{j+1}} = \overline{v\xi(ih,(j+1)h + vk)}$$

v can be expressed in terms of ξ through equations (1b), (1c); it is readily seen that when our assumption holds,

$$\overline{v(\xi - \overline{\xi})} = 0 \quad ;$$

thus

$$\overline{(v\xi)_{j+1}} \cong \overline{v'^2_{j+1}} \frac{\overline{\xi}_{j+3/2} - \overline{\xi}_{j+1/2}}{h} k \quad ;$$

where

$$v = \overline{v} + v', \quad \overline{v'} = 0 \quad .$$

Other terms can be evaluated in the same way, and an equation for

$\overline{(\xi_{j+1/2} - \overline{\xi}_{j+1/2})^2}$ obtained from

$$\partial_t \xi^2 = \text{div}(\underline{u}\xi^2) + \frac{2}{R} \xi \Delta\xi \quad ,$$

which also follows from (1a).

This method can be viewed as a way of summarizing the random vortex approximation. Rather than follow individual vortices, one divides their phase space into cells and applies a plausible assumption to the total vorticity per cell. The variables $\xi^n_{i,j+1/2}$ are linear functionals of the vorticity field. The possibility of approximating a functional of a field by a function of just exactly such linear functionals has been established in CHORIN [1973b]. The suggestion that such an approximation would be valid is apparently contained in the last footnote of HOPF [1952].

The approximation above has been applied to the problem of turbulent flow in a channel. The calculated means $\overline{\xi}$, $\overline{u}$, fluctuation levels $\overline{\xi'^2}$, $\overline{u'^2}$, $\overline{v'^2}$, as well as the pressure drop and other functionals of $\underline{u}$ are in excellent agreement with experiment. Detailed calculations will be published elsewhere.

4. CONCLUSION

The calculations briefly described above point out the fact that the marriage of probability theory with numerical analysis can be extremely fruitful. Its results allow one to solve problems unreachable by any other means, and should be the object of substantial further research.

ACKNOWLEDGMENT

This work was carried out while the author was an Alfred P. Sloan Research Fellow, with partial support from the Office of Naval Research under Contract USN-N00014-60-A-0200-1052.

REFERENCES

Chorin, A. J., "On the Convergence of Discrete Approximations to the Navier-Stokes Equations, Math. Comp., 23, 341 (1969).

__________, "Numerical Study of Slightly Viscous Flow," J. Fluid Mech., 57, 785 (1973a).

__________, "Accurate Evaluation of Wiener Integrals," Math. Comp., 27, 1 (1973b).

__________, "Gaussian Fields and Random Flow," J. Fluid Mech., 63, 21 (1974a).

__________, "Lectures on the Theory of Turbulence," Lecture Notes, Department of Mathematics, University of California, Berkeley (1974b).

E. Hopf, "Functional Calculus and Statistical Hydrodynamics, Arch. Rat. Mech. Anal., 1, 82 (1952).

M. Kac, Probability and Related Topics in the Physical Sciences, Interscience, New York (1959).

J. Marsden, "A Solution Formula for the Navier-Stokes Equations," Bull. Am. Math Soc., 80, 154 (1974).

Spline Smoothing of Spectra

Theo Gasser, Zurich

1. INTRODUCTION

To obtain derivatives of noisy curves is a problem of practical importance: Either the derivatives are needed explicitly or otherwise implicitly, when applying interpolation and approximation routines from numerical analysis. The superposed errors may impair a proper convergence of these algorithms, as soon as they are larger - by order of magnitude - than the discretization errors. A qualitative argument from Fourier analysis: The signal is usually concentrated in low frequency bands, the noise is approximately white. Since the transfer function for the first derivative is $2\pi i \cdot \nu$, for the second derivative $-(2\pi)^2 \nu^2$, then the total derivative of signal plus noise is largely the derivative of the noise. This article concentrates on smooth spectrum estimation: We want to arrive at good smoothness properties without introducing too much additional bias.

One approach to this problem is a least squares parametric fit, e.g. a polynomial fit. These fits are global, so that the behavior in one neighbourhood determines the function everywhere. We prefer a non-parametric smoothing procedure, since the spectrum should be a non-parametric device. I propose to apply cubic smoothing splines, which have both above mentioned features.

This new approach is compared with the classical estimate both in terms of smoothness quality and statistical quality. The classical estimate consists in smoothing the periodogram by a moving average.

2. SPECTRUM ESTIMATION

From a sample $\{X_t;\ t=0,1,\ldots,N-1\}$ of a stationary process X_t, we have to estimate its spectrum $f(\nu)$. The first three steps are identical for the two procedures:

A. Low-pass-filtering of the data, with a cut-off frequency not higher than $1/2\cdot\Delta t$ (Δt=discretization step). In the following we put $\Delta t=1$.

B. Tapering: To reduce the relative height of the side lobes, we multiply the data by a suitable function c(t) decaying smoothly to zero at both ends:

$$X_t^* = c(t/N)\cdot X_t\ ,\quad 0 \leq t \leq N-1$$

C. The FFT enables a highly efficient Fourier-transforming of the record for N a power of 2:

$$Y_q = \sum_{k=0}^{N-1} \exp(-2\pi i\frac{k\cdot q}{N})\cdot X_k^*\ ,\quad q=0,1,\ldots,N-1$$

Squaring yields the periodogram, the raw data in spectrum analysis:

$$Y_q = \frac{1}{N}\left|Y_q\right|^2$$

Because of the symmetry $\overline{Y}_q=Y_{N-q}$, we can restrict the evaluation to $q=0,1,\ldots,N/2$. The integer q stands for the frequency $\nu_q=q/N$.

D. The smoothing of the periodogram is done with a moving average:

$$\hat{f}_q = \sum_{k=-L}^{L} c_k\cdot I_{q-k}$$

$$\text{with:}\quad c_k \geq 0,\ c_k = c_{-k},\ \sum_{k=-L}^{L} c_k = 1$$

The high correlation for neighbouring values would make it wasteful to compute the spectrum with minimal discretization step $\Delta\nu = 1/N$ and the step shall be $\Delta\nu = \Delta Q/N$:

$$q = Q_o,\ Q_o+\Delta Q,\ldots,\ Q_o+(K-1)\cdot\Delta Q$$

$$K = \left\lfloor \frac{N/2 + 1 - Q_o}{\Delta Q} + 1 \right\rfloor$$

$$\lfloor r \rfloor = \text{largest integer} \leq r$$

Rosenblatt (1971) in his competent review on curve estimates has pointed out, that the choice of weights c_k has little influence on efficiency. This conforms well with my experience in spectrum analysis. I will therefore stick to the simple uniform weights $c_k=1/M$ with $2L+1=M$ called the window width.

The variance of $\hat{f}_q$ is given by

$$\frac{1}{M^2}\cdot\sum_{k=-L}^{L} f_{q-k}^2 \approx \frac{1}{M}\cdot f_q^2$$

This approximation holds for locally flat pieces only. The bias at frequency $\nu_q = q/N$ is:

$$b_q = f_q - \frac{1}{M}\sum_{k=-L}^{L} f_{q-k}$$

$$\approx -c\, f_q'' \quad , \quad c = \frac{1}{6}\left(\frac{M}{2\pi N}\right)^2$$

For a competent, partly historical review of spectrum analysis, see Tukey (1967).

3. SPLINES

Given a discrete function $\{(x_i,g_i);\ i=1,\dots,K\}$. A cubic interpolation spline is the solution to the following minimum problem:

(1) g_S is twice continuously differentiable.

(2) $g_S(x_i) = g_i$

(3) $\int_{x_1}^{x_K} (g_S''(x))^2 dx = \text{Min!}$

Higher order splines are defined in a similar way, but their marked oscillations may cause difficulties in practice (Greville, 1964). For smoothing splines, condition (2) has to be replaced:

$$(2')\quad Y_K = \sum_{i=1}^{K} \left(\frac{g_S(x_i) - g_i}{\sigma_i}\right)^2 \leq S\ , \qquad S>0$$

σ_i = standard deviation of g_i

The existence of such a g_S has been proved by Schoenberg (1964).

This leads to a non-parametric and local smoothing procedure, where the degree of smoothing is regulated by S. The case of normal independent g_i can give us a first guideline for the critical choice of S: $S \in (K-\sqrt{2K},\ K+\sqrt{2K})$.

Since splining incorporates a penalty for too rough a behavior of the estimate, it should be superior to the moving average estimate of section 3, regarding smoothness. The question is, if we have to pay heavily in terms of statistical quality. It is difficult to weigh these different factors by analytical reasoning. A Monte Carlo experiment should enable a better judgement. The computer program follows Reinsch (1967), with minor modifications.

4. CRITERIA

The judgement on the performance of a curve estimate depends on what we want to do with the estimate. The two statistical criteria used in this study are:

$$\text{MSE} = \frac{1}{K} \sum_{i=0}^{K-1} E\left[(f(\nu_i) - \hat{f}(\nu_i))^2\right]$$

$$\nu_i = (Q_o + i \cdot \Delta Q)/N$$

$$\text{MAXABS} = E(\max_{i=0,1,\dots,K} \left| f(\nu_i) - \hat{f}(\nu_i) \right|)$$

There is an optimal window width M=2L+1 with respect to MSE, i.e. an M which minimizes MSE. This optimum may be quite sensitive for a spectrum with accentuated intensity variations: For too low M, we have high variability and low bias, for too large M, we have low variability and high bias and of course improved smoothness. The qualitative behavior of the global criterion MSE and of the local MAXABS as a function of window width M is suprisingly similar.

I also propose a local and a global smoothness criterion, with heavy emphasis on the local one:

$$\text{GLOBSMOOTH} = \left(\sum_{i=1}^{K} (\hat{f}''(\nu_i))^2\right) / \left(\sum_{i=1}^{K} (f''(\nu_i))^2\right)$$

The ideal value of GLOBSMOOTH is 1; deviations indicate global undersmoothing resp. oversmoothing.

$$\text{LOCSMOOTH} = \frac{\max\limits_{i \in (1,K)} \left| \, |f''(\nu_i)| - |\hat{f}''(\nu_i)| \, \right|}{|f''(\nu_{max})|}$$

(ν_{max} = abscissa where numerator becomes maximal)

The ideal value of LOCSMOOTH is 0.

These four parameters are estimated from NRUN Monte Carlo realizations; the standard deviation is also computed.

5. EXPERIMENTAL RESULTS

The samples are generated by autoregressive-moving-average schemes of order (P,Q) with Gaussian white noise as input:

$$X_t = \sum_{k=1}^{P} A_k X_{t-k} + \sum_{k=1}^{Q} B_k Z_{t-k} + Z_t$$

Due to limitations of space, only two examples are discussed here:

1. P=4, Q=0: A_1=-.482, A_2=-.9472, A_3=-.3774, A_4=-.42

Realization length is N=128, and there are NRUN=100 samples from which the criteria are estimated (see table 1).

2. P=1, Q=0: A_1=.5

For this example I take N=512 and NRUN=40 (see table 2). The spectrum is given in fig. 1 resp. fig. 2, on equal and linear scale.

Splining a mildly pre-smoothed periodogram produced rather unsatisfactory results. This might be due to the non-homogeneity of the variance or to the large intensity variations occurring in spectrum analysis, or to both. These effects are reduced by a logarithmic transformation, which at the same time normalized the estimate. The procedure looks then as follows:

I. Pre-smooth the periodogram with a low $M=M_1$.

II. Take logarithms of this pre-estimate.

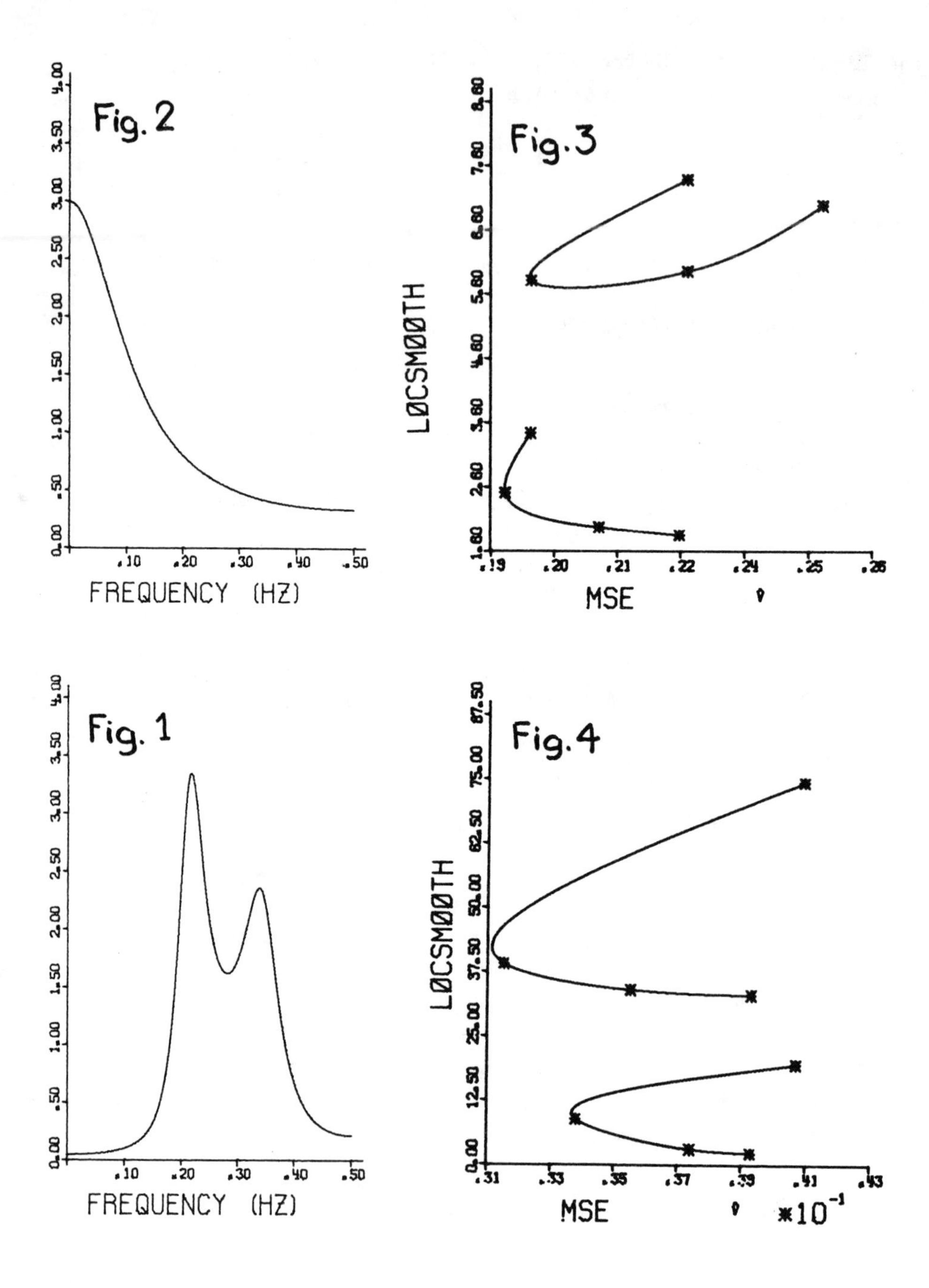
Fig. 2
FREQUENCY (HZ)
Fig. 3
LØCSMØØTH
MSE
Fig. 1
FREQUENCY (HZ)
Fig. 4
LØCSMØØTH
MSE
*10^-1

III. Spline the result.

IV. Exponentiate the estimate obtained by spline-smoothing.

The improvement by this procedure gave rise to a separate study, in which step III is replaced by ordinary periodogram-smoothing with a rectangular window. The results are excellent, particularly with respect to the sensitivity for a non-optimal choice of window width M. Some of the credit for the following results definitely goes to this new technique and not to splining.

TABLE 1 (Frequency grid: $Q_o = 0$, $Q = 3$, $K = 22$)

M	ESTIMATOR OF	MSE	MAXABS	GLOBSMOOTH	LOCSMOOTH
7	Mean	.2239	1.2814	1.48	7.40
	Stand Dev	.1290	.3875	.76	3.06
9	Mean	.1930	1.1782	.87	5.81
	Stand Dev	.1011	.3123	.46	2.20
13	Mean	.2007	1.2788	.40	5.86
	Stand Dev	.0673	.2837	.18	2.35
17	Mean	.2503	1.4375	.22	6.99
	Stand Dev	.0574	.2666	.11	3.13
M_1	Splined Estimate				
5	$S=.3(K-\sqrt{2K})=4.61$				
	Mean	.1929	1.1780	.64	3.44
	Stand Dev	.1089	.3692	.64	1.92
5	$S=.5(K-\sqrt{2K})=7.68$				
	Mean	.1879	1.2014	.33	2.51
	Stand Dev	.0863	.3232	.40	1.34
5	$S=.8(K-\sqrt{2K})=12.29$				
	Mean	.2065	1.3388	.14	1.97
	Stand Dev	.0732	.3218	.19	.70
5	$S=(K-\sqrt{2K})=15.37$				
	Mean	.2224	1.4481	.08	1.85
	Stand Dev	.0684	.3085	.11	.46

Fig. 3 gives a graphical comparison with curves representing LOCSMOOTH against MSE, parameterized by the smoothing parameter. The curves are obtained by spline interpolation through the tabled values, which are marked by crosses. If we want both a good statistical and a good smoothness quality, we should evidently take the splined estimate, the performance of which is given in the lower curve. Due to the correlatedness of the pre-estimate, one should S as small as $S=.5(K-\sqrt{2K})$ to avoid oversmoothing. The sensitivity to the choice of S is remarkably low, due to the logarithmic transformation.

TABLE 2 (Frequency grid: $Q_0=0$, $Q=7$, $K=37$)

M	ESTIMATOR OF	MSE	MAXABS	GLOBSMOOTH	LOCSMOOTH
41	Mean	.0411	.5615	59.33	74.59
	Stand Dev	.0349	.2742	33.36	26.17
69	Mean	.0317	.4559	21.02	39.05
	Stand Dev	.0319	.2539	10.52	13.77
83	Mean	.0357	.5033	14.64	34.04
	Stand Dev	.0324	.2330	6.65	12.72
93	Mean	.0395	.5297	12.63	33.10
	Stand Dev	.0323	.2160	5.00	9.65
M_1	Splined Estimate				
7	$S=(K-\sqrt{2K})=28.40$				
	Mean	.0409	.6128	9.60	19.58
	Stand Dev	.0374	.3126	22.14	27.78
7	$S=(K-.4\sqrt{2K})=33.56$				
	Mean	.0340	.4911	2.98	8.84
	Stand Dev	.0302	.2720	7.16	16.18
7	$S=(K+.6\sqrt{2K})=42.16$				
	Mean	.0376	.4608	.23	3.00
	Stand Dev	.0263	.1679	.56	4.90
7	$S=(K+\sqrt{2K})=45.60$				
	Mean	.0395	.4578	.11	2.24
	Stand Dev	.0254	.1531	.15	2.04

In this Markovian example, the improvement with regard to LOCSMOOTH brought about by the splined estimate is impressing: The optimum with regard to MSE is on the other side a bit higher (note that the origin is not at 0). A periodic spline might do better if we have substantial spectral mass at the ends of the interval $[0,1/2)$.

In the spline routine, an estimate of the variance is needed. A very accurate asymptotic formula of $\log\chi_k^2$ is given by Davis and Jones (1968):

$$\mathrm{Var}(\log\chi_k^2) \approx \frac{2}{k} + \frac{2}{k^2}$$

An additional advantage of the logarithmic transformation does materialize here: In the linear case, the variance is proportional to the square of the true spectrum, so that an iteration is needed.

From these and other examples not mentionned here, the following conclusions can be drawn:

A. Smoothing splines are an excellent device to obtain estimates with good smoothness properties.
B. Care has to be taken to reduce the true intensity variations and to homogenize the variance. The spline has otherwise a tendency to oversmooth the data, which results in a large bias at peaks.
C. A reasonable choice of the smoothing parameter still depends on the true shape of the underlying curve: For curves with high intensity variations, the smoothing parameter should be taken relatively small.
D. The technique is quite easy to implement and to apply and allows spline interpolation just after the smoothing process.

ACKNOWLEDGEMENT:

I would like to thank Prof. P.J. Huber for stimulating discussions and critical comments.

REFERENCES:

Davis, H.T., R.H. Jones (1968). Estimation of the Innovation Variance of a Stationary Time Series. J.Am.Statist.Ass. 63, 141-149

Greville, T.N.E. (1964). Numerical Procedures for Interpolation by Spline Functions. J.SIAM Numer.Anal., Ser. B 1, 53-64

Reinsch, Ch. (1967). Smoothing by Spline Functions. Numer.Math. 10, 177-183

Rosenblatt, M. (1971). Curve Estimates. Ann.Math.Stat., 42, 1815-42

Schoenberg, I.J. (1964). Spline Functions and the Problem of Graduation, Proc.Nat.Acad.Sci. USA, 52, 947-950

Tukey, J.W. (1967). Calculations of Numerical Spectrum Analysis. in: Spectral Analysis of Time Series, ed. B. Harris (J. Wiley, New York), 25-46

Statistical Analysis of Simulation Results by UNISIAS

M. Helm, Munich

INTRODUCTION

The evaluation of results is an old and difficult problem of simulation.Standard simulation packages like GPSS,SIAS,GASP etc. offer the user only mean values without any statistical analysis, so that misleading conclusions about the results may occur.Many efforts has been made to arrive at a more satisfactory treatment but the only one,which seems to us to be promising is that,which uses the powerful theory of time series analysis.
In this connection we have developed the system UNISIAS which supplements the GPSS-like simulation language SIAS with a time series package.Once a sample of data has been collected,the simu= lation is interrupted.The user has then the possibility to ana= lyze the data at the computer terminal.In using the resulting information he decides on whether to continue the simulation run. Thus UNISIAS opens to the user a "window",which allows him to observe and control the simulation in an interactive manner.
In the further sections we will briefly discuss the method and will give an example of the application of UNISIAS.

2. THEORY

The starting point of the present work are the papers by G.S. FISHMAN (1970).Many important elements of FISHMAN's method have been included,but some of his restrictions such as the postula= tion of stationary processes,the unsatisfactory treatment of the initial values and the absence of any analysis of residuals seem us to be to dangerous when applyed to practical simulation stu= dies.Therefore we have used the complete theory of time series as described by BOX and JENKINS (1970).The resulting system UNISIAS turned out to be flexible enough to analyze a broad class of time series generated by various simulation models.
The most important points of the theory are:

2.1 THE MODEL

At first we assume that all time series originating from simu= lation experiments may be represented by means of the following autoregressive model:

$$w_t=\alpha_1 w_{t-1}+\alpha_2 w_{t-2}\dots\dots\dots\dots\alpha_m w_{t-m}+\alpha_{m+1}+\varepsilon_t \qquad (1)$$

where

$$w_t=\nabla_{s_1\ s_2\dots\dots\ s_n} z_t \ . \qquad (2)$$

$$\{z_t\} \quad t=1,2,\dots\dots N \qquad (3)$$

is the time series to be analyzed.The filters ∇_{s_i} are defined by

$$\nabla_{s_i} z_t = z_t - z_{t-s_i} \qquad (4)$$

and are regarded as a simple transformation of the original

data.It is known,that the model (1)-(4) is valid for a very gene= ral class of time series.

At first we remove long-term effects such as trends,instabilities and periodic variations by applying the filters (2); then we use the model (1) to fit the filtered data w_t ,which still contain the short term effects.The coefficients α_i are determined by maximum likelihood methods.By investigating a great variety of time series we have ascertained the residuals of model ε_t contain no further structure (i.e. $\varepsilon_t \neq \beta_1 \tau_{t-1} + \dots \dots \beta_k \tau_{t-k}$) or moving average terms,if we allow high order models ($m \leq 12$).The advantage of pure autoregressive (A.R.) models over autoregressive-moving average models (A.R.M.A.) is that A.R.-models have an exact so= lution which can be formulated recursively.From the computational standpoint this is of great importance.

2.2 MODEL-IDENTIFICATION

Firstly we must determine the kind of the filters (2) to be applied and the order m of the model (1).This model-identifi= cation requires the following steps:

i)FILTERING OF DATA

The first question to be ask is:Are there any instabilities in the original series z_t? To answer this question,we proceed as follows:

By substituting the filtered data (2) in the model (1) we obtain the integrated model for the series z_t:

$$z_t = \delta_1 z_{t-1} + \delta_2 z_{t-2} + \dots \dots \dots \dots \delta_{m+s_n} z_{t-m-s_n} + \delta_{m+s_n+1} + \varepsilon_t \qquad (5)$$

In (5) the δ_i are simple linear combinations of the α_i. Equation (5) is the discrete analogy to an ordinary linear differential-equation of order $m+s_n$ and constant coefficients. Therefore the roots of the corresponding characteristic polynomial determine the stability behavior of the series z_t. Setting $z_t=x^t$ in (5) we obtain:

$$\varphi(x)=x^{m+s_n}-\delta_1 x^{m+s_n-1}-\ldots\ldots\ldots\ldots-\delta_{m+s_n}=0 \qquad (6)$$

If the roots of (6) are such that $|x_i|\leq 1$, the solutions z_t of (5) are stable, otherwise they are instable.

For testing the stability, we examine the maximal root (i.e. x_1) of the characteristic equation of our integrated model (5). If we find, that its value is nearly equal to 1, we use the filter $\nabla_1 z_t=z_{t+1}-z_t$ to remove instability. This procedure is repeated for the filtered series until all instabilities are removed.

Now we examine the series for periodical effects. If we measure the utilization of a computer every hour, we obtain a time series with period 24 and pronounced peaks. It is clearly of great in= terest to investigate various effects caused by such peaks. For example it is misleading to simulate such situations on the hypo= thesis, that the load will be permanently equal the maximum value. Thus , methods for handling such problems will be very important.

To determine the periods of the series, which has beforehand been detrendet, we calculate the spectrum. The cycles $s_1, s_2, \ldots\ldots s_k$ are then removed with the filter $\nabla_{s1\ s2 \cdots\cdots\ sk} z_t$ and the whole filter (2) is now defined. The filtered series is futher investi= gated.

ii) DETERMINATION OF THE ORDER OF THE A.R.-MODEL

At first we assume a low order (for ex. m=2) for our model (1) and calculate the coefficients α_i.It is the basic hypothesis of the theory,that the residuals ε_t are independent from each other.

If ε_t is a random series (white noise) the following relations hold:

$$E(\varepsilon_t)=0 \qquad (7)$$

$$E(\varepsilon_{t_1}\cdot\varepsilon_{t_2}) \begin{cases} =\sigma^2 & t_1=t_2 \\ =0 & t_1\neq t_2 \end{cases} \qquad (8)$$

whreby E(.) denotes the expected value.

However because the ε_t defined in (1) are residuals from a re= gression analysis,their independence can not be defined as in (7) -(8),but must be treated with special methods(BOX and PIER= CE (1970)).

Starting with a low initial order for m we test the resulting ε_t for independence.If the result is not satisfactory the order m is increased,until the residuals ε_t satisfy the assumtion of independence.

2.3 INITIAL VALUES

No time series generated by simulation is stationary because their initial values are chosen arbitrarily.Thus some bias in the entire time series will be induced. We have always a con= ditional ensemble(given the fixed initial values).

For these reasons we have to deal with a conditional algorithm when calculating the coefficients α_i.This algorithm enables us to treat the effects due to bias exactly without eliminating any value of our time series, as this is the case in conventional methods (FISHMAN 1971)).So precious computer time will be saved.

2.4 ITERATION

For identifying our model (1)-(4),we use in general 50-100 data points of the simulated process.As the simulation advances,the time series-model remains unchanged ,but the tolerance intervals of the coefficients α_i shrink.This implies that the precision of the results is increased.

UNISIAS calculates the sample size necessary to obtain the desired precision of the results.The user at the terminal indicates how many of the proposed values shold be simulated.If a continuation of the simulation run seems not reasonable it will be terminated by the user.

3. AN EXAMPLE

By using a simple simulation model we will demonstrate some possibilities,which UNISIAS offers for the analysis of simulated time series.

3.1 THE SIMULATION-MODEL

We consider a simplified rent-a-car office.Initially there are 20 cars available.The arrivals of the clients are random and the inter-arrival-times are exponentially distributed.If the client gets a car,he starts immediately for a trip.The travelling times are also assumed to be exponentially distributed (mean=480 minute so that the results may be checked against those arising from queuing theory.If no car is available the clients have to wait until a car is returned.For simplicity both the arrival and the service-processes are considered stationary. As a matter of course seasonal peaks in the arrival process may be of great interest. As mentioned above such periodicities may be treated without dif= ficulties using UNISIAS.

At this point we seek answers to two questions:

1) What is the threshold value of the arrival rate λ of clients beyond which the system becomes unstable?

2) What is the probability that a new client obtains a car without delay?

3.2 RESULTS

We have investigated two time series in the simulation model:

1) Waiting times of the clients.

2) Number of available cars.

The values of the series are measured every 30 minutes. Each series consists of 200 data points, which represents a reasonable sample size.

In Fig. 1 the maximum root x_1 is plotted against the utilization rate $\varrho = \lambda/(20.\mu)$ (see also eq.6).

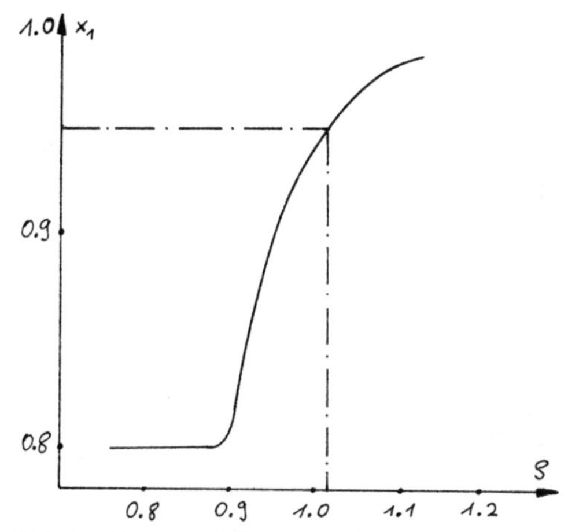

Fig.1) Determination of the instability point.

Beyond the value $\rho=0.9$ the root x_1 increase rapidly and for ρ greater than 1 the value $x_1=1$ is approached asymtotically. We regard a process with $x_1 \geqslant 0.95$ as instationary. For $x_1=0.95$ we have $\rho_c=1.04$ as the threshold beyond which the process becomes instable. Thus we have found a good aproximation of the theoretical value $\rho_c=1.00$.
In general by repeating a simulation experiment with different ran= dom numbers there results different time series, which are reali= zations of the same stochastic process. Thus for any time t we have the probability that n cars are available, i.e. p(n,t). These time varying probabilty densities estimated from a single time series (N=200) are shown in Fig. 2:

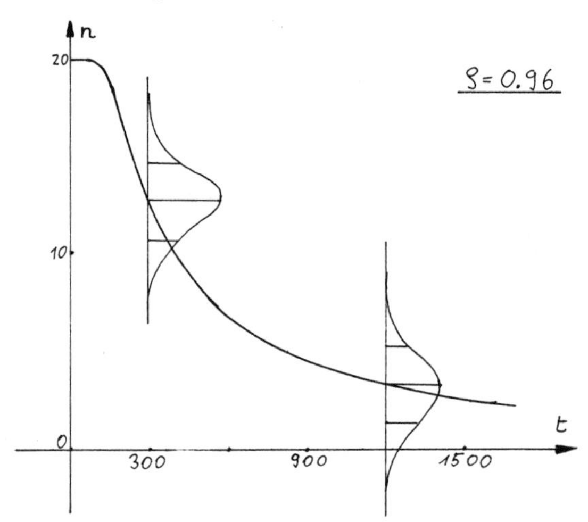

Fig.2) Probability, that n cars are available at time t.

We see from Fig. 2 that n is a continuous variable and that also negative values occur. This is done by the simplification, that z_t in (1)-(4) are continuous variables which can assume every real value. Otherwise our algorithmus would be very complicated.
To pass from continuos values to discrete and positive values for n we accumulate the probability mass of the distribution at the dis= crete points 0,1,2,......20.
For example we define the probability, that no car is available as:

$$P(0,t)= \int_{-\infty}^{0.5} p(n,t)dn \qquad (9)$$

The probability that at least one car is available is $\bar{P}(0,t)=1-P$. Its dependence on ϱ is shown in Fig 3:

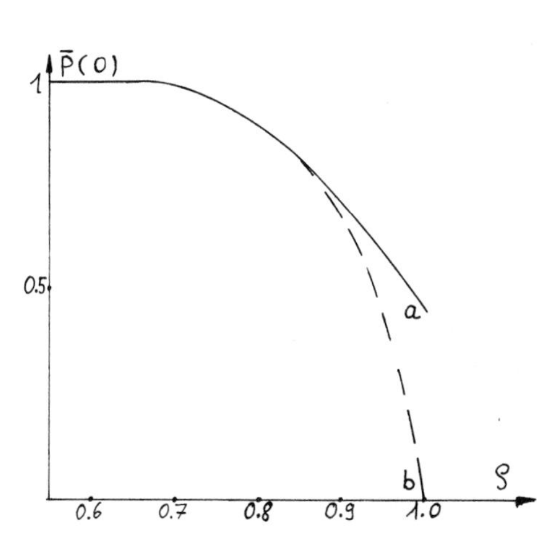

ig.3 Probability, that at least one car is available.

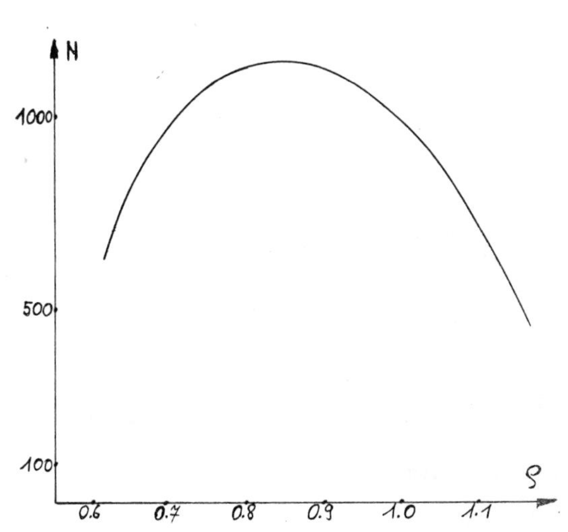

Fig.4) Necessary sample size.

We see, that up to $\varrho=0.95$ the agreement with theoretical values (b) is good. Beyond this value some discrepancies caused by the rela= tive small sample size N=200 occur.

Finally Fig. 4 shows the sample size necessary to determine the mean value of the available cars with a tolerance of ±1 car and a confidence level of 95%.

REFERENCES

BOX G.E.P. and G.M.JENKINS,Time Series Analysis and Control Holden-Day San Francisco (1970).

BOX G.E.P. and D.A. PIERCE,Distribution of residual autocorrelations in autoregressive-moving average time series models, JOUR.AMER.STAT.ASSOC.,Vol 64,pp 299-312,(1970).

FISHMAN G.S. ,Estimating sample size in computing simulation experiments,Management Science ,Vol.18,pp21-38 (1971).

FISHMAN G.S.,Bias considerations in simulation experiments OPER.RES. Vol.20,pp785-790 (1972).

FISHMAN G.S. and P.J. Kiviat,The analysis of simulation generated time series.Mangement Science Vol.13,pp 525-557 (1967).

Variance Reducing Two-Stage Procedures for the Monte Carlo Analysis of Markov Chains

J. Kohlas, Fribourg

1. INTRODUCTION

Variance reducing techniques for Monte Carlo computations have been introduced early in the history of the Monte Carlo method [KAHN, MARSHALL, 1953]. They have been developped systematically with respect to integral computations [HAMMERSLEY, HANDSCOMB, 1964; HANDSCOMB, 1968; HALTON, 1970]. As most Monte Carlo applications can be regarded as an estimation of expected values by artificial sampling, i.e. the computation of integrals of random variables, it has often been argued, that variance reducing integral computations are generally applicable.

These variance reducing methods apply to independent sampling of random variables. But there is a number of problems, especially in Operations Research, where sampling from a stochastic process rather than from independent random variables is necessary. An example is the Monte Carlo analysis of queueing systems [MOY, 1969; PAGE, 1965]. Variance reducing methods for such problems are far less developped, although some of the methods known for integral computations can be carried over, at least conceptually.

The simplest case of dependent sampling arises in connection with Markov chains. A variance reducing method for the evaluation of the expected value of a functional over a stationary Markov chain will be introduced in this paper. Although very simple, the problem is typical for a number of problems arising in O.R. This shows that it is possible to develop variance reducing methods for this type of problem.

In the second section the well known control variate method will be extended from the case of independent sampling to the sampling from Markov chains. This extension is straightforward. A control functional is used to control the estimation. This presumes of course the knowledge of an appropriate control functional.

The à priori knowledge of a control functional is in most cases a too restrictive assumption. It will therefore be shown in section three, how a control functional can be constructed in a first stage of the computations. This control functional can then be used in the second stage to reduce variance or to accelerate convergence of the estimate. Thus, a two-stage procedure will be introduced. The importance of multi-stage methods for Monte Carlo computations has already been noted by MARSHALL [1956].

In the fourth section finally, a result on the asymptotic efficiency of the two-stage procedure will be given.

2. A CONTROL VARIATE METHOD

2.1 Recapitulation of the Method

The control variate method for integral computations is well known [HAMMERSLEY et al., 1964]. Instead of integrals, the computation of finite sums will be considered. Thus, the sum

$$\theta = \sum_{i=1}^{n} f(i)p_i, \text{ where } 0 \leq p_i \leq 1, \sum_{i=1}^{n} p_i = 1 \tag{1}$$

is to be calculated. Using a crude Monte Carlo method, N independent samples X_j from a random variable X, defined by X=i with prob. p_i, are drawn using pseudo random numbers [HAMMERSLEY et al. 1964, KOHLAS, 1972]. θ can then be estimated by

$$\bar{\theta} = \frac{1}{N} \sum_{j=1}^{N} f(X_j) . \tag{2}$$

This estimator has standard deviation

$$\sigma(\bar{\theta}) = \sigma/\sqrt{N} \; ; \; \sigma^2 = \sum_{i=1}^{n} (f(i)-\theta)^2 p_i . \tag{3}$$

If a second sum

$$\Theta' = \sum_{i=1}^{n} f'(i)p_i \tag{4}$$

is known, and if $f' \cong f$, then variance can be reduced by estimating the difference $\Delta\Theta = \Theta'-\Theta$. This is done by sampling as above and estimating $\Delta\Theta$ by

$$\Delta\bar{\Theta} = \frac{1}{N} \sum_{j=1}^{N} (f'(X_j)-f(X_j)) \ . \tag{5}$$

An estimator for Θ is then given by $\hat{\Theta}=\Theta'-\Delta\bar{\Theta}$. Its standard deviation is

$$\sigma(\hat{\Theta}) = \sigma'/\sqrt{N} \ ; \ (\sigma')^2 = \sum_{i=1}^{n} (f'(i)-f(i)-\Delta\Theta)^2 p_i \ . \tag{6}$$

If $f' \cong f$, then $\sigma' < \sigma$ and the variance of the estimator $\hat{\Theta}$ is smaller than the variance of $\bar{\Theta}$. This is the control variate method. It should be noted, that a bad choice of the control summands f' could lead to an increased variance. It is generally true, that a bad application of variance reducing methods can lead to worse estimates than crude Monte Carlo.

2.2 Control Variate Method Applied to Markov Chains

In talking about Markov chains the terminology of CHUNG [1967] is used. Given are transition prob. p_{ij} of a Markov chain with n states $i=1,2,\dots,n$. These states are assumed to form one positive, recurrent class with period 1. There exists therefore an unique stationary distribution p_i, $i=1,2,\dots,n$, which satisfies the system of equations $\sum_{i=1}^{n} p_i p_{ij}=p_j$, $j=1,2,\dots,n$. Given is furthermore a realvalued function $f(i)$ of the states of the chain. This defines a functional over the Markov chain, see [CHUNG, 1967]. The stationary expected value

$$\Theta = \sum_{i=1}^{n} f(i)p_i, \text{ where } 0<p_i<1, \ \sum_{i=1}^{n} p_i=1, \ \sum_{i=1}^{n} p_i p_{ij}=p_j \tag{7}$$

of the functional is to be determined. Instead of solving the system of linear equations for the stationary prob. and summing up (7) to obtain Θ, a crude Monte Carlo method can be applied.

A sample X_j, $j=1,2,\ldots,N$ of the first N transitions of the chain is drawn using pseudo random numbers. This is done by choosing X_o arbitrarily and drawing $X_j=i$ with prob. p_{ki} if $X_{j-1}=k$ [KOHLAS 1972]. Θ can then be estimated by

$$\bar{\Theta} = \frac{1}{N} \sum_{j=0}^{N} f(X_j) . \tag{8}$$

In order to emphasize the dependence of $\bar{\Theta}$ from the sample size N we write sometimes $\bar{\Theta}_N$. The following results can be found in [CHUNG, 1967]:

$$\lim_{N \to \infty} E[\bar{\Theta}_N] = \Theta , \tag{9}$$

$$P[\lim_{N \to \infty} \bar{\Theta}_N = \Theta] = 1 , \tag{10}$$

$$E[(\bar{\Theta}_N - \Theta)^2] = \frac{\sigma^2}{N} + \frac{o(N)}{N} , \quad (\lim_{N \to \infty} o(N) = 0) , \tag{11}$$

$$\lim_{N \to \infty} P\left[\frac{\bar{\Theta}_N - \Theta}{\sigma/\sqrt{N}} \leq x\right] = \frac{1}{\sqrt{2\pi}} \int_{-\infty}^{x} e^{-v^2/2} \, dv , \tag{12}$$

where σ^2 is defined by the following quadratic form

$$\sigma^2 = \sum_{i=1}^{n} p_i (f(i)-\Theta)^2 + 2 \sum_{j=1}^{n} \sum_{k=1}^{n} a_{jk} (f(j)-\Theta)(f(k)-\Theta). \tag{13}$$

The coefficients a_{jk} will not be discussed further, see [CHUNG, 1967; KOHLAS, 1973].

This crude Monte Carlo method can be improved, if a control functional $f' \cong f$ over the same chain, together with its stationary expected value

$$\Theta' = \sum_{i=1}^{n} f'(i) p_i \tag{14}$$

is known. By sampling X_j as above, the difference $\Delta\Theta = \Theta' - \Theta$ can be estimated by

$$\Delta\bar{\Theta} = \frac{1}{N} \sum_{j=0}^{N} (f'(X_j) - f(X_j)) . \tag{15}$$

$\hat{\Theta} = \Theta' - \Delta\bar{\Theta}$ is then an estimator for Θ. If $(\sigma')^2$ is the value of (13) for the difference functional $f''=f'-f$, then it may be expected that $\sigma' < \sigma$, if $f' \cong f$. According to (11) a variance reduction is achieved in this case.

3. A TWO-STAGE PROCEDURE

3.1 Relative Values

A control functional f' will only rarely be known in advance. Some means are therefore to be provided to construct a control functional, if the control variate method is to be of any practical value. This problem will be considered in this section.

$\bar{\Theta}$ depends on the initial value X_o. In order to emphasize this, we write $\bar{\Theta}_N/h$, if $\bar{\Theta}$ is computed with initial state $X_o=h$ and N transitions. Now

$$E[N\bar{\Theta}_N/h] - (N+1)\Theta = \sum_{s=0}^{N} \sum_{i=1}^{n} (p_{hi}^{(s)} - p_i) f(i) = \sum_{i=1}^{n} \left[\sum_{s=0}^{N} (p_{hi}^{(s)} - p_i)\right] f(i) , \qquad (16)$$

where $p_{hi}^{(s)}$ are the s-stage transition prob., $p_{hi}^{(0)} = \delta_{hi}$. The series within brackets in (16) can be shown to converge absolutely [FRECHET, 1950]. Hence

$$\lim_{N \to \infty} \left\{E[N\bar{\Theta}_N/h] - (N+1)\Theta\right\} = \sum_{i=1}^{n} s_{hi} f(i) = s_h , \qquad (17)$$

with

$$s_{hi} = \sum_{s=0}^{\infty} (p_{hi}^{(s)} - p_i) . \qquad (18)$$

Furthermore

$$\Theta + s_h = \sum_{i=1}^{n} p_i f(i) + \sum_{i=1}^{n} \sum_{s=0}^{\infty} (p_{hi}^{(s)} - p_i) f(i) = f(h) + \sum_{j=1}^{n} \sum_{i=1}^{n} \sum_{s=0}^{\infty} p_{hj} (p_{ji}^{(s)} - p_i) f(i)$$

$$= f(h) + \sum_{j=1}^{n} p_{hj} s_j \quad . \tag{19}$$

(19) is a system of linear equations for Θ and the values s_h defined in (17). As solution of this system of equations Θ is uniquely determined and the values s_h are determined up to an additive constant. Solutions s_h of (19) are called relative values (of the initial states $X_o=h$), a name which is explained by (17). Relative values and system (19) play an important role in the policy iteration algorithm for the dynamic optimization of Markov chains [HOWARD, 1960].

In the next subsection it will be shown, how relative values s_h can be estimated by the Monte Carlo method. This can be used to construct a control functional. Suppose that s_h^* are estimates for s_h. $\bar{\Theta}$ is given by the crude Monte Carlo method (8). $\bar{\Theta}$ and s_h^* will not satisfy (19) exactly. Correction terms $\eta(h)$ have to be introduced

$$\bar{\Theta} + s_h^* = f(h) + \eta(h) + \sum_{j=1}^{n} p_{hj} s_j^* \quad . \tag{20}$$

$\eta(h)$ can of course be calculated from $\bar{\Theta}$ and s_h^* using (20). (20) implies on the other hand that $\bar{\Theta}$ and s_h^* are the <u>exact</u> stationary expected value and the <u>exact</u> relative values corresponding to the functional $f' = f + \eta$. Hence, if estimates for s_h can be obtained then a control functional can be constructed.

3.2 Estimation of Relative Values

As the relative values s_h are only determined up to an additive constant, a state i can be chosen arbitrarily and s_i can be put equal to zero, $s_i=0$. This state i will be hold fixed in the sequel. It follows from (17) that

$$s_h = \lim_{N \to \infty} \left\{ E[N\bar{\Theta}_N/h] - E[N\bar{\Theta}_N/i] \right\} . \tag{21}$$

If f(i) is interpreted as a payoff obtained when the chain reaches state i, then $N\bar{\Theta}_N$ is the sum of the payoffs obtained during the first N transitions and s_h is according to (21) the expected difference between the payoff-sums for a large number of transitions if the chain starts in state h and if it starts in state i. This leads to the following intuitive idea for the estimation of s_h: Suppose the chain arrives in state h. Sum afterwards all payoffs (including f(h) for the starting state h) until the chain reaches state i. This gives a sum U(h), which is of course a random variable. Suppose it took $\rho(h)$ transitions to reach state i; $\rho(h)$ is also a random variable. If from state i on summation of payoffs is once continued and once started from scratch, then the difference between the two sums will always be U(h), but the second sum has $\rho(h)$ summands less than the first one. For a large number of transitions the mean payoff in each transition is θ. For the $\rho(h)$ missing summands in the second sum can therefore be accounted by introducing a correction $\theta\rho(h)$. Thus, it can be conjectured, that

$$s_h = E[U(h)] - \theta E[\rho(h)] \quad . \tag{22}$$

That this is in fact true has been proven in [KOHLAS, 1973].

As all states are recurrent, state h and afterwards state i will be reached an infinity of times. Denote the sum U(h) attached to the ν-th time state h is reached by $U_\nu(h)$ and $\rho(h)$ accordingly by $\rho_\nu(h)$. In view of (22) it seems then plausible that s_h can be estimated by

$$s_h^* = \frac{1}{N(h)} \sum_{\nu=1}^{N(h)} U_\nu(h) - \bar{\theta}_N \cdot \frac{1}{N(h)} \sum_{\nu=1}^{N(h)} \rho_\nu(h) \ , \tag{23}$$

where N(h) is the number of times state h and afterwards state i is reached during the N transitions. It has in fact been shown in [KOHLAS, 1973] that s_h^* (23) is a consistent estimator for s_h.

3.3 Two-Stage Procedures

In a first stage of the computations to determine θ, the crude Monte Carlo method gives a first crude estimate $\bar{\theta}$ for θ. At the same time, the simulated sample X_j, j=1,2,...,N can also be used to obtain estimates s_h^* for s_h by (23). For the state i we put

$s_i^* = s_i = 0$.

After terminating the first stage, the corrections $\eta(h)$ are calculated using (20). In the second stage the control functional $f' = f + \eta$ is used to improve the estimate of Θ just as in the method described in subsection 2.2. It has to be reminded that now $\Theta' = \bar{\Theta}$, the crude estimate of the first stage.

The first stage of this two-stage procedure can of course be replaced by any other method to solve (19) approximately.

As both estimators $\bar{\Theta}$ and s_h^* of the first stage are consistent, it follows, that $\eta(h)$ can be made arbitrarily small with a prob. arbitrarily near 1. The variance of the estimator obtained in the second stage is then determined by (11), together with (13), applied to the functional $\eta(h)$. But (13) can be made arbitrarily small with $\eta(h)$.

Thus, for a sufficiently large sample size in the first stage, it is very probable, that the second stage will bring an advantage in terms of estimation variance, i.e. will be variance reducing.

4. STOPPING RULES AND ASYMPTOTIC RESULTS

It is, at least in principle, not the best method to fix the sample size of the first stage in advance. Preferable is a stopping rule, which decides on the basis of the results obtained during the computations, when to stop. Such a rule could be to stop, when (13) evaluated for the correction functional $\eta(h)$ becomes smaller than a given percentage of (13) evaluated for the original functional f. This would guarantee the variance reduction. Unfortunately (13) contains the unknown parameters Θ, p_i and a_{jk}, such that this rule has no practical value.

Other rules may be constructed. It is not at all clear, which rule is optimal in terms of the trade-off between computational effort required and accuracy of the result obtained. But an asymptotic result for certain stopping rules can be given.

Let τ be a stopping time (optional variable [CHUNG, 1967]) of the Markov chain. To each stopping rule corresponds a stopping time τ and σ_τ^2, defined as the value of (13) for the control functional η. To each stopping time τ and sample size N corresponds a truncated stopping time τ_N, defined by $\tau_N=\tau$, if $\tau \leq N$, $\tau_N=N$, if $\tau > N$. Put $\eta_{\tau_N}=f$, if $\tau_N=N$ and $\eta_{\tau_N}=\eta$, if $\tau_N < N$. N represents an upper limit on the sample size of the first stage.

Let now $\hat{\theta}_{N,M}$ be the estimator for θ obtained with the two-stage procedure, using a stopping time τ, truncated at N, with an overall sample size $M > N$,

$$\hat{\theta}_{N,M} = \frac{1}{\tau_N} \sum_{j=0}^{\tau_N} f(X_j) - \frac{1}{M-\tau_N} \sum_{j=\tau_N+1}^{M} \eta_{\tau_N}(X_j) \; . \tag{24}$$

If τ is a stopping time such that $P[\tau < \infty] = 1$, then

$$\lim_{N\to\infty} \lim_{M\to\infty} ME\left[(\hat{\theta}_{N,M}-\theta)^2\right] = E\left[\sigma_\tau^2\right]. \tag{25}$$

This result has been proven in [KOHLAS, 1973].

As an application of this result let τ' be the first time that (13) evaluated at the control functional η becomes smaller than $\varepsilon > 0$. τ satisfies the assumptions for (25). It follows therefore for this stopping rule

$$\lim_{N\to\infty} \lim_{M\to\infty} ME\left[(\hat{\theta}_{N,M}-\theta)^2\right] \leq \varepsilon. \tag{26}$$

REFERENCES

CHUNG, K.L.: Markov Chains with Stationary Transition Probabilities. Springer, Berlin-Heidelberg-New York, 1967.

Frechet, M.: Recherches théoriques modernes sur le calcul des probabilités. Vol. II: Méthodes des fonctions arbitraires, théories des événements en chaîne dans le cas d'un nombre fini d'états possibles. Gauthier-Villars, Paris, 195o.

HALTON, J.H.: A Retrospective and Prospective Survey of the Monte Carlo Method. SIAM Review, 12, No. 1, 1970.

HAMMERSLEY, J.M. and D.C. HANDSCOMB: Monte Carlo Methods. Methuen, London, 1964.

HANDSCOMB, D.C.: Monte Carlo Techniques: Theoretical. The Design of Computer Experiments. Edited by T.H. NAYLOR. Duke University Press, Durham, N.C., 1969.

HOWARD, R.A.: Dynamic Programming and Markov Processes. M.I.T. Press, Cambridge, Mass., 1960.

KAHN, H. and A.W. MARSHALL: Methods of Reducing Sample Size in Monte Carlo Computations. Op. Res., 1, 263-278, 1953.

KOHLAS, J.: Monte Carlo Simulation im Operations Research. Lecture Notes in Economics and Mathematical Systems, 63. Springer, Berlin-Heidelberg-New York, 1972.

KOHLAS, J.: Relative Werte und Monte Carlo Analyse von Kosten- oder Erlös-Strukturen auf Markoff-Ketten. Juris, Zürich, 1973.

MARSHALL, A.W.: The Use of Multi-Stage Sampling Schemes in Monte Carlo Computations. Symposium on Monte Carlo Methods. Edited by H.A. MEYER. Wiley, New York, 1958.

MOY W.A.: Monte Carlo Techniques: Practical. The Design of Computer Simulation Experiments. Edited by T.H. NAYLOR. Duke University Press, Durham, 1969.

PAGE E.S.: On Monte Carlo Methods in Congestion Problems, II: Simulation of Queuing Systems. Op. Res., 13, 300-305, 1965.

Efficient Estimation via Simulation of Work-Rates in Closed Queueing Networks

Stephen S. Lavenberg, San Jose

1. INTRODUCTION

Closed queueing networks, as considered in this paper, are composed of interconnected multi-server stages serving a fixed number of customers. The route followed by a customer in the network is described by a Markov chain over the stage names. Service times at a particular stage are independent and identically distributed random variables and service times at different stages are mutually independent. Such networks have found modeling application in a variety of areas including multistage production processes [Koenigsberg, 1958], machine repair [Barlow and Proschan, 1965], and, most recently, multiprogrammed computer systems [Gaver, 1967], [Buzen, 1971].

While analytic solutions for such networks can be obtained in certain special cases, e.g. [Gordon and Newell, 1967], [Lavenberg, 1973], in general these networks remain intractable analytically and/or unfeasible computationally. Thus, one is led to consider simulation of these networks. This paper illustrates the benefits of employing analytic results in simulation. An analytically derived relation between response variables is used to suggest improved estimators of these response variables and results on the stochastic structure of the system being simulated are used to provide a distributional theory for estimating confidence intervals.

The response variables considered in this paper are work-rates in closed queueing networks. The work-rate for a stage is defined as the long-run time-average of the total busy time for all servers at the stage. Work-rates provide a measure of resource utilization for the system being modeled. Using a known relation [Chang and Lavenberg, 1972] between work-rates at different stages, new work-rate estimators are proposed which are linear combinations of the straightforward estimators of work-rates at different stages. The coefficients of the linear combinations are constrained so that the new estimators are asymptotically strongly consistent. Previously, Lavenberg and Shedler [1973] investigated these new estimators in a 2-stage closed queueing network having a single server at each stage. They found that substantial reductions in confidence interval widths can be obtained using the new estimators. This paper extends the investigation of these estimators to more complex networks.

In section 2 closed queueing networks and the straightforward and new work-rate estimators are defined. Section 3 contains an analytic derivation of asymptotic confidence intervals for work-rates under the assumption that all service time distributions are exponential. The optimal coefficients for the new estimators are computed and numerical results are presented which demonstrate that substantial reductions in confidence interval widths are possible using the new estimators. In section 4 networks with non-exponential service times are considered whose stochastic structure is such that the work-rate estimators are asymptotically normal. A method for choosing a priori the coefficients for the new estimators is proposed. Confidence intervals, estimated via simulation using the t statistic, are presented and have substantially smaller widths using the new estimators in almost all simulations. In addition, comparisons of estimated coverages with theoretically predicted coverages are given in section 4.

2. DEFINITIONS AND NOTATION

The network consists of s interconnected service stages, named by the integers 1,2,...,s. For each i, stage i consists of a queue served by m_i identical parallel servers. There are N customers in the network. Customers enqueued at a stage are served on a first-come first-served basis. Service times at stage i are independent identically distributed non-negative random variables, each having non-zero finite expectation μ_i. Service times at different stages are mutually independent. Upon completion of its service at stage i the customer just served proceeds instantaneously to stage j with probability $p_{ij}\geq 0, \sum_{j=1}^{s} p_{ij}=1$. The transition matrix $P=(p_{ij})$ is assumed to be irreducible. Let $\underline{\pi}=(\pi_1,\pi_2,\ldots,\pi_s)$ be the unique probability vector which satisfies $\underline{\pi}P=\underline{\pi}$, where $\pi_i>0$, $1\leq i\leq s$, since P is irreducible.

Consider an arbitrary location of customers in the network at time zero with zero service time rendered. Let $W_i(t)$ be the total busy time for all servers at stage i in the time interval [0,t]. (The dependence on the initial location of customers is suppressed in the notation.) Then Chang ... et al. [1972] established the following results:

For each i, $1\leq i\leq s$,

$$U_i \equiv \lim_{t\to\infty} W_i(t)/t \quad \text{exists with probability 1,} \tag{1}$$

$$\frac{U_i}{U_j} = \frac{\pi_i\mu_i}{\pi_j\mu_j} \quad \text{for all } j\neq i\ . \tag{2}$$

U_i as defined in eq. (1) is called the work-rate for stage i. U_i is a degenerate random variable, i.e., a constant, whose value is independent of the initial location of customers in the network.

A straightforward estimator of U_i is $U_i(\tau)=W_i(\tau)/\tau$ where τ is the fixed time at which a realization of the simulation is stopped. It follows from eq. (1) that $U_i(\tau)$ is an asymptotically strongly consistent estimator of U_i. The relation in eq. (2) suggests the following as a new estimator of U_i:

$$U_i(\tau,\underline{\alpha}) = \sum_{j=1}^{s} \alpha_j(\pi_i\mu_i/\pi_j\mu_j)U_j(\tau)$$

where $\underline{\alpha}=(\alpha_1,\alpha_2,\ldots,\alpha_s)$ is a vector of constant coefficients and $\sum_{j=1}^{s}\alpha_j=1$. $U_i(\tau,\underline{\alpha})$ is also an asymptotically strongly consistent estimator of U_i. The use of this estimator is an example of correlated sampling. Correlated sampling methods, such as antithetic variates, have been used to achieve variance reduction in simulation, e.g. [Moy, 1971]. In the remainder of this paper we compare the accuracy of the new and straightforward estimators by comparing the widths of confidence intervals for work-rates obtained using these estimators. The choice of the vector $\underline{\alpha}$ will be discussed.

3. NETWORKS WITH ALL SERVICE TIMES EXPONENTIALLY DISTRIBUTED

It is assumed throughout this section that all service time distributions are exponential, i.e. if $F_i(t)$ is the service time distribution for stage i, then $F_i(t) = 1-\exp(-t/\mu_i), t\geq 0$. With this assumption the steady-state joint distribution of the number of customers in each stage is given by a well-known expression [Gordon ... et al., 1967] from which the work-rates can be computed. Thus, there is no need to estimate the work-rates via simulation. Nevertheless, the analytic results on confidence intervals derived in this section allow a preliminary comparison to be made of the accuracy of the new and straightforward work-rate estimators. Also, the optimal coefficients for the new estimators computed in this section will be used in section 4 when simulating networks with non-exponential service times.

3.1 Derivation of asymptotic confidence intervals

Let $n_i(t)$, $t\geq 0$, be the number of customers in stage i at time t and let $\underline{n}(t)=(n_1(t),n_2(t),\ldots,n_s(t))$. Assume that $\underline{n}(0)$ is given and adopt the convention $\underline{n}(t)=\underline{n}(t+)$. Then $\underline{n}(t)$ is a finite state irreducible Markov process over the state space $A=\{(n_1,n_2,\ldots,n_s): n_i\geq 0,\ 1\leq i\leq s,\ \sum_{i=1}^{s} n_i=N\}$. The state space has cardinality $|A| = \binom{N+s-1}{s-1}$. Let $\{e_k:k=0,1,2,\ldots\}$ denote epochs of transition in the Markov process, i.e. e_k is an epoch at which a customer completes service at a stage and instantaneously enters the next stage on its route. It is assumed that $e_0=0$. Let $\{r_k:k=0,1,2,\ldots\}$ denote those epochs for which $\underline{n}(r_k)=\underline{n}(0)$; note that $r_0=0$. The r_k are regeneration points [Smith, 1958] in the process $\underline{n}(t)$. The times between them, denoted by $\{Y_k:k=1,2,\ldots\}$ where $Y_k=r_k-r_{k-1}$, $k\geq 1$, are independent random variables identically distributed as a random variable Y, i.e. the Y_k form a renewal process. From the properties of $\underline{n}(t)$ it follows that $E[Y]<\infty$, $E[Y^2]<\infty$. Let $W_{ik}=W_i(r_k)-W_i(r_{k-1})$, $k\geq 1$. Observe that W_{ik}, $k\geq 1$, are independent random variables identically distributed as a random variable W_i. Thus, $W_i(t)$ is a cumulative process defined with respect to the sequence of regeneration points $\{r_k\}$. Furthermore, since $W_{ik}\leq m_i Y_k$ it follows that $E[W_i]<\infty$, $E[W_i^2]<\infty$. Thus, from cumulative process results [Smith, 1958]

$$U_i = E[W_i]/E[Y],$$
$$\mathrm{Var}[W_i(t)/t] \sim \sigma_i^2/t,$$

where $\sim$ denotes asymptotic equality for large t,

$$\sigma_i^2 = (E[W_i^2] + E[Y^2]U_i^2 - 2E[W_iY]U_i)/E[Y] \ , \tag{3}$$

and $(W_i(t)-U_it)/\sigma_i t^{1/2}$ is asymptotically normally distributed with mean zero and variance one, i.e.

$$\lim_{t\to\infty} \Pr\{(W_i(t) - U_it)/\sigma_i t^{1/2} \leq \gamma\} = \phi(\gamma) \ , \tag{4}$$

where

$$\phi(\gamma) = \int_{-\infty}^{\gamma} (2\pi)^{-1/2} \exp(-x^2/2)dx \ .$$

From eq. (4) it follows that for τ large and $\gamma\geq 0$

$$[U_i(\tau) - \gamma\sigma_i/\tau^{1/2},\ U_i(\tau) + \gamma\sigma_i/\tau^{1/2}] \tag{5}$$

is approximately a $100(2\phi(\gamma)-1)\%$ confidence interval for U_i.

Since $W_i(t)$, $1 \le i \le s$, are cumulative processes defined with respect to the same sequence of regeneration points, $\sum_{j=1}^{s} \alpha_j (\pi_i \mu_i / \pi_j \mu_j) W_j(t)$ is also a cumulative process with respect to this sequence of regeneration points. Thus, it can be shown that for τ large and $\gamma \ge 0$

$$[U_i(\tau,\underline{\alpha}) - \gamma\sigma_i(\underline{\alpha})/\tau^{1/2},\ U_i(\tau,\underline{\alpha}) + \gamma\sigma_i(\underline{\alpha})/\tau^{1/2}) \tag{6}$$

is approximately a $100(2\phi(\gamma)-1)\%$ confidence interval for U_i where

$$\sigma_i^2(\underline{\alpha}) = (\pi_i\mu_i)^2 \left(\sum_{j=1}^{s} \alpha_j^2 \sigma_j^2 / (\pi_j\mu_j)^2 + \sum_{j=1}^{s} \sum_{\ell \ne j}^{s} \alpha_j \alpha_\ell \sigma_{j\ell} / (\pi_j \mu_j \pi_\ell \mu_\ell) \right) \tag{7}$$

and

$$\mathrm{Cov}[W_i(t)/t, W_j(t)/t] \sim \sigma_{ij}/t,$$

$$\sigma_{ij} = (E[W_i W_j] + E[Y^2]U_i U_j - E[W_i Y]U_j - E[W_j Y]U_i)/E[Y] . \tag{8}$$

In order to compare the lengths of the approximate confidence intervals in eq. (5) and eq. (6), σ_i^2 in eq. (3) and σ_{ij} in eq. (8) are computed next. The Markov process $\underline{n}(t)$ is viewed as a semi-Markov process with exponential waiting times and semi-Markov process analysis techniques are used in the computation [Barlow... et al., 1965].

Let $n_{ik} = n_i(e_k)$. Then $\{(n_{1k}, n_{2k}, \ldots, n_{sk}) : k=0,1,2,\ldots\}$ is a finite state irreducible Markov chain over the state space A with initial state $\underline{n}(0)$. If $\underline{n} = (n_1, \ldots, n_i, \ldots, n_j, \ldots, n_s)$, $i \ne j$, let $\underline{n}(i,j) = (n_1, \ldots, n_i - 1, \ldots, n_j + 1, \ldots, n_s)$ provided $\underline{n}(i,j) \varepsilon A$. Let $q(\underline{n}, \underline{n}')$ denote the probability of transition from state $\underline{n}$ to state $\underline{n}'$ in the Markov chain. Let $\lambda(\underline{n}) = \sum_{k=1}^{s} \min(n_k, m_k)/\mu_k$. It is straightforward to show that

$$q(\underline{n},\underline{n}') = \begin{cases} \min(n_i, m_i) p_{ij} / \mu_i \lambda(\underline{n}) & , \underline{n}' = \underline{n}(i,j) \varepsilon A,\ i \ne j \\ \sum_{i=1}^{s} \min(n_i, m_i) p_{ii} / \mu_i \lambda(\underline{n}) & , \underline{n}' = \underline{n} \\ 0 & , \text{else.} \end{cases}$$

Let $v(\underline{n})$, $\underline{n} \varepsilon A$, be the unique solution to the equations

$$\sum_{\underline{n} \varepsilon A} v(\underline{n}) q(\underline{n}, \underline{n}') = v(\underline{n}') , \quad \underline{n}' \varepsilon A \tag{9}$$

$$\sum_{\underline{n} \varepsilon A} v(\underline{n}) = 1 , \tag{10}$$

i.e. $v(\underline{n})$, $\underline{n} \varepsilon A$, are the unique stationary probabilities of the Markov chain. It can be verified by direct substitution into eq. (9) that

$$v(\underline{n}) = (\lambda(\underline{n}) \prod_{k=1}^{s} (\pi_k \mu_k)^{n_k} / \beta(n_k, m_k))/C$$

where

$$\beta(n,m) = \begin{cases} n! & , 1 \le n \le m \\ m! m^{n-m} & , n > m \end{cases}$$

and C is a normalizing constant chosen so that eq. (10) is satisfied.

The time $e_{k+1} - e_k$ between successive epochs e_k and e_{k+1} is, given the states at these epochs, an exponentially distributed random variable which is independen of the times between previous epochs and of the states at previous epochs. Thus,

$\underline{n}(t)$ can be viewed as a semi-Markov process with exponential waiting times. Let $\Delta(\underline{n},\underline{n}')$ denote the time between successive epochs given that the state at the first of these epochs is $\underline{n}$ and at the second is $\underline{n}'$. It is straightforward to show that $\Delta(\underline{n},\underline{n}')$ is exponentially distributed with rate parameter $\lambda(\underline{n})$, i.e. for all $\underline{n}'$ such that $q(\underline{n},\underline{n}')>0$

$$\Pr\{\Delta(\underline{n},\underline{n}')<t\} = 1-\exp(-\lambda(\underline{n})t) \quad , \quad t\geq 0 \ .$$

Let $w_i(\underline{n},\underline{n}')$ denote the busy time for all servers at stage i between successive epochs given that the states at these epochs are $\underline{n}$ and $\underline{n}'$ respectively. Clearly, $w_i(\underline{n},\underline{n}') = \min(n_i,m_i)\Delta(\underline{n},\underline{n}')$ so that $w_i(\underline{n},\underline{n}')$ is exponentially distributed with rate parameter $\lambda(\underline{n})\min(n_i,m_i)$.

The random variables Y and W_i were defined earlier. In terms of the semi-Markov process Y is the first passage time from state $\underline{n}(0)$ to state $\underline{n}(0)$ and W_i is the busy time for all servers at stage i during this first passage time. Expressions for the expectations E[Y], $E[W_i]$, $E[Y_2]$, $E[W_i^2]$, $E[W_iY]$ and $E[W_iW_j]$ can be obtained using standard techniques and are given in the Appendix. It should be noted that U_i, σ_i^2 and σ_{ij} do not depend on the initial state $\underline{n}(0)$ even though the expectations do depend on $\underline{n}(0)$. Thus, any initial state can be used in the computation of U_i, σ_i^2 and σ_{ij}.

In order to minimize the width of the asymptotic confidence interval for U_i in eq. (6), $\underline{\alpha}$ should be chosen to minimize $\sigma_i^2(\underline{\alpha})$ in eq. (7). It is straightforward to show that the value of $\underline{\alpha}$ which minimizes $\sigma_i^2(\underline{\alpha})$ subject to the constraint $\sum_{i=1}^{s} \alpha_i=1$ is

$$\underline{\alpha}^* = \underline{1}D^{-1}/\underline{1}D^{-1}\underline{1}^T \ ,$$

where $D=(d_{ij})$ is an s by s matrix with entries $d_{ij}=\sigma_{ij}/(\pi_i\mu_i\pi_j\mu_j)$, $\underline{1}$ is a row vector of dimension s with all components equal to unity and $\underline{1}^T$ is the transpose of $\underline{1}$. The resulting minimum value of $\sigma_i^2(\underline{\alpha})$ is

$$\sigma_i^2(\underline{\alpha}^*) = (\pi_i\mu_i)^2/\underline{1}D^{-1}\underline{1}^T \ .$$

Note that $\underline{\alpha}^*$ minimizes $\sigma_i^2(\underline{\alpha})$ subject to $\sum_{i=1}^{s} \alpha_i=1$ for all i.

3.2 Numerical results

In tables 1-2 the straightforward estimators and the new estimators with optimum coefficients are compared by numerically comparing σ_i^2 and $\sigma_i^2(\underline{\alpha}^*)$ for two closed queueing networks. In the tables $\rho_i=\sigma_i^2(\underline{\alpha}^*)/\sigma_i^2$. The maximum reduction in the width of the asymptotic confidence interval for U_i obtained by using the new estimator is equal to $\rho_i^{1/2}$. The tables indicate that for a fixed level of confidence and a fixed stopping time substantial reductions in confidence interval widths are possible using the new estimators (reductions in width by factors of 2-5 are common, corresponding to reductions in stopping time by factors of 4-25 for a fixed width).

When estimating U_i via simulation using the estimator $U_i(\tau,\underline{\alpha})$, $\underline{\alpha}^*$ is not known. Either a value of $\underline{\alpha}$ which is believed to be near $\underline{\alpha}^*$ is chosen a priori or $\underline{\alpha}^*$ is estimated empirically from multiple realizations of the simulation. Thus, the increase in accuracy obtained in practice when using the estimator $U_i(\tau,\underline{\alpha})$ will not be as great as is indicated in the tables.

Observe from the tables that $U_i(\tau)$ and $(\pi_i\mu_i/\pi_j\mu_j)U_j(\tau)$ have in general unequal variances, i.e. $\sigma_i^2\neq(\pi_i\mu_i/\pi_j\mu_j)^2\sigma_j^2$. Although not shown in the tables, all computed values of σ_{12} were negative for the 2-stage network of table 1. (Only negative covariances were observed by Lavenberg...et al. [1973] for the 2-stage network with a single server at each stage and one service time distribution arbitrary.) However, for the more complex network in table 2 both positive and negative values of σ_{ij} were computed. Thus, it appears $\underline{\alpha}^*$ could have some negative components. The computed values of $\underline{\alpha}^*$ in the tables have only positive components, however. Also, from the tables the components of $\underline{\alpha}^*$ are such that stages having larger work-rates per server have larger components, i.e. if $U_i/m_i > U_j/m_j$, or equivalently from eq. (2) if $\pi_i\mu_i/m_i > \pi_j\mu_j/m_j$, then $\alpha_i^* > \alpha_j^*$.

μ_2	N	i	U_i	σ_i^2	α_i^*	$\sigma_i^2(\underline{\alpha}^*)$	ρ_i
1	2	1	.800	.160	.700	.016	.100
		2	.800	.800	.300	.016	.020
	4	1	.957	.063	.870	.030	.482
		2	.957	1.48	.130	.030	.020
	6	1	.989	.019	.950	.014	.733
		2	.989	1.80	.050	.014	.008
2	2	1	.600	.352	.467	.011	.030
		2	1.20	1.09	.533	.043	.039
	4	1	.778	.466	.492	.065	.139
		2	1.56	1.77	.508	.260	.147
	6	1	.846	.523	.497	.094	.179
		2	1.69	2.04	.503	.374	.183
4	2	1	.385	.422	.262	.003	.007
		2	1.54	.888	.738	.047	.052
	4	1	.475	.730	.132	.013	.018
		2	1.90	.479	.868	.215	.449
	6	1	.494	.893	.053	.007	.008
		2	1.98	.151	.947	.107	.708

Table 1.
Comparison of estimators - exponential service times
$s=2,\ m_1=1,\ m_2=2,\ \mu_1=1$

$$P = \begin{pmatrix} 0 & 1 \\ 1 & 0 \end{pmatrix}$$

μ_2	i	U_i	σ_i^2	α_i^*	$\sigma_i^2(\underline{\alpha}^*)$	ρ_i
1.11	1	.928	.907	.746	.394	.434
	2	.928	2.60	.090	.394	.152
	3	.464	7.99	.082	.098	.012
	4	.464	7.99	.082	.098	.012
2.22	1	.650	4.27	.246	1.02	.238
	2	1.30	19.8	.262	4.07	.205
	3	.650	17.5	.246	1.02	.058
	4	.650	17.5	.246	1.02	.058
4.44	1	.355	3.61	.055	.791	.219
	2	1.42	62.1	.324	12.7	.204
	3	.710	36.0	.310	3.16	.088
	4	.710	36.0	.310	3.16	.088

Table 2.
Comparison of estimators - exponential service times
$s=4,\ m_1=1,\ m_2=2,\ m_3=1,\ m_4=1,\ \mu_1=1,\ \mu_3=\mu_4=9\mu_2,\ N=6$

$$P = \begin{pmatrix} 0 & .9 & .05 & .05 \\ 1 & 0 & 0 & 0 \\ 1 & 0 & 0 & 0 \\ 1 & 0 & 0 & 0 \end{pmatrix}$$

4. NETWORKS WITH NON-EXPONENTIAL SERVICE TIMES

In this section it is assumed that each service time distribution function is either

(i) an arbitrary finite mixture of Erlangian distributions, i.e. a distribution of the form $\sum_{j=1}^{J} \delta_j F(k_j,t)$, $t\geq 0$, where $\delta_j\geq 0$, $1\leq j\leq J$, $\sum_{j=1}^{J} \delta_j=1$, k_j is a positive integer and $F(k,t)$ is an Erlang-k distribution with arbitrary finite mean, or

(ii) a bounded distribution

and at least one distribution is of type (i). The closed queueing networks having such service time distributions are in general intractable analytically. However, the following theorem for such networks, which has been proved by the author and whose somewhat lengthy proof will be published elsewhere, provides the basis for a distributional theory to be used in estimating confidence intervals for work-rates via simulation.

Theorem - For a closed queueing network with each service time distribution of either type (i) or type (ii) and at least one distribution of type (i), there exists an increasing sequence of random times $\{r_k:k\geq 1\}$, $r_1\geq 0$, which are regeneration points in the stochastic process describing the evolution of the network in time. Furthermore, the first two moments of the time between successive regeneration points are finite.

Using this theorem, it can be shown from cumulative process results [Smith, 1958] that $(U_i(t)-U_i)/(\mathrm{Var}[U_i(t)])^{1/2}$ and $(U_i(t,\underline{\alpha})-U_i)/(\mathrm{Var}[U_i(t,\underline{\alpha})])^{1/2}$ are asymptotically normally distributed with mean zero and variance one. Therefore, confidence intervals for U_i can be estimated using the t statistic from multiple independent realizations of a simulation as described below. (The theorem also provides sufficient conditions for estimating confidence intervals using the method proposed by Crane and Iglehart [1973]. This method is not pursued in this paper, however.)

Let $M>1$ be the number of independent realizations. Each realization is started at simulated time zero under fixed initial conditions and is stopped at simulated time τ. Let the superscript k refer to the value of an estimator observed on the kth realization. Let

$$\hat{U}_i(\tau) = \sum_{k=1}^{M} U_i^k(\tau)/M ,$$

$$\hat{V}_i^2(\tau) = \sum_{k=1}^{M} (U_i^k(\tau)-\hat{U}_i(\tau))^2/(M-1) .$$

Then $M^{1/2}(\hat{U}_i(\tau)-U_i)/\hat{V}_i(\tau)$ is asymptotically distributed as the standardized t statistic with M-1 degrees of freedom. Therefore, for τ large and $\gamma\geq 0$,

$$[\hat{U}_i(\tau)-\gamma\hat{V}_i(\tau)/M^{1/2},\ \hat{U}_i(\tau)+\gamma\hat{V}_i(\tau)/M^{1/2}] \tag{11}$$

is approximately a $100(2\theta_{M-1}(\gamma)-1)\%$ confidence interval for U_i, where $\theta_{M-1}(\gamma)$ is the distribution of the standardized t statistic with M-1 degrees of freedom. Similarly, for fixed $\underline{\alpha}$ let

$$\hat{U}_i(\tau,\underline{\alpha}) = \sum_{k=1}^{M} U_i^k(\tau,\underline{\alpha})/M ,$$

$$\hat{V}_i^2(\tau,\underline{\alpha}) = \sum_{k=1}^{M} (U_i^k(\tau,\underline{\alpha})-\hat{U}_i(\tau,\underline{\alpha}))^2/(M-1) .$$

Then, for τ large and $\gamma \geq 0$,

$$[\hat{U}_i(\tau,\underline{\alpha}) - \gamma\hat{V}_i(\tau,\underline{\alpha})/M^{1/2},\ \hat{U}_i(\tau,\underline{\alpha}) + \gamma\hat{V}_i(\tau,\underline{\alpha})/M^{1/2}] \tag{12}$$

is approximately a $100(2\theta_{M-1}(\gamma)-1)\%$ confidence interval for U_i.

In the above it is assumed that the vector $\underline{\alpha}$ is chosen a priori rather than estimated during the simulation. For a network with specified values of s, N, m_i, $1 \leq i \leq s$, P, μ_i, $1 \leq i \leq s$, and service time distributions $F_i(t)$, $1 \leq i \leq s$, $\underline{\alpha}$ is chosen equal to $\underline{\alpha}^*$, the optimum vector of coefficients computed in section 3 under the assumption that all service times are exponentially distributed. The estimation of $\underline{\alpha}$ during the simulation is not considered in this paper.

In tables 3 and 4, 90% confidence intervals for U_i obtained from using the estimators $\hat{U}_i(\tau)$ and $\hat{U}_i(\tau,\underline{\alpha}^*)$ are compared. The networks in these tables are intractable analytically. In each table a summary of results from 25 independent experiments is presented, where each experiment consists of five independent realizations. The columns headed by $\overline{\text{Int}_i}$ and $\overline{\text{Int}_i}^*$ contain the averages over the 25 experiments of the estimated confidence intervals for U_i in eq. (11) and eq. (12) respectively. The last column in the tables contains the number of experiments for which the width of the estimated confidence interval for U_i in eq. (12) is less than the width of the estimated confidence interval for U_i in eq. (11). Note that in almost all experiments the use of the new estimators results in confidence intervals having smaller widths. The tables demonstrate substantial reductions in confidence interval widths using the new estimators in practice.

μ_2	i	$\overline{\text{Int}_i}$	$\overline{\text{Int}_i}^*$	(see text)
1	1	(.929, .942)	(.932, .943)	16
	2	(.903, .998)	(.932, .943)	25
2	1	(.708, .747)	(.723, .740)	24
	2	(1.41 ,1.53)	(1.45 ,1.48)	25
3	1	(.420, .480)	(.449, .462)	25
	2	(1.78 ,1.87)	(1.80 ,1.85)	24

Table 3.

Estimated 90% confidence intervals for network described in Table 1
N=4, τ=1000, M=5, 25 experiments
Stage 1 service times: Hyperexponential with coef. of var. = 2
$F(t) = \varepsilon(1-\exp(-t/\gamma_1)) + (1-\varepsilon)(1-\exp(-t/\gamma_2))$, $t \geq 0$
$\gamma_1 = 4.0$, $\gamma_2 = .5$, $\varepsilon = .143$
Stage 2 service times: Constant

μ_2	i	$\overline{\text{Int}_i}$	$\overline{\text{Int}_i}^*$	(see text)
1.11	1	(.905, .927)	(.910, .925)	19
	2	(.899, .948)	(.910, .925)	25
	3	(.422, .495)	(.455, .462)	25
	4	(.420, .501)	(.455, .462)	25
2.22	1	(.619, .670)	(.633, .658)	21
	2	(1.25 ,1.34)	(1.27 ,1.32)	22
	3	(.587, .694)	(.633, .658)	25
	4	(.591, .704)	(.633, .658)	25
4.44	1	(.339, .386)	(.355, .377)	24
	2	(1.38 ,1.55)	(1.42 ,1.51)	22
	3	(.653, .802)	(.710, .754)	25
	4	(.657, .818)	(.710, .754)	25

Table 4. (Caption on next page)

Estimated 90% confidence intervals for network described in Table 2
$\tau = 4000$, $M = 5$, 25 experiments
Stage 1 service times: Hyperexponential with coef. of var. = 2 (see Table 3)
Stage 2 service times: Constant
Stage 3 service times: Erlang-2
Stage 4 service times: Erlang-2

In table 5 computed work-rates, average estimated confidence intervals and estimated coverages are presented for networks with all service times exponential. The results are based on 100 independent experiments of five independent realizations each. The estimated coverage (expressed as a percentage and denoted by $\hat{C}_i$ or $\hat{C}_i^*$) is the number of experiments for which the estimated confidence interval contains the computed work-rate divided by the total number of experiments. Note that the estimated coverages are close to the theoretically predicted coverage of 90%. These results support the validity of the confidence intervals in eq. (11) and eq. (12).

μ_2	i	U_i	$\overline{Int}_i$	$\hat{C}_i$	$\overline{Int}_i^*$	$\hat{C}_i^*$
1	1	.957	(.948, .963)	93%	(.951, .961)	95%
	2	.957	(.921, .994)	97%	(.951, .961)	95%
2	1	.778	(.759, .798)	95%	(.771, .786)	95%
	2	1.56	(1.52 ,1.59)	94%	(1.54 ,1.57)	95%
4	1	.475	(.451, .501)	91%	(.472, .479)	93%
	2	1.90	(1.88 ,1.92)	95%	(1.89 ,1.91)	93%

Table 5.
Estimated 90% confidence intervals and coverages for the network described in Table 1 - exponential service times, $N = 4$, $\tau = 1000$, $M = 5$, 100 experiments

ACKNOWLEDGMENT
T. C. Kelly wrote the program to compute the asymptotic confidence intervals in section 3. D. P. Gaver suggested the choice of the vector $\underline{\alpha}$ in section 4.

APPENDIX

The following expressions can be obtained in a straightforward manner using analysis techniques similar to those used in Barlow ... et al. [1965] to compute the moments of first passage times in semi-Markov processes.

$$E[Y] = \sum_{\underline{n}\in A} v(\underline{n})/\lambda(\underline{n})v(\underline{n}(0))$$

$$E[W_i] = \sum_{\underline{n}\in A} v(\underline{n})\min(n_i,m_i)/\lambda(\underline{n})v(\underline{n}(0))$$

$$E[Y^2] = 2\sum_{\underline{n}\in A} v(\underline{n})E[Y^{(1)}(\underline{n},\underline{n}(0))]/\lambda(\underline{n})v(\underline{n}(0))$$

$$E[W_i^2] = 2\sum_{\underline{n}\in A} v(\underline{n})E[W_i^{(1)}(\underline{n},\underline{n}(0))]\min(n_i,m_i)/\lambda(\underline{n})v(\underline{n}(0))$$

$$E[W_iY] = \sum_{\underline{n}\in A} v(\underline{n})(E[Y^{(1)}(\underline{n},\underline{n}(0))]\min(n_i,m_i)+E[W_i^{(1)}(\underline{n},\underline{n}(0))])/\lambda(\underline{n})v(\underline{n}(0))$$

$$E[W_iW_j] = \sum_{\underline{n}\in A} v(\underline{n})(E[W_j^{(1)}(\underline{n},\underline{n}(0))]\min(n_i,m_i)+E[W_i^{(1)}(\underline{n},\underline{n}(0))]\min(n_j,m_j))/\lambda(\underline{n})v(\underline{n}(0))$$

The expectations $E[Y^{(1)}(\underline{n},\underline{n}(0))]$ and $E[W_i^{(1)}(\underline{n},\underline{n}(0))]$ are computed as follows:

Let Q^0 denote the $|A|$ by $|A|$ matrix obtained from the matrix $Q = (q(\underline{n},\underline{n}'))$ by setting all entries in the column of Q corresponding to state $\underline{n}(0)$ equal to zero. Let $Z = (z(\underline{n},\underline{n}')) = (I-Q^0)^{-1}$. Then,

$$E[Y^{(1)}(\underline{n},\underline{n}(0))] = \sum_{\underline{n}'\in A} z(\underline{n},\underline{n}')/\lambda(\underline{n}')$$

$$E[W_i^{(1)}(\underline{n},\underline{n}(0))] = \sum_{\underline{n}'\in A} z(\underline{n},\underline{n}')\min(n_i',m_i)/\lambda(\underline{n}').$$

REFERENCES

Barlow R. E. and F. Proschan, Mathematical Theory of Reliability, John Wiley and Sons, Inc., New York, 1965.

Buzen J. P., "Queueing Network Models of Multiprogramming," Ph.D. Thesis, Division of Engineering and Applied Physics, Harvard University, 1971.

Chang A. and S. S. Lavenberg, "Work-Rates in Closed Queueing Networks with General Independent Servers," IBM Research Report RJ 989, San Jose, Ca., 1972 (to appear Opns. Res.).

Crane M. A. and D. L. Iglehart, "Statistical Analysis of Discrete Event Simulations," Proceedings of the 1974 Winter Simulation Conference, 513-521, 1974.

Gaver D. P., "Probability Models for Multiprogramming Computer Systems," J. ACM 14, 423-438, 1967.

Gordon W. J. and G. F. Newell, "Closed Queueing Systems with Exponential Servers," Opns. Res. 15, 254-265, 1967.

Koenigsberg E., "Cyclic Queues," Opnal. Res. Quart. 9, 22-35, 1958.

Lavenberg S. S., "The Steady State Queueing Time Distribution for the M/G/1 Finite Capacity Queue," IBM Research Report RJ 1150, San Jose, Ca., 1973 (to appear Mngt. Sci. Th.).

Lavenberg S. S. and G. S. Shedler, "Derivation of Confidence Intervals for Work-Rate Estimators in Closed Queueing Networks," IBM Research Report RJ 1297, San Jose, Ca., 1973.

Moy W. A., "Variance Reduction," Chapter 10 in Computer Simulation Experiments With Models of Economic Systems, T. H. Naylor, John Wiley and Sons, Inc., New York, 1971.

Smith W. L., "Renewal Theory and Its Ramifications," J. Roy. Statist. Soc. B 20, 243-302, 1958.

Use of the Diffusion Approximation to Estimate Run Length in Simulation Experiments

T. Moeller and **H. Kobayashi**, New York

ABSTRACT

This paper presents an application of the diffusion process approximation to the statistical analysis involved in simulation experiments of queueing systems. The autocovariance function of the queue size process for both a GI/G/1 queue and a cyclic queueing system are obtained. Run length is predicted and the variance (hence confidence intervals) of estimates of performance such as average queue size, queue size distribution, and server utilization are computed using the approximate autocovariance function. These techniques are used in a simulation experiment of a cyclic queueing model of a multi-programmed computer system.

1. INTRODUCTION

The analysis of outputs of simulations of queueing processes is often made difficult because of the high degree of serial correlation in the output time series. If the autocorrelation function is positive over some range of time as is the case in many queueing processes, a variance estimate made on the assumption of independent and identically distributed outputs may be a serious underestimation of its true value.

In order to overcome such a difficulty, a number of techniques have been developed. The method of blocking of the output samples is the most widely used technique to achieve nearly independent sample outputs. Also commonly used is the method of independent replications of the experiment.

FISHMAN [1971] has proposed a method in which regression analysis is performed during the simulation to dynamically estimate the autocorrelation in the output. This estimate is then used to determine when to halt the simulation. CRANE and IGLEHART [1974] recently discuss the concept of regeneration cycles to obtain groupings of the output which are independent and identically distributed, whereby the variance of the sample means can be calculated in a straightforward manner. Other methods of analysis of variance which take the autocorrelation into account are given in FISHMAN and KIVIAT [1967].

An exact expression for the autocorrelation function of a queueing process, however, is known only for the M/M/1 case, MORSE [1955], who obtained the solution as an infinite series of modified Bessel functions.

For simultations of queueing systems we propose here the use of approximate analytical solutions. The diffusion approximation of the queue size process of a GI/G/1 system and a two stage cyclic queueing system is used to estimate the autocovariance functions of the processes. They, in turn, are used to calculate the run length required to achieve a prespecified value for the variance of the sample mean of the quantity one wishes to estimate.

2. GI/G/1 QUEUEING SYSTEM

Consider a GI/G/1 system with FCFS (first-come, first-served) queue discipline. Let us assume that the interarrival times and service times are both independently and identically distributed with means and variances given by (μ_a,μ_s) and (σ_a^2,σ_s^2), respectively.

Let Q(t) represent the number of customers in the system (i.e., those in service or in queue) at time t. A realization of Q(t) is a random step function with vertical jumps of magnitude one at instants of customer arrivals departures from the system.

If the traffic density $\rho=\mu_a/\mu_s$ is close to unity, then the server is rarely idle, i.e., Q(t) is seldom near the barrier Q=0 and it is well justified to replace Q(t) with an approximate continuous process X(t), t>0. In this case, X(t) is a diffusion process with a reflecting barrier at X=0. For a discussion of the validity of this approximation, see COX and MILLER [1965], KOBAYASHI [1974], and NEWELL [1971].

For a stochastic process X(t) which approximates the queue size process of a GI/G/1 system, the conditional probability density function $p(X_0,X:t)$ of X(t) given that $X(0) = X_0$, satisfies the Fokker-Planck equation:

$$\frac{\partial}{\partial t} p(X_0,X;t) = \frac{\alpha}{2} \frac{\partial^2}{\partial X^2} p(X_0,X;t) - \beta \frac{\partial}{\partial X} p(X_0,X;t) \tag{1}$$

with the boundary conditions

$$\frac{\alpha}{2} \frac{\partial}{\partial X} p(X_0,X;t) - \beta\, p(X_0,X;t) = 0 \text{ at } X = 0 \tag{2}$$

where

$$\alpha = \frac{C_a}{\mu_a} + \frac{C_s}{\mu_s}, \quad \beta = \frac{1}{\mu_a} - \frac{1}{\mu_s}$$

and

$$C_a = \frac{\sigma_a^2}{\mu_a^2}, \quad C_s = \frac{\sigma_s^2}{\mu_s^2}. \tag{3}$$

The solution to Eqn. (1) is given by

$$p(X_0,X;t) = \frac{\partial}{\partial X} \{\Phi(\frac{X-X_0-\beta t}{\alpha t}) - e^{\frac{2\beta X}{\alpha}} \Phi(\frac{X+X_0+\beta t}{\alpha t})\} \tag{4}$$

where $\Phi(\cdot)$ is the integral of the unit normal distribution, i.e.,

$$\Phi(X) = \frac{1}{\sqrt{2\pi}} \int_{-\infty}^{X} e^{-t^2/2}\, dt. \tag{5}$$

Using the property that the process X(t) is, in equilibrium, covariance stationary, the autocovariance function of a queue size process Q(t) can be approximated by

$$R(t) = \sum_{k=0}^{\infty} \sum_{m=0}^{\infty} [k - E\{X\}][m - E\{X\}]\, \hat{p}(k)\hat{p}(m,k;t) \tag{6}$$

where $\hat{p}(m,k,t)$ is the discrete conditional probability density function corresponding to $p(m,X;t)$ and is given by

$$\hat{p}(m,k;t) = \int_{k}^{k+1} p(m,\xi;t)\,d\xi, \quad k=0,1,\ldots \tag{7}$$

and

$$\hat{p}(k) = \lim_{t\to\infty} \hat{p}(m,k;t). \tag{8}$$

3. CYCLIC QUEUEING SYSTEM

Let $Q_1(t)$ represent the queue size of the first server of a cyclic queueing system, shown in Fig. 1. The number of jobs N remains constant and the jobs circulate in a closed loop of two servers. Each server follows the FCFS (first-come, first-served) discipline and when service is completed at one server, the job instantaneously moves to the queue of the other server. If the service process at each server is governed by a general probability distribution with means (μ_1,μ_2) and variances (σ_1^2,σ_2^2), and then the $Q_1(t)$ process is a discrete-valued random process with reflecting barriers at $Q_1=0$ and $Q_1=N$.

If the number of jobs N in the system is sufficiently large and the ratio μ_1/μ_2 is close to unity, then $Q_1(t)$ seldom reaches the barrier and it is appropriate to approximate $Q_1(t)$ by a diffusion process q(t). We then derive a normalized diffusion process $y(\tau)$ according to the transformations

$$\tau = t \Bigg/ \frac{\mu_1(C_1 + C_2\rho)}{(1-\rho)^2} \tag{9}$$

and

$$y = q \Bigg/ \frac{C_1 + C_2\rho}{1 - \rho} , \tag{10}$$

and we obtain the corresponding Fokker-Planck equation in KOBAYASHI [1974]:

$$\frac{\partial}{\partial\tau} p(y_0,y;\tau) = \frac{1}{2} \frac{\partial^2}{\partial y^2} p(y_0,y;\tau) - \frac{\partial}{\partial y} p(y_0,y;\tau) \tag{11}$$

with boundary conditions

$$\frac{1}{2} \frac{\partial}{\partial y} p(y_0,t;\tau) - p(y_0,y;\tau) = 0 \text{ at } y = 0 \text{ and } y = b \tag{12}$$

where those parameters which appeared in Eqns. (9) - (12) are defined by

$$b = \left| \frac{N+1}{\frac{C_1 + C_2\rho}{1-\rho}} \right| , \quad C_1 = \frac{\sigma_1^2}{\mu_1} , \quad C_2 = \frac{\sigma_2^2}{\mu_2}$$
$$\rho = \frac{\mu_1}{\mu_2} < 1. \tag{13}$$

Note that τ represents a normalized time and y, the normalized queue size.

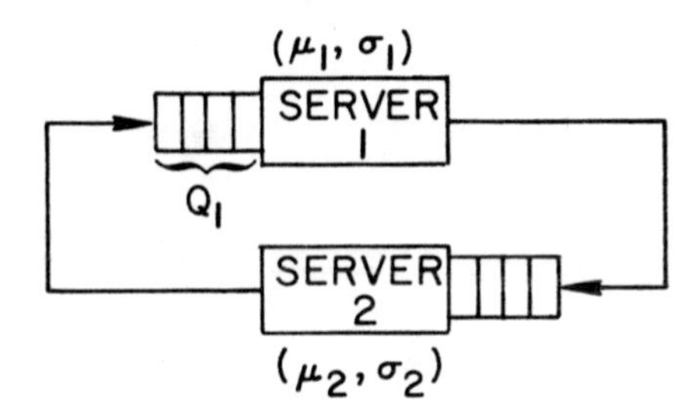

Fig. 1 A two-stage cyclic queueing system

Using the conditional probability density function $p(y_0,y;\tau)$ and its steady state probability density function $p(y) = \lim_{\tau\to\infty} p(y_0,y;\tau)$, the autocovariance function for the normalized queue size process is given by

$$R(\tau) = \int_0^b [y(s) - E\{y\}]p(y(s))dy(s) \int_0^b [y(s+\tau) - E\{y\}]\cdot p(y(s+\tau), y(s);\tau)\, dy(s+\tau) \tag{14}$$

A solution to the boundary value problem given by Eqns. (9) and (10) is given in SWEET and HARDIN [1970]

$$p(y_0,y;\tau) = \frac{2e^{2y}}{e^{2b}-1} + e^{y-y_0} e^{-\frac{\tau}{2}} \sum_{n=1}^{\infty} \phi_n(y_0)\phi_n(y) e^{-\frac{\lambda_n^2\tau}{2}} \tag{15}$$

with

$$\phi_n(y) = \sqrt{\frac{2\lambda_n^2}{b(\lambda_n^2+1)}} \{\cos\lambda_n y + \frac{1}{\lambda_n}\sin\lambda_n y\}$$
$$\lambda_n = \frac{n\pi}{b}, \quad n = 1,2,... \tag{16}$$

Using an approximate solution for $p(y_0,y;\tau)$ defined by Eqn. (15) in the expression for $R(\tau)$ yields an estimate of the autocovariance function

$$R(\tau) = \sum_{n=1}^{\infty} a_n^2 e^{-\frac{(\lambda_n^2+1)\tau}{2}} \tag{17}$$

where

$$a_n = \left[\frac{2}{e^{2b}-1}\right]^{\frac{1}{2}} \int_0^b \xi\, e^{\xi}\, \phi_n(\xi)\, d\xi \,. \tag{18}$$

3.1 Mean Value of Queue Size

The sample mean of queue size is given for an observation of length T by

$$\bar{y} = \frac{1}{T} \int_0^T y(t)dt \,. \tag{19}$$

The variance of $\bar{y}$ may be used to estimate how close the sample mean is to its population mean $E\{y\}$. The sample covariance is given in terms of the auto-covariance $R(\tau)$ as

$$\mathrm{var}(\bar{y}) = \frac{1}{T} \int_{-T}^{T} (1 - \frac{|\tau|}{T})\, R(\tau)\, d\tau \,. \tag{20}$$

Since $R(\tau)$ given by Eqn. (17) approaches zero sufficiently fast as $\tau \to \infty$, we obtain the following asymptotic result

$$\lim_{T\to\infty} \int_{-T}^{T} (1 - \frac{|\tau|}{T})\, R(\tau)\, d\tau = 2 \int_0^{\infty} R(\tau)d\tau \,. \tag{21}$$

For the case where $R(\tau)$ is approximated by Eqn. (17) the assumption for Eqn. (21) holds and thus

$$\mathrm{var}(\bar{y}) = \frac{1}{T} \sum_{n=1}^{\infty} \frac{4\, a_n^2}{\lambda_n^2+1} \tag{22}$$

Thus, an estimate of simulation run length T^* required to obtain a specified level of variance V^* in the sample mean is

$$T^* = \frac{1}{V^*} \sum_{n=1}^{\infty} \frac{4\, a_n^2}{\lambda_n^2+1} \tag{23}$$

3.2 Queue Size Distribution Function

Consider an estimation of the distribution function for the queue size process y(t) of the cyclic queueing system. The first-order distribution function of y(t) is given by

$$F(y) = \Pr\{y(t) \leq y\} , \quad 0 \leq y \leq b . \tag{24}$$

For the process y(t) define a 1-0 valued process $Z_y(t)$ by

$$Z_y(\tau) = \begin{cases} 1 \text{ if } y(\tau) \leq y \\ 0 \text{ if } y(\tau) > y . \end{cases} \tag{25}$$

The expected value of Z_y is thus the value of the distribution function $F(\cdot)$ at point y:

$$E\{Z_y(t)\} = \Pr\{y(t) \leq y\} = F(y) . \tag{26}$$

For a simulation of run length T, define the sample estimate of the distribution function F(y) by

$$\bar{F}(y) = \frac{1}{T} \int_0^T Z_y(t)\, dt . \tag{27}$$

The variance of this sample value is given by

$$\mathrm{var}\{\bar{F}(y)\} = \frac{1}{T} \int_{-T}^{T} (1 - \frac{|\tau|}{T}) [G(y,\tau) - F^2(y)] d\tau \tag{28}$$

where

$$\begin{aligned} G(y,\tau) &= E\{Z_y(t+\tau) Z_y(t)\} \\ &= \Pr\{X(t+\tau) \leq y, \quad X(t) \leq y\} \\ &= \int_0^y p(\xi) \int_0^y p(\xi,\eta;\tau)\, d\eta\, d\xi . \end{aligned} \tag{29}$$

Since $R(\tau)\to 0$ sufficiently fast as $\tau\to\infty$, we may write for sufficiently large T

$$\mathrm{var}\{\bar{F}(y)\} = \frac{2}{T} \int_0^\infty [G(y,\tau) - F^2(y)] d\tau . \tag{30}$$

Using the diffusion process approximation for $p(\xi,\eta;\tau)$ we have an estimate for the variance of $\bar{F}(y)$ given by

$$\mathrm{var}\{\bar{F}(y)\} = \frac{1}{T} \sum_{n=1}^{\infty} \frac{4\, b_n^{\,2}(y)}{\lambda_n^{\,2} + 1} \tag{31}$$

where

$$b_n(y) = \left[\frac{2}{e^{2b}-1}\right]^{\frac{1}{2}} \int_0^y e^{\xi} \phi_n(\xi)\, d\xi . \tag{32}$$

Using Eqn. (31) an estimate of run length T_y^* may be obtained for the desired level of confidence V_y^* in the variance of the sample distribution point $\bar{F}(y)$:

$$T_y^* = \frac{1}{V_y^*} \sum_{n=1}^{\infty} \frac{4\, b_n^{\,2}(y)}{\lambda_n^{\,2} + 1} . \tag{33}$$

3.3 Simulation Experiment

As an example of the run length and confidence interval prediction techniques we take a cyclic queue model (Fig. 1) of a multiprogrammed computer system. Programs (jobs), in the system wait for service at the CPU (server 1), then after an I/O request they wait for service at the I/O device (server 2). It is assumed that there is a constant number of jobs N in the system and that the CPU has an exponential service time distribution and the I/O device has a five stage erlang service time distribution. For this example, N=10, $\rho=\mu_1/\mu_2=.8$ and $\mu_1=4.0$, $\mu_2=5.0$, $\sigma_1=4.0$, $\sigma_2=\sqrt{5}$. Since $\bar{q} = ((C_1 + C_2\rho)/(1 - \rho))\bar{y}$ (where $\bar{y}$ is given by Eqn. (19)) is essentially a linear sum of a random variables y(t), $0 \le t \le T$, the random variable $\bar{q}$ is asymptotically (i.e., as $T\to\infty$) normally distributed. Thus a confidence interval for $\bar{q}$ for sufficiently large T can be derived using the relation

$$\Pr(\ |\bar{q} - E\{q\}| \le L \cdot \sqrt{(\mathrm{var}\ \bar{q})}\ ) \ge 1 - C \tag{34}$$

with L such that $\Phi(L) = \frac{C}{2}$, $0\le C\le 1$. Thus the confidence interval is given by $2\, L\sqrt{(\mathrm{var}\ \bar{y})}$ and the confidence level is 100(1-C)%.

The simulation was programmed using the SIMPL/I language [IBM, 1972] and run on a IBM 370/158 computer. The experiment is initialized by placing five jobs in each queue and beginning service to the first job in each queue at time zero. The stopping condition was determined from the run length estimate given by Eqn. (23) together with a variance of sample mean such that the confidence interval is .2 and the confidence level is 90%.

The sample mean queue size for the first server and the sample queue size distribution $\{\bar{F}(q)\}_{q=1}^{10}$ are generated from the simulation, and confidence intervals are calculated for $\{\bar{F}(q)\}_{q=1}^{10}$ based on the run length T* = 62,180 (in the same units as μ_1) and using Eqn. (31) and Eqn. (34). Table 1 summarizes the results of the simulation.

Of special interest is the fact that the sample value of utilization for the CPU (server 1) is given by $1 - \bar{F}(0)$. The confidence interval for this sample utilization is therefore given by the confidence interval for $\bar{F}(0)$.

$\bar{q}$ = 1.92, 90% confidence interval of .2

With $\bar{q}$ the sample mean queue size

q	$\bar{F}(q)$	δ_i	$[\bar{F}-\delta_i, \bar{F}+\delta_i]$
0	.245	.002	[.243, .247]
1	.530	.005	[.525, .535]
2	.712	.008	[.704, .720]
3	.826	.011	[.815, .837]
4	.895	.014	[.881, .909]
5	.938	.016	[.922, .959]
6	.965	.017	[.948, .982]
7	.981	.016	[.965, .997]
8	.992	.014	[.972, 1.000]
9	.997	.008	[.887, 1.000]
10	1.000	0	---

Table 1 Estimates of Queue Size Distribution Function For Cyclic Queue System

(N=10, ρ=.8, Run Length = 62,180)

4. CONCLUSIONS

The techniques of run length prediction or confidence interval prediction may be applied to other systems for which an expression for the autocovariance function is known. Here a cyclic queueing system and a GI/G/1 were studied. The estimates for run length and confidence intervals take into account the serial correlation of the output time series. An extension of the method to a simulation of a more general queueing network is currently being investigated.

REFERENCES

[1] Cox, D., Miller, Theory of Stochastic Processes, John Wiley and Sons, New York, 1965.

[2] Crane, M., and Iglehart, D., "Simulating Stable Stochastic System I: General Multi-server Queues", JACM, Vol. 21, No. 1, pp. 103-113, 1974.

[3] Fishman, G., "Estimating Sample Size in Computing Simulation Experiments", Management Sci., Vol. 18, No. 1, pp. 21-38, 1971.

[4] Fishman, G, and Kiviat, P., "The Analysis of Simulation Generated Time Series", Management Sci., Vol. 13, No. 7, pp. 525-557, 1967.

[5] IBM, Simpl/I Program Reference Manual, SH 19-5060-0, White Plains, New York, 1972.

[6] Kobayashi, H., "Application of the Diffusion Approximation to Queueing Networks, Part II", to appear in JACM, July 1974.

[7] Morse, P, "Stochastic Properties of Waiting Lines", J. Op. Res. Soc. Am., Vol. 3, No. 3, pp. 255-261, 1955.

[8] Newell, G, Applications of Queueing Theory, Chapman and Hall Ltd., London, 1971.

[9] Sweet, A. L., and Hardin, J. C., "Solutions for Some Diffusion Processes with Two Barriers", J. Appl. Prob., Vol. 7, pp. 423-431, 1970.

Time Series Analysis and Simulation

Peter Naeve, Berlin

1. INTRODUCTION

Since the stochastic nature of time series has become the main subject of research in time series analysis simulation techniques have been among the famous tools of time series analysts. Already the now classical papers by Slutzky [1937] and Yule [1926] use simulation experiments as basis for their reasoning.

Applying the theory of weak stationary processes to time series analysis creates a lot of methodical problems which can be solved in part only by simulation. Roughly one can classify the applications of simulation experiments as follows. First there are problems the theoretical background of which is worked out but one has not yet been able to deduce procedures which can be applied in empirical studies. Secondly problems where one hopes to find some hints by simulation experiments which relations would play a prominent rôle in explaining the theoretical background.

In the remainder of this paper we will discuss some examples of both kinds.

2. WHAT IS THE RIGHT DIVISOR WHEN ONE ESTIMATES THE COVARIANCE FUNCTION

Let x_t stand for the time series being the observations of the stochastic process[1] X_t . It is well known that

$$c(\tau) = \frac{1}{d} \sum_{t=1}^{n-|\tau|} x_{t+\tau} x_t \qquad (1)$$

is an estimator for the unknown covariance function[2] of X_t . There is

1) Unless otherwise stated by stochastic process we mean a weak stationary stochastic process having the ergodic property.

2) Strictly speaking we are estimating the covariance coefficients for $\tau = 0, \pm 1, \pm 2, \ldots$, we follow common practice to use the word covariance function for them.

quite a debate among time series analysts as how to choose d. It is easily seen that $d = n - |\tau|$ in eq.(1) will produce an unbiased estimator whilst $d = n$ comes out with a biased one. This makes some authors prefer the divisor $d = n - |\tau|$.

On the other hand applying some straight forward arithmetic to

$$Q(\mathcal{C}, u) = \frac{1}{2n} \sum_{j=0}^{n-1} \{(\sum_{i=1}^{n} x_i u_k)^2 + (\sum_{i=1}^{n} (-1)^{\left[\frac{i+j}{n}\right]} x_i u_k)^2\} \qquad (2)$$

at which $k \equiv i + j - 1 \pmod n$ and $[r]$ stands for the largest integer equal to or less than r one gets

$$Q(\mathcal{C}, u) = \sum_{i=0}^{n-1} \sum_{j=0}^{n-1} c(i-j)\, u_i\, u_j \geq 0 . \qquad (3)$$

$c(i-j)$ is as defined in eq.(1) with divisor $d = n$. As eq.(3) shows in that case the estimator is a positive definite function.

If one replaces $c(i-j)$ in eq.(3) by the estimator with divisor $d = n - |\tau|$ it is easy to construct a counter example showing that in this case $Q(\mathcal{C}, u)$ is not positive definite. Parzen [1964] claims positive definitness to be the essential attribute of the covariance function for estimating spectra; therefore he recommends to use $d = n$ even though accepting a biased estimator.

If one tackles the question of the right divisor via simulation experiments one soon notices that this is a problem of minor practical importance. The following table presents a typical section of results one gets by simulation experiments.

Table 1

Power spectrum computed from $c(\tau)/c(0)$ using Tukey-window

frequency $\pi j/20$	divisor $d=n$	divisor $d=n-\lvert\tau\rvert$	circular	theoretical spectrum
j = 2	.593	.599	.593	.650
3	.487	.488	.488	.498
4	.359	.356	.359	.333
5	.233	.228	.232	.185
6	.130	.125	.129	.077
7	.063	.059	.061	.017

The time series was generated by a moving average of five with equal weights .2 over random numbers uniformly distributed over (0,1). "Circular" means that in computing $c(\tau)$, $\tau \neq 0$ always n values were used by putting $x_{t+\tau} = x_i$ for $t + \tau > n$ and $t + \tau \equiv i \pmod n$. The covariance function was estimated up to a lag of $m = 10$. The time series were $n = 200$ points long.

The table shows - as further results do - that the estimated spectra based on various methods of defining $c(\tau)$ are relative close together compared with their deviation from the theoretical values of the power spectrum.

The results are not due to the choice of the lag window as can be seen in table 2. There is shown a section of the results of the same simulation experiment which one gets using the lag window

$$\begin{cases} 1 - \frac{6\tau^2}{m^2} \left(1 - \frac{|\tau|}{m}\right) & 0 \leq |\tau| \leq \frac{m}{2} \\ 2 \left(1 - \frac{|\tau|}{m}\right)^3 & \frac{m}{2} \leq |\tau| \leq m \end{cases} \tag{4}$$

as proposed by Parzen.

Table 2

Power spectrum computed from $c(\tau)/c(0)$ using Parzen-window

frequency $\pi j/n$	divisor $d=n$	divisor $d=n-\lvert\tau\rvert$	circular	theoretical spectrum
j = 2	.554	.560	.554	.650
3	.465	.467	.465	.498
4	.359	.358	.359	.333
5	.254	.251	.253	.185
6	.163	.159	.162	.077
7	.097	.094	.096	.017

In both tables frequency intervals centered around very low frequencies are omitted. The reason for this is a methodical problem which shall be demonstrated now.

3. CORRECTION FOR MEAN

If the process X_t has an expectation value $E[X_t] = \mu \neq 0$ this value would be unknown in most cases prior to empirical investigations. For an ergodic process X_t one has the unbiased estimator $\bar{x} = \frac{1}{n} \sum_{t=1}^{n} x_t$ for μ but it can be shown (e.g. Jenkins et.al. |1968|) that the use of data points $x_t - \bar{x}$ which are corrected for mean produces an additional bias of order $\frac{1}{n}$ if one takes an estimator for $C(\tau)$ as given by eq.(1).

Estimating the power spectrum this bias will result in an underestimation of the spectrum for low frequencies.

As can be seen from simulation studies made by Granger and Hughes [1968] and König and Wolters [1972] this bias can be very disturbant especially for short time series.

The following table lists some results of a simulation experiment by König et. al.[1972]. The estimating procedure used Parzen window with a lag up to 24. The series was 200 data points long.

Table 3

frequency $\pi j/n$	theoretical spectrum	no correct. for mean	correction for mean	Fishman procedure
j = 0	1.768	36.43	1.129	1.240
1	1.561	24.68	1.058	1.131
2	1.156	7.411	0.889	0.907
3	0.810	1.404	0.746	0.747
4	0.573	0.784	0.637	0.637
5	0.420	0.545	0.478	0.478

The last column presents corrected values following a proposal by Fishman[1969].

4. FURTHER METHODICAL DIFFICULTIES

It is evident that in practical application of spectral theory the prerequisite that one knows the covariance function for all real τ is never met. Moreover the time series at hand will be so short that one cannot rely on the asymptotic properties of the various estimators as proposed in the literature. So one has to face questions like:
What is the proper truncation point when estimating the covariance function?
Which window should be chosen when using an indirect estimating procedure?
For which frequency intervals should one take estimates?
How varies the bias of the spectral estimators with frequency?
One looks for procedures which enable those to get the right answer in their concrete data situation who would apply spectral theory. It revealed that simulation experiments (Jenkins et.al.[1968], Naeve [1969], König et.al.[1972]) can lead to a lot of rules as how to proceed. These rules have proved to be appropriate in numerous simulation studies and in many investigations of empirical time series.

Especially the book by Jenkins et.al. |1968| can be highly recommended to all who would apply spectral analysis.

5. THE NEED FOR A SPECTRAL SIMULATOR

The simulation studies mentioned so far were mostly designed to work like this. One produces a large number of simulation runs and computes spectral quantities by taking some sort of average. In comparing these quantities with their theoretical values one reaches some final conclusions or rules. A typical example was demonstrated in section 2. As has been mentioned one can get valuable procedures for practical applications going along this way.

Designing lectures on time series analysis one should make use of simulation techniques in quite another way. The wide spreading of computers and the installation of interactive computer languages give the opportunity to develop program systems which could take over a large part of the students practical training in time series analysis.

Such a system would confront the student with simulated time series. In a first step and with the help of built in procedures the student should try to find estimation values for covariance function, power spectra etc. which he considers to be "good". In a second step he then can compare his estimates with the theoretical values. Variation of the parameters for the estimation procedures will show him the influence of the various estimators on the final estimation values. If he goes through these steps for some time series generated by the same stochastic process he can get a feeling as how the finiteness of the time series will influence the results.

Training the students in this way could stop the bad habit just to use the program on time series analysis implemented at the computer center and then interpreting the results in a mix-up of methodical effects and the characteristics of the underlying process.

6. INVESTIGATION OF PHASE RELATIONS

Up to now only problems in univariate spectral analysis had been solved by simulation experiments. It is not necessary to point out that methodical problems will multiply in multivariate spectral analysis. Here too simulation studies can help to clarify the situation and do help to solve some of these problems.

Vicarious the problem of estimating the phase spectrum of two stochastic processes X_t and Y_t shall be demonstrated. The phase spectrum is especially valuable in economic applications for it reflects the temporal relations among economic variables.

Take for instance the saving relation of the Harrod-Domar growth model

$$S_t = s\,Y_{t-1} . \tag{5}$$

Interpreting saving S_t and income Y_t as stochastic processes it follows from eq.(5) that their phase spectrum is

$$\Phi(\omega) = \omega . \tag{6}$$

Applying this theoretical concept on empirical time series one has to answer the question if, due to the relative shortness of the time series, one would be able to estimate the phase spectra so that relations like eq.(6) and more complex ones can be detected.

Various simulation studies [Naeve 1968, 1969] show that under certain circumstances this can be done. The studies were based on pairs of time series which were constructed to have a phase function of one of the following types

$$\begin{aligned} \Phi(\omega) &= c && \text{fixed angle lag} \\ \Phi(\omega) &= a\omega && \text{fixed time lag} \\ \Phi(\omega) &= a_1\omega + a_2 && . \end{aligned} \tag{7}$$

Let X_t be the first process of the pairs then the second one will be denoted in the sequence of eq.(7) X_t^c , X_{t-a} , $X_{t-a_1}^{a_2}$. In addition pairs of the form

$$X_t \;,\; \alpha X_{t-b}^{a} + \beta X_{t-d}^{c} \tag{8}$$

were investigated. The procedures to generate such pairs are described in the paper by Naeve [1968].

Two typical results will be presented here. In the phase and Argand diagram the broken line stands for the graph of theoretical values.

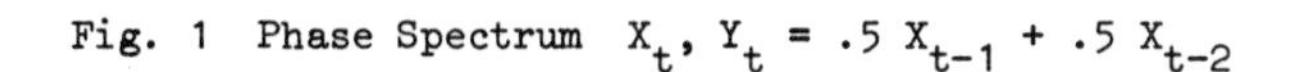

Fig. 1 Phase Spectrum X_t, $Y_t = .5\ X_{t-1} + .5\ X_{t-2}$

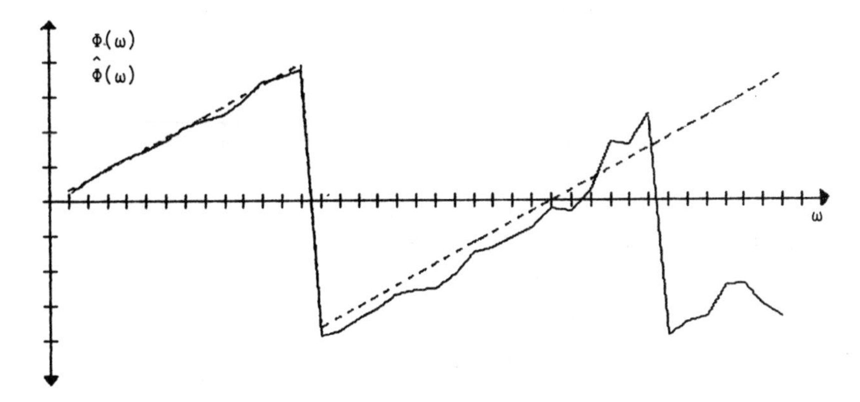

Fig. 2 Argand Diagram X_t, $Y_t = .5\ X_{t-1} + .5\ X_{t-2}$

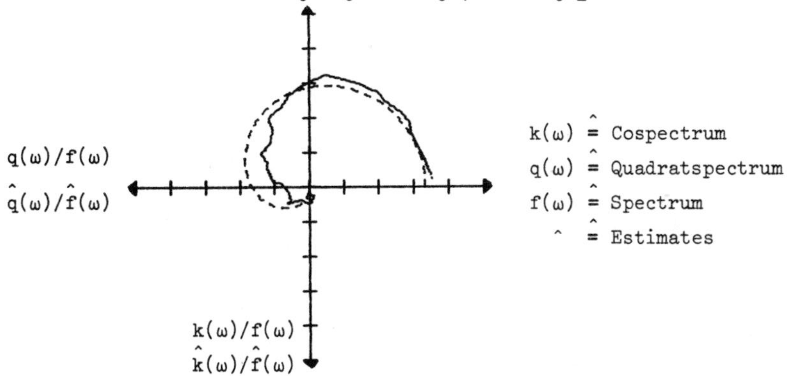

It seems possible to detect simple relations between time series. But even for relations of the type $\Phi(\omega) = a_1\omega + a_2$ it is often not easy to determine a_2. Relations as in eq.(8) are very difficult to spy. One has to take great care when one is interpreting phase functions computed in empirical applications. Identification of phase relations requires a great deal of experience, needless to stress the usefulness of training programs as pointed out in section 5.

7. CROSS SPECTRAL ANALYSIS IN THE PRESENCE OF FEEDBACK

It can be shown [Granger 1964] that cross spectral analysis is less appropriate if feedback is present in the system generating the time series to be studied. But how can one identify feedback without knowledge of the generating system? In two papers Garbers [1970, 1971] offers a rule of thumb. I quote from his latest paper: "Vorausgesetzt zwischen zwei Variablen besteht überhaupt ein Zusammenhang, so dominiert in dieser Beziehung dann eine Richtung, wenn die aus den Originalwerten geschätzten "Kohärenzen" und "Phasen" mit den aus den 1. Differenzen berechneten weitgehend übereinstimmen. Diese Dominanz ist schwächer, wenn die "Phasen" kaum, die "Kohärenzen" jedoch weitgehend übereinstimmen. Besteht diese Ähnlichkeit schließlich weder für die "Phasen" noch für die "Kohärenzen", so existiert zwischen den Variablen ein ausgeprägter "Feedback"." [1)]

Due to theoretical considerations there is little reason to accept this rule of thumb, for one can show [Jenkins et.al.1968] that phase spectra and coherence spectra remain unaltered by linear prefiltering operations, - and computing first differences is just such kind of filtering. However, there are two aspects of practical importance which might be in favour of Garbers' suggestion. His empirical series are in some sort of feedback relation, provided the economic reasoning is valid. Also it is wellknown [Jenkins et.al 1968, Naeve 1969] that one has to make special efforts to discover the structure of the underlying processes if only short series are at hand, - and economic time series must be considered to be very short according to the theory of weak stationary random processes. Combining these two aspects, one might suspect that they cause the difference between the spectra computed from the original series and their first differences respectively.

Since Garbers substantiates the existance of feedback relations between time series strictly on a qualitative basis, it seems to be a natural step to look out for a definition of feedback that allows its measurement prior to spectral procedures.

[1)] Translated quotation: "On the assumption that two variables are connected with each other, one direction of this mutual influence is dominating if the coherence and phase spectra, estimated from the original time series, correspond extensively to those spectra computed from the first differences. This dominance is of less weight if the phase spectra hardly correspond, although the coherences still do fairly well. Finally, there is confirmed feedback between the variables if neither the phases nor the coherences correspond to one another."

We will adopt the definition of feedback and causality in the book of Granger [1964]. Granger's definition seems to fit best into spectral theory: both resting on second moment properties. So if we have two processes labelled j and k we take C(k,j) as a measure of the strength of the feedforward relation from process k to process j where C(k,j) is defined as

$$C(k,j) = 1 - V_j(j,k) / V_j(j) . \qquad (9)$$

$V_j(j)$ is the prediction error variance of the best linear predictor of the process j using only past values of this process whereas $V_j(j,k)$ stands for the corresponding variance if one uses the past values of both processes. If one interchanges j and k one gets the measure C(j,k) for the feedforward relation from process j to k . The product C(k,j)C(j,k) will be taken as measure for feedback between process j and process k .

The following table summarizes some results. It can be seen that there is on the ground of the adopted definition of feedback, no evidence for Garbers' rule of thumb. The first pair of time series shows results which are in full agreement with his rule of thumb, but the second pair brings quite the opposite results one should expect.

Table 4

Time Series	1→2	2→1	1<->2	difference in phase	difference in coherence
1 Auftragseingang Industrie 1955-1968 2 Umsatz Industrie	yes	yes	yes	yes	yes
1 Kurzfristige Kredite an Nichtbanken 1957-1968 2 Bankenliquidität 2. Grades	no	no	no	yes	yes

j → k means feedforward from j to k ; j <-> k means feedback relation.

Now one could rise the question whether the proposed procedure would work at all in practical applications as predicted by theoretical considerations. A natural way to find it out would be to start simulation experiments.

Simulation experiments made by Birkenfeld [1973] prove two things. First the adequacy of our tools in feedback situations and secondly that Garbers' rule is wrong. Birkenfeld experimented with simulated time series generated by processes with known feedback relations. In the presence of feedback his results never revealed any considerable deviations between the coherence spectra or phase spectra estimated from the original time series and their first differences. Beyond that, he was always able to redetect the feedback relations applying a procedure based on Granger's definition.

8. CONCLUSION

It has been shown by various examples that simulation is a valuable aid in preparing for empirical studies on time series analysis. But prior to any simulation experiment one thing should be made distinct. If one checks on methods all facts deduced from theory must be correct. This seems obvious but unfortunately one can find counter examples.

For instance Schips and Stier [1973] spent quite a lot of effort to invent a theory about the possibility to estimate the phase spectrum in dependence of the type of the power spectra. They back their results by simulation studies. Unfortunately they have overseen just one small point: their theoretical phase function is wrong. If one tackles the problem using the correct phase function all difficulties are gone with the wind. It was just much ado about nothing.

REFERENCES

Birkenfeld, W. — Zeitreihenanalyse bei Feedback-Beziehungen; Schäffer, K.A., P.Schönfeld, W.Wetzel, Physica-Verlag, Würzburg, 1973

Birkenfeld,W., P.Naeve — Programme zur bivariaten Zeitreihenanalyse; Institut für Quantitative Ökonomik und Statistik, Freie Universität Berlin, Berlin, 1974

Fishman, G.S. — Spectral Methods in Econometrics Rand Corporation, Santa Monica, 1968

Garbers, H. — Chapter 2 of Neuere Entwicklungen auf dem Gebiet der Zeitreihenanalyse, Wetzel, W., Vandenhoeck & Rupprecht, Göttingen, 1970

Garbers, H. — Zur Spektral-und Kreuzspektralanalyse stationärer stochastischer Prozesse. Münzner, H., Wetzel, W., Physica-Verlag, 1971

Granger, C.W.J. — Spectral Analysis Of Economic Time Series. Princeton University Press, Princeton, 1964

Granger,C.W.J., A.O.Hughes — Spectral Analysis Of Short Series - A Simulation Study. J.R.S.S. Vol.131 pp. 83-99, 1968

Jenkins,G.M., D.G.Watts — Spectral Analysis And Its Applications, Jenkins,G.M., E.Parzen, Holden-Day, San Francisco, 1968

König,H., J.Wolters — Einführung in die Spektralanalyse ökonomischer Zeitreihen, Verlag Anton Hain, Meisenheim am Glan, 1972

Naeve, P. — Phasenbeziehungen ökonomischer Zeitreihen, Institut für Höhere Studien und Wissenschaftliche Forschung Wien, Wien, 1968

Naeve, P. — Spektralanalytische Methoden zur Analyse von ökonomischen Zeitreihen, Münzner,H., W.Wetzel, Physica-Verlag, Würzburg, 1969

Parzen, E. — An Approach To Empirical Time Series Analysis, Radio Science, Vol. 68 pp.937-951, 1964

Schips,B., W.Stier — Einige Bemerkungen über die Rolle der Spektralanalyse in der empirischen Wirtschaftsforschung Schweiz.Zeitschrift f.Volkswirtschaft u.Statistik, pp. 233-243, 1973

Slutzky, E. — The Summation Of Random Causes As The Source Of Cyclic Processes. Econometrica Vol.5 pp.105-146, 1937

Yule, G.U. — Why Do We Sometimes Get Nonsense-Correlations Between Time Series? J.R.S.S. Vol.89, pp.1-64, 1926

Simulation Package for Computer Operating System

Shafeek I. Saleeb and Mokhtar Boshra Riad, Cairo

ABSTRACT

The main objective of this paper is the construction of a basic simulation model for the disc-based system GEORGE 2E used to run the ICL 1905E computer of the Cairo University, which can be used to study the computer system performance, to find the optimal policy for running the computer to give the maximum CPU productivity and also to determine the relative merits in the utilization of the central processor of adding additional central memory, replacing the disc with faster models or controlling the type and specifications of the jobs run.

1. INTRODUCTION

The operating system (OS) is essentially needed to run the modern high speed, multiprogrammed computers in order to save the time lost by the human intervention during the computer operation, and to accurately interpret the users' requirements.

The system considered here is the GEORGE 2E used to run the ICL 1905E computer of 64k words. It comprises 3 programs [1] : the Input program to transfer the input from cards to disc, the Central Module to look after the execution of the program, its input|output is from and to disc and the Output program to transfer the results from disc to printer.

2. THE SYSTEM DESCRIPTION AND THE JOB GENERATION

Fig.1 shows the components of the computer system to be simulated the Central Processing Unit (CPU), the Central Memory (CM) and the movable head disc. The card-to-disc and disc-to-printer operations are of little effect on the CPU utilization as they are overlapped with the program's execution [2,3 and 4]. The figure shows the flow of the jobs through the system and the processing steps where it is controlled by 3 queues (CM, CPU and Disc queues) are treated in the simulator as linked lists with two elements per entry (Job number (JN) and time of entry in the queue T_{in}). Service is provided to the jobs according to the first-in, first-out discipline. Statistics are gathered on the queues' behaviour by means of the entry and removal algorithms outlined in the following equations:

<u>Insert entry in queue</u>:

1. Insert JN & TIME in the

<u>Remove entry from queue</u>:

1. Unlink entry at head of

queue list
2. $\Sigma TQ = \Sigma TQ + (TIME-T_{last}).Q$
3. $Q = Q + 1$
4. $Q_{max} = MAX [Q, Q_{max}]$
5. $N = N + 1$
6. $T_{last} = TIME$

the queue list
2. $\Sigma TQ = \Sigma TQ + (TIME-T_{last}).Q$
3. $Q = Q + 1$
4. $\Sigma T_W = \Sigma T_W + (TIME-T_{in})$
5. $W_{max} = MAX [W_{max}, (TIME-T_{in})]$
6. $T_{last} = TIME$

Where: Q is the current queue length, with maximum value Q_{max}, N is the queue entry counter, TIME is the current time value, T_{last} is the time of last entry|removal, W_{max} is the max. waiting time in the queue, ΣTQ is the length time product accumulator and ΣTW is the waiting time accumulator.

At the end of the simulation run the following values are computed for each queue:

Mean queue length = ΣTQ|TIME
Mean waiting time = ΣTW|N

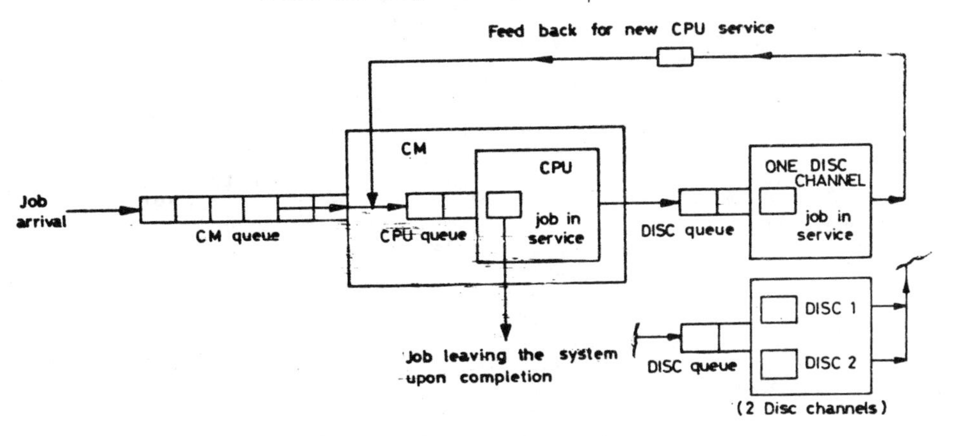

Fig.1: The Operating System : GEORGE 2E

The arrival of a job in the system is marked by the generation of a set of values for the following characteristics:

1. Job inter-arrival time i.e. the interval between the arrival of successive jobs.
2. Central memory space requirement.
3. Total CPU mill time requirement.
4. Number of I|O requests and I|O request length.
5. Inter-request time i.e. the interval between a job's assignment to the CPU and its issue of an I|O request.

A statistical study and analysis of the job's requirements and the job generation techniques are given in detail in [4].

3. THE SIMULATOR DESIGN

The simulator which we designed is an extension to the BASYS simulator designed by MacDougall [3], to be compatible with the GEORGE 2E O.S. and the ICL software. In the simulator, a job is represented by an entry in a job table which contains the job's characteristics. As the job moves through the system - enters queues, is assigned to the CPU, etc - its movement is reflected by moving a pointer to this job table entry (job number).

The progress of the job through the system is marked by the occurance of a series of events. These events correspond to transition points between operations or activities, they represent a change of state. The significant events considered in the simulator are the following:- E1: marks the arrival of a job in the system; E2: a job's request for central memory space; E3: a job's request for the central processor; E4: release the CPU by a job issuing an I|O request; E5: marks the initiation of processing for an I|O request; E6: marks the completion of processing for an I|O request and E7: marks the completion of a job (the job release the CPU and the CM). Scheduling of the events is facilitated by the use of a linked event list of three element entries, (Job number, Event identifier and Event time).

The simulation program (FORTRAN), [4], is developed so that there is a one-to-one correspondance between the events of the simulation model and the simulator routines. With this philosophy, each simulator "event routine" essentially does two things (1) it performs the operations whose initiation corresponds to the occurance of this event; (2) it predicts, for the job for which the operation was performed, which event is to occur next and at what time it is to occur. An event list of three entries is used to keep these values and to schedule the events.

In the simulator, job processing is a quasi-parallel operation, reflecting the true simultaneity of the system being simulated. The predicted duration of an operation is added to the value of simulated time at which the operation was initiated to obtain the point in simulated time at which an event (the end of that particular operation) is to occur.

There are instants where the next event time for a job cannot be predicted and so no entry for this job appears in the event list. This occurs, for example, when a job finds a facility busy and enters a queue. At the time at which this event occurs it is impossible to predict the time at which this job will leave the queue and then assigned to the facility. When a facility is released, a job is selected from the queue according to the appropriate queue discipline and scheduled for the event which will result in its assignment to the facility. In the simulator, every job in the system is represented by the appearance of its job table entry pointer (job number) either in the event list or in the queue list, but not in both.

4. THE SIMULATOR DESCRIPTION

The simulation program consists of the following segments, Fig.2.

1. The MASTER segment is the initialization routine of the simulator it establishes the system parameters (the CM size,

the mean inter-arrival time of jobs, the disc transfer rate, the maximum number of streams allowed to run, the number of jobs in the batch, etc) and the job mix parameters, and triggers the simulation run by scheduling the arrival of the first job. The master segment was developed in such a manner to allow new simulation runs with different system parameters.

2. Subroutine PRINTQR: In this subroutine the statistics about the jobs run and about the queues are accumulated and then printed.

3. Subroutine SCHD: This is the scheduler of the simulator, the basic steps in event scheduling are the following:
   (a) The scheduler removes the entry at the head of the event list. This entry specifies an event time T, and event identifier E, and a job table pointer JN.
   (b) The scheduler advances the clock (TIME) to the event time T specified in the first step, because this time represents the earliest of all the scheduled events to occur.
   (c) The scheduler transfers control to the event routine designated by the event identifier E.
   (d) The event routine performs the required processing for the job, determines its next event and event time, and inserts the event identifier, event time, and job table pointer in the event list. It then returns control to the scheduler.

   The scheduler also scans the CM queue list searching for the first waiting job for which the available free space in the CM, is sufficient and removes this job (if exists) from the CM queue, and allocates CM space to it. The next event for this job is event 3 and is scheduled to occur at a time equal to the current time plus the time required to relocate the CM.

4. Subroutine JOBCH: This subroutine samples randomly the job characteristics: CM size, CPU mill time and the volume of disc I|O records, from the appropriate frequency distributions.

5. Function TIA: This function draws the value of the job inter-arrival time or the job CPU ←→ Disc inter-request time from the negative exponential distribution.

6. Subroutine UT: This subroutine is used to give the TIME values during the simulation run. On each call it updates the TIME value by adding the difference (ICT-IST) between the current time value ICT and the previously measured value IST in the previous call. Either ICT or IST is the output of the subroutine TIME which gets the time value.

7. Subroutine ORDEREL: This subroutine orders the entries in the event list in ascending order with respect to the event time.

8. Subroutine REMOVE: This subroutine removes the evant just ex-

ecuted from the event list.

9. Subroutine QUEUE: The function of this subroutine is to insert entry in queue if the job finds the requested facility busy, or to remove entry from queue when the facility is free and became available. Also it performs the operations described in sec.2 concerning the queue entry and removal algorithms.

10. Subroutine CHECKQ: This subroutine checks the queue number K when the corresponding facility is free. If the queue has entries it removes the entry at the head of the queue list and gives the job number in NJN and sets QEMPTY=.FALSE. If the queue has zero entry count, it outputs with QEMPTY=.TRUE.

11. Subroutine EVENT 1: Event 1 marks the arrival of a job in the system. The subroutine EVENT 1 determines the characteristics of this job either by sampling the appropriate distributions through calling the subroutine JOBCH, or from the data read in the master segment. It stores these characteristics in the job table and other arrays. The following job characteristics are generated:
    1. CM space required.
    2. Processor mill time required.
    3. Number of records read|written.
    4. Mean inter-request interval (computed by dividing the job's processor mill time by the number of I|O requests).

    The record size is assumed to be constant for all the jobs. The next event for the arrived job is scheduled by inserting the job table pointer (JN), the event designator (E) and the event time (T) in the event list. In our basic model, there is no delay between the time at which the job arrives and the time at which it requests central memory. Therefore, the job transfers directly to EVENT 2 routine. Also the subroutine EVENT 1 predicts the next arrival time of a new job (JN').

12. Subroutine EVENT 2: Event 2 is the occurance of a job's request for CM space. The CM space requirement of the job (from the job table entry JN) is compared with the available CM free space. If sufficient space is available, it is allocated to the job; otherwise, the job is entered in the CM queue. It is assumed that allocation of CM space to the job may require relocation of other jobs to provide sufficient contiguous CM space. The next event for this job (event 3) is scheduled to occur at a time equal to the current time plus the time (TR) required to relocate the CM.

13. Subroutine EVENT 3: Event 3 corresponds to a job's request for the central processor. If the CPU is free, it is reserved for the requesting job; otherwise the job is entered in the CPU queue. The number of records to be read or written and the mean interval between I|O requests were established when the job is arrived. If the record count has not been reduced to zero, a sample from a distribution of the negative exponential–

form with this mean value is used to determine the time at which the next I|O request is issued. The record count is decremented and event 4 is scheduled for this time. If the record count has been reduced to zero or the remaining CPU time for this job has been reduced to zero, event 7 is scheduled.
It was observed that the GEORGE 2E logging facilities record the maximum CM size for the job during its running. If the core size of the object program, after compilation and consolidation, is small then the maximum core size will be equal to the size of the compiler or the consolidator whichever is larger. We have the sizes of the FORTRAN disc compilers as follows:-
XFAE = 11008 words $\simeq$ 12000 words } when using intervals in
XFAT = 18176 words $\simeq$ 20000 words } a frequency table.
between a sample of 10250 jobs run we have:-
2329 jobs requiring a maximum core of 12000 words.
6072 jobs requiring a maximum core of 20000 words.
But, this maximum core size happens only during the compilation time which is less than 20 seconds in average, and the job then occupies a smaller core size during its execution time. Therefore, we can assume that after the compilation process, the job core size is a fraction p of the maximum size. We assume the range of p to be from 0.50 to 1.00. Hence, we can generate a uniformly distributed random number for p in the assumed range, or simply (as we did in our simulator) a constant value for p can be assumed, say p=0.60. Then the shrinked core size will be computed by multiplying p by the maximum core size.

14. Subroutine EVENT 4: Event 4 is the release of the CPU by a job issuing an I|O request. Upon the requestor's release of the CPU, the CPU queue is examined; if there are waiting jobs, one is selected from the queue (the first entered in the queue) and scheduled for event 3. The job releasing the CPU is scheduled for event 5. Both events are scheduled to occur at a time equal to the current time plus an overhead time (TO) required to process the request.

15. Subroutine EVENT 5: Event 5 marks the initiation of processing for an I|O request. If the disc is busy, the job is entered in the disc queue. If the disc is free, it is reserved for the job and the request processing time (TD) is computed in the master segment upon the initialization of the simulation run, as follows:-

TD = TP + TL + TT** R

where: TP is the disc positioning time of the heads (average seek time).
TL is the disc latency time (half a revolution time in average).
TT is the disc transfer time per word.
R is the request length in words.

The next event (event 6) is scheduled to occur at a time equal

to the current time plus TD.
In this subroutine, we assumed an overhead of 5 seconds charged to each job on its first I|O request. This overhead caters for the delay caused by the activity of the O.S. central module overlaid from disc.

16. Subroutine EVENT 6: Event 6 is the completion of processing for an I|O request. The disc is released and the disc queue checked for waiting requests; if there are waiting requests, the first entered is removed from the queue and scheduled for event 5. — The job releasing the disc is scheduled for event 3. Here again, the overhead time will be considered, and both events are scheduled to occur at a time equal the overhead time (TO) plus the current time.
17. Subroutine EVENT 7: Event 7 marks the completion of a job. The CPU is released, and if there are jobs in the CPU queue, one is removed and scheduled for event 3 at time equal to the current time. Also the CM space allocated to the job is now released.

N.B.: The complete details of the simulator, the program listing and description and examples of the simulation results are given in [4]

5. SIMULATION EXPERIMENTS

Many simulation experiments can be performed using the simulator such as studying:-

1. The stability of the simulator through observing the queues' characteristics.
2. The effect of the load factor (mean job mill time|mean inter-arrival time of jobs) on the CPU productivity.
3. The effect of the batch size required to be executed during the run and the number of streams on the CPU productivity.
4. The effect of the CPU speed and the type of jobs run on the CPU productivity.
5. The effect of the memory size and the number of streams on the CPU productivity.
6. The effect of the jobs' requirements from the CM space and the number of streams on the CPU productivity.
7. The effect of the disc speed and the number of disc channels on the CPU productivity specially when running business jobs.

Results of these and other simulation experiments which can be easily performed are given in [5]

REFERENCES

1. ICL: "GEORGE 1 & 2 Operating Systems, 1900 Series" . Technical Publication No.4229, ICL, London, 1970.
2. LACOUTURE, M.B. and POIX, A.: "Un Simulateur de Système" . Revue Francaise d'Automatique Informatique, Recherche Operationnelle, 6eme Annee, Sept.1972, pp.27-38.
3. MACDOUGALL, M.H.: "Computer System Simulation:An Introduction". Computing Surveys, Vol.2, No.3, Sept. 1970, pp.191-208.
4. RIAD, M.B.: " Computer System Simulation for the Scientific Computation Center, Cairo University". M.Sc. Thesis, Cairo Univ. 1973
5. SALEEB, S.I. and RIAD, M.B.: " Use of Simulation for Operating System Optimization". IFIP Congress 74.

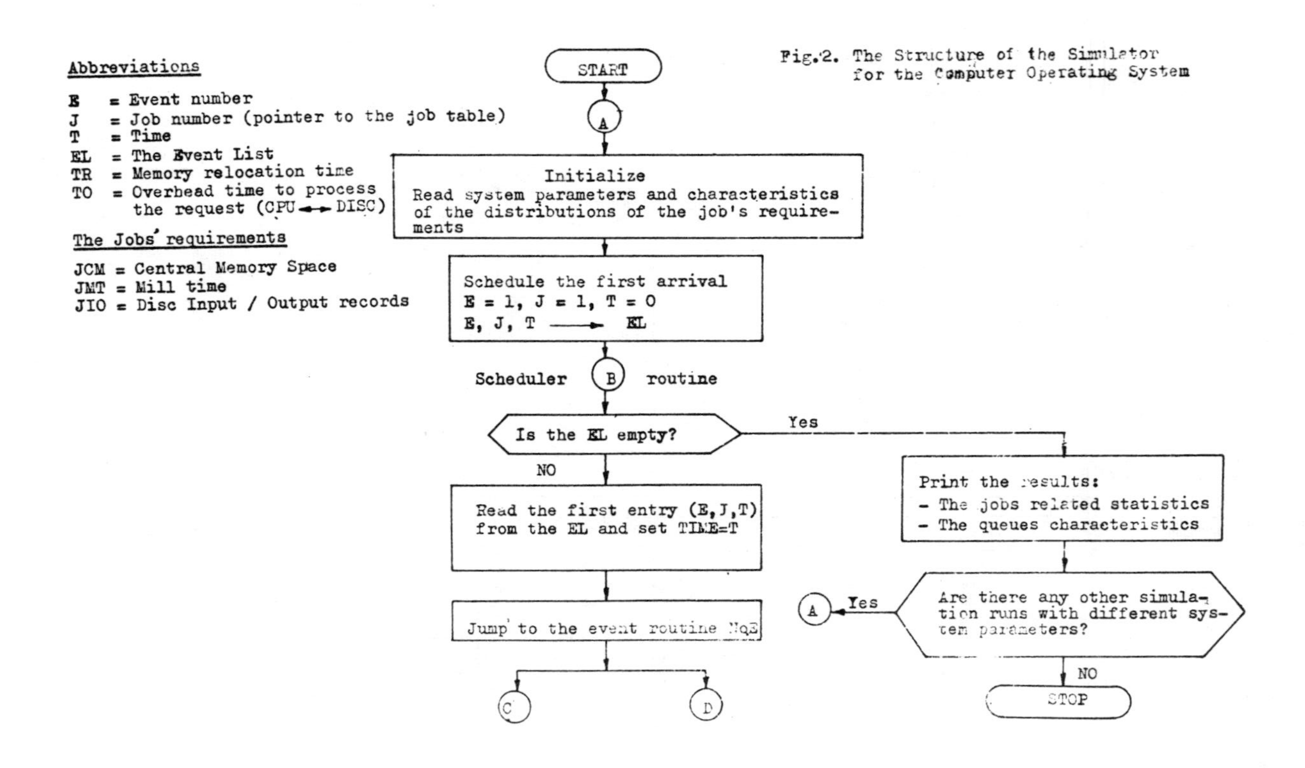
Abbreviations
E = Event number
J = Job number (pointer to the job table)
T = Time
EL = The Event List
TR = Memory relocation time
TO = Overhead time to process the request (CPU ↔ DISC)
The Jobs' requirements
JCM = Central Memory Space
JMT = Mill time
JIO = Disc Input / Output records
START
A
Initialize
Read system parameters and characteristics of the distributions of the job's requirements
Schedule the first arrival
E = 1, J = 1, T = 0
E, J, T ⟶ EL
Scheduler
B
routine
Is the EL empty?
Yes
NO
Read the first entry (E,J,T) from the EL and set TIME=T
Jump to the event routine NoE
C
D
Print the results:
- The jobs related statistics
- The queues characteristics
Are there any other simulation runs with different system parameters?
Yes
A
NO
STOP

Fig. 2. The Structure of the Simulator for the Computer Operating System

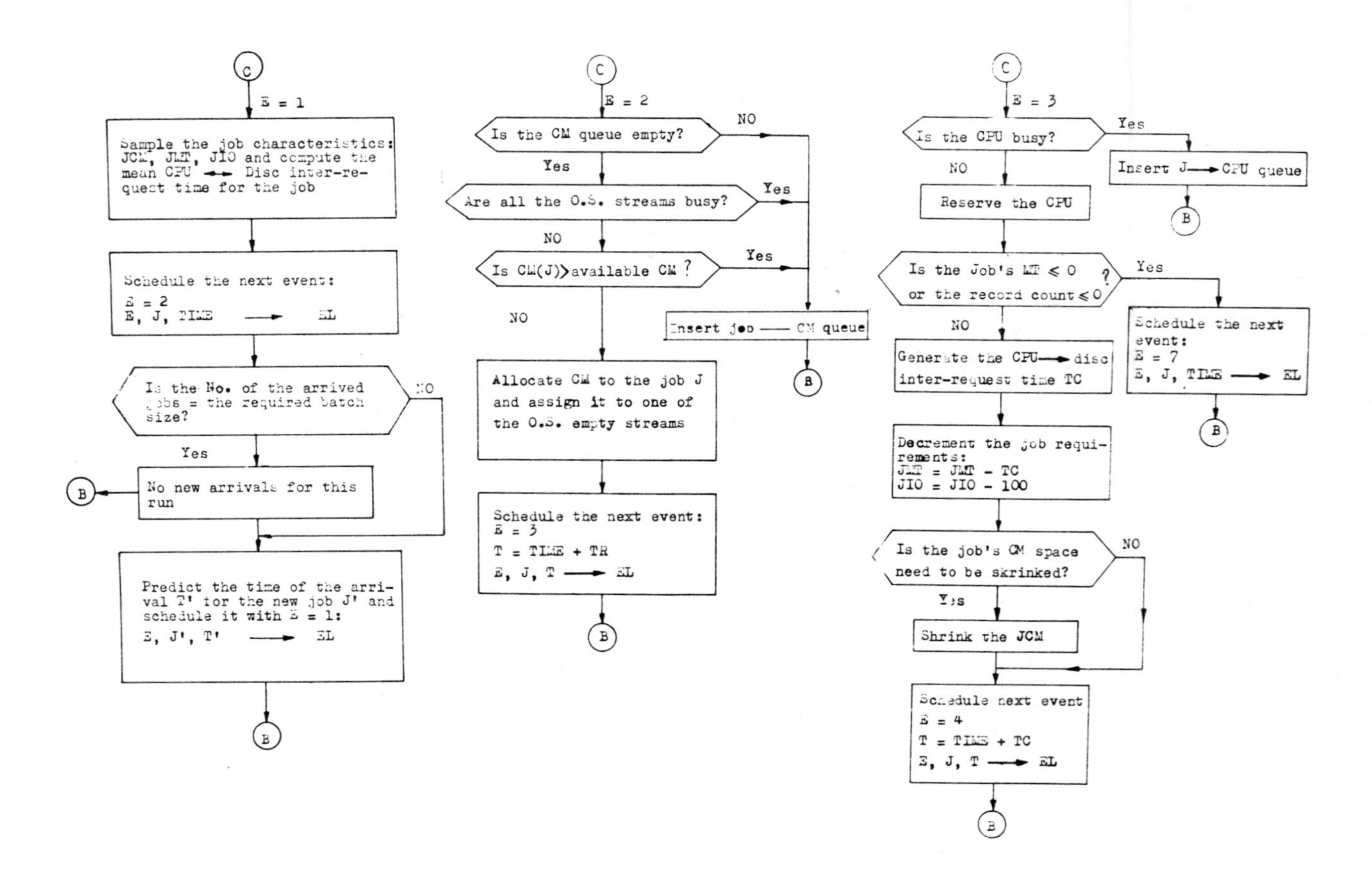
C
E = 1
Sample the job characteristics: JCM, JMT, JIO and compute the mean CPU ↔ Disc inter-request time for the job
Schedule the next event:
E = 2
E, J, TIME ⟶ EL
Is the No. of the arrived jobs = the required batch size?
Yes
NO
No new arrivals for this run
B
Predict the time of the arrival T' for the new job J' and schedule it with E = 1:
E, J', T' ⟶ EL
B
C
E = 2
Is the CM queue empty?
Yes
NO
Are all the O.S. streams busy?
Yes
NO
Is CM(J) > available CM ?
Yes
NO
Insert job —— CM queue
B
Allocate CM to the job J and assign it to one of the O.S. empty streams
Schedule the next event:
E = 3
T = TIME + TR
E, J, T ⟶ EL
B
C
E = 3
Is the CPU busy?
Yes
NO
Insert J ⟶ CPU queue
B
Reserve the CPU
Is the Job's MT ≤ 0 or the record count ≤ 0 ?
Yes
NO
Schedule the next event:
E = 7
E, J, TIME ⟶ EL
B
Generate the CPU ⟶ disc inter-request time TC
Decrement the job requirements:
JMT = JMT - TC
JIO = JIO - 100
Is the job's CM space need to be skrinked?
NO
Yes
Shrink the JCM
Schedule next event
E = 4
T = TIME + TC
E, J, T ⟶ EL
B

Simulation Modelling for Analysis of Optimum Sequencing in the Shop-Floor

Ernoe Stauder, Budapest

ABSTRACT

This paper presents a simulation modellig method to select a production control policy for optimal or near-optimal manufacturing of jobs and job operations. The objective is to optimize some production factors with the "most intelligent policy". Job operations are sequenced so as to minimize total idle times on machines while completing individual jobs according to specific lead-times. Jobs are received at the shop-floor with a predefined arrival rate for each job. The number of operations per job varies between jobs with a fixed routing sequence. Job operations are sequenced dynamically between operations, in front of each machine /machine group/. The output of the simulation runs contains the following statistical data: flow time per job type, cumulated waiting time per machine group, average utilization of machine group, queuing time and number of waiting jobs in front of each machine /machine group/. Different priority scheduling policies are used to evaluate the effect on the investigated production factors. Some experiments have also to be done in a stochastic production environment. /Constant operation time versus stochastic operation time. Constant arrival rate versus stochastic arrival time./ The paper discusses the results from experiments of 20 jobs, minimum 58 machines grouped into 16 work centers and all relevant information.

1. THE BASIC MODEL

This paper will attempt to set up a generalized dynamic simulation model of a shop-floor to obtain statistics for each work center and various overall plant statistics ba parts such as run times and waiting times. By changing some basic information, the production process and the entire system can be analyzed. The model can be used to justify different production sheduling and control polices by computer simulation.

In this paper, the interpretatron of the detailed data as results, the reaching of steady state and the length of the simulator run are left to the reader's own judgement. In addition, there is no "feed-back" or "feed-forward" control within the plant, i.e. the established control policy is forced on the shop-floor at each work center. The representative model is runing with a sample of 20 different type of jobs, on 16 different work centers. The following assumptions are made and based on the real shop-floor environment of any typical non-assembly line manufacturing process.

1. To produce a part, a number of operations have to be completed in a predefined sequence. The documentation is set up for routing, with all job operation times as standards of scheduling.

2. Routing sequence of each job, or each part is fixed and rigid.

3. Job or operation sequence for each work center is dynamic and changes from time to time.

4. The sample size has been extrapolated to represent the full loading rate of the plant by assuming 100% work center efficiency factors.

5. "Operation time" is the period during the work center is seized by the selected part. This period consists of the set up, the machining and the post operation time.

6. To simplify the modelling, we ignore the inter-operation time which is usually needed to transport the part from one work center to the other.

7. The possible discrepancies are assumed to be zero, and this allows rework.

8. Variable number of jobs on a variable number of machines can be handled ith the model.

9. The number of operations per part varies between parts.

10. Each machine belongs to a specific machine group, and the number of machines are fixed during a simulation run, each machine being capable of handling one release at a time.

11. The number of machines per machine group varies between machine groups.

12. Each operation must be performed at a specific machine group. Replacement is not allowed.

13. The lot sizes are fixed and remain constant throughout the simulation. Each standard job operation time is an average operation time which is needed to work on the part at a specific work center.

14. The average cycle time of each part is based on historical data. The inter-arrival rate of a new part-order is gained from it and this is used to generate a new incoming part for operations in the simulation model.

2. CHARACTERISTICS OF THE SHOP FLOOR

When investigating the shop-floor control we have to describe the environment and the behaviour of the system in which the product moves ahead until it is finished. The flow of the product is prearranged by the routing sequence as mentioned before. The work center assigned to each operation of the part is arranged in the plant as a shop-foor and each of the unique work centers must be available for operations. The work center can service the reques of the operation provided it is not seized by another operation. If the assigned work center is busy, the arriving part has to queue up for service. As a rule, the product can arrive in the system at any point or at any work center, then following a predefined routing sequence within the shop-floor it leaves the plant to be kept in inventory.

In this sense, the shop-floor seems to be a network system with as many queues as the number of work centers. Each work-center can be described as a multi-server station with a definite number of servers. As the number of operations or requested services depends on the arrival rate of each product type and their routing, the analytical study of such a system is fairly difficult with mathematical formulas.

Other useful measures of behaviour of the system would be to make a simulation model of the shop-floor with all respected characteristics of work centers and products - and let's run. We can measure the system with the following:

A. The average number of parts in the queues in front of each work center

B. The average waiting time spent in queue.

C. The average throughput time of a product spent in the system /waiting in queues and job operation times/.

D. The utilization of each work center.

E. The maximum length of queue in front of each work center.

The queueing theory addresses itself to the problems of idleness and congestion. Many applications are described and published on this topic. The theory is most successful where the Poisson process operates /assumed Poisson arrival, servicing/, because the mathematical analyses are less complex. The results of queueing theory are applicable to the most complex network of waiting lines and facilites, assuming that the network receives units or items in Poisson fashion, from outside the network, and that each facility /or node/ operates in Poisson fashion upon these units, discharging them in streams towards other nodes or completely out of the system. However, some real queueing networks can not be analyzed merely on mathematical bases, if the system follows other than Poisson or Erlang disciplines. Nevetherless, certain measures of behaviour of the system might remain the same, namely:

L_q - the average number of items in the queue

W_q - the average waiting time spent in queue

W_L - the average time spent in system /queue and service at facility/, the throughput time

L_{max} - the maximum number of items in queue

R - the utilization coefficient of the servers $/0<R<1/$

Part of this paper discusses some empirical results of a simulation model, describing a shop-floor system as a queueing network with queueing nodes as so many work centers as are used for routing products throughput the shop-floor.

3. THE ACTIVE MODEL

The problem of scheduling with sequencing as the sole control technique of the in-process batches is now examined. A feasible state of the system can be imagined as consisting of all the decisions that load the available job operations on available machines. For each point in time at which a decision is to be made, a priority value is assigned to each activity competing for an available resource. Smaller value of priority indicates lower priority. Thus, the activity or job operation with the highest priority value has the available resource assigned to it.

Let us suppose that operation A_1 of batch H_1 and A_2 of batch H_2 are both competing for the use of a machine of the defined work center at time t. Let us introduce f, as a priority function. Then f /A_1, H_1, t/ and f /A_2, H_2, t/ are computed, and activity A_1 is loaded if f /A_1, H_1, t/ $>$ f /A_2, H_2, t/. The priority function f should reflect the current status of the system, and induce local decisions to sequence compatibly with the overall objectives of management. Such decisions can be easily computed.

For an operation A of batch H, ready to be loaded at time t, we introduce the following variables associated with this operation and this batch:

M - expected duration of operation A, based on the standard labor

S - batch slack /difference between due date and earliest possible completion date/ at time t

V - value of batch /related to the total cost/ at time t

We now define a priority function f^*, with arguments M, S and V. Without specifying this function, we can state the general properties required of it: f^* increases with M and S and decreases with V. This is the reverse the priority value mentioned above; the higher the value of f^* the lower the priority value in control.

Heuristically, this means:

$$\frac{\partial f^*}{\partial M} \geq 0, \quad \frac{\partial f^*}{\partial S} \geq 0, \quad \frac{\partial f^*}{\partial V} \leq 0.$$

Smaller values of M tend to give an operation higher priority, thus tending to shorten the average waiting time for job operation in queue and reduce the average number of units in queue. Ordering by the use of M alone as the shortest operation discipline.

A developed priority function formula can be expressed as

$$f_i^* = a_1 M_i + a_2 S_i + a_3/V_i$$

for which operation i precedes operation k if $f_i^* < f_k^*$. In our investigation we examine the effect of a different control policy selecting only one argument to compute the priority function ignored the parameter value of a_1,

a_2 and a_3. It should be kept in mind that we want to optimize a production system in its day-to-day operation. The chief goal is to improve actual system performance. Some of the overall goals of interest in sequencing operations are:

- Minimize the values of waiting time
- Maximize the utilization of manpower and/or equipment
- Minimize the throughput time of batches
- Reduce the length of the waiting operations in queues.

The active model was run with different priority control argument to obtain the effect of each control policy.

4. EXPERIMENTAL RESULTS.

The control method described was programmed for the IBM/370 and several experiments and model runs were performad by use of a simulated job shop that insorporated simplified factory data. A set of runs was performed for validation with a time period of 60 000 time unit in minutes. After the validation runs, we simulated the job shop with 300 000 time units in minutes equivalent to 5000 fabrication hours which period was long enough to accept all statistics as the result of steady state. The arrival of a shop order was generated separately for each type of product. We used a standard cycle time of order issue to calculate an actual inter-arrival time after the formula:

$$T_i^* = T_i \left(0.5 + 0.5\, E/t/ \right)$$

where the notations were

T_i^* - actual inter-arrival time for the next shop-order of product type $\underline{i}$

T_i - the average inter-arrival time measured for long period

$E/t/$ - normalized exponential distribution function with mean of 1, and the actual value was established by using a random number generator.

To get the actual operation time we used the standard labor with the relevant assignement of work center and assumed that the service time has followed some exponential distribution which could be given by the formula:

$$M_{ij}^* = M_{ij} \left(0.9 + 0.1\, E/t/ \right)$$

where the notations were

M^{*}_{ij} - actual operation time required to work product i at work center j

M_{ij} - standard labor of product i at work center j

E/t/ - as defined above.

We used a different random number generator for T^{*}_{i} and M^{*}_{ij}.

We ran the model for three different control policies. We considered the priority argument separately to control the sequencing within the shop-floor and a the simulation model for optimizing whith minimum, maximum operation time, then with maximum accumulated value. We collected data of waiting time, at each work center, the throughput time of each product type, the sum of total output as total value of the entire production period and some total penalty of waiting time. In comparison to the change control policy based on a non-changing mathematical and structural model, the least penalty cost was achieved by control policy with maximum value priority. The least waiting time was achieved by priority control of the shortest operation time descipline. The results are shown in Table 1. It should be noted that the production output was almost the same in all three cases. Nevetherless, the cumulated waiting time gives a great difference. The maximum is in case of priority policy by longest operation discipline, the minimum is in case of shortest operation discipline and the difference is about 14 %.

Table 1. Experimental results

Control policy	Total waiting time of released batches 1/	Total waiting time of ordered batches 1/	Total output value 2/	Total penalty cost 2/
Priority M_{min}	705505	708800	24879021	1365714
Priority M_{max}	799654	801902	24870784	1548718
Priority V_{max}	739508	742204	24877656	1242583

1/ in minutes

2/ in unit of money

5. PROGRAMMING CONSIDERATION

The simulation language used for modelling and programming is GPSS/360. This was chosen because of the very good capabilities of GPSS to describe all kinds of job flow, queueing and time interdependent situations within a job-shop and because of the graphic output available with GPSS/360.

The data, which the user has to provide for GPSS, was entered in the form of variable cards, storage definition cards for each work centers with fixed number of machines, the operation time of job type by machine in matrix format, the arrival rates per job via savevalue initialization cards. To describe distributions the definition of certain functions may also be necessary, and these were supported by function cards. The length of the simulation run was controlled by the number of time units.

Although the program is relatively short, its logic is fairly complicated and no more than a brief mention of one of its basic programming solutions can be given here.

All the jobs in the simulation are carried out regarding the fixed routing sequence. All job operations follow the same procedure at each work center effected by them, hence one generalized set of blocks is required which can be reused for the same activities of a procedure on transaction as job operation at a work center. This set of blocks was the nucleus of the program giving all the statistical data on every work center: queueing statistics, utilization statistics. Waiting and flow time distributions were collected on each job type from this set, too.

Because of our aim to investigate the effect of different production control policies, as many different and separate model runs as policies and assumtions were established. The great deal of statistical printout was very useful to evaluate the dynamic behaviour of the shop-floor. The limits of this paper do not allow for illustrating all features of GPSS used extensively during the simulation modelling.

REFERENCES:

A.B.Calica: Production order sequencing.
IBM Systems Journal. vol.4. No.3 225-240. /1964/

R.W.Conway and W.L.Maxwell: Network scheduling by the shortest possible discipline. Operations Research 10, No.1. 51-73 /1963/

S.Gorenstein: Parameter values for sequencing control. IBM Systems Journal. vol. 4. No.3. 241-249 /1964/

C.W.Merrian: Optimization theory and the design of feedback control systems. McGraw-Hill Book Company, New York, Chap.3 and 4. /1964/

T.L.Saaty: Elements of queueing theory with applications. McGraw-Hill Book Company, New York /1961/

QUEUE IDENTIFIER	NUMBER OF OPERATIONS	AVERAGE QUEUE LENGTH	MAXIMUM QUEUE LENGTH	ZERO ENTRIES	PERCENT ZEROS	AVERAGE QUEUE TIME
LATH1	3109	.113	5	2672	85.9	77.725
LATH2	2277	.070	4	1997	87.7	75.853
LATH3	2200	.124	3	1772	80.5	87.053
DRIL1	2317	.158	4	1726	74.4	80.663
DRIL2	1896	.124	4	1490	78.5	91.987
DRIL3	1472	.416	5	723	49.1	166.630
MILL1	2600	.172	4	1984	76.3	84.176
MILL2	2958	.186	5	2423	81.9	104.448
MILL3	2025	.244	6	1542	76.1	152.062
MILL4	1193	.025	3	1092	91.5	76.237
GRIN1	1963	.309	5	1349	68.7	151.024
GRIN2	1174	.155	3	917	78.1	182.062
GRIN3	1295	.060	3	1122	86.6	104.716
HEAT1	2654	.175	6	2144	80.7	103.084
GALV1	1519	.039	3	1402	92.2	102.076
INSP1	4381	.096	5	3595	82.0	36.749

PRIORITY WITH VALUE

W.CENTER IDENTIFIER	AVERAGE LOAD	W.CENTER UTILIZATION	AVERAGE QUEUE LENGTH	MAXIMUM QUEUE LENGTH	ZERO ENTRIES	AVERAGE QUEUE TIME
LATH1	3.147	.629	.113	5	2672	77.725
LATH2	2.477	.619	.070	4	1997	75.853
LATH3	2.750	.687	.124	3	1772	87.053
DRIL1	1.878	.626	.158	4	1726	80.663
DRIL2	1.789	.596	.124	4	1490	91.987
DRIL3	1.464	.732	.416	5	723	166.630
MILL1	2.634	.658	.172	4	1984	84.176
MILL2	4.448	.741	.186	5	2423	104.448
MILL3	3.966	.793	.244	6	1542	152.062
MILL4	1.562	.520	.025	3	1092	76.237
GRIN1	3.096	.774	.309	5	1349	151.024
GRIN2	2.021	.673	.155	3	917	182.062
GRIN3	1.709	.569	.060	3	1122	104.716
HEAT1	5.240	.748	.175	6	2144	103.084
GALV1	3.127	.625	.039	3	1402	102.076
INSP1	1.512	.504	.096	5	3595	36.749

PRIORITY WITH VALUE

MATRIX FULLWORD SAVEVALUE 4

ROW	COLUMN 1	2	3	4	5	6	7
1	252	38870	251	38870	1278560	65739	5093
2	204	22811	202	22811	1048063	29941	5188
3	321	25403	319	25403	1344447	28634	4214
4	337	23317	333	23191	1152038	21056	3459
5	287	47079	285	46899	1270514	55540	4457
6	289	36352	287	36352	926886	31225	3229
7	196	47412	193	47096	1779450	102015	9219
8	250	49970	248	49393	1356089	91065	5468
9	164	50972	162	50433	1276420	96123	7879
10	200	50236	199	50081	1198872	92944	6024
11	157	31411	156	31308	1150506	70572	7375
12	171	31317	169	31054	1353757	72127	8010
13	256	32726	255	32726	1225057	34511	4804
14	238	32581	235	32518	1314417	26514	5593
15	198	22758	197	22758	865870	20963	4395
16	221	37756	221	37756	1186298	38490	5367
17	193	40997	192	40963	1167210	71289	6079
18	197	34848	197	34848	1320561	67729	6703
19	138	60058	136	59985	1326611	163104	9754
20	144	25330	143	25063	1336030	63002	9342

PRIORITY WITH VALUE

TOTAL VALUE: TOTAL WAITING PENALTY:

24877656 1242583

PRODUCT TYPE	RELEASED BATCHES	AVERAGE THROUGHPUT TIME	S.DEVIATION OF THROUGHPUT TIME	CUMULATED WAITING TIME
1	251	3280.278	209.062	38870
2	202	2631.564	186.000	22811
3	319	2274.093	121.437	25403
4	333	2453.249	147.875	23191
5	285	2560.501	247.125	46899
6	287	2345.661	174.500	36352
7	193	4215.039	229.312	47096
8	248	3105.649	202.562	49393
9	162	3872.185	320.000	50433
10	199	3444.914	254.437	50081
11	156	3471.717	207.000	31308
12	169	3821.295	234.500	31054
13	255	2600.694	196.500	32726
14	235	2838.110	237.937	32518
15	197	2737.913	181.750	22758
16	221	2894.018	228.312	37756
17	192	3156.750	252.875	40963
18	197	3435.573	218.625	34848
19	136	5114.117	386.000	59985
20	143	4617.054	230.875	25063

PRIORITY WITH VALUE

Some Simulations of Targeted Orthogonal Rotation of Factors – 1. Random Factors and Real Hypotheses –

N. A. Van Hemert, M. Van Hemert, and J. J. Elshout, Amsterdam
and Utrecht

1. INTRODUCTION

One is continually confronted with criticisms of factor analysis for its implied subjectivity. This criticism is heard particularly where targeted rotation is used to test hypotheses concerning (1) the number of factors as well as (2) the factor loadings of the variables, i.e., the kind of factors. By targeted rotation it is meant that an empirical factor matrix is rotated to as close a match as possible to an hypothesized factor (target) matrix that sums up the hypotheses under investigation and which has the form of an ideal solution typically showing maximal simple structure. Since Horn's (1967) demonstration the dependence of a researcher on this kind of rotational procedure, also known as Procrustes rotation, has been severely criticized because of the subjective character of the approach.

In Horn's simulation experiment seventy-four random scores for each S (N=300) were generated, arbitrarily named, intercorrelated, factored and rotated obliquely into a least squares approximation to an hypothesis factor matrix, containing twenty-one factors and defined on the basis of a theory and results obtained in earlier studies of Horn (see: Horn and Cattell,1966). If a variable given a particular name was supposed to help define a given factor, it was given a non-zero loading of .2, .3, .4, .5 or .6 in the hypothesis target matrix; otherwise a zero was recorded.

A short explanation is perhaps appropriate. In his ability research Cattell (1971) propagates the use of so-called 'hyperplane stuff'. That

is, in a battery of intelligence tests a number of variables is included from which it is expected that they are uncorrelated with the intellectual factors searched for, so that these variables can define a hyperplane perpendicular to which the factors are located. The idea, then, is that the locations of a factor in different investigations will be about the same regardless of the specific variables used in each investigation to measure this factor; i.e., it should be a means to insure a higher degree of factorial invariance. Here these 'hyperplane stuff' variables will be called H-variables whereas the variables intended to measure a factor will be referred to as F-variables.
In Horn's target matrix twelve F-variables had two, fourty-two F-variables had one and twenty H-variables had no hypothesized loadings different from zero, totalling sixty-six target loadings $\geq |.20|$, sixty-five of which were positive.
From his results Horn concluded:

"The resulting factors were found to be quite interpretable and to have high hyperplane counts, although communalities and factor loadings were low. There was thus indication that if an investigator were willing to interpret relatively low loadings if these seemed to 'make sense', he needn't bother to gather actual data: random variables may be labeled arbitrarily and pushed into solutions that make quite 'good sense'."

He then continues, however:

"The intercorrelations among factors determined in this way tended to be large, however, thus suggesting that this kind of information can be used to indicate the invalidity of a factorial solution arrived at by subjective rotation."

This suggests that Horn did not intend his conclusions also to pertain to the orthogonal case.
In spite of a reaction of Guilford and Hoepfner (1969) who judge Horn's demonstration to be a "caricatured example of a 'procrustean' rotation" and in spite of Horn's remark concerning the resulting factor intercorrelations his results are used, even by Horn himself, as arguments against targeted orthogonal rotation of real data (e.g. Horn,1970; Carroll,1972; Hakstian,1972; Eysenck,1973; Nijsse,1973).
There seem to be reasons enough, then, to ask two questions: (1) Are Horn's results also to be expected from targeted orthogonal rotation of random data? And, more interesting, (2) are his results valid as an argument against targeted orthogonal rotation of real data? This last question will be the subject of the second part of this paper. To pro-

vide for an answer to the first question in this part Horn's experiment will be replicated for the orthogonal case.
In view of the factor intercorrelations reported by Horn and given the fact that in orthogonal rotation the angles of separation between factors are fixed at ninety degrees one might expect, for instance, (1) a lesser number of 'high' positive loadings which might be even lower than those reported by Horn. In view of the fact that all but one of the target loadings are positive one might also expect (2) a larger number of 'high' negative loadings which might be of an absolute higher value. Furthermore, in view of the simple structure properties of oblique solutions one might expect (3) a less good simple structure. Moreover, one might also expect (4) a larger number of 'high' loadings of the H-variables which more outspoken than in Horn's study will have to function as a kind of 'garbage can' variables.

2. PROCEDURES

With the exception of the kind of rotation the procedures were identical with Horn's. That is, seventy-four random variables each based upon an N of 300 and drawn from a normal distributed population of real numbers were separately generated. The names assigned to the variables were the same as Horn's. As he, however, reports but the names of those variables which either had an hypothesized loading or had a loading $\geq |.20|$ in his solution the names of fourteen variables were unknown. Because these names also are not given in the paper Horn refers to (Horn and Cattell,1966) these fourteen H-variables were assigned arbitrary names like Social Class, Self Esteem, and the like.
The intercorrelations of the seventy-four variables were factored using the method of principal axes with iterative communality estimates. Twenty-one factors were extracted and iterations continued till each of the communalities differed less than .01 from corresponding estimates in the immediately preceding solution. The resulting communalities ranged from .10 to .46 with a mean value of .264; so 26.4% of the total variance of the variables was accounted for.
Subsequently, instead of the Hurley-Cattell (1962) oblique Procrustes procedure which was used by Horn, Cliff's (1966) orthogonal rotational procedure was used to obtain a least squares fit between the obtained random factors and the hypothesized target matrix which was identical

with Horn's. The coefficients of congruence between each rotated factor and its corresponding target factor ranged from .35 to .64 with an overall coefficient between the rotated solution and the target matrix of .46 which in view of the positive loadings in the target matrix is extremely low, but not to be evaluated by means of a statistical test. The variance accounted for by the rotated factors ranged from .77 to 1.03 per factor with a mean value of .9306 which is very low also.

3. RESULTS

Concerning the magnitude of the 'high' loadings in both studies some means were calculated (see Table I).

Mean of:	Horn:	We:
positive loadings of +.20 or larger	+.301	+.280
negative loadings of -.20 or smaller	-.224	-.233
positive loadings of +.30 or larger	+.390	+.378
negative loadings of -.30 or smaller	none	none

Table I: Means of some 'high' loadings in Horn's and in this study.

Although the differences in Table I are in the expected directions (i.e., in this study lower 'high' positive loadings and in absolute value higher 'high' negative loadings) they are not very impressive. Concerning the resulting simple structure, defined as the proportion of loadings within the $\pm$.10 hyperplane, the expectation is that the results of this study should show a less good simple structure than Horn's solution. Horn reports a proportion of .60 or better (i.e., higher) for all factors whereas for one factor we found .53 and for the remaining twenty factors a range from .65 to .78 with a mean proportion for all twenty-one factors of .70. Although it may be that the resulting simple structure in this study is a little bit worse (or rather less good) than in Horn's study it cannot be denied that it is to be considered a rather good one and, as such, it is more or less against the expectation. Simple structure as such, without further evidence, seems to be a rather poor argument in favor of the validity of a solution.

Concerning the other expectations first of all some terms will be defined: 'high' loadings, as did Horn, will be defined for the moment as loadings $\geq |.20|$; 'hits' will be defined as high loadings in places where high loadings of the same sign were hypothesized; 'misses' will be defined as low loadings (i.e. $< |.20|$) in places where high loadings were hypothesized and 'extra's' will be defined as high loadings of F-variables on other factors than their hypothesized one(s). Then, the results of both studies can be summarized as presented in Table II.

Number of:	Horn:	We:
Hits (misses = 66 - hits)	50(16)	40(26)
Extra's	22	40
High loadings of F-variables	72	80
High loadings of H-variables	7	39
High loadings (total)	79	119
High negative loadings F-variables	13	17
High negative loadings H-variables	4	25
High negative loadings (total)	17	42
High positive loadings F-variables	59	63
High positive loadings H-variables	3	14
High positive loadings (total)	62	77

Table II: Numbers of hits, misses, extra's and of high positive and negative loadings of F- and H-variables in Horn's and in this study. ('High' means $\geq |.20|$)

From Table II it can be noted first of all that the total number of high loadings in this study is much higher than in Horn's. Expressed as proportions of the total number of loadings (i.e. 1554) this difference is significant ($z=2.94; p \leq .0033$; two-tailed).
Concerning the numbers of high positive loadings, expressed as proportions of the total number of loadings, the difference is not in agreement with the expectation: Horn's proportion is even lower though not significantly so ($z=1.30; p \leq .1936$; two-tailed).
Concerning the numbers of high negative loadings, expressed as proportions of the total number of loadings, the difference is in agreement with the expectation: the proportion in this study is much higher and very significantly so ($z=3.29; p \leq .0006$; one-tailed). Even in arguing that these numbers should be expressed as proportions of the total number of high loadings this difference remains significant though, of course, less ($z=2.06; p \leq .0188$; one-tailed). Accepting this argument, however, one should also express the numbers of high positive loadings as proportions of the total number of high loadings in which case the difference is in the expected direction (Horn's proportion is much higher then) and of course at the same level of significance as the dif-

ference between the numbers of high negative loadings expressed as proportions of the total number of high loadings.
From Table II it can readily be noted, however, that all these differences are to be attributed to the H-variables. There are no appreciable differences concerning these numbers for the F-variables in these particular studies. It will be clear that the expectation concerning the numbers of high loadings of the H-variables is strongly supported: in this study this number is nearly six times the number Horn reported. Even if expressed in terms of proportions of the total number of high loadings this result is highly significant ($z=3.73; p \leq .0001$; one-tailed).

From Table II some further comparisons can be made. Concerning the numbers of hits and misses the difference doesn't seem very impressive and expressed in terms of proportions of the number of target loadings (i.e. 66) this difference is barely significant ($z=1.87; p \leq .0307$; one-tailed). In comparing these numbers one should, however, take into account that there is a difference in total number of high loadings in both studies and if there are more high loadings there is a better chance on hits. Expressed as proportions of the total number of high loadings the difference in numbers of hits comes out to be very significant: Horn's proportion is much higher ($z=4.11; p \leq .0001$; one-tailed). Even in arguing that the hits and misses pertain to the F-variables only and that therefore one should express the numbers of hits as proportions of the numbers of high loadings of the F-variables this difference remains significant though, of course, less so ($z=2.43; p \leq .0075$; one-tailed).
Concerning the numbers of extra's the difference is rather appreciable. Expressed, however, as proportions of the total number of high loadings this difference is not significant ($z=.86; p \leq .3898$; two-tailed). Arguing as above, however, and expressing these numbers as proportions of the numbers of high loadings of the F-variables this difference also comes out to be significant ($z=2.44; p \leq .0146$; two-tailed). If, moreover, the high loadings of the H-variables are interpreted as extra's (i.e. as 'wrong' loadings and why shouldn't it be done?), combining them with the extra's and expressing the resulting numbers (29 versus 79) as proportions of the total number of high loadings this difference is very significant ($z=4.11; p \leq .0001$; two-tailed).
From Table II it can readily be noted that Horn arrived at a much bet-

ter solution than we did. Concerning the numbers of hits, extra's and high loadings of the H-variables the expected frequencies are 66, 0 and 0; Horn obtained 50, 22 and 7, which perhaps is not too bad. We, however, obtained 40, 40 and 39, which gives an impression of a random or chance distribution. This is, of course, exactly what it is and accordingly our solution should make this impression.

In view of these results we assume that nobody will be surprised that, by taking loadings $\geq |.20|$ to be significant (and, of course, not only if they "make sense"), we judge our solution to be uninterpretable even if it is allowed to rationalize after the fact[1]).

4. DISCUSSION

It will be clear that in view of the results the very first conclusion has to be that Horn's demonstration, even if one is convinced that it is valid as an argument against targeted oblique rotation (which, by the way, doesn't go for us), is <u>not</u> valid at all as an argument against targeted orthogonal rotation.

In the second place we think Horn makes a mistake in arguing:

> "In a factor analysis based upon a sample of 300 subjects, as represented in this study, it would not at all be unusual for an investigator to base his interpretations of factors on loadings of about .20 or .25 and larger."

We can readily agree that a correlation of .20 based upon an N of 300 is significant beyond the .01 level. But the argument that because of this any factor loading $\geq |.20|$ can be considered to be significant at this level also, is contradicted by the findings of both studies. That is, in these particular studies where, moreover, all but one of the loadings in the target matrix are positive, Horn's percentage of loadings $\geq |.20|$ equals 5.1% and in our study this percentage even amounts to 7.7%. We do not think that anybody will agree that taking a factor loading $\geq |.20|$ to be significant is solid enough as a defense against a Type I error. In these particularly designed studies even taking a loading $\geq |.30|$ to be significant is not very conservative as an action with regard to the defense of a Type I error: Horn obtains 1.7% and we 1.6% of loadings $\geq |.30|$. Only if we take loadings $\geq |.40|$ these

---

[1]) For those who are interested a listing per factor of the target loadings and the loadings $\geq |.20|$ of the variables in both studies is available from: N.A. Van Hemert, Psychologisch Laboratorium, Weesperplein 8 (room 743), Amsterdam, Holland

percentages in both studies lower to .6% which is acceptable.
But even if, according to the well known rule of thumb, we take loadings $\geq |.30|$ to be significant it is clear that both solutions are uninterpretable. Concerning Horn's study this is already to be understood from his remarks that indicate the necessity of taking "relatively low loadings" (i.e. "about .20 or .25") to be significant ("if these seemed to 'make sense'"!). Concerning this study it can suffice to say that then there are only two factors with two significant hits, which is required by another well known rule of thumb for the interpretation of a (common) factor, three factors with zero significant hits and sixteen factors with one significant hit. Even in accepting the argument, that psychologists can rationalize almost anything after the fact and taking into account in the interpretation also the extra's and the high H-loadings we only count six common factors, defined each by two variables. That is, of the twenty-one hypothesized factors more than 70% doesn't even come out to be a common factor. In this case Table II reduces to Table III (the number of target loadings reduces to 58).

Number of:	Horn:	We:
Hits (misses = 58 - hits)	26(32)	20(38)
Extra's	0	2
High loadings of F-variables	26	22
High loadings of H-variables	0	3
High loadings (total)	26	25

(There are no high negative loadings)

Table III: Numbers of hits, misses, extra's and of high loadings of F- and H-variables in Horn's and in this study. ('High' means $\geq |.30|$)

Concerning the numbers in Table III there is hardly any difference between both studies and in both studies the number of hits is outnumbered by the number of misses, even more so in the orthogonal case. Once again the conclusion has to be that Horn's results cannot legitimately be used as an argument against targeted rotation of factors, let alone targeted orthogonal rotation of factors.
Finally, once again defining 'high' loadings to be $\geq |.20|$ something can be noted about the relation between the percentage of hits and the number of target loadings a factor receives in the hypothesis target matrix as well as the magnitude of these target loadings. These per-

centages are presented in Table IV.

Mean % of hits with:					
Number of targets:	Horn:	We:	Magnitude of targets:	Horn:	We:
1 (2 factors)	100.00	100.00	.2(8 variables)	0.00	37.50
2 (8 factors)	100.00	84.25	.3(12 variables)	58.33	25.00
3 (4 factors)	83.33	41.67	.4(21 variables)	85.70	57.10
4 (3 factors)	66.75	58.25	.5(19 variables)	100.00	84.20
6 (4 factors)	58.33	54.18	.6(6 variables)	100.00	100.00

Table IV: Mean percentage of hits related to the number and to the magnitude of target loadings.

From Table IV it is clearly to be noted that the lesser the number of hypothesized loadings for a factor and the greater the magnitude of the target loadings the better the chances on hits become. It is suggested, for instance, that one shall take at least a minimum of four variables for each factor. The experiment, however, was not designed to answer this kind of questions and on subjects like these the interested reader is referred to Humphreys et al. (1969). For a critique of the use of 'hyperplane stuff' the reader is referred to Humphreys (1967).
The final conclusion of this part of this paper is, that whether one considers loadings $\geq |.20|$ or loadings $\geq |.30|$ to be significant Horn's results are certainly not valid as an argument against targeted orthogonal rotation, not even in the case of random data.
The question, however, remains in what way targeted rotation can force a solution upon a given set of real data. That is, is such a solution to be discriminated from a random solution, i.e. from a solution based on random hypotheses? For the orthogonal case this question will be dealt with in the second part of this paper.

5. SUMMARY

Although Horn's (1967) results pertained to oblique rotation of random data they are used as arguments against targeted orthogonal rotation of real data. To check for the validity of these arguments in this study Horn's experiment was replicated with the exception, however, that the rotation was orthogonal. That is, seventy-four random normal-deviates for an N of 300 were generated, named according to Horn, intercorrelated, factored and rotated orthogonally into a least squares approximation to a target matrix which was identical with Horn's.
In comparing the results, taking loadings $\geq |.20|$ to be significant,

some significant differences were noted. With regard to the magnitude of the significant loadings as well as to the resulting simple structure in both solutions the differences were not very impressive. They mainly pertain to the total number of significant loadings and to the numbers of hits, misses and extra's, as well as to the numbers of significant negative loadings and, above all, of significant loadings of variables for which no loading different from zero was hypothesized. As a consequence of these differences the orthogonal solution was uninterpretable. Apparently there are some real differences between Procrustean oblique and orthogonal rotation.

If only loadings $\geq |.30|$ are considered to be significant then these differences are of only minor importance. In that case, however, Horn's solution is uninterpretable, which goes, perhaps even stronger, also for the orthogonal solution.

The conclusions are: (1) whether one considers loadings $\geq |.20|$ or loadings $\geq |.30|$ to be significant, Horn's results are <u>not</u> valid as an argument against targeted orthogonal rotation, and (2) a large N is not sufficient as an argument for not keeping to the rule of thumb that for a loading to be significant it should be $\geq |.30|$. If one sticks to this rule and to the other rule of thumb which requires at least two significant loadings for a factor to be interpreted, then, a Procrustean (whether oblique or orthogonal) rotation of random factors typically leaves one with an uninterpretable solution.

REFERENCES

Carroll, John B., Stalking the wayward factors: book review of Guilford and Hoepfner (1971), Contemporary Psychology,6,321-324, 1972

Cattell, Raymond B., Abilities: their structure, growth and action, Houghton Mifflin Cy., Boston,1971

Cliff, N., Orthogonal rotation to congruence, Psychometrika,31,33-42, 1966

Eysenck, H.J., The measurement of intelligence, Medical and Technical Publishing Co. Ltd., Lancaster, England,1973

Guilford, J.P., The nature of human intelligence, McGraw-Hill, New York,1967

Guilford, J.P. and Ralph Hoepfner, Comparisons of Varimax rotations with rotations to theoretical targets, Educ. and Psych. Measurement,29,3-23,1969

Guilford, J.P. and Ralph Hoepfner, The analysis of intelligence, McGraw-Hill, New York,1971

Hakstian, A. Ralph, Book review of Guilford and Hoepfner (1971), Educ. and Psych. Measurement,32,211-215,1972

Horn, John L., On subjectivity in factor analysis, Educ. and Psych. Measurement,27,811-820,1967

Horn, John L., Book review of Guilford (1967), Psychometrika,35,273-277,1970

Horn, John L. and Raymond B. Cattell, Refinement and test of the theory of Fluid and Crystallized intelligence, J. Educ. Psych., 57,253-270,1966

Humphreys, Lloyd G., Critique of Cattel's "Theory of Fluid and Crystallized intelligence: A critical experiment", J. Educ. Psych., 58,129-135,1967

Humphreys, Lloyd G., D. Ilgen, D. McGrath and R. Montanelli, Capitalization on chance in rotation of factors, Educ. and Psych. Measurement,29,259-271,1969

Hurley, J.R. and Raymond B. Cattell, The Procrustes program: Producing direct rotation to test a hypothesized factor structure, Behavioral Science,7,258-261,1962

Nijsse, M., Creativiteit en de meting er van bij kinderen: een literatuur overzicht, Ned. Tijdschr. Psych.,28,477-502,1973

Some Simulations of Targeted Orthogonal Rotation of Factors – 2. Real Factors and Random Hypotheses –

M. Van Hemert, N. A. Van Hemert, and J. J. Elshout, Amsterdam and Utrecht

1. INTRODUCTION

In the first part of this paper it was demonstrated that the results of Horn's (1967) simulation experiment are not valid as an argument against targeted orthogonal rotation of factors. This demonstration, which was an orthogonal 'replication' of Horn's experiment, pertained to the rotation of random factors to real hypotheses. In our opinion, however, the only light that the analyses of random data can shed upon the problem of rotation is whether the conventional lines of defense against a Type I error are solid enough. Much more interesting is the case of real data. That is, we should be concerned not with the 'informed' analysis of random data, but with the rotation of real factors to random hypotheses. What would happen, for instance, if we randomly permuted the names of the variables in a particular study, revised the target matrix accordingly and left the factor matrix intact? Is it to be expected, then, that targeted rotation can force a solution upon the set of given real data in such a way that there will not be an appreciable difference in fit between, on the one hand, a realistic target matrix and the factors rotated to this target, and, on the other hand, a random target matrix and the factors rotated to this random target? This paper intends to provide for an answer to this question in the orthogonal case.

In trying to do so a second independent variable which sometimes is considered to be relevant in targeted rotation will be introduced. In his very critical review of Guilford and Hoepfner (1971) Hakstian (1972)

refers to Horn (1967,1971) and, among others, states that given a large ratio of retained factors to variables (i.e. a small variable/factor ratio) virtually any set of data can be fitted to any hypothesis. This might be in agreement with the suggestion given in the first part of this paper that one shall have to take at least four variables for each hypothesized factor. So, the second question is a kind of specification of the first one: can a small variable/factor ratio be of decisive influence on the fit between a target matrix and factors rotated to this target?
To obtain evidence on these questions it was decided to do at least two simulation experiments in each of which an unrotated factor solution of real data is rotated orthogonally to a least squares fit with a realistic target matrix as well as with a large number of random target matrices generated by means of a computer program, SIMRANTAR, especially written for these experiments and to be described in the next section. Of each rotated factor solution measures of resemblance will be computed concerning its resemblance to the corresponding target matrix. In one of these simulations the variable/factor ratio is high (i.e. 5.17); in the other one this ratio is low (i.e. 1.62). Because of the results obtained from this latter experiment a planned third simulation with a medium variable/factor ratio (i.e. 2.76) was not carried out.

2. THE COMPUTER PROGRAM SIMRANTAR[1])

In this and the next sections we will define the following matrices:

A = an empirical factor matrix
REALTAR = a realistic target matrix
RANTAR = a random target matrix
RANROT = RANTAR rotated to REALTAR
F = A rotated to REALTAR
X = A rotated to RANROT

Following Cliff's (1966) orthogonal rotational procedure SIMRANTAR rotates a given A to a given REALTAR. After this first rotation the program generates a given number of RANTAR's in which the hypothesized loadings of the variables are equal to those in REALTAR, but where the factor, on which the loading (taken to be the square root of the commu-

[1]) The computer program SIMRANTAR is written by the authors in ALGOL 60 and was processed on the EL-X8 computer of the Mathematical Centre in Amsterdam.

nality) is hypothesized, is decided upon randomly.
In doing so, however, the program requires each factor in RANTAR to have at least a minimum number of target loadings. This means that one can prevent the occurrence of zero vectors in RANTAR; i.e. one can require the number of factors with one or more hypothesized loadings in REALTAR and in every RANTAR to be equal. This means that if a given variable/factor ratio is equal to 1.0 and if the desired minimum of loadings per factor in RANTAR is equal to 1 (which means that in this case every factor in RANTAR will have exactly one hypothesized loading) every RANTAR will give a rotational result which is identical to the result obtained from the rotation of A to REALTAR (i.e. F and all X's will be identical).
Because sometimes some variables have hypothesized loadings on two factors in REALTAR, SIMRANTAR can be instructed to generate RANTAR's in which a number of variables is hypothesized for two factors (in which case each hypothesized loading of the variable concerned is taken to be the square root of half of the communality). Depending on an input number the number of variables with two hypothesized loadings in the RANTAR's can vary from none to all variables with all possibilities in between. Because of this feature it is possible to have in every RANTAR approximately the same number of variables with two hypothesized loadings as in REALTAR. <u>Which</u> variables will have two loadings in any RANTAR, however, is completely randomized.
The measures of resemblance used by SIMRANTAR in comparing two matrices are the coefficient of factorial similarity (abbreviated to CFS) as well as the sum of squared deviations (abbreviated to SSD) of corresponding elements of the factors. These measures are computed for each comparison of corresponding factors as well as for the compared matrices as a whole treating all factor columns in a matrix together as one column. (If the resulting CFS for a factor is negative, the rotated factor in the comparison is reflected to give the coefficient a positive sign.)
The measures of resemblance are computed to decide upon the resemblance of F to REALTAR. For each of the RANTAR's two comparisons are made. This is done because the degree of correspondance between REALTAR and any RANTAR will influence the degree of resemblance between X and RANTAR. Hence, to be able to control for this possibility for every RANTAR

the measures concerning its resemblance to REALTAR are computed. To make this comparison meaningful it is necessary that every RANTAR be rotated to a least squares fit with REALTAR. After having performed this rotation SIMRANTAR computes the CFS's and the SSD's to decide upon the resemblance between RANROT and REALTAR. (It will be clear that RANROT is not essentially different from RANTAR as initially generated by SIMRANTAR.) After having rotated A to RANROT SIMRANTAR, finally, computes the measures of resemblance between X and RANROT. (Computation of these measures between, on the one hand, A rotated to RANTAR and, on the other hand, RANTAR will, of course, give identical results.)

3. THE SIMULATIONS AND THEIR RESULTS

3.1 Introduction.

In designing the experiments the most important problem, of course, was the construction of REALTAR. It would, of course, be nonsense to take the final factor solution of the selected studies to be REALTAR: evidently this would result for each study in a perfect resemblance between REALTAR and F. As one of the selected studies stems from the Aptitude Research Project (ARP) of Guilford (1967) c.s. there was no problem concerning the construction of REALTAR for this study. This is because in any ARP study the hypotheses that identify the factors on which any variable will have a loading are stated explicitly. Thus, in these studies this problem can be solved easily and objectively. That is, in the case of the ARP study subjected to SIMRANTAR the REALTAR was taken to be a zero matrix except for those elements where loadings of variables were hypothesized in which case this loading was equal to the square root of the communality of the variable concerned. Perhaps it has to be stated explicitly that this procedure in constructing REALTAR maximizes the chances to get a 'negative' answer to the questions posed (i.e. in fact the 'positive' answer that by means of targeted rotation, indeed, virtually any set of data can be fitted to any hypothesis). This is because in an ARP study typically about half of the variables are 'new' ones, that is, factor analyzed for the first time in their Structure of Intellect (SI) career (for a description of the SI theory and the SI model see Guilford (1967) and Guilford and Hoepfner (1971)). But even although for a number of variables a secondary (or even a primary) loading on another factor as well as , or instead

of, on the hypothesized one, may sometimes be expected and accounted for in SI terms (before any rotation is performed) the loadings of these variables are hypothesized uniquely to be on one factor and one factor only. (One can imagine that when an investigator knows that his target matrix will be compared to a number of random targets, this target will perhaps be made a little less 'ideal' and a bit more 'realistic'.)
With regard to the second study selected to be subjected to SIMRANTAR the REALTAR was constructed in a somewhat different way, which is described in section 3.2.
Another problem was the minimum number of hypothesized loadings any factor in RANTAR should have. It was decided that this minimum number of loadings should be equal to the number of loadings of that factor in REALTAR which had the least number of loadings.

3.2 Simulation experiment I.
The study selected for having a high variable/factor ratio is "a classic in factor analysis literature" (Harman,1970) and has been used extensively to illustrate different methods of factor extraction and rotation. The initial data (the scores of 145 seventh and eighth grade school children) were gathered by Holzinger and Swineford (1939). The names of the twenty-four variables, their descriptive statistics and their intercorrelations are given by Harman (1970). As these variables can be considered to be tests of intelligence, it was decided to construct a REALTAR as if these variables were SI tests. This is not to say that the battery can be considered to be a battery typical of an ARP study; on the contrary, in SI terms this battery would be 'rather messy', containing tests that obviously correlate with more than one SI factor.
On the basis of the names and, as far as possible, the content of these tests it was decided that their intercorrelations could be accounted for by six factors (not to be considered as SI factors, but rather as kinds of 'mixtures' thereof). This means that the variable/factor ratio is equal to 4.00. Moreover, because of the heterogenity of the tests, it was decided that seven of the twenty-four tests should be hypothesized to be loaded on two factors. This makes the variable/factor ratio 5.17. The numbers of hypothesized loadings on the factors in REALTAR are: one factor with eight, two factors with six, one factor with

five and two factors with three loadings.
Although Harman gives the first ten principal components he gives but one solution with more than five factors, which, however, is a bifactor solution. So, to obtain six principal factors it was decided to use the method of principal axes with iterative communality estimates. The first six factors accounted for 54% of the total variance of the variables.
SIMRANTAR was instructed to generate 250 RANTAR's in each of which every factor had at least three hypothesized loadings and in each of which about seven of the twenty-four variables had hypothesized loadings on two factors, and to do the rotations and comparisons described in section 2.
The CFS between REALTAR and F was equal to .8486 and the SSD was equal to 3.9268. Concerning the 250 RANTAR's the CFS's ranged from .5001 to .6685 and the SSD's ranged from 12.9353 to 8.5798. The frequency distribution of all 250 CFS's is presented in Figure 1 (the distribution of the SSD's which gives an analogous picture is omitted).

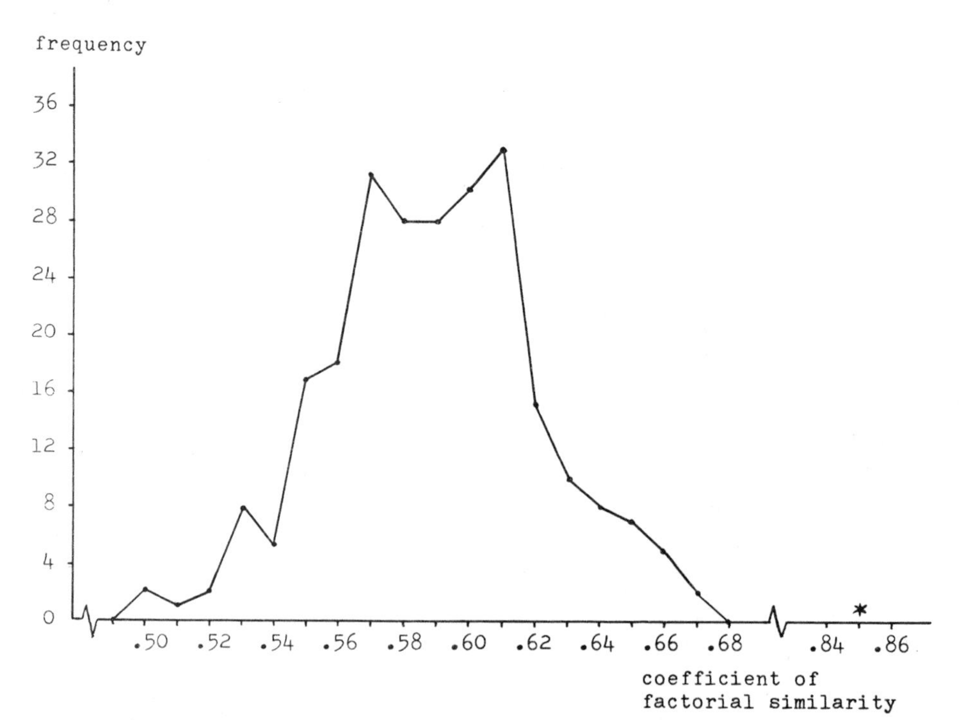

Figure 1: The distribution of CFS's obtained from the rotations to 250 random target matrices in Experiment I (* = the CFS in the case of rotation to the realistic target matrix).

3.3 Simulation experiment II.

In this study, selected for having a low variable/factor ratio, the data of an ARP study of Merrifield, Guilford and Gershon (1963) were used. The original study had thirty-three variables, but in a reanalysis (Guilford and Hoepfner,1971) sixteen of the original variables were reduced to eight variables. This was done because these sixteen variables were originally eight tests which were split each into two parts. To demonstrate common factors by means of two parts of the same test, however, is a questionable procedure. So, each two corresponding parts were combined to one test. Furthermore, four variables in the original analysis were removed from the battery because these tests were not really SI measures. This left twenty-one variables and the thirteen hypothesized factors mentioned in Guilford and Hoepfner (1971) which means a low variable/factor ratio of only 1.62. Five of these factors were singlets, i.e. represented each by only one variable. The remaining eight factors were represented by two variables each. The REALTAR was based on the hypotheses in the original study and was arrived at in the manner described in section 3.1.

As the correlation matrix was available only from the original study it was of order 33. By using the formula for computing the correlation between a variable and an unweighted sum (where the correlations of this variable with each of the components of the unweighted sum are known) and by omitting the correlations of the four omitted variables we arrived at a correlation matrix of order 21.

The first thirteen factors, resulting from a principal factor analysis with iterative communality estimates, accounted for 48% of the total variance. Together with the REALTAR they were input to SIMRANTAR, which generated 120 RANTAR's in each of which every factor had at least one loading and every variable exactly one loading, and which did the rotations and comparisons described in section 2.

The CFS between REALTAR and F was equal to .7330 and the SSD was equal to 5.3989. The CFS's of the 120 RANTAR's ranged from .6008 to .6942 and the SSD's ranged from 8.1491 to 6.2424. The frequency distribution of all 120 CFS's is presented in Figure 2 (the distribution of the SSD's which gives an analogous picture is omitted).

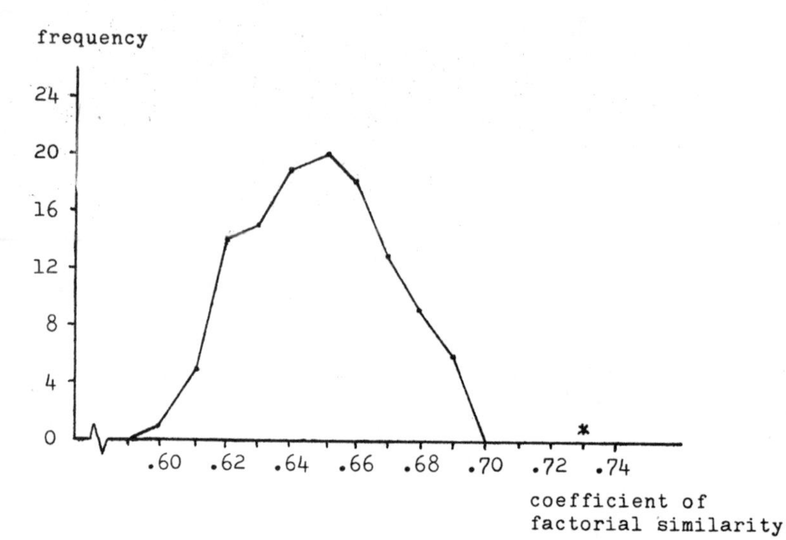

Figure 2: The distributions of CFS's obtained from the rotations to 120 random target matrices in Experiment II (* = the CFS in the case of rotation to the realistic target matrix).

Even with a low variable/factor ratio the best random target matrix can not be considered to be as good as a realistic target matrix. Although the differences are not as sharp as in the former simulation, they are clear enough. Doing another simulation experiment with a medium variable/factor ratio seemed rather superfluous.

4. DISCUSSION

From these results it will be clear that the answer to the first question posed has to be a definite 'no'. We must conclude that it is not true that targeted rotation will produce such a good fit for any hypothesis tested that to discriminate between random hypotheses and hypotheses that describe the data well is impossible. In both our studies the coefficient of similarity for the realistic hypotheses lies far outside the range of coefficients obtained for the random hypotheses. That the separation is clearer in the first study than in the second one can be explained from the way the hypotheses were arrived at; bet-

ter hypotheses should lie farther out.
As was to be expected a weak research design in terms of the variable/factor ratio will make the distribution of coefficients for the random targets shift to the higher values. It should be noted, however, that even in the second study with its extremely low variable/factor ratio this effect did not obscure the randomness of the random hypotheses[2]. This answers the second question posed. Even a very low variable/factor ratio is <u>not</u> sufficient to make the discrimination between different hypotheses on the basis of their fit impossible.
The influence of the research design on the CFS distribution, however, points to the problem of the evaluation of the magnitude of a CFS. An investigator who wants to know if his theory, as operationalized in his target matrix, is better than a 'random theory' is not helped very much by computing the CFS between his target and the factors rotated to this target. In view of this, the results of our experiments suggest a practical use of SIMRANTAR. That is, a researcher who feels uneasy about the CFS obtained for his hypothesis and who wants to know whether he has outdistanced the 'random theorist' in this particular situation, if at all, can let the computer generate the distribution of CFS's achieved by his competitor. It will be clear that this might be a rather costly affair.
Still costlier would be another practical use suggested for SIMRANTAR. That is, the construction of tables of 'significance' for CFS's obtained under particular circumstances. Very probably the magnitude of the CFS not only depends on the number of factors and of variables, but also, for instance, on the variable/factor ratio, the number of subjects on which the correlations and, hence, the factors are based, the 'evenness' with which the variables are spread across the factors as well as the occurrence of negative loadings in a target matrix. By systematically varying these variables SIMRANTAR might be used eventually in the construction of such tables. It will be clear, however, that this Monte Carlo research, if at all possible, implies an enormous

---

[2]) Moreover, it must be stated that, because we dealt with tests of intelligence, <u>no negative</u> loadings were allowed in the random target matrices. Had we left the decisions about the signs of the loadings in the random target matrices to the 'random theorist', i.e. to SIMRANTAR, in both studies mean CFS's of zero could have been expected for the random hypotheses.

amount of work and of computer time. At the same time the usefulness of such tables is undeniable.
To finish this paper the final conclusion is that targeted orthogonal rotation as such is not to be considered as a questionable procedure. If one wants to criticize a theory which uses targeted orthogonal rotation to provide for supporting evidence for its hypotheses, this theory should be criticized on other grounds than because of its use of this particular technique. The second conclusion is that even a very low variable/factor ratio is not sufficient as an argument against the use of targeted orthogonal rotation. It is certainly not true that in the case of a low variable/factor ratio virtually any set of data can be fitted to any hypothesis by using targeted orthogonal rotation of factors.

5. SUMMARY

In the first part of this paper it was demonstrated that a Procrustean rotation of random factors will result in an uninterpretable solution. The arguments against Procrustean rotation, however, concern targeted rotation of real data. One more argument reads that using a Procrustean rotation and "given a large ratio of retained factors to variables ... virtually any set of data can be fitted to any hypothesis".
To provide evidence of the validity of these arguments in the orthogonal case two simulation experiments were done, in each of which real data were rotated orthogonally to a least squares fit with (1) a 'realistic' target matrix, as well as (2) a large number of randomly generated target matrices in which the communalities of the variables were kept constant. The factors on which the loadings of the variables in the random target matrices were hypothesized were decided upon randomly. The number of hypothesized loadings for the random factors as well as for the variables in the random matrices were about equal to these numbers hypothesized in the realistic target matrix. In the first simulation experiment a high variable/factor ratio (5.17) was used whereas in the second experiment this ratio was low (1.62).
In both studies the resemblance between the empirical factors rotated to the realistic target matrix and this latter matrix was greater than the resemblance between the empirical factors rotate to any of the (250 and 120) random target matrices and these latter matrices.
It is concluded that the proposition that using a Procrustean rotation

any set of data can be made to fit virtually any hypothesis, even if given a low variable/factor ratio, is not valid as an argument against targeted orthogonal rotation of real data.

ACKNOWLEDGEMENT

We are indebted to Prof. Dr. Fred N. Kerlinger for his comments on an earlier draft of this paper.

REFERENCES

Cliff, N., Orthogonal rotation to congruence, Psychometrika,31,33-42, 1966

Guilford, J.P., The nature of human intelligence, McGraw-Hill, New York, 1967

Guilford, J.P. and Ralph Hoepfner, The analysis of intelligence, McGraw -Hill, New York, 1971

Hakstian, A. Ralph, Book review of Guilford and Hoepfner (1971), Educ. and Psych. Measurement,32,211-215, 1972

Harman, Harry H., Modern factor analysis, Univ. of Chicago Press, Chicago, 1970

Holzinger, K.J. and F. Swineford, The stability of a bi-factor solution, Supplementary Educational Monographs,48, Univ. of Chicago, Dep. of Educ., Chicago, 1939

Horn, John L., On subjectivity in factor analysis, Educ. and Psych. Measurement,27,811-820, 1967

Horn, John L., The structure of intellect: Primary abilities. Chapter VI-B in R.M. Dreger (ed.), Multivariate personality research: Contributions to the understanding of personality in honor of Raymond B. Cattell, Claitor, Baton Rouge, Louisiana, 1971

Merrifield, P.R., J.P. Guilford and A. Gershon, The differentiation of divergent-production abilities at the sixth grade level, Reports from the Psychological Laboratory, Univ. of Sthrn. California,27, 1963

A Comparison of Simulated Results with Those Obtained by Steady State Formulae

K. Venkatacharya, Benghazi

1. INTRODUCTION

In recent years, concepts of Renewal Process and Markov Renewal Processes have been applied to anumber of fields including human reproduction[Perrin and Sheps, 1964]. The human reproductive process can be treated as a Markov Renewal Process. The term'event' is used in Renewal Process to mean various occurrences such as a conception; the'risk of event' to mean the monthly chance of conception (MCC) or fecundability; and the 'dead times' to mean the various nonsuceptible (to the risk of conception) periods associated with abirth. Thus the results of one system are applicable to the other under identical assumptions.

Making use of the theory of Markov Renewal Process a number of results pertaining to human reproductive process are obtained. These include the expected number of live births--ELB--(expected number of renewals) and the expected interval between (live) births or events -- ELBI--(recurrence times).

It is known that the results of Markov Renewal Process are valid under certain assumptions. The reliability of the use of the theoretical results to practical situations depends upon the sensitivity of these results to deviations in the assumptions underlying the theory of Markov Renewal Process. The effect of deviations in some of these assumptions on steady state results have been examined earlier by some authors.

The three important assumptions under which the expected values of various statistics connected with a renewal process are derived are :

(a) The probabilities of occurrence of various 'events' such as a conception, live birth termination etc, are constant i.e the probabilities such as MCC do not vary between women.

(b) The probabilities do not vary over time or order of event i.e the distribution of biological variables such as MCC remain constant with age and/or birth order of women .

(c) The process continues for infinitely long period of time so as to reach a steady state.

Under the above assumptions we obtain ELBI as

$$\text{ELBI} = m + \frac{q+\pi w}{p'} = \frac{mp' + \pi w + q}{p'} \qquad \dots (1)$$

where $p'=(1-\alpha)p$, p being MCC and α the probability of pregnancy wastage. $\pi = \alpha p$ and $q = 1-p$. m and w are total nonsuceptible periods associated with a live birth and a pregnancy wastage respectively. g is the gestation period leading to alive birth conception. The values of g,m and w are measured in months which approximates an intermenstruum.

Further under the above assumptions we obtain the mean number of live births [Sheps et al, 1969] for t months of (marital) duration as

$$\text{ELB}(t) \doteq \frac{t-g}{\mu^{(1)}} + \frac{\mu^{(2)}+\mu^{(1)}}{2[\mu^{(1)}]^2} - \frac{\mu^{(0)}}{\mu^{(1)}} \doteq \frac{t-g}{\mu^{(1)}} \qquad \dots (2)$$

where $\mu^{(1)}$ and $\mu^{(2)}$ are the first two factorial moments about zero of live birth intervals, and $\mu^{(0)}$ the mean interval between consummated marriage and first live birth. Eq. (2) can be put explicitly as

$$\text{ELB}(t) = \frac{(t-g).p'}{mp'+\pi w+q} \qquad \dots (3)$$

From eqs.(1) and (3) we obtain

$$\text{ELB}(t) = \frac{(t-g)}{\mu^{(1)}} = \frac{(t-g)}{\text{ELBI}} \qquad \dots (4)$$

The inverse relation between ELB and ELBI is valid only if t is sufficiently large (assumption -a). If t is small then estimates of birth intervals get biased and hence eq.(4) will not be valid. Later reference to this aspect will be made.

2. PROBLEM

The effect of relaxing assumptions (a) and (c) on ELB and ELBI has been discussed by some authors [Sheps, 1964; Henry, 1965]. In case the gestation period and postpartum amenorrohea (PPA) vary for women only -- not over timeor age --the eqs. (1) and (2) can be modified to yield ELBI and ELB in the heterogeneous case. The expected number of live births can be obtained by multiplying the left hand member of eq.(2) by the respective probability density functions and integrating the product with respect to the variables used in the product. It was found that the expected number of live births for the heterogeneous case very well approximated by using eq.(2) with the mean values (averaged over women) of PPA , w and g instead of their fixed values. In the case of ELBI also the same result holds.

However if p is distributed between women, the expected number of live births for the heterogeneous case is found to be smaller than the value obtained through the eq.(2) by using mean values of p instead of p itself [Perrin and Sheps, 1964]. But the expected mean live birth interval for heterogeneous case is found to be greater than the value obtained through (1) using mean value of p (averaged over women) instead of p [Sheps, 1964].

There isa second direction in which the biological variables in general and MCC in particular can vary, that is variation by age and/or birth order. In the present paper we are interested in obtaining results on ELB and ELBI by relaxing assumption (b) while retaining the other two (in case of steady state formulae assumption- c is needed but in simulation model -- to be introduced subsequently-- assumption - c will be ralaxed). No study appears to have been made to the author's knowledge about the impact of deviations of MCC with duration on ELB and ELBI.

The mathematical derivation of equations similar to (1) and(2) for ELBI and ELB when MCC is assumed to vary by age is difficult.

Even when such mathematical relations are obtained they are expected to be complicated functions which need numerical approximation through computers.

Alternatively we can examine the difference between the value of ELB obtained after taking into account the actual variations in MCC by age and the value of ELB approximated by using mean MCC values (averaged over duration) in formula (2). In the present paper the ELBI and ELB are obtained through a Monte Carlo model which allows for variations of gestation and PPA between women , and variations in MCC over duration. The later values are treated as the estimates of the true value of ELBI and ELB when assumptions a, b, and c are relaxed. These values are then compared with those obtained by steady state results using (1) and (2) with the mean values of MCC (averaged over duration), gestation and PPA (averaged over women).

3. SIMULATION MODEL

The simulation model generated events such as marriage, occurrenceof a conception, length of gestation, type of conception outcome, length of PPA etc. depending on the input probabilities. Making use of the ages of occurrence of the above events in a woman's simulated history, various fertility measures can be computed.

It is well known that fertility measures based on simulated histories are subject to sampling errors. In order to minimize the sampling errors, for each combination of input values 1000 women histories were generated which has been established as a good sample size [Roy and Venkatacharya ,1970].

In the present model three levels of MCC (F_0,F_1 and F_2)are considered. These levels reflect the fecundity of high and low fertility populations. The pregnancy wastage is taken at two levels (P_0andP_1). The PPA following a live birth is taken at three levels (A_0,A_1 and A_2). Only means of the distributions are shown in the Table 1. The choice of the parameters is made on the experience of some developed and developing countries.

Since we have three levels of MCC, three levels of PPA and

two levels of pregnancy wastage , in all eighteen variable combinations are possible. For the purpose of the present paper only results corresponding to P_0 level are presented. The results for P_1 are quite similar.

The simulated results on ELB are obtained for each variable combination at 60,120,180,240 and 360 months of marital duration corresponding to 5,10,15,20 and 30 years. The corresponding value of ELB using eq.(2) are derived by using the mean values of MCC, gestation and PPA (shown in Table 1) at each of the five marital durations mentioned above.

4. RESULTS

An examination of Table 2 indicates that the ELB obtained through the simulation model and those obtained through the use of steady state eq.(2) are quite close. For the high and medium MCC levels in most of the casesELB of steady state formula are more than the corresponding results obtained through the simulation model.For low MCC this pattern does not hold in all cases. The main observation that can be made from this table is that eq.(2) making use of mean MCC (over duration) give very close results to those of the simulation model where MCC varied over duration.

For duration of 60 months in the low MCC case the difference between the two estimates of ELB (Table 2) is large. However, this difference decreases fast from and after 120 months of duration. This is due to the duration of 60 months being insufficient for ELB to stabilize to steady state results.

Thus in cases where MCC is not constant over the reproductive span, the ELB could be estimated through eq.(2) using mean MCC values(over duration) of the three patterns considered in this paper.The results on ELB for other patterns of MCC also gave the same type of observations.

The results on ELB from Table 2 also indicate that the use of eq.(2) may result in slight overestimation compared to simulation results. Experiments with continuosly increasing pattern of MCC-- though unrealistic -- also indicated slight overestimation by eq.(2). However, the magnitude is small for moderate values of MCC. But if MCC increased to very high values the overestimation was significant. If MCC on the other hand decreased by duration significantly then eq.(2) is found to slightly underestimate ELB, but only after certain duration, during which period it gave overestimation.

One remark on the three patterns of MCC used in the paper is worth mentioning. Though MCC varied over duration, the mean MCC-- Table 1 -- by duration fluctuated within a narrow range. Also the MCC increases in the early period and decreases in the later period.Asa result of these characteristics of fecundability the eq.(2) gives very satisfactory results.

One of the interesting applications of this result lies in the fact that the eq.(2) can be used (with mean values of biological variables) to readily estimate ELB for complete or incomplete reproductive span. The eq.(2) can also be used for validation of the results of a simulation model.

Now we turn to the study of ELBI.Table 2(II) presents ELBI generated through simulation model and also by using eq.(1). It can be easily noticed that the values of ELBI obtained in the simulation model are always smaller than those estimated through the steady state formula. The difference between the two values--termed as error -- could be due to two reasons:

(i) Use of mean MCC instead of a duration pattern of MCC,and

(ii) Assuming infinite time for reproductive process in the derivation of the steady state results, where as a finite period is used for the simulation model.

It is difficult to specify the contribution of each of the above factors because they are not independent. For example , as the MCC values decrease -- such as at terminal durations--the mean MCC decreases. Simultaneously, the truncation error increases with the decrease in MCC [Sheps et al,1970; Venkatacharya, 1969]. The same thing is discovered in Table 2(II), where the error increases as the level of MCC decreases from F_2 to F_0 .

The error due to truncation is knowntobesignificant even if MCC were to be constant over time [Sheps et al,1970;Venkatacharya, 1969]. Thus with MCC varying over time we should expect this pattern of truncation error to hold. From Table 2(II) it is evident that

TABLE 1

Input values in the simulation model

(a) Monthly chance of conception--MCC:

Level of MCC	Duration in months (x)	Values of MCC
F_o	1- 30	.0049+ .00049 (x- 1)
	31- 78	.0199+ .00004 (x- 30)
	79-138	.0646+ .00004 (x- 78)
	139-198	.0673- .00043 (x-138)
	199-234	.0413- .00041 (x-198)
	235-360	.0266-. 00020 (x-234)
F_1	1- 60	.05 + .0025 (x- 1)
	61-120	.20
	121-300	.20 - .00083 (x-120)
	301-360	.05 - .00042 (x-300)
F_2	1- 60	.10 + .0025 (x- 1)
	61-180	.25
	181-240	.25 - .00083 (x-180)
	241-300	.20 - .0017 (x-240)
	301-360	.10 - .0008 (x-300)

Mean MCC values at various durations (months)

	60	120	180	240	360
F_o	.016	.034	.043	.041	.033
F_1	.120	.160	.165	.155	.114
F_2	.170	.210	.223	.236	.214

(b) Pregnancy wastage: (In the simulation model still births and abortions are taken seperately)

P_o : 15% P_1 : 25%

(c) Gestat on(distributed among women) leading to a	Mean (Months)	Variance
live birth	9.6	.46
still birth	8.2	.60
abortion	3.7	1.40
(d) Postpartum amenorrohea following a live birth:		
Level A_o	10.8	46
Level A_1	14.7	32
Level A_2	20.1	94

(e) The mean total nonsuceptibility associated with a live birth (m) corresponding to the three levels of PPA are 20.4,24.3 and 29.7 months respectively.The mean total nonsuceptibility for a pregnancy wastage (w) is 7 months.

TABLE 2

Expected number of live births for pregnancy wastage (P_o), three levels of MCC, three levels of PPA and five durations of simulation model (SIM) and of steady state formula (SS).

Duration (months)		LEVEL OF MCC								
		F_2			F_1			F_o		
		LEVEL OF PPA								
		A_o	A_1	A_2	A_o	A_1	A_2	A_o	A_1	A_2
		I: Expected Live Births (ELB)								
60	SIM	2.13	1.87	1.72	1.85	1.64	1.53	.66	.66	.61
	SS	2.13	1.92	1.71	1.91	1.74	1.56	.57	.55	.53
120	SIM	4.56	3.98	3.52	4.16	3.63	3.24	2.18	2.04	1.87
	SS	4.56	4.02	3.48	4.25	3.79	3.30	2.08	1.96	1.86
180	SIM	6.96	5.99	5.22	6.42	5.55	4.87	3.71	3.46	3.10
	SS	6.95	6.10	5.28	6.46	5.72	4.95	3.70	3.44	3.17
240	SIM	9.49	8.06	6.95	8.48	7.34	6.36	4.90	4.53	4.05
	SS	9.40	8.21	7.01	8.50	7.52	6.50	4.80	4.47	4.09
360	SIM	13.5	11.6	9.8	10.8	9.5	8.3	5.8	5.4	4.8
	SS	13.8	12.1	10.3	11.6	10.4	9.0	6.3	5.9	5.5
		II: Expected Live Birth Intervals(ELBI) (for duration 360 months)								
	SIM	24.5	28.9	33.3	28.9	33.0	37.1	44.3	47.3	52.1
	SS	25.7	29.9	35.3	30.5	34.7	40.1	55.8	60.0	65.4
	% Error	4.9	3.5	6.0	5.5	5.2	8.1	25.9	26.8	25.5
		III : Estimates of ELB using ELBI(SIM) of II and eq.(4) with duration 360.								
	SIM	14.69	12.46	10.81	12.45	10.91	9.70	8.13	7.61	6.91
	SS	14.01	12.04	10.20	11.80	10.37	8.97	6.51	6.00	5.50

the percentage error in ELBI in the simulation model ranges from 4 to 30. A significant part of this error should be due to truncation. Thus the steady state eq.(2) for ELB gives reasonably accurate estimates (or slightly overestimates) even when MCC varies over duration, but the eq.(1) for ELBI gives overestimates.

From eq.(4) we note that under the assumptions (a),(b) and (c) ELB and ELBI are inversely related. However, results of the simulation model indicate that this relation need not hold when assumption (c) is relaxed. But where truncation errors are small, this relation approxiamtely holds-- a fact that can be verified from Table 2(III).

REFERENCES

Henry,L. French statistical research in natural fertility. In Mindel C.Sheps and J.C.Ridley (Eds). Public Health and Population change- Current Research Issues, Pittsburg University Press, 1965.

Perrin E.B and M.C.Sheps. Human reproduction : A stochastic process. Biometrics, 20,28-45, 1964.

Roy T.K and K.Venkatacharya. An application of analysis of variance technique to Monte Carlo data on human reproduction. Sankhya-B, 32,1970.

Sheps M.C. On the time required for conception. Population Studies, 28,85-97,1964.

Sheps M.C.,J.A.Menken and A.P.Radick. Probability models for family building: An analyatical review. Demography,6,161-183,1969.

Sheps M.C.,J.A.Menken.,J.C.Ridley and J.W.Lingner. Truncation effect in closed and open birth interval data, Jour.Amer. Stat. Assn, 65,678-693,1970.

Venkatacharya.K. An examination of a certain bias due to truncation in the context of simulation models of human reproduction, Sankhya,31,397-412,1969.

SUBJECT GROUP E

Software Packages

The BOON System – A Comprehensive Technique for Time-Series Analysis

K. G. Beauchamp, Cranfield, **P. Kent**, Berkshire, **S. E. Torode**,
E. Hulley and **M. E. Williamson**, Cranfield

1. INTRODUCTION

The collection and analysis of data series taken at equal intervals of time, distance or some other measure is common to many fields of research; for example, stock market prices [Godfrey, 1965], ocean tides [Hasselman, Munk and MacDonald, 1963], large pulses in electricity grid systems [Fox and Kent, 1970], magnetic and gravitational anomalies in the ocean floor [Niedell, 1965], the population of swans on a reservoir [Holgate, 1966] or the unstable behaviour of coffee futures due to the time lag in growing new trees when the price is high to replace trees cut down when the price is low. Despite this very wide range of application the analysis of all these problems has much in common. This analysis is quite complex and often requires very sophisticated computing techniques. Since many of the users have little or no computing experience it is essential to provide them with a system of programs that is easy to understand and to use.

The BOON system [Cranfield Institute of Technology, 1973] allows users to manipulate series by issuing a set of simple commands. Commands exist to read, plot, calculate the Fourier transform or the spectrum of a series, etc. The user assembles a number of such commands to form his BOON program. The BOON monitor program then checks all his commands and issues subroutine calls to perform the necessary manipulations. The BOON program also takes complete control of the data handling and file management, so allowing the user to concentrate wholly on the analysis of his data.

2. PROCESSING OF A DISCRETE SERIES

Much of the data acquired from research experiments is continuous in form and a number of constraints have to be accepted when the data is converted into a discrete series for the digital computer. These constraints arise because:-

(a) the data is sampled
(b) the data is quantised
(c) the record length is finite
(d) the frequency sample is finite

Some of the implications of these constraints are referred to in the techniques for discrete series manipulation discussed below.

Full consideration of the properties of sampled and quantised signals in this context is given elsewhere [Beauchamp, 1973b].

2.1 Pre-Processing

A pre-requisite of several analytic processes is that the data have zero-mean value and unit variance. This is the case for example with normalised autocorrelation. Again where a finite length record is taken from a much longer record, tapering will need to be carried out to modify the abrupt beginning and ending of the finite record. These constitute pre-processing operations needed before analysis can be attempted. A common pre-processing problem is the need to improve the dynamic range for analysis of small signals superimposed on a slowly varying base line or trend. An example is shown in Figure 1 in which trend removal, a zero-mean operation and tapering have all been carried out on the data.

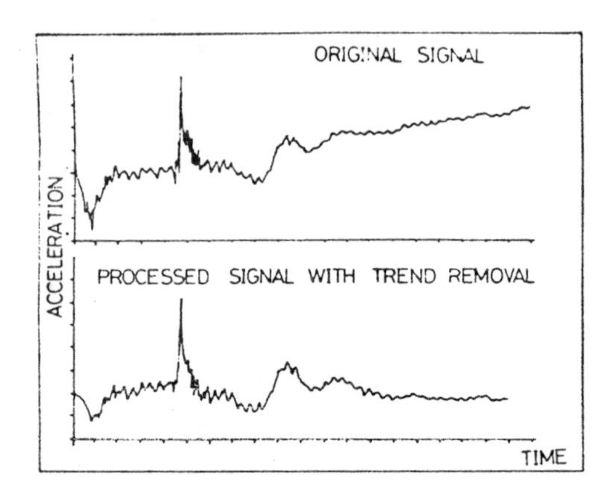

Figure 1 Pre-Processing Operations on a Transient Signal

2.2 Statistical Analysis

Statistical algorithms are required to enable mean, rms value, standard deviation and variance of a data series to be evaluated. Others carry out probability density calculations, probability distribution and peak probability analysis. Statistical tests are needed to determine the stationarity of the series. One form of test produces a running mean and variance from which a decision can be made concerning the quasi-stationary period over which short-term spectral analysis can be attempted.

2.3 Transform Programs

Domain translation using an orthogonal tranformation forms an important operation in processing a time series. This is because the characteristics of a time series can be expressed by a far smaller series of numbers when translated in this way. Additionally, transformation techniques are employed in filtering and in correlation.

It is well known that considerable economy in computing time can be obtained if a special ordering routine, commonly known as a fast transform, is employed [Cooley and Tukey, 1965]. Using the fast Fourier transform (FFT) for example, only $2N\log_2 N$ complex mathematical operations are required to transform N values compared with N^2 operations if earlier methods are used. Alternative transforms based on the Walsh series [Walsh, 1923] or the Harr series [Haar, 1910] enable even faster transform calculations to be carried out since the mathematical operations reduce to additions and subtraction only, and in the case of the Haar transform, a reduced number of these.

The choice of transform depends on the origin of the data series. Where the series can be related to a real physical system, synthesis of the waveform from a given number of sinusoidal functions is possible. This is the case for many laboratory experiments and here the use of Fourier, rather than Walsh or Haar transformation is likely to minimise errors in any form of subsequent analysis [Beauchamp, 1973a]. A rectangular based signal is generally the result of technology generated systems (eg communication coding systems) and some analytic advantages may be obtained if the Walsh or Haar transform is used.

A number of subroutines have been written to carry out the FFT, the fast Walsh transform (FWT), and the fast Haar transform (FHT). The latter two have a speed advantage over the FFT of about eight times when programmed in FORTRAN.

The basic transforms are used in spectral analysis, correlation, and digital filtering, although the three transformations can only be regarded as alternatives for digital filtering. Correlation is not easily possible using the Haar or Walsh transform and slightly different methods are required to use them for spectral analysis [Beauchamp, 1972]. An FFT routine for spectral analysis must permit the analysis bandwidth to be specified. A similar routine using the FWT is also possible. Figure 2 shows a comparison between results obtained for spectral analysis of the same transient signal using both systems.

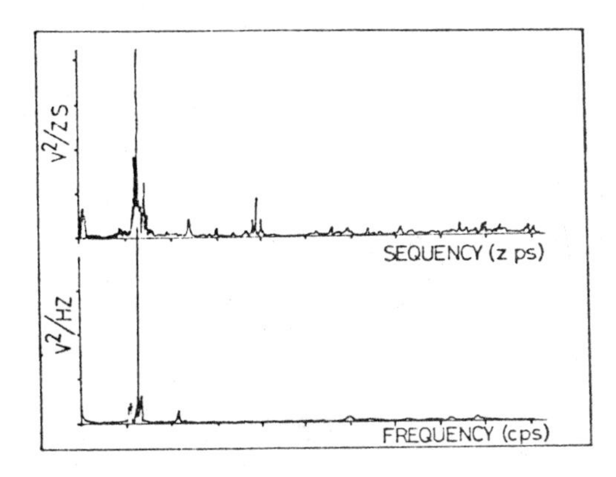

Figure 2 Spectral Analysis Using the Fourier and Walsh Transform

The FFT can also be used as a basis for correlation and convolution. The auto- and cross-spectrum (phase and amplitude characteristics) are obtained as auxiliary outputs from the correlation routines.

2.4 Digital Filtering

Filtering is one essential operation common to many analytical processes. Often the fundamental characteristics of a process can be obtained simply by selecting appropriate frequency regions and measuring the contents found there. Another use of the digital filter is to smooth a data series in the time or frequency domain.

We can classify digital filters into two broad categories, recursive and non-recursive, which have quite distinct characteristics. The non-recursive filter is stable and has zero phase shift. Providing the required filter response is known, either directly or as a transfer function, there is no difficulty in realising its characteristics using a convolution method. In this method the convolution of two time series is obtained indirectly as the inverse transformation of the product of their Fourier transforms. Techniques of FFT manipulation are available to permit series of unlimited lengths to be filtered [Beauchamp and Williamson, 1973]. Recursive filters are implemented directly from a summation which includes a feedback term. Instability is possible and zero phase shift will not necessarily be obtained. However, the operation of recursive filtering takes less computing time and requires a smaller number of parameters than the non-recursive case. It would therefore be chosen for many operations. Wiener filtering has been implemented which allows the user to select his transform (Fourier, Walsh, or Haar).

In order to develop a flexible recursive digital filter a cascaded model is used where the signal is passed through a sequence of recursive filters. Each elemental filter is simple in design but the composite filter can have complex performance characteristics. The required filter configurations (low-pass, high-pass, band-pass, band-stop) can be implemented directly from the specification of the filter parameters. An applications example is shown in Figure 3 which shows the extraction of a periodic signal from a composite wave-form giving information not readily apparent from inspection of the original record.

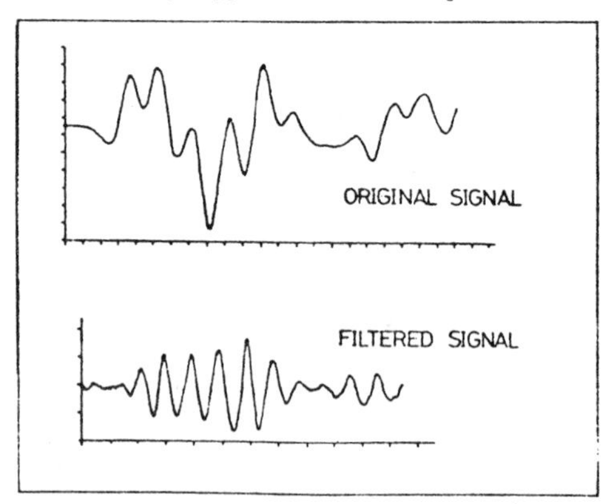

Figure 3 Digital filtering

3. A SIGNAL PROCESSING SYSTEM

An original version called BOMM was written in the early 1960s [Bullard, 1966] to help in the analysis of geophysical data. This was written for the IBM 7090 computer. Later versions were available on the CDC 1604 and 3600 machines. From 1964 the Atlas Computer Laboratory provided a computing service to University and Government financed research workers via its Atlas computer. A copy of the BOMM program became available to the Atlas users in 1967. This was used until the closure of the Atlas computer in 1973.

In 1972 it was decided to offer similar facilities for users of the new Atlas Laboratory computer, an ICL 1906A. As the Cranfield Institute of Technology had developed a set of programs for analysing time series using more modern numerical techniques a contract was placed to develop a BOMM-like system based on these programs. The result of this co-operation is the BOON system (<u>BOM</u>M <u>O</u>n <u>N</u>ineteen hundreds).

The first version of BOON was released in June 1973. Work is now in hand to develop a Mark 2 version having much better error checking facilities and graphical output.

3.1 The BOON Command System

The user directs his analysis by means of BOON commands issued in the order in which they are to be obeyed. Each command consists of an identity name followed by a list of qualifying parameters. The commands fall into six groups which:

(a) control data channels but have no effect on the data 'series', eg opening and closing magnetic tape

(b) control input and output of series on standard peripherals

(c) control graphical representation of series

(d) generate new series from parameters supplied in the command or by operation on series already present, eg function generation, pre-processing, filtering, calculation of spectra

(e) calculate only one value from a series, eg standard deviation

(f) control the flow of the system, eg looping

As a series is input or created, it is given a 'name' by the user. The series is referred to by name in any subsequent commands.

A complete example of the use of BOON for an investigation into the effect of pulses in Electricity Grid Systems is given in the Appendix.

3.2 BOON Structure

The BOON system is written in FORTRAN segments controlled by a 2-stage overseeing routine. The first stage unpacks and checks each command and writes the unpacked version to a system file. The second stage uses the finished system file as input for the execution of the commands. This allows easy enhancement of the system.

A command is read and unpacked and default parameters, if any, are supplied. A check is made firstly for incompatibilities within the parameter list (eg ensuring that a finish time is greater than a start time) and secondly that the series named in the command will exist. The unpacked commands are retained on backing store ready for stage two. If, however, an error is detected, an appropriate diagnostic is printed and further commands are checked but not added to the system file. The execution stage may then be entered.

The time taken to execute the BOON commands will depend upon the number of commands and the length of the series. As this varies from run to run, it was decided to leave the use of the dump and restart facility to the users' discretion.

3.3 Use of Backing Storage

BOON allows series, once created, to be used in any order so they are stored on random access devices for speed of retrieval. The user may specify that a series is not to be retained on backing store in which case it is only available to the next command. This can be useful in cases such as pre-processing immediately after input, where the original series is no longer required. Obviously, some series are too big to be totally resident in core and BOON reserves an area for ones that are only required by the following command.

3.4 Peripheral Handling

Data may be input from cards or magnetic tapes and may come from a variety of sources so each series may vary in format and size. Similarly, output must be flexible to meet each user's requirements. This has been achieved by allowing users the option of defining their own FORTRAN formats in the input/output commands. Although this will require some knowledge of FORTRAN on the part of the user, it was felt that this would be a reasonably flexible method. This enables the series to be input or output in a form suitable for communication with other programs.

3.5 Graphical Presentation

Graphical output may be sent to a lineprinter, a plotter or a micro-film recorder. Magnetic tapes may be produced for the latter two enabling plotting to take place at any installation having the appropriate hardware, whilst a rough graph may be obtained on the lineprinter.

There are three kinds of photographic output: 16 mm, 35 mm film and paper, different intensities being available to distinguish curves. Long series may be plotted on cinefilm. More sophisticated facilities are available on the plotter and micro-film recorder. These include:

(a) linear or logarithmic scales in either or both the x, y directions

(b) choice of heading and x, y titles

(c) off-set curves with hidden line suppression (3-dimensional effect)

There is provision for several curves per graph. Each curve may have its points joined by:

(a) straight lines
(b) curves
(c) a step function/histogram

or be marked only by a symbol of the user's choice.

An example of graphical output is given in Figure 4. This shows the three dimensional display of short-term spectral analyses, using hidden line suppression to enhance the clarity of presentation.

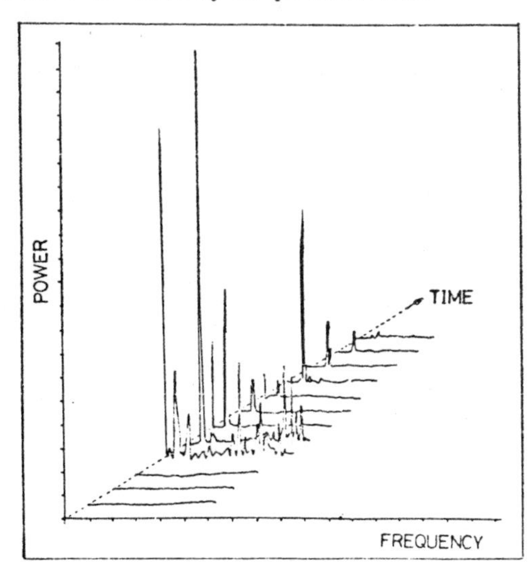

Figure 4 Three dimensional display of short-term spectral analysis using hidden line suppression

4. CONCLUSIONS

Processing of discrete data series is a corollary to very many research projects, (not all of them scientific). The computational tools required are often developed 'ad hoc' and much duplication of effort is apparent.

The BOON system described in this paper is an attempt to make widely available a unified system of processing routines, and simplify implementation so that the further duplication is avoided. Its ease of extension to include new procedures when required has resulted in a versatile system capable of meeting future user demands. Further extension to include interactive operation is under consideration and would provide the user with not only a rapid analytical tool but also a means of studying the methods of analysis themselves as suggested elsewhere [Freedman, 1970].

APPENDIX

This concerns a study undertaken to determine the behaviour of large inter-connected electricity transmission/generation systems when subjected to heavy pulsed loads. The pulses were produced by repetitively exporting electricity to France via a cross-channel link. In this way, a train of 10 to 25 square waves was produced having amplitudes of 80 to 160 M watts and repetition rates of 0.25, 0.33 and 0.5 Hz. The effect of the pulses on the electricity grid systems was measured at three places in the UK and also at CERN in Switzerland. The results were printed at intervals of 0.5 seconds with an accuracy better than 0.01%.

It was necessary to know the form and amplitude of the disturbances induced in the electricity grid system. If such loads caused no resonances in the system and were adequately damped out by the generators, then high energy accelerators, which demand large pulses of electricity, could be connected directly to the electricity grid system and thus avoid costly flywheel stabilisation systems.

The analysis of the data proceeded as follows:-

The original series was read in and plotted (Figure 5) using commands:

```
READF LPH 15 1 256 0 0.5 (15F5.0)      (inputs series of length 256 and
Data Series                             named LPH)
PLOT LPH
```

The auto spectrum was then computed and plotted using the commands:

```
AUTOS LPH SPECT                         (series named SPECT created by this
PLOT SPECT                              command)
```

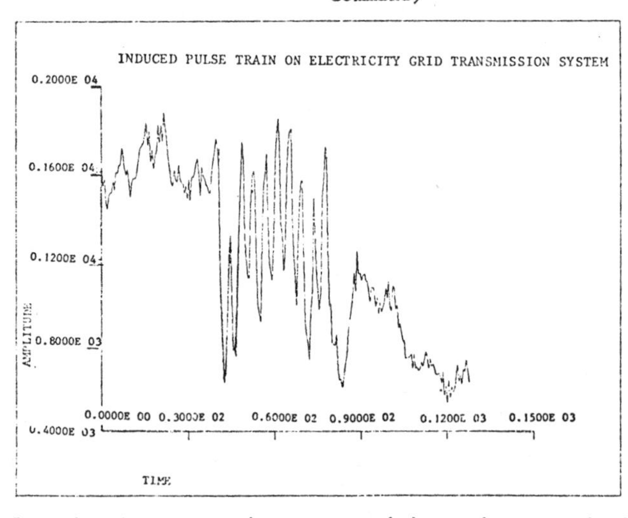

Figure 5 Induced Pulse Train on Electricity Grid Transmission System

The results at this stage could not be used directly since the low frequency component dominated the spectrum preventing the pulse frequency component from being seen. To overcome this the logarithmic transform of the spectrum was computed, smoothed and plotted viz:

SARIT 1 SPECT LGSPEC NOTAPE	(computes the logarithmic transform of SPECT and names the results LGSPEC which is not written to backing store)
DESF LGSPEC 1 18.0 20.0 20 NOTAPE	(designs and applies a filter to series LGSPEC and overwrites original series with the filtered series of the same name)
PLOT LGSPEC	

The low frequency component was removed and the spectrum recomputed to enable the spectral peak at 0.25 Hz to be seen.

DESF LPH 2 .1 .5 20 PH NOTAPE	(creates new series PH which is LPH filtered)
AUTOS PH PHSPEC NOTAPE	
PLOT PHSPEC	

The three components low frequency, high frequency and pulse frequency were then separated from the original series and plotted on the same graph (Figure 6).

DESF LPH 1 0.0 .15 20 L	(low-pass filter)
DESF LPH 2 .35 .5 20 H	(high-pass filter)
DESF LPH 3 .15 .35 20 P NOTAPE	(band-pass filter)
PLOT L H P	

The list of commands were terminated by:

END

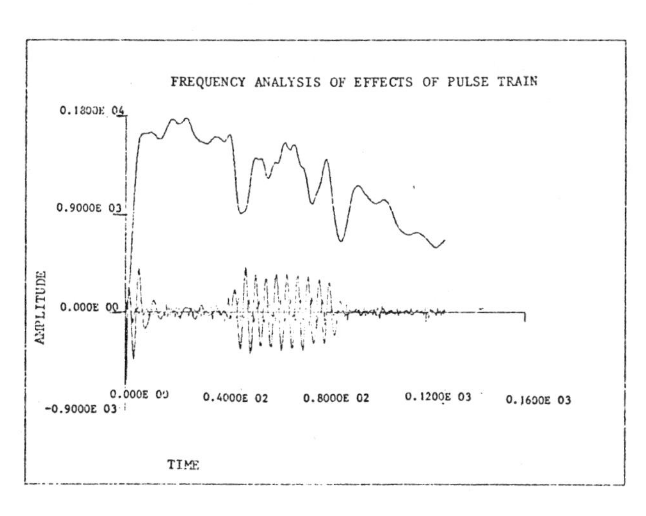

Figure 6 Frequency Analysis of Effects of Pulse Train

The start of pulsing at time 42 can be clearly seen in the graph of P, the end of pulsing occurs at time 86. There is also a noticeable fall in the low frequency component at the start of pulsing as the generators compensate for the extra load. There is a corresponding rise after the pulsing has stopped. By using BOON the analysis of this data is almost trivial, without such a system the analysis would require a major effort by the research workers involved.

REFERENCES

Beauchamp, K G, The Walsh Power Spectrum. Proc NEC, 27, Chicago USA, 377-382, 1972.

Beauchamp, K G, Waveform Synthesis Using Fourier and Walsh Series. Hatfield Colloquium 'Theory and Applications of Walsh Functions', 1973a.

Beauchamp, K G, Signal Processing. George Allen & Unwin, 1973b.

Beauchamp, K G and M E Williamson. Digital Filtering for System Analysis using Fourier and Walsh Techniques, IEE Conference 'The Use of Digital Computers in Measurement', York University, 1973.

Bullard, E C et al. A User's Guide to BOMM, A system of Programs for the Analysis of Time Series, the Institute of Geophysics and Planetary Science La Jolla, California, 1966.

Cooley, J W and J W Tukey. An Algorithm for the Machine Calculation of Complex Fourier Series, Math Comp, Vol 19, 297, 1965.

Cranfield Institute of Technology, The Computer Centre, Cranfield and the Applications Software Group, Atlas Computer Laboratory, Chilton. A User's Guide to BOON - Mark 1 (BOMM on the 1906A), 1973.

Freedman, R A, Interactive Signal Processing. Computer Graphics Symposium, Brunel University, 1970.

Fox, J A and P Kent, Power System Pulse Loading Studies on the Atlas Computer. Proceedings of Fifth Universities Power Engineering Conference, University College of Swansea, 1970.

Godfrey, M D, An Exploratory Study of the Bi-Spectrum of Economic Time Series. J of the Royal Stat Soc, Series C, Vol XIV, No 1, 48-69, 1965.

Haar, A, Zur Theorie Der Orthogonalen Funktionen Systeme, Math Ann 69, 331-371, 1910.

Hasselman, K, W Munk and G MacDonald, Bi-Spectrum of Ocean Waves, Time Series Analysis, ed M Rosenblatt, J Wiley & Sons, New York, 125-139, 1963.

Holgate, P, Time Series Analysis Applied to Wildfowl Counts, App Stats, Vol XV, No 1, 15-23, 1966.

Neidell, N S, A Geophysical Application of Spectral Analyses. J of the Royal Stat Soc, Series C, Vol XIV, No 1, 75-88, 1965.

Walsh, J L, A Closed Set of Orthogonal Functions, Ann Journ Math 55, 5-24, 1923.

MAD – The Analysis of Variance in Unbalanced Designs – A Software Package

Gale Rex Bryce and Melvin W. Carter, Provo

1. INTRODUCTION

The acronym which gives this paper its title, stands for the modified abbreviated Doolittle procedure which was the basic algorithm in the original versions of the package, and not for the state of mind often engendered by the prospect of analyzing data from unbalanced designs. In this paper we shall refer to any design in which the usual orthogonality between any pair of factors has been lost as an unbalanced design. For those interested in further study of the classical methods used in the analysis of these designs we cite Yates' (1934) paper in which most of the methods still used today were first suggested, and Bancroft's (1968) text in which the basic methods are applied to a variety of data. For a more modern approach to the problem consult Searle (1968), Federer and Zelen (1966), and Speed and Hocking (1974).

MAD was developed at Brigham Young University Provo, Utah by the authors over a period of several years. It was developed in response to our needs as statistical consultants to a wide variety of users in the University community. While we have attempted to keep the package relatively simple to use, a certain amount of statistical expertise will be necessary to make of it the powerful analysis tool that it can be. However, once the basic philosophy behind the theory which is applied in MAD is well understood most applied statisticians will have little trouble in using it to great advantage.

Two basic principles have been kept in mind in developing the theory utilized in the MAD package. First, when one obtains a sum of squares associated with a particular factor in a linear model for an unbalanced design the expectation of that sum of squares becomes of primary and fundamental importance. This is obvious due to the aforementioned loss of orthogonality and the possible resultant contamination of the sum of squares with extraneous factors which will tend to invalidate tests of hypotheses to a greater degree than either nonchi-squaredness or non-independence which are the more commonly recognized problems of unbalanced designs. The second principle is that we will seek that sum of squares which has the minimum possible number of components. We shall use as our guide in deciding what constitutes this "minimum number" of components, the expected mean square for the same factor from a balanced version of the unbalanced design under investigation.

Few statisticians will question the appropriateness of the first of these basic principles. However, most who have analyzed complex unbalanced designs using the least squares or fitting constants method will be well aware of the problems inherent in satisfying the second principle. In this paper, we shall show that the theory implemented in MAD not only allows us to satisfy these tw basic principles, but as a side benefit allows the calculation of other quantities of considerable interest to the practitioner.

In the remainder of this paper we shall give primary attention to the theory which is implemented in the MAD package and hold illustrative examples a minimum. For those interested in numerical examples, Francom (1973) present eighteen worked examples taken from the literature which cover a wide range of experimental designs in both the balanced and unbalanced case. In the followi section we shall present the general theory associated with matrix factorizati methods. In section three we will apply the theory particularly to the method used in MAD. In the final section we shall review some of the I/O features of the package.

2. FACTORIZATION TECHNIQUES

As was mentioned earlier the basic algorithm used in the original version of MAD was the abbreviated Doolittle technique which factors a square symetric matrix into the product of upper and lower triangular matrices. An equivalent and mathematically more appealing algorithm is the Cholesky or square root fac orization which factors a square symetric matrix into the product of a lower triangular matrix and its transpose. Suppose we consider the linear model

$$\underline{y} = X\underline{\beta} + \underline{\varepsilon} \tag{1}$$

where X is matrix of known constants $\underline{\beta}$ is a vector of fixed unknown constants and random variables and $\underline{\varepsilon}$ is a vector of independent normally distributed errors. If we consider the matrix

$$C = \begin{bmatrix} X'X & X'\underline{y} \\ \underline{y}'X & \underline{y}'\underline{y} \end{bmatrix}$$

where a prime denotes the transposition of the matrix, then the Doolittle opera tions yield $C = A'B$, where $B = \text{diag}^{-1}(A)A$, and diag (A) is a diagonal matrix made up of the diagonal elements of A. The Cholesky factorization yields

$$C = T'T \tag{2}$$

with the relationship between the two factorization methods being $T = \text{diag}^{-\frac{1}{2}}(A)$

The principle application of these methods has been in the full rank case. However, Rohde and Harvey (1965) have shown that with slight modification (the M of MAD) they are applicable to singular matrices. It is well known that othe useful quantities may be obtained from the forward solutions (e.g. various sums of squares) and Gaylor et. al. (1970) have shown that the forward solution "... can be used to obtain the expected mean squares for the analyses of variance in the general case with fixed finite or random effects for any model."

For the convenience of the reader, we shall review here the basic distribu tion theory associated with the factorization (2). A detailed derivation, including variances and covariances of sums of squares from a general mixed model, is given in Bryce (1974). Let us define our linear model (1) as follows

$$\begin{aligned} \underline{y} &= X\underline{\beta} + \underline{\varepsilon} \\ &= \left[F \mid R \right] \begin{bmatrix} \underline{\Phi} \\ \underline{\rho} \end{bmatrix} + \underline{\varepsilon} \end{aligned} \tag{3}$$

$$= \left[F_1 | \dots | F_f | R_1 | \dots R_r\right] \begin{bmatrix} \underline{\Phi}_1 \\ \vdots \\ \underline{\Phi}_f \\ \underline{\rho}_1 \\ \vdots \\ \underline{\rho}_r \end{bmatrix} + \underline{\varepsilon}$$

where

X is a known matrix of constants with F_i, $i=1,\dots,f$ and R_j, $j=1,\dots,r$ being columnwise partitions of X associated with the f fixed factors and the r random factors in the model;

$\underline{\Phi}_i$ is a fixed vector of unknown constants $i=1,\dots,f$;

$\underline{\rho}_i$ is a vector of normally distributed random variables with $E(\underline{\rho}_i) = \underline{0}$, $V(\underline{\rho}_i) = \sigma_i^2 I_i$ and $Cov(\underline{\rho}_i, \underline{\rho}_j) = 0$, $i=1,\dots,r$; $j=1,\dots,r$; $i \neq j$ with I_i, $i=1,\dots,r$; being identity matrices of appropriate order:

and

$\underline{\varepsilon}$ is a vector of normally distributed random variables with $E(\underline{\varepsilon}) = 0$, $V(\underline{\varepsilon}) = \sigma^2 I$ and $Cov\ (\underline{\rho}, \underline{\varepsilon}) = 0$.

From this we note that $\underline{y} \sim N_n(F\underline{\Phi}, RVR' + \sigma^2 I)$ where $V = \text{blockdiag}(\sigma_1^2 I_1, \sigma_2^2 I_2, \dots, \sigma_r^2 I_r)$. Any of the algorithms yielding a Cholesky factorization can be thought of as premultiplication by a full rank matrix which is the product of elementary lower triagular matrices corresponding to row operations. Let Q be such a matrix, then

$$\begin{matrix} Q & C & = & T \\ \begin{bmatrix} Q_1 & 0 \\ Q_2 & Q_3 \end{bmatrix} & \begin{bmatrix} X'X & X'\underline{y} \\ \underline{y}'X & \underline{y}'\underline{y} \end{bmatrix} & = & \begin{bmatrix} T_1 & T_2 \\ 0 & T_3 \end{bmatrix} \end{matrix}$$

and $T_1 = Q_1 X'X$ and $T_2 = Q_1 X'\underline{y}$. From these considerations it is relatively easy to show that

$$T_2 \sim SN(T_F \underline{\Phi}, T_R V T_R' + \sigma^2 H) \tag{4}$$

where $SN(\cdot)$ stands for the singular normal distribution and $T_F = \left[T_{F_1} | T_{F_2} | \dots | T_{F_f}\right]$ with $T_{F_i} = Q_1 X' F_i$, $i=1,\dots,f$; with an analogous definition for T_R. The matrix H of (4) is a diagonal matrix with ones and zeros on the diagonal, the zero's corresponding to null diagonal elements in T_1. The singularity of the distribution (4) is more of a notational convenience than a distributional reality. It arises from the fact, that if the normal equations form a singular system of equations, redundancies will be indicated by null rows in the T matrix of (2). Thus, if a null element occurs on the diagonal of T in a partition associated with say T_R, then the corresponding rows of T_F, H and T_2 will be null and the marginal distribution for that element will be degenerate.

It is well known that the sum of squares for the ith factor in the model (3), adjusted for previous factors and ignoring subsequent ones is found by summing the squares of the appropriate elements of T_2. To be more precise let $T_2' = (\underline{t}_1', \underline{t}_2', \dots, \underline{t}_p')$ be partitioned conformable to the $p = f + r$ factors in the model, where the ith factor may be either fixed or random. We also partition T_1 in an analogous manner by both rows and columns i.e.,

$$T_1 = \begin{bmatrix} T_{11} & T_{12} & \cdots & T_{1p} \\ 0 & T_{22} & \cdots & T_{2p} \\ 0 & & \vdots & T_{pp} \end{bmatrix}$$

and since H is a diagonal matrix we partition it as $H = \text{diag}(H_1, H_2, \ldots, H_p)$ again conformable to the p factors in the model. In addition we define N_F and N_R as being the set of indices taken from 1,2,...,p which correspond to fixed and random factors respectively. With this notation in mind consideration of the marginal distribution of $\underline{t}_i$ allows us to obtain the expected value of the ith sum of squares as

$$E(\underline{t}_i{}'\underline{t}_i) = \text{tr}(H_i)\sigma^2 + \sum_{\substack{j \in N_R \\ j \geq i}} \sigma_j^2 \ \text{tr}(T_{ij}{}'T_{ij}) + \sum_{\substack{j \in N_F \\ j \geq i}} \ \sum_{\substack{k \in N_F \\ k \geq i}} \underline{\phi}_j{}'T_{ij}{}'T_{ik}\underline{\phi}_k \quad (5)$$

Several important observations can be made relative to the expectation (5). First we note that the expected value is uncontaminated by factors prior to the ith factor ie., we have the sum of squares for the ith factor adjusted for all previous factors in the model. Secondly, we note that other than the unknowns eqn. (5) is only a function of partitions of the Cholesky factorization of X'X. Finally, we note that the final term in eqn. (5), will contain bilinear and quadratic forms in all of the unknown fixed effects ordered after the ith factor in the model.

Thus we have, at least conceptually, a means of satisfying the first of our basic principles, that of obtaining the expected value of every factor associated sum of squares. The fact the ith sum of squares is adjusted for all factors ordered prior to it in the model might lead us to the conclusion that by judicious choice of reorderings we could obtain for each factor in the model a sum of squares with a minimum number of components and thus satisfy the second principle as well. However, if one takes X of (1) to be the usual design matrix of zeros and ones, then any ordering of the model which entails ordering an interaction factor prior to any of its interacting main effects or prior to a lower order interaction containing a subset of the interacting factors will result in the clearly erroneous conclusion that the adjusted sum of squares is identically zero! Some authors (e.g. Bock (1963) and Harvey (1970)) have attempted to overcome this problem by a reparametrization to obtain a full rank model. This has been successful with respect to the fixed model but can be generalized to the mixed model for only a few rather specialized cases. However, Bryce (1970) and Lawson (1971) have investigated a reparametrization method which appears to be applicable to any model without restriction. This method is implemented in the MAD package.

3. MATRIX TRANSFORMATIONS

Rohde and Harvey (1965) point out that the Doolittle method, and by implication the Cholesky decomposition "... is merely a systematic method for obtaining the Fourier coefficients needed in applying the Gram Schmidt orthogonalization process to the columns of X." Thus, we can visualize the Cholesky operations as sequentially orthogonalizing the columns of X ie., the second column is made orthogonal to the first, the third is made orthogonal to the first two etc. Clearly, if any column is a linear combination of prior columns in such a procedure it will become a null vector, which by definition is orthogonal to any vector.

The implications of such a process with respect to X of (1) defined as a design matrix, are obvious. The submatrix associated with any interaction factor in the model will simply be cartesian products of the columns of the submatrices associated with the interacting main effects. Thus, when an interaction is ordered before its interacting main effects we will obtain the sum of squares due to fitting all the main effects in that interaction and all possible lower order interactions of those main effects i.e., a "subclass" sum of squares. While such a sum of squares may be useful it is of no interest if we were hoping to obtain the sum of squares for a main effect adjusted say, for its interaction with another main effect. Thus the unavailability of certain desirable sums of squares when using the design matrix approach is obvious.

This characteristic of the Gram-Schmidt process is not totally detrimental however. It becomes in fact a distinct advantage when one considers a completely random model. It is well known, in this case, that for the balanced design the expected mean square for say, a main effect, will contain the variance components of interactions of that main effect with all other main effects and interactions in the model. Thus, for the random case we are forced by the nature of our process and the fact that X of (1) is a design matrix to consider only those sums of squares having the minimum possible number of components.

In the fixed model, however, we have a dilema. In the balanced case, under the usual assumptions, all effects are adjusted for all other effects in the model. This implies that we would want to reorder our model repeatedly so that we could obtain the sum of squares for each factor in the model adjusted for all other factors (be they main effects or interactions). Clearly, from the foregoing discussion, this cannot be done. This, along with the resultant savings in the size of the matrices which must be dealt with, has motivated the use of reparametrization techniques to obtain full rank matrices. For example, the fixed model would be reparmetrized as follows:

$$\begin{aligned} y &= Z\,\underline{\gamma} + \underline{\varepsilon} \\ &= ZA^{-}A\,\underline{\gamma} + \underline{\varepsilon} \\ &= F\,\underline{\Phi} + \underline{\varepsilon} \end{aligned} \qquad (6)$$

where Z is nxm design matrix of rank p, $\underline{\gamma}$ is an mx1 vector of unknown constants and A is an mxp matrix of full column rank, A^{-} being a generalized inverse of A. This is the form of reparameterization suggested in both Bock (1963) and Harvey (1970) with slight variations in the definition of A^{-}.

MAD also handles the fixed model with a reparametrization of the form (6). The major difference in its approach and that found in the papers cited is that A is chosen to be a matrix of orthogonal contrasts of the levels of the various factors. This results in a rather simple and unique form for the generalized inverse of A i.e.,

$$A^{-} = (A'A)^{-1}A'.$$

Since A'A is a diagonal matrix the transformed vector of estimable linear functions

$$\underline{\Phi} = A^{-}\underline{Y} = (A'A)^{-1}A'\underline{Y}$$

is expressed as the same orthogonal linear contrasts of the elements of $\underline{Y}$ as were used in transforming the design matrix to a matrix of full rank. This choice of A^{-} leads to a rather convenient simplification since we are able to immediately determine which estimable linear functions of the parameters will be estimated given a particular transformation of the design matrix. On the other hand if we are interested in a particular set of estimable linear functions of the parameters we know immediately which transformation of the design matrix should be used to estimate them.

Obviously in this full rank case we are free to reorder the model in anyway we desire. MAD provides the capability of reordering the factors in a model with a simple control card included in the input stream. In addition to providing the capability to reorder factors (sets of columns) MAD has provision for reordering individual degrees of freedom (single columns) within factors. If the user wishes to obtain estimates of and tests for each individual degree of freedom adjusted for all other effects in a fixed model i.e., every individual degree of freedom ordered last in the model, this may be done with a single control card. The user may select his own set of comparisons for fixed main effects or default to the standard set supplied by MAD. Interaction comparisons are automatically formed by taking Kronecker (or direct) products of the submatrices associated with the sets of comparisons of the interacting main effects. Of course nonorthogonal comparisons can be used, but one will have to explicitly solve for the appropriate transformation matrix to be used in transforming the design matrix i.e., if B' is the set of nonorthogonal comparisons desired then let $A = B(B'B)^{-1}$.

Thus, in order to satisfy the two basic principles of obtaining the sum of squares with the minimum number of components and their expected values, we must use the design matrix for the random model and a full rank reparametrization for the fixed model. But what of the most general case, the mixed model? In MAD the nature of the factors (fixed or random) are allowed to dictate the form of the associated coefficient matrix. Thus, in eqn. (3) R will be a design matrix and F will be a full column rank matrix with some partitions corresponding to reparametrizations of fixed effects and others to concomitant variables in the model if they exist.

Several important consequences follow from this definition of the coefficient matrix in the mixed model. The first is that we are now able to satisfy for any model, whether fixed, random or mixed, the final principle of obtaining the sum of squares with the minimum number of components or, in the context of our original definition, that sum of squares having only those components which are present in a balanced version of the unbalanced model under investigation. The only requirement necessary to this result being that $F \neq RK$ i.e., that the transformed coefficient matrix for the fixed factors not be in the column space of the design matrix of the random factors. This requirement will be satisfied by the class of transformation matrices we have chosen. Not only have we satisfied the principle, but in fact by using a coefficient matrix defined as above and a matrix factorization technique which uses a Gram-Schmidt process, we have a procedure which will not allow us to obtain sums of squares in an unbalanced design which are inappropriate in terms of an analogous balanced design. This is true since if an inappropriate reordering of the model is chosen, e.g., a random interaction before its interacting main effects, obvious descrepancies will occur in the degrees of freedom for the misordered effects.

The second consequence of our definition of the coefficient matrix occurs with respect to interactions between fixed and random factors. As was previously indicated with respect to interactions among fixed effects the transformation matrix for an interaction is simply the Kronecker product of the transformation matrices of the interacting main effects. If we think of random factors as being "transformed" by identity matrices of appropriate order then the definition of the transformation matrix for a fixed-random interaction is the obvious extension of the fixed-fixed interactions i.e., the Kronecker product of a matrix of orthogonal comparisons (the fixed effect) and an identity matrix of appropriate order (the random factor). Higher order interactions are just the logical extension of this Kronecker product transformation concept. Thus, we retain in our definition of fixed-random interactions the concept of a factor which is indeed fixed in one direction and random in another, rather than arbitrarily assigning such interactions to be either fixed or random. In addition, since the coefficinet matrix for say a fixed main effect will now be in the column space of the coefficinet matrix of its interaction with a random factor, we will not be able to reorder the fixed effect after the interaction but the random factor can be reordered after both.

This topic has recently come under considerable discussion because of variations in the definitions given to various factors in the mixed model and the subsequent discrepancies in expected mean squares (see e.g. Hocking (1973)). The controversy revolves largely around the inclusion or exclusion of the interaction component in the expected mean square for the random factor. In the balanced case it is excluded; but many algorithms for finding expected mean squares in the unbalanced case include it. Hartley and Searle (1969) refer to this situation as "a discontinuity in mixed model analysis." MAD resolves this discontinuity by simply taking into account the essential nature of the interacting factors and the result is that one can obtain expected mean squares in the unbalanced case having the same components (although clearly not the same coefficients) as would be obtained in the balanced case.

Thus we have outlined here, and implemented in MAD, a procedure by which the two basic principles stated in the introduction can be realized. With MAD one can analyze any model, whether fixed, random, or mixed, balanced or unbalanced, univariate or multivariate or with concomitant variables. In the unbalanced case one can obtain sums of squares whose expectations contain only those factors contained in the expectation of the same sum of squares of an analogous balanced model. Finally using equation (5) MAD supplies the expectation for every sum of square for each ordering of the model factors. (Actually since for fixed effects equation (5) contains quadratic forms in the unknown effects the trace of the matrix of the quadratic form is printed indicating presence or absence of the factor in the sum of squares.) To the knowledge of the authors there exists no computer package which approaches the generality and flexability of MAD

4. INPUT/OUTPUT FEATURES

MAD was written with the non-computer oriented researcher in mind. Thus, the details of its inner workings outlined in the previous section are necessary only to an understanding of how various quantities are obtained, not to understand what quantities are given.

Input is greatly simplified by the fact that the user need only specify, in standard linear model notation, the model to be analyzed along with the nature and maximum number of levels of each main effect. From this information and the subscripts defining the origin of each data point MAD automatically generates the transformed matrices discussed in the previous section. As an example consider the data from Chakravarti et. al. (1967: page 352) in which he considers a two-way crossed model with three levels for each factor. If we consider observer fixed and subjects random, the input to MAD would be of the form

```
MODEL     Y(IJ) = O(I) + S(J) + OS(IJ) + E
EFFECT      O       FIXED   I = 3   OBSERVERS
EFFECT      S       RANDOM  J = 3   SUBJECTS
```

Two subscripts would be associated with each data point or points if we were considering the same model with a multivariate response or with concomitant variables. From this information the package automatically generates the independent variables and forms the sums of squares and cross products matrix.

This matrix is stored for the duration of the analysis with appropriate portions being assembled according to the orderings of the factors selected by the user. In Table 1 we present sample output from the results of three orderings of the model associated with the Chakravarti data. The analysis of variance tables have been compressed but are identical in format to those produced in MAD except for additional identification information supplied by the package. From the ordering (a) we would obtain the mean squares and their expectations for the fixed factor (O - observers) and for interaction. We note also that the mean square for the random factor (S - subjects) is contaminated by the fixed effect as well as the interaction component. Thus, if a test or an estimate of that component is desired the reordering (b) will be necessary. In this way we can obtain sums of squares analogous to those of an equivalent balanced model. The ordering (c) is included to illustrate the results of a misordering of the model. Clearly, from the analogy with the balanced model the ordering (c) would be inappropriate, which a glance at the output from MAD makes very obvious.

In addition to the capability of performing any analysis of variance, MAD provides aids to interpretation of the analysis. In the less than full rank case the solution to the normal equations can be stated as

$$\hat{\underline{\beta}} = (X'X)^{-}X'\underline{y} \tag{7}$$

where $(X'X)^{-}$ is any generalized inverse of $X'X$. Although this solution is not unique,

$$\hat{\mu} = \underline{x}'\hat{\underline{\beta}}$$

is, provided $\underline{x}$ is in the column space of X' i.e., provided $\underline{x} = X'\underline{\ell}$ for some vector $\underline{\ell}$. Thus we can uniquely estimate any cell mean or combination of cell means for a given design. MAD performs this estimation procedure for any factor or combination of factors averaged across the levels of the remaining factors in the model. For the fixed model the estimated standard error for each mean is also provided. In designs in which a complete cell or subclass is empty MAD allows for estimation of cell means excluding the empty cell or if the user desires a value can be estimated for the empty cell and included in the marginal totals.

The estimates given by eqn. (7) can also be used to obtain the vector of residuals $\underline{y} - \hat{y} = \underline{y} - X\hat{\underline{\beta}}$ which is obviously unique. This has led to a version of MAD currently under preparation by Mr. Del Scott of Pennsylvania State University, University Park, Pennsylvania which provides extensive plotting capabilities for examing residuals and to aid in the interpretation of interactions etc.

TABLE 1. SAMPLE OUTPUT - Analyses of variance for various orderings of a mixed model.

(a) Model: $y_{ijk} = \mu + S_j + O_i + OS_{ij} + e_{ijk}$

SOURCE	DF	SUM OF SQRS	MEAN SQRS	EMS COEFFICIENTS		
				S	O	OS
MEAN	1	76936.				
S	2	20662.	10331.	7.318	0.122	0.091
O	2.	508.38	254.19	0.0	7.196	2.446
OS	4	709.91	177.48	0.0	0.0	2.375
ERROR	13	1204.0	92.615			

(b) Model: $y_{ijk} = \mu + O_i + OS_{ij} + S_j + e_{ijk}$

SOURCE	DF	SUM OF SQRS	MEAN SQRS	EMS COEFFICIENTS		
				O	OS	S
MEAN	1	76936.				
O	2	872.23	436.12	7.318	2.490	0.122
OS	4	1264.2	316.04	0.0	2.399	0.070
S	2	19744.	9872.1	0.0	0.0	7.057
ERROR	13	1204.0	92.615			

(c) Model: $y_{ijk} = \mu + S_j + OS_{ij} + O_i + e_{ijk}$

SOURCE	DF	SUM OF SQRS	MEAN SQRS	EMS COEFFICIENTS		
				S	OS	O
MEAN	1	76936.				
S	2	20662.	10331.	7.318	0.091	0.122
OS	6	1218.3	203.05	0.0	2.399	2.399
THE TERM ABOVE HAS 2 DEGREES OF FREEDOM TOO MANY. YOU REORDERED A TERM BEHING ITS ERROR.						
O	0	0.0	0.0	0.0	0.0	0.0
THE 2 DEGREES OF FREEDOM LOST ARE CONFOUNDED WITH THE TERMS ABOVE THIS MESSAGE						
ERROR	13	024.0	92.615			

5. LIMITATIONS AND IMPLEMENTATION

After four years of steady use by all of the consulting statisticians at Brigham Young University and at some seventeen other installations across the United States we have found very few problems which could not be analyzed with MAD. Those which we have been unable to analyze have been simply too large for our computing facilities. However, Francom (1973) gives a simple way of overcoming this difficulty when a design includes nested factors with many levels. The other major problem which we have encountered is with the accumulation of round-off errors when multiple covariate analysis is being performed in very large models i.e., cases where X'X is of the order of 100 or more.

MAD is currently operational on both the IBM 360/65 and IBM 7030 at Brigham Young University. Except for two subroutines in assembly language it is written in the FORTRAN IV language. The assembly language routines are used in the object time allocation of memory and in obtaining the time and date. The basic load module requires about 48K bytes of storage with the actual amount of memory required depending upon the size of the model being considered. In order to conserve storage, the package is heavily overlayed. Peripheral devices necessary to a basic analysis include the card reader, printer and a sequential file. If the estimated means options is called for or a multivariate analysis is being performed an additional random access device will be required.

For further information contact the authors at Department of Statistics, Brigham Young University, Provo, Utah 84602.

6. REFERENCES

Bancroft, T. A. Topics in Intermediate Statical Methods Iowa State University Press; Ames, Iowa (1968).

Bock, R. D. "Programming Univariate and Multivariat Analysis of Variance" Technometrics 5: 95-117; (1963).

Bryce, G. R. A Comparison of Pseudo F-Statistics with Regard to the Split Plot Repeated-Measures Design Unpublished Ph. D. Dissertation; University of Kentucky (1974).

Bryce, G. R. A Unified Method for the Analysis of Unbalanced Designs Unpublished Masters Thesis; Brigham Young University (1970).

Chakravarti, I. M., R. G. Laha, and J. Roy Handbook of Methods of Applied Statistics, Vol I. John Wiley and Sons; New York (1967).

Federer, W. T. and M. Zelen "Analysis of Multifactor Classifications with Unequal Numbers of Observations" Biometrics 22: 525-552; (1966).

Francom, S. F. A Multiple Linear Regression Approach to the Analysis of Balanced and Unbalanced Designs Unpublished Masters Theses; Brigham Young University (1973).

Gaylor, D. W., H. L. Lucas, and R. L. Anderson "Calculation of Expected Mean Squares By the Abbreviated Doolittle and Sqare Root Methods" Biometrics 26: 641-655; (1970).

Hartley, H. O. and S. R. Searle " A Discontinuity in Mixed Model Analysis" Biometrics 25: 573-576; (1969).

Harvey, W. R. "Estimation of Variance and Covariance Components in the Mixed Model" Biometrics 26: 485-504; (1970).

Hocking, R. R. "A Discussion of the Two-Way Mixed Model" The American Statistician 27: 148-152; (1973).

Lawson, J. The Validity of a Unified Method for the Analysis of Unbalanced Designs Unpublished Masters Thesis; Brigham Young University, Provo, Utah; (1971).

Rohde, C. A. and J. R. Harvey "Unified Least Squares Analysis" J. Amer. Statistical Association 60: 523-527; (1965).

Speed, F. M. and Hocking, R. R. "Computation of Expectations, Variances, and Covariances of ANOVA Mean Squares" Biometrics 30: 157-169; (1974).

Yates, F. "The Analysis of Multiple Classifications with Unequal Numbers in the Different Classes" J. Amer. Statist. Assoc. 29:51-66; (1934).

Supporting P-STAT on 7 Types of Computers

Roald Buhler, Princeton

1. INTRODUCTION

The purpose of this paper is to present some of the programming techniques which have become necessary to permit a large statistical system to run on a variety of computers. The goal is to be able to send, for each supported computer, a Fortran source file and (where necessary) an overlay structure deck, both in a form that will directly work without modification. It is the author's belief that this is the only way in which large, constantly improving systems can reliably be run on a worldwide basis. Before going into the technical problems of portability, it may be useful to provide a brief description of the P-STAT system itself.

2. THE P-STAT SYSTEM

P-STAT is a computing system for file manipulation and statistical analysis of social science data. It began and grew rapidly on the IBM 7090/94s from 1963 to 1966. The first user's manual of some seventy-nine pages was printed in February of 1964. It slowed down somewhat from 1966 to 1970 but has been growing and improving rapidly since 1970.

It is written in very clean Fortran. The current version has about 48,000 Fortran statements, one third of which are comments, and three small machine dependent subroutines. The user's manual has 185 pages. About 50 computing installations run P-STAT on IBM 360s, 370s, Sigma 7s, PDP-10s, Univac 1108s and, most recently, Burroughs 6700s. It is not yet running on Honeywell 6000s or on CDC 6400-6600s, although an appropriate source has compiled on each of those computers. The special problems posed by the overlay aspects of these systems are discussed in a later section.

P-STAT works with the conventional rectangular files of cases (rows) by variables (columns). Both rows and columns have 16-character labels. Missing data are permitted in data files and are accepted in some reasonable manner by many of the programs. Files can optionally be in double precision.

Its capabilities include correlation, regression, factor analysis, discriminate analysis, MANOVA, plots, histograms, cross-tabulation and frequency distribution. It has an extensive set of file manipulation operations, including various merge, match, collate, sort and update commands. It can be used in either batch or interactive modes. It is, in effect, partly competitive and partly complementary to systems like SPSS, SAS, BMD, OSIRIS and DATA-TEXT. A more complete description of

P-STAT is available in its user's manual [Buhler, 1973a] and in papers by Buhler [1973b] and by McLean [1973].

3. REASONS FOR A PRE-PROCESSOR

If you have written a large program for the IBM/360, for example, and a University in Australia with a Univac 1108 wishes to make your program available to their users, you have two choices. (A third, ignoring the letter, is not unknown in computing circles.) You can give them your 360 oriented Fortran source and let them go ahead and change it, perhaps substantially, to run on an 1108. The alternative is to be able to generate on your 360 a source which will compile and run on their 1108. The first approach, sending the original source, works reasonably well when the program is both well debugged and not being improved. If, however, major improvements occur yearly (as in P-STAT) that approach is a disaster.

Several such conversions were done with earlier versions of P-STAT; each took several skilled man-months at the receiving end. Once that was done, the communication of error-fixing was found to be additionally time-consuming. (Some bugs affect a single statement, others cause changes in many subroutines.) When a considerably improved (and changed) version arrived a year later, the entire effort had to be repeated from scratch.

It became clear from these experiences that control and responsibility of the source must stay with the author. If so, one approach would be to maintain a separate source file for each supported computer. At 50,000 statements each, and probably 10 supported computers in another year or two, that would never work. The keypunching and control aspects would be overwhelming.

The better approach is to have one generalized source with codes punched to indicate those statements which need modification for one computer or another. Then, a pre-processor reads that source and, given control parameters for a specific computer, produces a compilable source for that computer.

The first effort by the author in this pre-processor direction was begun in 1971 [Buhler, 1973 c]. It supported 4 computers (360, Sigma 7, 1108 and PDP-10) and worked well enough to warrant writing a faster and much more general pre-processor. This was done in early 1974. The new pre-processor is itself about 2000 Fortran statements. It took a month or so to write; changing the P-STAT source from the old to the new pre-processor control codes took about three months.

The next section of this paper will describe in some detail the Fortran portability problems that are controlled by this pre-processor.

4. PRE-PROCESSOR CAPABILITIES

P-STAT (as of June, 1974) has 48,566 source statements. 6,479 of them (about 13%) are subject to pre-processor action because they have pre-processor codings in columns 1 through 5. The rest are perfectly normal Fortran statements. The codes fall into 3 general classes: alphabetic codes before signs, alphabetic codes after signs and other codes.

4.1 Alphabetic codes before signs

These codes identify cards which are affected by the choice of P-STAT size, use of single or double precision, the form of encode-decode syntax and the form of alphabetic compares.

(i) P-STAT size

In order to fit well in different sized memories or regions, P-STAT is available in six sizes. From the pre-processor's point of view, the 6 sizes are...

Size Name	360/370 Region	Number of Variables	Common (Words)
TINY	170k	150	5,000
SMALL	200k	300	10,000
MEDIUM	230k	500	15,000
LARGE	270k	750	23,000
XLARGE	420k	1,500	50,000
JUMBO	680k	3,000	100,000

The codes are...T,S,M,L,X and J.

The following is typical pseudo Fortran going into the pre-processor...

```
T       COMMON    A( 4300), X( 150 )
S       COMMON    A( 8500), X( 300 )
M       COMMON    A(12500), X( 500 )
L       COMMON    A(19000), X( 750 )
X       COMMON    A(42500), X(1500 )
J       COMMON    A(85000), X(3000 )

T       MAX =  4300
S       MAX =  8500
M       MAX = 12500
L       MAX = 19000
X       MAX = 42500
J       MAX = 85000
```

The pre-processor, in this instance, is purely a <u>selecting</u> program. If the medium size is desired, an M is punched on a pre-processor control card, which is read by the pre-processor telling it what to do. It will then discard any cards with size controls (TSMLXJ) other than M. For an M card it simply blanks the M and keeps that card. The pre-processor output is the compiler input. The above 12 cards would become...

```
        COMMON    A(12500), X( 500)
        MAX = 12500
```

There are currently 2,420 cards with size codes. Most of them are

DIMENSION or COMMON statements.

(ii) Precision

"G" is used to indicate statements to be included only if an entirely single precision source is wanted. They are omitted if double precision is wanted. Similarly, an "H" card is included in double and omitted in single. Double precision is desirable in many situations on the 32 and 36 bit computers but not on the 60 bit computers (6400 - 6600).

```
H       DOUBLE PRECISION  C1, PROD, SQPROD
                ----
G       C1 = 1.
H       C1 = 1.DO
                ----
G       SQPROD =  SQRT  ( PROD)
H       SQPROD = DSQRT  ( PROD)
```

The pre-processor must be told "G" or "H". If "G", a single precision source is generated. Note that "REAL*8" is not general, while "DOUBLE PRECISION" works on 360s and on the others.

The Burroughs 6700 has a 48 bit word with some extra tag bits, one of which signals a double precision data element. The use in P-STAT of both single and double precision files is handled by many subroutines which receive the data row, for example, as an argument. They then look at a variable in common, which is set to 1 or 2 to indicate the precision of the file, to decide how to move the data around or such. This approach is utterly impossible on the B6700. It cannot be faked in this manner. If the compiler notes, while compiling the entire source, that the first reference to PSRROW has a single precision first argument, it considers in error any calls to PSRROW with a double precision first argument. I had thought that keeping track of single and double precision areas was my business; Burroughs thinks it is their business and they win. Fortunately, 48 bits in single precision is quite good and that is the way a B6700 source is generated.

This, incidentally, is a good example of the unexpected benefits of a generalized pre-processor. The single-double coding was put in with Control Data in mind, it turned out to be crucial for generating a Burroughs source. The P-STAT source has 43 G cards and 118 H cards. The larger number of H cards is for "DOUBLE PRECISION" specification statements.

(iii) Encode-decode syntax

The letter E signals an ENCODE or DECODE statement. Every Fortran supported so far has ENCODE-DECODE except IBM. Within IBM, P-STAT uses a modified version of FIOCS which treats unit 99 as an internal buffer to write into and read out of (using formatted I/O). For example, assume one has read a card in CX in 80A1 format and one now wishes to pack it in CC in 20A4 format. If QQED is dimensioned 33, the following works on the Sigma 7, 1108, PDP-10 and 6400.

```
E        ENCODE  (   80,40,QQED    )  CX
      40 FORMAT  ( 80A1  )
E        DECODE  (   80,60,QQED    )  CC
      60 FORMAT  ( 20A4 )
```

Obviously a machine language pack-unpack subroutine is faster, but this illustrates the idea. There are 247 encodes or decodes now in P-STAT. Honeywell 6000s do not require the maximum character count (the 80 in the encode statement), so one would expect this to work for Honeywell...

```
         ENCODE    ( 40, QQED )  CX
```

but no, Honeywell reverses the arguments...

```
         ENCODE    (  QQED, 40 )  CX
```

Also, Honeywell Fortran prefers QQED to be in character mode. Burroughs 6700 Fortran within the last year has begun to support this feature, they also use the Honeywell arguments but use write-read instead of encode-decode...

```
         WRITE    (  QQED, 40 )  CX
```

The IBM form uses KKOUT as a unit number in ENCODE situations, and KKIN in DECODE situations. Both are set to 99.

```
         WRITE    ( KKOUT, 40 )  CX
```

All <u>but</u> Univac permit more than 132 characters. Sigma 7 requires a list in the encode; i.e., one cannot use encode-decode to replace DATA statements on a Sigma 7...

```
         ENCODE    (  26, 60  QQED )
      60 FORMAT    ( 26HABC...Z)
```

The modification for IBM, Honeywell and Burroughs is one of the few instances where the pre-processor is actually doing things to the contents of the statement.

(iv) <u>Alphabetic compares</u>

An "A" indicates an alphabetic compare using .EQ. or .NE. . These compares ane worrisome everywhere and P-STAT has a problem with Burroughs in this area.

```
         DATA    BLANK    /    1H /
         DATA     DASH    /    1H-/
A        IF ( BLANK .EQ. DASH) GO TO 70
```

The .EQ. causes an arithmetic subtract on the B6700. The two literals do differ but in a manner that, after an arithmetic subtract, the result is considered to be zero and off to statement 70 we go. The logical operators work properly for all computers except Burroughs so far. For those, the A is dropped and the statement left as is. For Burroughs the statement is changed into a logical function call.

```
         IF (  PPS   (  BLANK,  DASH  ) )  GO TO 70
```

PPS returns true if they are the same. PPD is used if the operator was .NE., it returns true if the arguments are different. Those two functions, PPS and PPD, are supplied with the Burroughs source and omitted otherwise. There are 477 alphabetic compares now in P-STAT.

4.2 Alphabetic codes after signs.

Signs can occur as pre-processor codes in columns one to five. There are currently some 27 codes that can come after a sign. There are 2186 such cards in P-STAT, 98% of them follow + signs, the rest follow - signs. The pre-processor begins by initializing those 27 options so that the + cards for each of them are rejected (and the - cards accepted). The pre-processor controls for each computer include a set of options for which the + cards should be accepted and the - cards rejected. Unless indicated, there are only a few of each of these types of cards.

(i) Vertical bar

```
+B    DATA    BAR / 1H| /
-B    DATA    BAR / 1HI /
```

One or the other of these statements will be accepted.

(ii) Date

```
+D    CALL    PSDATQ
```

This call survives the pre-processor only if a machine dependent date (like 10/24/74) subroutine is available and the D option was therefore selected.

(iii) EOF

```
-E    READ    ( JT, 40, END=200) X
+E    READ    ( JT, 40 )  X
+E    IF      ( EOF  (JT) .NE. 0 )  GO TO 200
```

Control Data Fortran does not allow "END=200", for example, but supports instead an "EOF" function. Thus, CDC should get the +E cards and all other computers get the -E cards. There are 8 such sets of cards. Note that an unsigned E means encode, a signed E indicates EOF.

(iv) Define file

+F is used to indicate cards to be included if the define file option is needed. So far, 360s have used it, the others have not. There are 26 references to +F or -F in the source.

(v) Glitches

These are extremely specific one time problems. Two characters are used.

+G1 is for the Honeywell 6000. It causes QQED (see ENCODE) to be 132 characters rather than 33 words. That avoids many compiler warnings.

+G2 is for 360/370 O.S. using multi-region overlays. All names of subroutines in other regions are cited in EXTERNAL statements (with +G2 in columns 1-3) located in the first region's root segment.

+G3 used if an interface to OSIRIS files is there (yes on 360s, no on others).

+G4 used on 360s to reference a bit-modifying subroutine in a MANOVA program....

```
+G4   IF  ( X .NE. 0. ) X = PSRND (X)
```

(vi) Integer mode

If desired, a source can be generated such that all alphabetic variables are in integer mode. This protects against some clever Fortran normalizing a supposedly real but actually character value while it moves it.

```
+I     INTEGER  DOLLAR, BLANK, CHAR
+IC
```

If selected, that causes those 3 variables to be in integer rather than the default real mode. There are 869 such cards, indicating perhaps that statistical systems do a lot of character type decoding of commands. The above example shows another feature: if a "C" occurs after a signed code, the "C" is moved into column one if the test passes. Thus, comments can be included or deleted, either for spacing (as the above) or to explain something. 577 comments are subject to such tests.

(vii) Logical

This is specific to the "A" code for alphabetic compares. If the PPS or PPD functions are used, they must have been declared logical at the start of the subroutine. There are 240 of these.

(viii) Machine dependent

As with G, an M is a two character code. It refers to a Machine dependent statement. "+MU" is specifically for Univac 1108, "-MB" is for all machines except Burroughs.

A code can occur in the first card of a subroutine or function. If so, and the test fails, the entire subroutine or function is omitted. This is most often done with G or H (precision), +F (define file), +T (CPU time), +D (date) and +M situations. Currently, 56 of the 546 decks have a first card test.

(ix) Real

This control, +R, is the reverse of +I. It causes all character variables to be in real mode. Programmers are known to be a little compulsive; I cannot imagine a use for this but it was symmetric with +I so in it went, 175 cards so far.

(x) Scratch files

In a P-STAT run, use may be made of 3 particular scratch files, 21, 22 and 23. If +S1, +S2 and +S3 cards were included during pre-processing, these will indeed be scratch files involving I/O. If -S1, -S2 and -S3 cards were included, they become 3 labelled commons taking an extra

5000 words but avoiding the I/O. The latter is advantageous for a memo rich PDP-10 and clearly better for any paging machine. There are 114 +S and 19 -S statements. Most of the +S's do I/O.

(xi) Time

It is nice to report CPU time used by each P-STAT command within a run. That is usually not attempted in the first try at supporting a new computer, but is added in a later pre-processing after a subroutine which can access CPU time on that computer has been written. Such a subroutine for Honeywell would have +T+MH in columns 1-5 of the first card. In other words, signed tests can concatenate.

(xii) Underlining

Underlining must be made to go away if a computer does not support eithe the underscore character or the 1H+ format code. All such WRITE and FORMAT statements have +U codes. There are 72 of them.

(xiii) Word size

Whenever word size aspects occur, a set of 4 statements occur, for example...

```
+W4   8   FORMAT  ( 23A4 )
+W5   8   FORMAT  ( 18A5, A2 )
+W6   8   FORMAT  ( 15A6, A2 )
+W0   8   FORMAT  ( 9A10, A2 )
```

The "8" is just a statement number. There are surprisingly only 6 such sets in the system.

4.3 Other codes

R and Y are the most important. They indicate a range of statements to be excluded if the test on the R card fails....

```
RH
      double precision statements
YH
             ...or...
R+F
      define file statements
Y+F
```

These are quite useful. They allow the statements to be omitted at compile time if a feature is not wanted. 24 such ranges are in use.

Statement numbers sometimes get in the way.

```
, 170
+W4 . FORMAT (   etc.
+W5 . FORMAT (
```

If a period is found in column 1, the statement number is saved by the pre-processor. The next accepted card with a period in column 5 will be given that statement number.

A slash in column 5 causes the codes on that card to be continued to the next card.

```
-S2 /
   900 RETURN
```

This causes "900 RETURN" if -S2 is selected, both cards are omitted for the +S2 option.

A final, major use of the pre-processor is to insert sets of labelled common cards. These are indicated by a + in column 1 and the name of the labelled common in 5-10. There are 15 possible labelled commons and 676 references to them. They are read by the pre-processor and held in core for insertion as the source is processed. They have an extensive amount of pre-processor codes themselves, and are selected and modified according to the options of the particular run as they are read.

5. OVERLAY PROBLEMS

The pre-processor handles the across-machine Fortran problems quite well but it does not help on overlay portability, partly because the author preferred to keep overlay structuring independent from the source code.

The TINY size of P-STAT, when run in the ordinary manner, takes about 900k of core. Unless one has a paging computer like the 370 running VS or the B6700, this size program is intolerable, particularly when many parts of the system are totally unused in any given run. The operating systems of non-paging computers usually provide the capability of overlaying parts of one's program. There are two fundamental styles, implicit and explicit. This subject is covered with additional detail in [Buhler, 1973c].

5.1 Implicit overlays

In this form, a Fortran program simply says "CALL SUBX" without knowing if SUBX (1) is in the same overlay segment as the calling program, (2) is in another segment currently in core, or (3) is in a segment not in core that must be brought in. (A segment is simply a group of one or more subroutines that are brought into core together.) IBM 360s and 370s running O.S., 1108s, Sigma 7s and PDP-10s (using a new loader now in field test) all require an overlay definition deck to be submitted to the loader, but the Fortran itself is clean. The 360 overlay deck is about 700 cards. Each of these computers requires its overlay deck in a totally different syntax from the others. This 700 card deck was converted by hand to the Sigma 7 overlay language once, and very quickly afterwards a program was developed to do it automatically. This program now generates overlay decks for the above computers quite nicely. The problem is under control for these computers.

5.2 Explicit overlays

In this situation, the Fortran program itself calls an operating system subroutine requesting the loading of some needed segment. This must be done before a call to a subroutine in that segment can be made. This is true for Honeywell and CDC (using SCOPE 3.3). For Honeywell (and probably CDC as well) the problem actually is worse.

Consider segment W, with several segments below it including segment X, and below X are segments Y and Z. Of these, only W is now in memory, and it wants to call a subroutine in Z. In an implicit overlay system, you just call it. The implicit system, at that moment, also realizes that X needs to be in core if Z is to be called from W and, if X is not there, it brings it in. In an explicit overlay system, a call must be made asking for X and another asking for Z. In addition, Honeywell allows many levels of segments, CDC allows only 3. Placing these explicit loading calls into the source is an unattractive solution.

It seems possible that the best way to solve the overlay problem for Honeywell and CDC is to write a post-processor. This post-processor would read the overlay structure for Honeywell, for example, and would read the pre-processor output. It would be able to check each "CALL" against the structure and, if it is a potential segment-loading call, generate into the source a call to an interfacing subroutine with the nodal values (of the segment in the tree structure) as arguments. That interfacing subroutine would keep track of what segments are in core, and would generate the actual loading calls only when necessary. This has two constraints; function calls cannot cause loading of a lower segment, and calls which can cause loading cannot occur as part of logical IF statements. As it happens, the P-STAT source had no instances of either of these.

6. CONCLUSIONS

Will these problems go away by themselves?

The overlay situation will probably keep improving. The PDP-10 will soon have a reasonable, implicit-type overlay. CDC has improved its overlay in a newer system, SCOPE 3.4 . The change from 360/O.S. to 370/V.S. makes overlay go away on IBM paging hardware. The trend is clearly towards overlays causing less trouble in the next few years (unless some of the smaller manufacturers grow somewhat and re-create the cycle).

The pre-processor problems seem unlikely to improve. Word size, precision, trade-offs of I/O versus memory, character sets, etc., show no signs of standardizing. Will the rate of change in software slow down so that yearly conversions become unnecessary? Surely not, in the case of P-STAT the list of improvements to be made grows each year. Improvement simply creates demand for more improvement. The utility of pre-processors seems to be considerable for at least the rest of this decade, and probably much longer.

REFERENCES

(1) Buhler, R.,
"P-STAT - a computing system for file manipulation and statistical analysis of social science data," Princeton University Computer Center, 1973a.

(2) Buhler, R.,
Computer Science and Statistics: "Proceedings of 7th Annual Symposium on the Interface", W. Kennedy, Ed., Iowa State University, 1973b, 283-286.

(3) McLean, L., ibid, 1973, 287-290.

(4) Buhler, R., ibid, 1973c, 181-188.

Interactive Management for Time Series

Giorgio Calzolari, Pisa

1. INTRODUCTION

A very important element in order to study the structure of a national economy is the organization of a data file consisting of the observations, over time, on the variables which characterize the economy.

The main purpose of the package presented in this paper is to offer researcher a supply of information, which can be increased revised and updated during the research.

While the application of some form of computer processing can be great help whenever data must be manipulated in a repetitive way, an interactive approach, by allowing a direct dialogue between the user and the computer and, making it possible for a large number of users to have simultaneous access to the same information, facilitates this iterative procedure very much.

Data are handled globally by a package, called IMTS (Interactive Management for Time Series), running under the operating system CP-67/ CMS. By means of this package, it is possible to use data both from the common central file and from a private single user file; these data are accessed using only the abbreviated name of the series to be used. The user has therefore to know only the abbreviated names of the series, whose complete list may be obtained by issuing a simple command at the terminal.

The research capabilities supplied by the IMTS package are analysis, transformation and regression of stored data. As regards analysis and transformation of stored data, algebraic, not algebraic, logical operators can be applied as described in sections 4. 1 to 4. 6.

The estimation of the parameters of a model can be performed by means of three different methods: Ordinary Least Squares, Two Stage Least Squares and Limited Information Single Equation. They are described in section 4. 7.

These features can be used independently or in any combination by means of the IMTS language, which provides a single command for each application: it is therefore possible to write regression equations as well as algebraic expressions defining transformations of the economic variables.

For the standard commands no programming skill is required; the researcher can devote virtually all his attention to problem analysis.

Moreover, any user with some experience of Fortran and/or Assembler languages, can create special functions (additional estimation methods, particular cross-sections studies, etc) by means of subroutines which, after having been compiled, can be used immediately in the IMTS environment, in conjunction with any other standard or user's operator.

The research capabilities of IMTS are supported by many other , such as the on-line formal errors recovery procedure (see section 4. 9), the Macro facility (see section 3), the on-line plotter (see section 4. 7), the linkage with the simulation package (DMS command, see section 4. 7 and ref.), etc.

The IMTS system was originally designed for use in econometrics, but it is expected to be applicable to many other fields such as management science or social sciences.

2. CONNECTIONS BETWEEN THE PACKAGE AND THE OPERATING SYSTEM

As mentioned in the introduction, the IMTS package operates under the control of the operating system CP-67/CMS. This is general purpose time-sharing system for the IBM System/360: the CP (Control Program) creates the time-sharing part of the system to allow many users to access the computer simultaneously. The CMS (Cambridge Monitor System) provides the conversational part of the system to allow a user to monitor his work from a remote terminal.

CMS interprets a simple command language typed at the user's remote terminal. One of its purposes is to provide the user with various file-handling facilities, referring to data stored on disk, cards, or magnetic tape.

The description of the technical features of this operating system is beyond the scope of this paper; what we want to emphasize here is that, taking into account the interactive features of CP-67/CMS and suitably operating on the main storage of the computer, the package and the operating system are linked very closely together to perform any operation in a very effective way.

The IMTS environment must be entered by issuing a command in the CMS environment; in the turn, any CMS command may be executed while operating in the IMTS environment. In particular, all the commands dealing with "File Creation, Maintenance and Manipulation" of CMS, and with utility programs of IMTS may be issued without leaving the IMTS environment.

3. FILE MANAGEMENT

The input/output routines of the package provide for the maintenance of four classes of file.

The first is made up of one single file: the file of permanent time series available to all users.

The second is also made up of only one file, containing the abbreviated names of all the permanent time series. It is the directory.

The third class of files is made up of all the files containing the private time series of individual users, or the time series obtained as intermediate results of one procedure, for use in subsequent ones.

The fourth class includes only one file in which the commands to be executed are stored as and when they arise. It is used both for correcting formal errors in commands, and for carrying out burdensome sequences of commands, which may be stored once and for all and then recalled as and when needed (macro facilities).

Each series, in the central or in a private file, holds the following :

- the complete name of the time series;
- the abbreviated name of the series;
- the data source;
- the unit of measurement;
- a code to indicate whether the data are annual, semi annual, quarterly or monthly;
- the first year;
- the last year;
- the first quarter (or half-year or month);
- the last quarter (or half-year or month);
- the number of data in the series;
- the numerical data.

The data must be complete from beginning to end (missing data are not allowed).

4. ACCESS TO THE FILE AND DATA PROCESSING

Issuing of commands and printing of results are carried out by the user through the terminal. The user must issue at the terminal the command to be executed, eventually followed by the names of the time series (from the central file or private series: in the event of the same names, the private series is the first to be searched for), numerical constants, operator codes and function names. Generally speaking a command must be contained in one line (nevertheless, the continuation on one or more subsequent lines is allowed); the program decodes and executes each command-line as soon as it is written.

After executing a command, the program prints the results on the terminal, but does not store these results. Should the user wish to retain these results (and this is only possible when the results constitute a time series or when a single numerical result is involved, thus excluding plotter output, for example), it is necessary to write the key word FILE at the end of the data string, following it up immediately - without any intermediate space - with the name of the file to be created: for example FILEGNP, the results of the command go into the FILE GNP P1, (CMS file identifier) and may be used in all subsequent operations merely by referring to the GNP time series.

The following operations may be carried out on time series and on any constant :

(i) Operations and functions executed by routines held in the storage

.AND.,.OR.,.LT.,.LE.,.EQ.,.NE.,.GE.,.GT.

that are logical dyadic operators.

.NOT.

that is a logical monadic operator.

+	addition
-	subtraction
*	multiplication
/	division
**	raise to power

that are algebraic dyadic operators.

LOG	natural logarithm
LOG10	decimal logarithm
EXP	exponential (base e)
SIN	sine (Radians)
COS	cosine (Radians)
ATAN	arc tangent (Radians)
SINH	hyperbolic sine
COSH	hyperbolic cosine
ABS	absolute value
MAX	maximum value
MIN	minimum value
MEAN	mean value
VAR	variance
INT	integer part
RV	rate of variation
DIF	first difference
LAG(nn)	data lagging
ONE(nn/mm)	selection of one single item of data from a series
SEL(nn/mm,ll/kk)	selection of part of a series
CPR(nn)	compression of a series

that are monadic operators.

IF(THEN,
ELSE)

that are for discontinuities and decision rules.

(ii) Operations and functions carried out by special routines.

PLOT	in line plotter
OLS, OLSR	ordinary least squares estimate
TSLS	two-stage least squares estimate
LISE	limited information single equation estimate

DMS connection to DMS/2

user defined functions

4.1 Operator priority

Highest priority is given to the following functions: LOG, LOG10, EXP, SIN, COS, SINH, COSH, ATAN, ABS, MAX, MIN, MEAN, VAR, INT, RV, ONE, SEL, CPR, IF, OLSR. In the event of functions being chained, priority goes from right to left (the internal function being executed first).

These are followed, in order of precedence, by: raising to power, multiplication and division, in order from left to right, then addition and subtraction, in order from left to right. Then comparison operations in order from left to right. Then the logical operator .NOT. At the end, the logical operators .AND. and .OR. in order from left to right. In order to vary priority, parenthesis may be used.

4.2 Arithmetic dyadic operators

+, -, *, /, **

The operator must be written between the two operands to which it refers. No special separator signs are necessary.

The execution of these operations is subjected to the following rules. Between two constants, the result is a new constant. Between a constant and a series, the result is a new series with the same characteristics (starting date and finishing date, division into periods) as the original series. The operation between two series is allowed only if both series are divided into the same periods (for example, both with monthly data or both with annual data), in which case the result is a series limited to the period common to both the original series (for example, the sum of two series of annual data - one from 1953 to 1972 and the other from 1951 to 1970 - is a series of annual data from 1953 to 1970).

4.3 Logical dyadic operators

.AND. , .OR. , .LT. , .LE. , .EQ. , .NE. , .GE. , .GT.

As for the arithmetic dyadic operators, the result may be a constant or a series, whose numerical values may be 0. (FALSE) or 1. (TRUE).

For the .AND. and .OR. operators, the operation can have any value, but they are tested only to control if the value is 0. (FALSE) or ≠ 0. (TRUE).

4.4 Monadic operators

(i) EXP, LOG, LOG10, SIN, COS, ATAN, SINH, COSH, ABS, INT, .NOT.

The operator symbol must precede the variable (or constant) to which it is to be applied. It is possible to chain several operators, without using separator signs between them.

When applied to a constant, these functions transform it into a new constant. Applied to a series, they transform it into a new series with the same characteristics as the original one (starting date, finishing date, division into periods).

(ii) MAX, MIN, MEAN, VAR

They return only one numerical value, which has all the characteristics of a costant (there are no dates, etc.). However, if they are related to the operator CPR (compression of a series in conformity with another operator), they return a series (see CPR).

(iii) ONE(nn/mm)

Extracts from the series the numerical value corresponding to the year and the month (or quarter or half-year) indicated. For example, ONE(67/11)'SERIES1' returns the value for the month of November and for the year 1967 (SERIES1 must have monthly data); ONE(67)'SERIES1' returns the first (or only) item of data for the year 1967. The numerical value returned has all the characteristics of a constant.

(iv) LAG(nn)

This operator shifts the data of a time series (in one direction only) for as many positions as are indicated by nn (from 0 to 99). For example, LAG(4) 'SERIES1', where SERIES1 (annual data) starts from 1961 and ends in 1969, returns a series starting from 1965 and ending in 1969, with the original value for '61 assigned to '65 etc.

(v) SEL(nn/mm, ll/kk)

Selects from the series the values included between the year and the month (or quarter) indicated first, and that indicated second. For example, SEL(55/10, 66/2)'SERIES1' returns a series with monthly data (like SERIES1) included between October 1955 and February 1966.

(vi) RV

Calculates the variation rates of a series, taking into account the difference between one data and the preceeding,whatever the periodicity. The results are in percentage form.

4.5 Composite operators

CPR(nn)

The function CPR(nn) (compression of a series according to another operator) must be used together with one of the operators MAX, MIN, MEAN, VAR, +, *, : it must never be used alone. This function gives a series with characteristics as illustrated in the following examples: MAX CPR(1) 'SERIES1' (where SERIES1 has monthly data) gives a series of annual values (1 per year) equal to the maximum value among the monthly values for each year; MEAN CPR(4) 'SERIES1' gives a series of quarterly values (4 per year) equal to the mean from among the values for each quarter in the series; 0. +CPR(2) 'SERIES1' gives a series of half-yearly values (2 per year) equal to the sum of all the values for each half year in the series indicated. In these cases, the operators MAX, MIN, MEAN, VAR give a series instead of a constant.

4.6 The IF operator

Discontinuities and decision rules can be analyzed by means of the IF operator. It is used together with the pseudo-operators THEN and ELSE as illustrated in the following examples :

(i) IF ('SERIES1'.LT. 'SERIES2') THEN'SERIES3'ELSE'SERIES4'

First of all the program performs the operation with highest priority (.LT. because it is enclosed in parenthesis); the result is a series, limited to the period common to SERIES1 and SERIES 2 and with the same division into periods, with numerical values 0. and 1. This resulting series should have the same division into periods as SERIES3 and SERIES4 in order to perform the operation for the period common to three series, selecting for every year the value from SERIES3 if the corresponding value of the logical calculated series is nonzero (TRUE) or from SERIES4 if it is 0. (FALSE). These values build the resulting series.

(ii) IF('SERIES1'.LT. 1000.)THEN'SERIES2'ELSE IF(178.LT. 'SERIES3') THEN'SERIES4'ELSE 250.

In this example, comparisons are made between series and constants (every value of the series with the constant) and the resulting values are selected from series or from the constant. Each ELSE refers to the last unclosed THAN; each THEN must be closed by an ELSE.

4.7 Special functions

This group includes the on-line plotter routines, estimate routines (OLS, OLSR, TSLS, LISE), the routine for connection to DMS/2, and user-defined functions.

(i) OLS (ordinary least squares)

Multiple linear regression by the ordinary least squares method is executed by means of the SSP (Scientific Subroutine Package) library routines (see ref.). On the basis of the ordinary least squares method, is possible with these routines to calculate the standard deviations and means of both dependent and independent variables, simple and multiple correlation coefficients, and also to calculate regression coefficients. Variance analysis is also executed and, on request, the residue table (differences between the observed and interpolated values is printed.

On user's request, it is possible to specify the output unit (6=teletype terminal, 8= high-speed printer) when printed output of the results is required.
Example: OLS('SERIES1', RV('SERIES2'+'SERIES3'), 'SERIES4'); SERIES1 is the dependent variable (written first in the list of variables separated by commas), while the other two, that is the series resulting from the RV('SERIES2'+'SERIES3') operation and 'SERIES4' are the independent variables.

(ii) OLSR

It performs the same operations as OLS, but instead of printing a table of statistics, it gives, as a result, the series of the residuals of the linear regression. It can be so used as each other function. For example: 'SERIES1'-OLSR('SERIES1', 'SERIES2'); the result is the series of the interpolated values of SERIES1.
PLOT('SERIES2', OLSR('SERIES1', 'SERIES2'))(see. PLOT) plots at the terminal the residuals against the explaining variable SERIES2.
Among the various single-equation methods for estimating systems of simultaneous equations, TSLS and LISE are considered hereunder.

(iii) TSLS (two-stage least squares)

The TSLS instruction ('SERIES1', 'SERIES2', ...) sorts all the endogenous and predetermined variables (lagged endogenous and exogenous) involved in the system to be estimated: all the endogenous variables are defined first, then the predetermined variables. After preliminary control of the total number of variables sorted, by means of specification of the number of endogenous and predetermined variables, the program executes the OLS estimate of the 'reduced form' (first stage). Then (the second stage), for each structural equation (at most as many as the endogenous variables), the user must insert the specifications (number of variables and sorting index) needed for sorting the subset of variables involved in the equation. For each equation, the estimate of the coefficients and the estimate of the standard error referred to each coefficient are supplied.

(iv) LISE (Limited information single equation)

The operational characteristics of the LISE command are the same specified for TSLS: however, even if the user is not aware of this, it should be noted that the 'reduced form' estimate applies only to the 'standard error' estimate, while for estimating the coefficients of structural equations, in the iterative process which leads to the maximization of the variance ratio, the OLS estimate of the same equation is taken as the point of departure.

(v) PLOT

The PLOT command permits the printing on the terminal of a graph

in points for certain (dependent) variables versus another (independent) variable. The scale is automatically adjusted on the basis of the minimum and maximum values of the variables.
Example: PLOT('SERIES1', 'SERIES2', RV'SERIES3'*MAX'SERIES3'/ 100.);'SERIES1' is the independent variable (abscissa), while 'SERIES2' and the series resulting from the RV'SERIES3'*MAX'SERIES3'/100 operation are the dependent variables (ordinates).
In conversational mode the user is asked to choose between two different types of on-line plotter that is the normal plotter, with numerical values and scales which takes considerable time to print, or the high-speed, qualitative plotter, without numerical values and scales which expecially in the case of a single dependent variable, permits a considerable saving in time.

(vi) DMS

The DMS command ('SERIES1', 'SERIES2'. . . .) allows the selection of a certain number of time series, writing their names and numerical values on the FILE DMS P1, according to the format required by the DMS/2 input; the data are, however, arranged variable by variable, while DMS/2 requires them arranged year by year. It is therefore necessary to use the CMS SORT command which permits re-arrangement of the file in date order. In this case, series of different lenghts are not cut as this operation is handled by DMS/2.

4.8 User defined functions

The user himself can create special functions of interest for him, according to the general philosophy of the IMTS. He must write a program, whose FILENAME must be the name he wants to give the function, as a subroutine with a standard list of parameters. All the necessary operative information can be easily obtained by entering a standard subroutine, which is available for any user. After a correct compilation, this function is immediately available with no other formalities. User defined functions are treated as the other functions for the priority.

4.9 Formal error recovery procedure

During the operations for the coding and decoding of commands some types of formal errors are likely to occur in the program as, for example, number of open parenthesis different from number of closed parenthesis quotes opened but not closed to contain names of series, names of series exceeding 8 characters, and so on.

After reporting the type of error and the name of the routine containing the error (indication normally not essential for the user, but very useful for the system engineer in setting up new functions), the program calls the EDIT program of the CMS and applies it to the file in which each command is stored before decoding.

Using the usual EDIT requests the user corrects the given command, cancels from the file any other records not required, gives the FILE command, which retains the file content and returns control to the main program, then, if required, gives the GO command to repeat operation of the corrected command. If this still contains formal errors, the procedure starts again.

Other types of errors are also contemplated as, for example, time series which are not present either in the central file or in the user's private file operations which do not yield any result (for example, arithmetic operations on two incompatible series, one with monthly data and one with annual data, or on two compatible series without a common period) divisions by zero, logarithms of negative numbers, and so on. In the case of all these errors, which are not in the formal error category, the correction procedure is not called automatically. Should he think it advisable, however, the user may edit the command file and then proceed as for formal errors.

5. UTILITY PROGRAMS

The category of utility programs includes all routines in the operating system accessible to the user (EDIT, COMBINE, LISTF, SORT, APL etc.) and certain special programs which are external to the main program but are accessible from within it (CREARCH, NEW). The CMS routines are accessed from within the program by means of the usual CMS command. As the special programs are held on the same disk as the central file in module format, the are called by typing only their name.

(i) CREARCH

This program was used at the outset to create the central file, and it is used whenever new time series are to be added to this file. It may be used only by the responsible for the file maintenance, and the file, according to the program philosophy, may be read contemporaneously by more than one virtual machine, but may be modified by only one.

(ii) NEW

This is a program which permits the creation of a file containing a new time series in user's private file, without any problem. It reads the data as and when they are written and then stores them in the required file, in correct format. Once it has been created and completed, the new series may be added by the central file becoming available for all the users.

REFERENCES

I. B. M. (CP-67/CMS) Version 3 CMS Program Logic Manual - GY20-0591-0

I. B. M. (CP-67/CMS) Version 3 User's Guide - GH20-0859-0

Corsi P., A. Stajano, An interactive programming system to solve econometric non linear model - CSP010/513-3505

I. B. M. System/360 Scientific Subroutine Package Version III - Programmer's Manual - H20-0203

Kmenta J., Elements of Econometrics - The Macmillan Company, New York

EXAMPLE (see Kmenta op. cit., p. 564)
(with formal error)

```
--->
ols('qt', 'pt', 'ft', 'time'
(D2)ERROR : PARENTHESIS OR THEN-ELSE DO NOT BALANCE
(AMDM1) ERROR RECOVERY PROCEDURE
USE EDIT COMMANDS.
DEFAULT TABS SET.
EDIT:
p
OLS('QT', 'PT', 'FT', 'TIME'
c./' /')/
n
EOF
file
--->
go
OLS('QT', 'PT', 'FT', 'TIME')
THERE ARE 20 DATA FOR EACH SERIES, FROM 1922 TO 1941:1 EACH YEAR
1 SIMBOLIC UNIT (6/8)
2-9 NAME
10 RESIDUALS (0/1)
-........-
6supply   0
MULTIPLE REGRESSION.....SUPPLY
VARIABLE     CORRELATION   REGRESS.    STD. ERROR        COMPUTED
   NO.          X VS Y       COEFF.     OF REG. COEF.    T VALUE
    2          0.09810      0.16037       0.09488         1.69014
    3          0.68057      0.24813       0.04619         5.37226
    4          0.15004      0.24830       0.09752         2.54622
DEPENDENT
    1
INTERCEPT                          58.27541
MULTIPLE CORRELATION               0.80920
R**2                               0.65481
STD. ERROR OF ESTIMATE             2.40509
STD. DETERMINANT                   0.88357
SOURCE          D.F.    SUM OF SQUARES    MEAN SQUARES    F VALUE
REGRESSION       3         175.5632          58.5211       10.11697
RESIDUAL        16          92.5512           5.7844
TOTAL           19         268.1144
                 DURBIN WATSON STATISTIC = 2.1097
```

SAS – A Software System for Statistical Data Analysis[1])

S. Christeller, A. Meystre, U. Ballmer, and G. Glutz, Basle

1. INTRODUCTION

Of course it is not possible to describe a software system and demonstrate its abilities on a few pages. To do this, one would have to write a book of some hundred pages. Nevertheless, it will be tried in the following sections to show the guidelines, which we were following in building up our own system as well as its essential features.

The experience with our own special purpose programs showed that the different input modes and the varying amount of data caused difficulties in applying several different programs to the same data, so frequent adaptions were necessary. For several reasons we had still more difficulties with generally available programs: no possibility of handling automatically missing values in calculations and transformations of data, no filtering of data, differences in programming languages, insufficient documentation and unknown reliability.

So it seemed reasonable to create a general purpose system with a main frame, within which the different statistical problems could be solved, and which provides a great number of datahandling facilities and permits a practically unlimited range of applications.

The applications of the system should be as simple as possible and make efficient use of the chosen datastructure. Therefore, the existing programming languages as FORTRAN or PL/1 e.g. are not suitable for a frequently used application system. Thus a metalanguage had to be constructed, which consists of single simple instructions and a set of rules for using them: the SAS-language.

1) Not to be confused with an other system of the same name (Barr 1971)

To facilitate the writing and reading of sequences of instructions, the instructions have all a standardized format, for each instruction one named datacard is used.

In order to have a control over the performed operations a protocol is produced, which provides also hints and error messages (cf. also section 7).

2. THE DESIGN OF A STATISTICAL APPLICATION SYSTEM

The designer has to face quite a number of problems:

- kind and frequence of statistical problems
- sophistication and needs of the users
- data organisation and varying amount of data
- ease and rapidity in applications
- control of inputdata and readability of outputdata
- available computersystems and facilities
- documentation and maintenance of the system
- datahandling facilities and statistical routines
- costs for development, documentation, maintenance and applications
- programming languages
- reliability of results
- possibilities of addition of new data to already existing datasets
- filtering of data
- running the system on other computersystems
- the handling of missing values
- already existing systems.2)

Although this list is not complete [3], it is obvious that the relative importance of the above mentioned problems determines in a large degree the layout of a system. Of course the reliability must not be questioned (cf., e.g. Francis, 1973). There is further a strong trend to receive the results more quickly, more readable, with less of paper and at a lower cost

As the following sections will show, the facilities for handling the data and the need for a sufficiently selfexplanating output blow up a system considerably and form together one group of problems, whereas the statistical problems form the second one.

2) cf. for instance Dixon [1967], Nie [1970] , Armor [1972], Cooley [1962, 1971]

3) for a more detailed discussion see Muller [1970], Milton [1969]

3. DATA ORGANISATION

The data organisation should suit the needs of more complicated data structures. In clinical trials the following situation arises fairly frequently: for each patient there exists one basic report, several intermediate reports, one final report.

For one observed or controlled patient there may be a great number of variables, up to hundreds, which may be all different or occur repeatedly in groups during the course of time. The number of reports is generally varying, because the duration of stationary or ambulant treatment differs from patient to patient.

This fairly general situation together with the available computer-system and programmerteam suggested the following approach:

- the patient data are arranged in a (m×n)-matrix, each row corresponding to one patient and each column to one variable
- explanation of the meaning of the variables by the aid of a datalist containing also the meaning of codes for qualitative variables
- the storage of the datamatrices together with the corresponding datalist for repeated use of the same data
- use of FORTRAN IV for the programs and a meta-language for the applications of the system in order to avoid unnesessary compilations and linkages.

4. DATAMATRIX AND DATALIST

Previous experience showed us that the data matrix should contain only numerical values and that the dimension of the matrix should be fairly large.

For ease of repeated reference to the same data, the datamatrix and the datalist are stored on external storage under a single code, so that data may be referenced by matrix-code and variablenumber.

This approach makes it possible that one user may own a whole series of datamatrices and several users may use the system simultaneously.

The datalist contains not only all relevant information concerning the meaning of the variables, but also the ranges of the values. This indication is needed for plausibility checks.

The arrangement of the data in a matrix causes of course an asymmetry because there is a difference in handling data in a column, i.e. a variable, or data in a row, i.e. data belonging to the same item.

5. CREATION OF THE DATAMATRIX

In a well planned study there exists always a questionaire, which can be directly used for punching the data or recording it on tape, perhaps after coding some qualitative variables. The questionaire also indicates the format of the data, which can be easily specified and serves as a base for the datalist. The input data may be read from cards, disk or tape.

In general it is advisable to check the data for completeness and redundancy before creating the datamatrix.

6. CHECK OF THE DATA IN THE DATAMATRIX AND CORRECTIONS

As already mentioned the given ranges of the variables in the datalist are used to detect impossible or implausible values and in a preliminary output these values are marked. With SAS-instructions the faulty values can be directly corrected in the datamatrix. Unfortunately this procedure has a serious disadvantage: the original data remain uncorrected. Nevertheless it cannot be emphasized enough, that bad coding of data costs a lot of time and money [Dempster, 1971].

7. IDENTIFICATION OF OUTPUT

There is a possibility to print a title on every page of the output. The title or parts of it can be easily changed.
The data and the daytime of the run is also printed together with a pagenumber, because the systemmessages and the jobcontrol are usually thrown away in order to diminish the amount of paper. This procedure is not only helpful in the discussion and interpretation of results after the run, it permits also a quick and unambiguous reference in a later point of time and prevents a confusion of apparently looking alike outputs.

The interpretation of the output is further improved by the possibility of inserting comments in the main printout and the protocol of the used SAS-instructions. To find a certain computed result quickly in the main output, the protocol gives the respective page number.

If calculations are performed under restrictions they are specified on the output as well as the identifications of the occuring variables.

8. OPTIONS IN THE INSTRUCTIONS

Most instructions can be used with one or more options, which produce different or more ample output. Contrary to the rule that the user should code as less as possible, a deviation from the minimal standard output must be always explicitly demanded. Otherwise there will be a waste of paper and computertime, because most users are too lazy to exclude non wandted information.

9. DATAFILTERS, LOOPS AND SWITCHES

In many cases we have to examine the data under varying restrictions. E.g. a set of questions has to be answered separately for males, females and both together and that the results have to be reported for the different investigators (not knowing in advance how many investigators were involved and on how many cases each one of them reports) and only those investigators are interesting, which report on at least ten cases.

To solve problems like this, the metalanguage provides the facilities of datafiltering, writing loops and asking switches, which means that one can build up boolean expression of different complexity, write do, if and goto statements without using a FORTRAN compiler.

10. TRANSFORMATION OF VARIABLES, COMPUTATION WITH VARIABLES AND CREATING NEW OR AUXILLIARY VARIABLES

The system provides a number of standard transformations as ln, log, exp, $x^{m/n}$, trigonometric functions and their inverses, the possibility to perform elementary arithmetical calculations without using a FORTRAN compiler.
There are further possibilities to create new variables according to given criteria, e.g. to classify people in groups of over-, normal- and underweight or to create random variables for different distributions.

All the transformed, calculated or generated variables can be added to the datamatrix and the datalist may be updatet. Missing values are automatically handled correctly.

11. STATISTICAL ANALYSIS

As shown in the previous sections, the data handling features play an important role in SAS, but they are only means to an end: statistical analysis. Therefore, the system provides a set of instructions for standard statistical computations:

- descriptive statistics
- printplotter graphics
- cross classification tables
- parametric and non-parametric tests
- sorting algorithmes
- random number generators
- standard statistical distributions
- multivariate techniques.

Evidently it is not possible nor feasible to provide more than a limited number of standard procedures. So only the frequently used ones are implemented in the system. But their is the possibility to insert special statistical programs as so called user-programs, which can be used by a user-instruction, in an environment with a considerable amount of datahandling facilities. If the procedure is needed frequently enough, it may be incorporated as a standard procedure.

The system may also be a help in the solution of rarely occuring problems, insofar as the bulk of calculations may be done within the system and the final calculations on a deskcomputer.

12. FREQUENCY OF THE USE OF INSTRUCTIONS

Because there are many user of the system, it might be difficult to decide if an instruction is used frequently or not. To get a clear picture of the situation, the system provides a statistic of the use of every SAS-instruction. This information gives also very useful hints for optimizing the performance of the system.

13. WORKING WITH THE SYSTEM

To day about 140 SAS-instructions are contained in the system. Thus, to apply it efficiently a user needs a considerable knowledge of the SAS-language; but the applications can be carried out with an absolute minimum of technical knowledge about computers and computing algorithmes.

As a mnemotechnic aid we have little reference booklets, which indicate for each instruction the essential and optional details.

As it is the case with every far developped system, a casual user will not be able to make great profit of its possibilities. Therefore, it is better to have a special team, which is thoroughly acquainted with it and translates the problems into the SAS-language. This approach has the further advantage, that in the course of time, the team acquires an ample experience to the benefit of the clientel and the further development of the system.

14. AN EXAMPLE

The following example will summarize some of the feature of SAS. It supposed that the variables mentioned between DATALIST ans DATAEND have been recorded for 1487 persons and that the data have been recorded on tape.

First an overall description of the population is desired in regard to age in five-year groups from 20 to 70 years and sex(AKAP2)as well as a frequency distribution of the treatment modes (AKAP3, INITKAP initialize the fields for these calculations).

Then follows a calculation of an "ideal" bodyweight (in variable 8) according to the comment under C. The partition of the sample in normal- and overweighted persons is done in an obvious way: the first NUMMER initializes variable 9 with blanks, SETBOOL sets a condition, viz. if variable 4 is greater or equal to variable 8, variable 9 is set to one, i.e. there is overweight, whereas in the opposite case variable 9 is set to zero. The statements following DATEIN update the datalist and NOSUBST clears the last restriction. MASTORE stores the datalist and the (1487x9)-datamatrix with matrixcode EX on disk. MASORT sorts the datamatrix according to the identification and the two following instructions produce the output of the datalist and datamatrix.

Further some basic descriptive statistics of age, bodyweight, bodylength and ideal weight are wanted (minimum, maximum, median, number of cases, mean, standard deviation) grouped according to sex, weight-class and pretreatment. The necessary calculations are performed after the next SETBOOL in the LOOP. The SETBOOL contains three restrictions R1, R2 and R3, which are connected by "and".

```
TITEL     1      *****
TITEL     2     * SAS *   E X A M P L E
TITEL     3      *****
TAPEIN   EX 321  1487      7
DATALIST
  1  O                  IDENTIFICATION
  2          15     80AGE IN YEARS
  3  N        1      2SEX
  3  C                  1MALE           2FEMALE
  4          35    110BODYWEIGHT IN KG
  5         140    200BODYLENGHT IN CM
  6  N        1      3MODE OF TREATMENT
  6  C                  1AMBULANT       2STATIONARY     3AMB. & STAT.
  7  N        1      3PRETREATMENT
  7  C                  1NO             2PRAEP. A       3PRAEP. B
DATAENDE
C        (F7.0,F3.0,F1.0,2F3.0,2F1.0)
INITAKAP
AKAP2         2   3     20     70       5
AKAP3         6   1   2   3
C        PARTITION ACCORDING TO WEIGHT IN TWO
C        GROUPS: NORMAL AND OVERWEIGHT
COMPUTE  V  8=V5-100
NUMMER        9
SETBOOL  S=R1
SUBST         4 >=V  8
NUMMER        9       1
SUBST         4 < V  8
NUMMER        9       0
DATEIN   EX
DATALIST
  8                     IDEAL BODYWEIGHT, BROCCA-INDEX
  9  N                  WEIGHT-CLASS
  9  C                    0NORMAL         1OVERWEIGHT
DATAENDE
NOSUBST
MASTORE  EX 28                                         1203.5(M13500) EX
MASORT        1
DATOUT
MATOUT
C        BASIC DESCRIPTIVE STATISTICS ACCORDING
C        TO SEX, WEIGHT-CLASS & PRETREATMENT
SETBOOL  S=R1*R2*R3
LOOP      A 12
SUBST         3 =     .
CHANGE   19     111222#
SUBST         9 =     .
CHANGE   19     0@061@06
SUBST         7 =     .
CHANGE   19     123#
BASIC          I  2  4  5  8
LOOPEND   A
```

15. TECHNICAL ASPECTS

SAS is run on an IBM 370/155 under OS and needs 240 K bytes of core storage. The maximum number of elements of the datamatrix which may reside in core is limited to 15000 under the further restriction that the product of the number of rows of the datamatrix and the number of variables in corestorage is $\leq$ 15000 and the number of variables in core is ≤ 50.

To make good use of the capacity of the highspeed printers, we decided to use the full range of an IBM 1403 printer, i.e. to exploit all the 132 printpositions and the maximum linenumber, practically 68 lines.

16. REFERENCES

Armor, D.J. and A.S. Couch: DATA-PRIMER, an introduction to computerized social data analysis, New York, 1972.

Barr, A.J. and J.H. Goodnight: Statistical Analysis System, Rayleight: Student Supply Store, North Carolina State, University, 1971.

Cooley, W.W. and P.R. Lohnes:
a) Multivariate Procedures for the Behavioural Sciences, J. Wiley, New York, 1962
b) Multivariate Data Analysis, J. Wiley, New York, 1971.

Dempster, A.P.: An Overview of Multivariate Data Analysis, Journ. of Multivariate Analysis 1, 1971, pp 316-346.

Dixon, W.J., ed., BMD-Biomedical Computer Programs, Berkely, Calif., University Press 1967.

Francis, I.: A Comparison of Several Analysis of Variance Programs, JASA 68, 1973, pp 860-865.

Milton, R.C. and J.A. Nelder. ed.: Statistical Computation, Academic Press, New York, 1969.

Muller, M.E.: Computer as an Instrument for Data Analysis, Technometrics 12, 1970, pp 259-293.

Nie, N. et al.: Statistical Package for the social Sciences, Mc Graw-Hill, New York, 1970.

An Econometric Program System – EPS

Hans-Jörg Haas and Georg Erber, Berlin

1. INTRODUCTION

The 'ECONOMETRIC PROGRAM SYSTEM' (EPS) is an integrated system of programs for the analysis of economic data by econometric methods such as OLS (Ordinary Least Squares), GLS (Generalized Least Squares), TSLS (Two Stage Least Squares) and LISE (Limited Information Single Equation). It combines a high flexibility in data organisation, definition of models, modification of data and model and data transformations. EPS is an attempt to depart from the numerous 'one purpose' programs which have been used up to now in econometric practice. The program contains not only the operations necessary for the estimation and testing of a model but also those necessary for the preliminary preparation of the data:

- setting up of a databank
- preparatory data analysis
- definition of an econometric model
- modification of the model and data in use
- estimation and testing of the model
- prediction using the reduced form
- storage of model and data for later access.

Routines offered by the system enable the user to define within the model most of the familiar data transformations used in econometric models, e.g. lags, differences and basic arithmetic transformations. Theoretical variables such as trends and dummies and all the variables which may be derived by the above transformations of the original data will be generated automatically during the estimation process; after estimation, they will

be deleted.
EPS can only be used to full advantage by interactive processing from remote terminals. The main requirement which influenced the system design was the need for easy handling of continuous alteration of model specification and data set. An interactive system seemed necessary since the specification of a model is an iterative process. The econometrician first specifies a model using a priori information. He then tests this model using empirical data. If it seems from the results of his tests that the model is not a good one, he must develope an alternative hypothesis. He must then test his new hypothesis. And so on. Since in general, the decision process is simple, the effective response time is small, and a large number of alternative hypothesis are needed for any one particular problem, an interactive system is obviously more effective than traditional batch processing.

2. META-LANGUAGE

EPS is easily operated by a meta-language based on econometric terminology. The language constructed is very similiar to that of SPSS, the Statistical Package for the Social Sciences.[1] A user familiar with econometric terms will have little difficulty in learning the language of the program. After a few hours he will be able to use most commands with facility. Elements of this language are commands such as the following:

COMMAND	ABBREV.	COMMENT
NO. OF STEQ	N,S	Number of stochastic equations
ADD IDENTITY	A,I	Add an identity equation to the model
GLS	---	The parameters will be estimated by Generalized Least Squares
LIST MODEL	L,M	The whole model will be listed
READ DATA	R,D	Start data input
LIST ENDOGEN	L,E	All endogenous variables will be listed

The meta-language consists at present of more than 70 commands; most of them may be abbreviated to speed up input. Associated with the commands are parameters for precise specification of the intended operations, e.g. ADD IDENTITY Y(T) = C(T) + I(T) . The setting up of the databank and the model, and the specification of the estimation methods and tests are all done by means of this meta-language. A number of LIST- and DUMP-commands (especially if the program is run interactively) enable the user to obtain information about the parameters, the structure of the model and the data.

3. SYSTEM FILES

Databank and model form together a system. The system may be written in any state on a storage medium for later access. That means not only that complete models or a prepared databank can be retrived later, but also that any step of model- and data modification can be documented so that there is the possibility to go back to every earlier version of a model and data.

4. THE DATABANK

4.1. Setting up of a databank

The data are organized by EPS in a databank. Original data can be read in any format-specification from every machine readable input medium such as disk, tape or card.
The basic data element is the variable. Variables are described by

- the variable name
- the variable type (stock or flow).

The variable name is necessary to allow further references to the variable; the definition of the variable type allows the proper handling of stocks and flows in time aggregation processes.
The databank should contain only original data. Variables which can be derived by simple arithmetic transformations (logarithms, lags, differences, exponentiations) from the original data and will only be used in the model need not be defined.

(i) example

```
COMMENT       SETTING UP OF A DATABANK
VARIABLES     K,CAPITAL STOCK/
              C,CONSUMER EXPENDITURE/L,LABOUR INPUT/
              Y,NATIONAL INCOME/
              I,INVESTMENT;
STOCKS        K,L;
NOBS          48
FORMAT        (3X,5F6.2)
READ DATA     (I=DATA,DO=P,AGGR=4)
SAVE SYSTEM
FINISH
```

The data input is realized by a sequence of commands. The VARIABLES-command specifies the names and labels of the five variables K,C,L,Y and I; the two variables K and L are stocks, the rest will be treated as flows. The number of observations (NOBS) per variable and the format-specification of the data is given, the READ DATA-command starts the input process. In the above example, the input medium is a file named 'DATA' (I=DATA), the data organisation is 'period by period' (DO=P) instead of 'variable by variable', and the 48 quarterly values per variable will be aggregated over 4 periods (AGGR=4) to twelve annual values per variable. The SAVE SYSTEM card writes the databank on a storage medium. The FINISH-command marks the physical end of the program.

4.2.Modification of the databank

Additional variables may be read into the databank through the ADD VARIABLES-command; the command DELETE VARIABLES removes variables from the databank.

Further extension of the databank could be done by the creation of new variables through arithmetic operations on variables already in the databank.

(i) example

```
COMPUTE      KCOEF = K/Y; or
COMPUTE      Y     = MVAV4 (Y);
```

The first command creates the variable KCOEF (capital coefficient); the second results in the replacement of the variable Y by its moving averages of fourth order.

4.3. Preparatory data analysis

For a first overview over the variables, simple uni- and bivariate analysis may be performed. The user has the possibility of calculating statistics such as mean, standard deviation, kurtosis and skewness for each variable in the databank. Pearson correlation coefficients are offered, and graphs of the variables against time may be plotted.

4.4. Filter

To reduce further the number of variables in the databank there exists a number of filters which the user may specify. A filter is defined as an operator that will be applied to all variables in the databank. All variables taken from there during the estimation process of a model have to go through the specified filters. The original variables in the databank are therefore not altered; if the filters are switched off, the original data can be used again.
Following filters are available:

(i) AGGREGATE

The use of this filter results in an aggregation of all variables over a specified number of periods (with respect to the variable type STOCK or FLOW).
Example:There are monthly data in the databank, but the econometrician wants to build a model with quarterly data. By the command AGGREGATE = 3, all variables used will be aggregated over three periods from monthly to quarterly data.

(ii) S-ADJUST

By this filter, seasonal adjustment on the variables is performed. At the present stage of program development only the method of moving averages is available; we hope to be able to offer spectral analytical methods in the near future.

(iii) DIFFERENCE

The result of this filter is the replacement of all variables by their own differences of a specified order.

5. MODEL MANAGEMENT

5.1. Model definition

A model consists of a number of equations which are divided into two groups, stochastic equations and identities. EPS can handle only linear equations, but it provides the user with possibilities to linearize intrinsic linear equations by log-transformations or exponentiations.
The model may be defined in a notation very similiar to that used in economic literature. All familiar variable transformations, such as logarithms, lags, differences may be applied within the model definition.

(i) example

The databank created in 3.1. will be used and the following model will be defined:

$$C_t = a + b\, C_{t-1} + c\, Y_t + u_t \quad \text{(Brown's consumption function)}$$

$$Y_t = I_t + C_t \quad \text{(Identity)}$$

The following deck has to be prepared to define the model and to estimate the parameters of the consumption function by the method of Two Stage Least Squares:

```
COMMENT       RETRIEVE DATABANK, DEFINE MODEL, ESTIMATE
GET SYSTEM
NO. OF STEQ   1
NO. OF IDENT  1
READ MODEL    C(T) = C(T-1) + Y(T);
              Y(T) = I(T) + C(T);
COMMENT       MODEL IS DEFINED, SAVE IT (WITH DATA)
SAVE SYSTEM
COMMENT       CHOOSE ESTIMATION METHOD
TSLS
COMMENT       START ESTIMATION PROCESS
GO
FINISH
```

5.2. Model modification

Since in general, economic theory is very vague in specifying the form of an economic equation, for any economic problem in view the econometrician has to test several alternative hypothesis and to choose the best one. That means that the specification process of an econometric model is not finished with the first version of a model - it starts there. Equations may be altered or deleted, new ones may be added to the model, old equations may be replaced by new ones. EPS allows all these operations.

(i) example

The Modigliani consumption function

$$C_t = a + b \left(\max_{i=1}^{t-1} Y_t\right) + c\, Y_t + u_t$$

will replace Brown's consumption function in the above model by the following command:

```
ALTER STEQ    1, C(T) = MDGL(Y(T)) + Y(T);
```

The first stochastic equation in the model is altered.

5.3. Theoretical variables

Up until now we have only used empirical variables or their derivates in our examples. But the econometrician often introduces dummy-variables into his model. EPS recognises four types of them.

(i) The constant term

Since nearly all econometric equations are assumed to be nonhomogeneous and therefore have to be estimated with a constant term, EPS automatically inserts this variable into every stochastic equation, so that it does not have to be defined explicitly. If the user wants to specify an equation as homogeneous, he has to insert into the list of the explaining variables the variable NOCONST.

(ii) Trendvariables

These variables may be defined by the code TREND(k), where k is the order of the trend. The command ADD STEQ I(T) = Y(T-1) + TREND(1) ; adds the stochastic equation ($I_t = a + b\, Y_{t-1} + c\, t + u_t$) to the model.

(iii) Dummy-variables

Dummy-variables may also be set for every period explicitly, so as to take account of wartime and other such phenomena. These variables have the code DUMMY. The variable DUMMY(4-8,12-19) is set to 1 in the periods 4 to 8 and 12 to 19, and set to 0 in all other periods.

(iv) Seasonal dummies

Seasonal variation may be dealt with on the one hand by seasonal adjustment of the original variables (see S-ADJUST - filter), on the other by defining seasonal dummies in the equations of the model. By the command SEASON=n, n seasonal variables will blended into every equation of the model. The command SEASONR does the same except that there are now linear restrictions underlying the season variables, so that the estimated parameters sum to zero.

6. ESTIMATION METHODS AND TESTS

At the present stage of program development the following estimation methods are offered:

OLS	Ordinary Least Squares
GLS	Generalized Least Squares, with respect to - autoregressive residuals up to fourth order - heteroscedasticity
TSLS	Two Stage Least Squares
LISE	Limited Information Single Equation

The whole model or selected equations for any specified subperiod may be estimated. All problems which arise in respect to the subperiod by the use of lags or moving avarages (these operations reduce the number of observations) will be solved by the program itself.

If one of the quasi-simultaneous methods LISE or TSLS is chosen, an algorithm will be started to examine the structure of the model and to classify all variables as endogenous or predetermined. All necessary data matrices for the estimation process are built up automatically by the program.

The tables which result from the estimation processes give all information necessary to judge the quality of an equation. Coefficients, standard error, t-values and their levels of significance are computed, and confidence intervals for the coefficients are calculated. All familiar tests such as coefficients of determination, Durbin-Watson and von Neumann for autocorrelation, Goldfeld-Quandt for homoscedasticity and others are performed. If estimating by LISE, Basman's F-test for iveridentification is also carried out. All test-statistics are associated with their level of significance or (e.g. Durbin-Watson) with their upper/lower limits. By meams of several commands, additional tests are done: predictive tests (Wald's Janus coefficient, Theil's inequality coefficient), and Chow's test on the equality of the coefficients in two relations. It is also possible to detect autoregressive processes in the residuals of higher order than one. Plots of the original and the fitted values are offered, and of the residuals to enable visual detection of autocorrelation or heteroscedasticity.

7. PROGRAMMING NOTES

7.1. Programming language

To facilitate transfer to other machines, EPS is primarily written in FORTRAN; subroutines for memory allocation and file-handling use COMPASS, the assembly language of the CONTROL DATA 6000/7000 computers.
The system is running at Berlin Free University on a CD CYBER 72; work is already in progress to adapt EPS to a TR 440 and an IBM 360.

7.2. Program organisation

To minimize the amount of core and to get a reasonable response time if EPS is run interactively, the program has an overlay structure up to second order. Data is organized 'variable by variable', the model 'equation by equation' on disc files, not in core. This reduces the common area and thus the main-overlay to a minimum.
EPS is a dynamic program; small and medium-sized models require much less core than larger ones.

8. APPLICATIONS

The simplicity of operating the system makes it possible to use

EPS for university courses in econometrics. Therefore the gap between methodological education on the one hand and practical experience with estimation methods on the other may be closed more easily by means of the program system. The students are enabled to gain experience in practical econometric work without being confused by the difficulties of programming and the properties of input/output organisation of models and data as would be the case with less developed programs.

9. FURTHER DEVELOPMENT

We are aware that there are many requirements which cannot be fulfilled by the program in its present stage of development. We feel, however, that the program is a step in the right direction. The further extension of the program system (new estimation methods, tests, filters, restrictions, non-linear regression, etc.) should be oriented towards the requirements of practical work. Therefore a broad discussion between users and system designers regarding the criteria for further development must bestarted.

REFERENCES

(1) Nie,N.;Bent,D.;Hull,C.: Statistical Package for the Social Sciences - SPSS, New York 1970

Genstat – A Statistical System

J. A. Nelder, Harpenden

0. HISTORY

I began Genstat in 1965 in Adelaide, South Australia, in collaboration with the Division of Mathematical Statistics, C.S.I.R.O., and the first version was developed on the CDC 3300 computer. This used the data matrix as the basic structure, much as SPSS does today. The present version, developed at Rothamsted by twelve members of the Statistics Department is much more general, and has been in use since 1971.

1. FORM OF THE SYSTEM

Genstat is written in Fortran, except for a few assembler routines, and currently contains about 275 subroutines involving about 25,000 lines of executable code. It exists as a compiled and link-edited program, presenting the user with an interpretive language for running his jobs. This language is 'compiled' into pseudo-object code, which in turn drives the various interpreters in the system. A single array in blank common serves to hold the directory of current structures, their attributes and values, and the pseudo object-code. Main components of the system for the user are

Data input
Output - both tabular and graphical
Basic data operations
Regression
Analysis of designed experiments
Multivariate analysis
Storage and retrieval

Before describing these facilities, we shall look at the data structures the system recognises and at the control language.

2. DATA STRUCTURES

Genstat recognises the following data structures, shown in geometrical form

```
Heading           ''THIS IS A LIST OF GENSTAT DATA STRUCTURES''

10.9       1  A  12.4          SMITH     2     X1
Scalar     2  B   8.9          BROWN    13     X2
           3  C  10.2          JONES     5     X3
           .  .    .             .       .      .
           .  .    .             .       .      .
           .  .    .             .       .      .
           n  C  13.3            .       .      .

           Data matrix         Name   Integer Pointer
           components             Other Vectors
```

```
  0 0 . . . 0        0                          0
  0 0 . . . 0        0 0                          0
  0 0 . . . 0        . . .                          .
  . . . . . .        . . . .                          .
  . . . . . .        0 0 . . 0                          .
  . . . . . .        x x . . x x                          0
  0 0 . . . 0

  Rectangular        Symmetric matrix           Diagonal
    matrix           and SSP structure           matrix
```

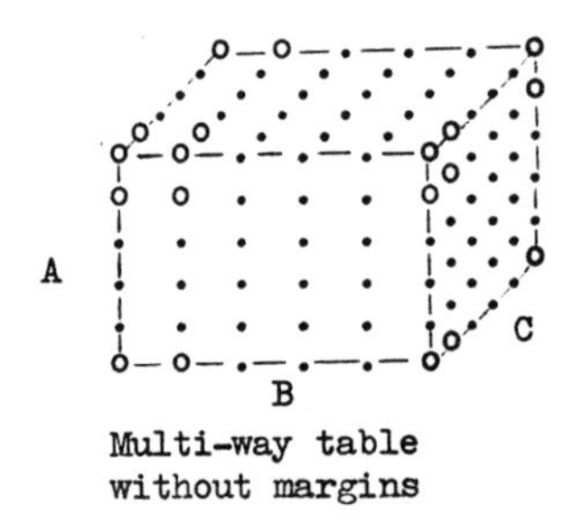

Multi-way table without margins

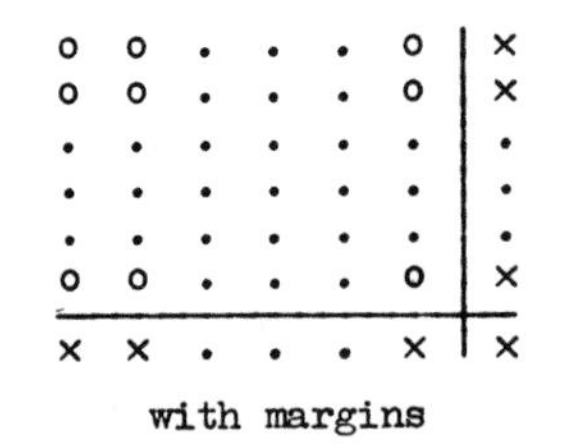

with margins

All structures can be assigned values in the program or have values read in; they all have default output formats and all can be stored on backing store and subsequently retrieved. All output from algorithms uses these standard structures, and this output can be referenced in subsequent directives.

3. THE GENSTAT LANGUAGE

The language consists of a sequence of directives, each introduced by a system word enclosed in primes, thus

'PRINT' X, Y, Z

The directive acts like a subprogram name in Fortran, but the parameter sequence has a much more flexible structure. To each Fortran parameter corresponds a list in Genstat, with semicolon (;) as the separator of lists. Further many lists can be named, and such named lists can be written in any order. Optional lists at the end may be omitted.

Example:

'ANOVA' Y(1...5);RES = R(1...5)

asks for the analysis of variance for 5 variates Y(1) ...Y(5) with residuals being saved in new variates R(1)...R(5).

Note how lists may appear in suffices so that

A(1,3,6) ≡ A(1), A(3), A(6)

Number lists may be compacted by various devices including

Progressions: 1,3...9 ≡ 1,3,5,7,9

Pre-multipliers: 2(1,3) ≡ 1,1,3,3

Post-multipliers: (1,3)2 ≡ 1,3,1,3

Thus 2(1...3)2 ≡ 1,1,2,2,3,3,1,1,2,2,3,3

3.1 Compile-time substitutions

The directive

'SET' S = X,Y,Z

means that subsequent occurrences of S are compiled as X,Y,Z, thus providing a macro facility within lists. Sets of complete directives can be named as macros and invoked later in the program. The 'SET' directive allows substitutions of sub-expressions in expressions or of format elements in formats.

3.2 Branching and loops

Branching is by the directive

'GOTO' label [* logical expression]

Unconditional jumps use the simple label, conditional jumps the additional logical expression as in

```
'GOTO' L1 *(X.GT.O.AND.Y.LE.Z)
```

Labels are treated as scalars and may be reset dynamically. Loops differ from those of Fortran in two ways; first the index variable denotes a reference to an identifier, not a value, and secondly up to 10 loops can be run in parallel, with the longest list determining the length of the loop. Thus in

```
'FOR' A = X,Y,Z; B = U,V
        ...
'REPEAT'
```

A and B in the three traverses refer to the pairs (X,U), (Y,V) and (Z,U) respectively. Dummy identifiers, such as A and B can also be set by the ASSIGN directive to refer to the ith element of an identifier list.

3.3 Interleaving of compilation and execution

An advanced feature of the language is the complete flexibility in the use of the compiler. Not only may a job consist of several blocks, each first compiled and then executed, but the compilation or execution of a block may be halted while an inner block is compiled and executed. The effect is to allow dynamic settings of formats, structure dimensions etc, the information being obtained either from external data or internal calculation in the program.

3.4 Compilation technique

The list structure, the full generality of which is not shown here, forms syntactically a one-one precedence grammar (with small exceptions caused by using round brackets for enclosing suffices etc.) and a syntax analyser parses all lists. As usual, however, the 'back-end' of the compiler, which deals with the semantics is much the largest section.

4 INPUT/OUTPUT

General input-output facilities are a vital part of any statistical system. In Genstat the READ directive allows fixed or free (or mixed) formats, and serial or parallel input. There are numerous options controlling the symbols for missing values, the treatment of blank fields and of errors, etc. Formats allow one line to contain many units or many lines to contain one unit.

PRINT gives tabular output. All structures have a default layout with automatic page overflow and labelling. Parallel or serial output are alternatives, and formats control field width and no. of decimal places. Special options allow the omission of labels, and, for tables, rearrangement by permuting factors etc.

5. ALGORITHMS

5.1 Basic data operations

These include the formation of the values of derived structures by (i) arithmetic operations and (ii) logical rearrangement. Arithmetic operations are exemplified by

'CALCULATE' A(1...3) = (X,Y,Z)/C

Note that the operands may be lists, as elsewhere in the system, and also that the identifiers may refer to vectors, matrices, tables etc. with item-by-item operations implied. Logical operators return real values 1. for true and 0. for false. Besides the standard functions, many statistical functions such as MIN, MAX, MEAN, VAR, are available together with random-number and log likelihood functions. Matrix functions are available for matrix operands, and there is a complete table calculus so that the expression

$$a_{ijk} = b_{ij} + c_k$$

can be calculated by the simple 'CALC' A = B + C.

Logical rearrangement is done by the EQUATE directive, which allows arbitrary subsets of a set of structures, defined by a format, to be assigned to arbitrary subsets of another set. Thus, for example, a partitioned matrix may be constructed from its constituents, or a subset of a data matrix extracted from the whole. There are many other uses.

5.2 Regression

The basic algorithm uses a symmetrical sweep operator on the SSP matrix, which on hexadecimal machines must be kept in double precision. It allows both quantitative and qualitative terms (variates and factors), with optional weighting and omission of intercept. Basic operations on the model are

FIT ADD DROP SWITCH TRY BEST WORST MINIMIZE

These allow respectively a model to be fitted, terms to be added, terms to be dropped, terms to be switched (i.e. added if absent or dropped if present), and the effect of adding a term to be assessed in advance, while the last three deal with sequential fitting in the sense of finding the best unfitted term, dropping the worst fitted term, or the better of the two. All components of the output from a regression can be either printed or named and saved as new structures, or both. In consequence, iterative weighted regression can be easily programmed within the control language, thus allowing e.g. the generalised linear models of Nelder and Wedderburn (1972) to be fitted.

5.3 Analysis of designed experiments

For this the recursive algorithm of Wilkinson (1970) is used. The model is specified in three parts (i) block structure, (ii) treatment structure and (iii) covariates. The block structure defines the random elements in the model, the treatment structure the systematic effects, and the covariates are used if an analysis of covariance is required. (i) and (ii) use structure formulae involving operators for compound terms (.), conjunction (+), exclusion (-), crossing (*), and nesting (/). The following examples are self-explanatory

A*B ≡ A + B + A.B

A/B ≡ A + A.B

A*B*C-A.B.C ≡ A + B + C + A.B + A.C + B.C

(A + B).C ≡ A.C + B.C

The algorithm copes with all generally balanced designs with an arbitrary number of error terms, and arbitrary confounding, subject only to the condition of general balance. It deals with missing values (including missing blocks etc.) and covariance in a quite general way. Submodels are allowed with individual d.f. defined by orthogonal polynomials or user-defined contrasts. Again output from the analysis can be extracted and put into named structures, or printed or both. It is difficult to indicate in this short space the full generality of this algorithm, which deserves careful study.

5.4 Multivariate analysis

The algorithms provided are intended to help the exploration of multivariate data. Clustering can be determined from a similarity matrix by any of five methods, after which nearest neighbours, most typical elements etc. can be displayed. Similarity matrices can combine information from many kinds of variate, both quantitative and qualitative, and the minimum spanning tree can be printed out.

Basic matrix operations for multivariate analysis are the latent (eigen)-root-and-vector and singular-value decompositions (Golub, 1970) Directives are provided for these and for principal components, canonical variate analysis, factor notation, principal coordinate analysis (Gower, 1966) (metric scaling), and Gower's rotational fitting technique(1971). Two algorithms which classify by optimizing a criterion are also available.

6. STORAGE AND RETRIEVAL

These facilities allow the user to define sub-files as sets of structures, and files as sets of sub-files. There is a work-file, which is active when a job is running and user files which are permanent. All sub-files are self-defining; a request to save a structure (e.g. a table) automatically causes all structures (e.g. lists of classifying factors and their level names) to be stored with it. All these structures are thus subsequently retrieved by referring only to the original table. The directory of structures in a sub-file can be merged with the current core directory in two different ways and re-naming is allowed to avoid clashes of identifiers. The directive DISPLAY allows the directory and pointer structure of a file to be printed out.

7. DOCUMENTATION

The user can refer to five types of document:

(i) The prospectus. This outlines the facilities and availability of the system.

(ii) The reference manual. This gives a full formal description of the language and all the directives.

(iii) User's guides. These give informal introductions to various aspects of the system. At present seven have been published, and one more is planned.

(iv) Worked examples. These exist as files on the computer and give annotated examples including the output.

(v) Notice board. This is issued with each release of the system, and gives current restrictions and known faults.

8. AVAILABILITY

Genstat is available on the ICL System 4 range from Rothamsted Experimental Station, Harpenden, Herts., AL5 2JQ, and on the IBM 360/370 range from the Edinburgh Regional Computing Centre, James Clerk Maxwell Building, Mayfield Road, Edinburgh, EH9 3JZ. An annual licence to use is issued at rates depending on the status of the institution concerned. Conversions are in progress for the CDC 7600 and ICL 1900 series. The program requires a minimum partition size of about 150K bytes.

REFERENCES

Gower, J.C. Some distance properties of latent root and vector methods used in multivariate analysis. Biometrika, 53, 325-338, (1966).

Gower, J.C. Statistical methods of comparing different multivariate analyses of the same data. In: "Mathematics in the Archaeological and Historical Sciences". Editors; Hudson, F.R., Kendall, D.G. and Tartu, P. Edinburgh: University Press, 138-149, (1971).

Golub, C.H. and Reinsch, C. Handbook series linear algebra. Singular value decomposition and least squares solutions. Numer.Math., 14, 403-420, (1970).

Nelder, J.A. and Wedderburn, R.W.M. Generalized linear models. Jl R.statist.Soc. A, 135, 370-384, (1972).

Wilkinson, G.N. A general recursive procedure for analysis of variance. Biometrika, 57, 19-46, (1970).

APL as a Notational Language in Statistics

Bent Rosenkrands, Denmark

1. INTRODUCTION

Iverson's 'A Programming Language' (APL) is more than just one of the many programming languages we are using today for communication with computers. Often referred to as Iverson's notation, it is also a formula language, a notation, whose simple function concept and syntax make it suitable for description of statistical problems. The fact that the notation itself is also directly programming in systems such as *APL*\360, *APL.SV*. and several others makes it so much more attractive to use.

Over the years, other writers have dealt with the suitability of APL for statistical tasks. Let me mention just a few. Smillie stands behind STATPACK [1969b] with several other interesting publications, including [1969a], [1970] and the lecture 'APL and Statistics, Programs or Insight?' [1971]. From Yale University Anscombe [1968] contributed his Technical Report on APL for Statistical Computing, and Snyder spoke on a series of APL functions for data analysis and non-parametric statistics at the APL Conference in Copenhagen [1973]. The writer of this article contributed the uses of APL for statistical graphs [Rosenkrands, 1971].

2. BRIEF COMMENTS ON APL

A few characteristics of the APL function concept should be emphasized very briefly. For a more detailed description, see [IBM, 1973a-b].

2.1 APL Function Concept

The APL functions may be divided into primitive functions and defined functions.

- The primitive functions, each of which has a special symbol, are classified as either scalar or mixed.
  - Scalar functions are defined on scalar (that is individual) arguments and are extended to arrays in five ways: element-by-element, reduction (f/), scan (f\ ), inner matrix product (f.g) and outer matrix product (∘.f).
  - Mixed functions are all other primitive functions.

- Defined functions are defined by the user himself, and may contain primitive as well as defined functions. These are denoted in the same way as variables by an arbitrary string of letters and digits, except that the first character must be a letter.

Whereas most other programming languages use a function definition of the form y=f(a,b,c,...), APL uses y=a + b or, generally, y=a f b as a pattern - that is, same appearance for primitive as for defined functions, since the two function types share the following characteristics:

- There may be 0 (defined functions only), 1 or 2 arguments.
  1 argument (may be an array) is placed to the right of the function symbol or function name (monadic function).
  2 arguments (may be arrays of different types) are placed one at each side of the function symbol or function name (dyadic function).

- The function value found may be an arbitrary array.

Most of the symbols for the primitive functions may denote either a monadic function or a dyadic function. *The symbol always denotes a dyadic function if possible*, that is, it will take a left argument if one is present.

In addition, all defined functions may use global variables.

2.2 APL Syntax

It is in this respect that APL deviates most obviously from other programming languages. The rule determining when to evaluate the individual functions - primitive as well as defined functions - is based exclusively on their position in the expression. In [Falkoff et al., 1973], this is expressed in the following rule:

> Every function takes as its righthand argument the entire expression to its right up to the right parenthesis of the pair that encloses it.

Or, briefer: A compound APL expression is evaluated from right to left.

It will now be easy also to make a clear rule for necessary parenthesizing in APL expressions:

> Parentheses are necessary only around the left argument of functions, and only if this argument is an expression.

So there can be no doubt at all about the sequence in which compound APL expressions must be evaluated: 5×3+2 evaluates to 25, namely 5×(3+2), whereas 2+3×5 is equal to 17, derived from 2+(3×5). And this applies not only to arithmetic expressions but also to logical expressions, branch statements, input and output statements, expressions with user defined functions, and mixtures of these. If *SQRT* stands for square root, 9+*SQRT* 16 equals 13, whereas *SQRT* 16+9 evaluates to 5.

3. ELEMENTARY STATISTICAL EXPRESSIONS

3.1 Using Primitives only

In section 3 and 4 we will limit ourselves to looking at vectors. The following examples illustrate how elegantly arithmetic and logic may be combined in one expression.

```
OBS ← . . .            Data entry
OBS                    Values of OBS typed out
+/OBS                  Sum of OBS values
ρOBS                   Number of elements in OBS
(+/OBS)÷ρOBS           Mean value of OBS
(⌈/OBS)-⌊/OBS          Range of OBS, that is max(OBS) - min(OBS)
+/OBS>0                Number of OBS's greater than 0
∨/OBS=2                Do any OBS's equal 2?
∧/OBS<5                Are all OBS's less than 5?
(OBS<5)/OBS            Select all OBS's which are less than 5
(OBS<5)/ιρOBS          Select the subscripts for all OBS's
                       which are less than 5
(~OBS∊(⌈/OBS),⌊/OBS)/OBS         Select all OBS's which differ
                       from greatest OBS and smallest OBS
```

We see that there is a consistent APL notation for many of those concepts for which we often use an ordinary verbal description instead of an unambigous notation.

3.2 Using Defined Functions

All APL symbols and APL expressions may, if required, be replaced by a defined function named by the user. For a start, let us define the following functions:

```
SUM X   is defined by +/X
SIZE X  -     -     - ρX
MAX X   -     -     - ⌈/X
MIN X   -     -     - ⌊/X
```

The mean value of OBS - let us call it OBSMEAN - may be expressed as

OBSMEAN ← (SUM OBS) ÷ SIZE OBS

and the Sum of Squares of Deviations as

*OBSSSD ← SUM (OBS - OBSMEAN)*2*

whereas the range is obtained by

OBSRANGE ← (MAX OBS) - MIN OBS

It is appropiate to emphasize that although these APL expressions are 'notations', they are also ordinary programming statements.

3.3 Successive Creation of a Statistical Function Library

A user-defined APL function may contain other defined functions. We utilize this advantage in the gradual creation of a function library. We assume that the previously mentioned SUM and SIZE and SQRT (denoted by *X*0.5*) are available.

The mean value:

R ← MEAN X defined as *R ← (SUM X) ÷ SIZE X*

The sum of squares of deviations:

R ← SSD X defined as *R ← SUM (X - MEAN X)*2*

The variance:

R ← VAR X defined as *R ← (SSD X) ÷ (SIZE X) - 1*

The standard deviation:

R ← STD X defined as *R ← SQRT VAR X*

The sum of products of deviations:

R ← X SPD Y defined as *R ← SUM (X - MEAN X) × Y - MEAN Y*

The covariance:

R ← X COVAR Y defined as *R ← (X SPD Y) ÷ (SIZE X) - 1*

The correlation coefficient:

R ← X CORR Y defined as *R ← (X SPD Y) ÷ SQRT (SSD X) × SSD Y*

We now have 10 general statistical APL functions available for practical statistical calculations and also as a basis for further development.

4. GRAPHICAL METHODS

This subject has been elaborated elsewhere - for example, in [Rosenkrands, 1971], from which the following two simple examples have been taken.

4.1 Display of Statistical Functions

The normal distribution, in traditional mathematical notation expressed as

$$y = \frac{1}{\sqrt{2\pi}} e^{-\frac{x^2}{2}}$$

may be written as follows in APL

```
Y←(÷(○2)*÷2)×*-0.5×X×X
```

if only the primitive functions of APL are used. But for APL purposes it is equally correct to write

```
Y←(1÷SQRT 2×PI)×EXP -0.5×X×X
```

or why not define a function GAUSS

```
R ← GAUSS X as R ← (1 ÷ SQRT 2×PI) × EXP -0.5×X×X
```

and produce a curve through the points by means of the functions DRAW and VS (versus) from the public library.

```
X ← ARITHSER ¯4  0.2  4
Y ← GAUSS X

10 40 DRAW Y VS X
```

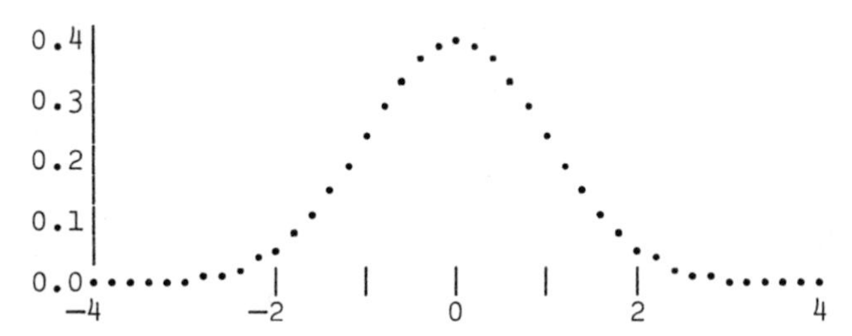

where ARITHSER generates the vector ¯4 ¯3.8 .. 0 0.2 ... 3.8 4. DRAW's left argument is an approximate dimension of the graph in number of lines (y axis) and number of strokes (x axis).

4.2 The Histogram

The creation of a histogram from a given observation vector consists of two parts:

- the administrative part, in which the observations are grouped and counted and
- the graphical part.

These two parts correspond to the two APL functions INTERVAL and HISTOGRAM, which may be used in the way the following example shows:

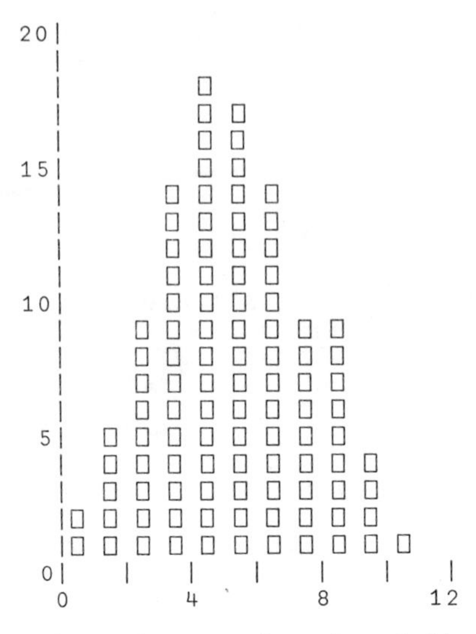

The function OF is a dummy, that is, defined as

```
     ∇R ← OF A
[1]  R ← A∇
```

INTERVAL's table representation clearly appears from

```
      (2 4 6 8 10 12) INTERVAL 3
 0   0
 3   1
 6   2
 9   1
12   2
```

This is one more example that justifies APL's right-to-left syntax.

5. MATRIX CALCULUS IN STATISTICS

All the primitive scalar functions denoted as 'f' and 'g' in the following examples are extended to arrays.

- Element by element corresponds to the examples used for the vectors.
- Reduction can for a matrix proceed along the first coordinate or along the second coordinate; for example: +/[1] *M* yields the column totals of the matrix M, and +/[2] *M* will result in the row totals of M. Reduction f/A is defined for any primitive scalar dyadic function on any array. See the following functions MEANS and RANK.

- Inner product: M1+.×M2 is the notation for the traditional matrix product of matrices M1 and M2. A similar definition applies to 'a f.g B', where f and g are both dyadic. See the following functions SPD and RANK.
- Outer product of two arrays X and Y is denoted by X ∘.f Y and yields an array, found by applying f to every pair of components of X and Y. See the following functions CORR and RANK.

Also several of the primitive mixed functions may be extended to arrays, depending on their meaning, of course. By compression it may, for example, be specified which coordinate of an array the selection is applied to. For example, the submatrix MS in the following expression contains the rows of matrix M, which has third column elements greater than 3000:

```
MS ← (M[;3] > 3000) /[1] M
```

5.1 CORR on a Matrix Argument

Let us once more find a function CORR. In section 3.3 the arguments for the dyadic CORR were the two vectors X and Y; this time, let us generalize by defining a monadic CORR whose argument is an arbitrary observations matrix and the result the matrix of the simple correlation coefficients. Moreover, let us develop the function as a kind of top-down programming:

R ← CORR X	defined as R ← (SPD X) ÷ SQRT (SSD X)∘.×SSD X
R ← SSD X	defined as R ← 1 1 ⍉ SPD X (the dyadic transposition 1 1 ⍉ results in the diagonal values).
R ← SPD X	defined as R ← (⍉DEV X) +.× DEV X
R ← DEV X	defined as R ← X - (⍴X)⍴ MEANS X (here we have to reshape the vector of the mean values to form a matrix with the same dimensions as matrix X).
R ← MEANS X	defined as R ← (+/[1] X) ÷ (⍴X)[1]

5.2 The Spearman Rank Correlation Coefficient

From a calculation and notation point of view the most interesting aspect is how to rank a vector of given values.
Let us look at APL's mixed function ⍋, called 'grade up'. The expression ⍋V (V is a vector) produces the permutation which would order V, that is, V[⍋V] is in ascending order. Since ⍋V is a permutation vector, then ⍋⍋V is the permutation inverse to ⍋V, that is, the ranks of the V values.
The matrix resulting from V∘.=V is multiplied by the vector of ranks, and the resulting vector is divided by the vector of the number of occurrences (derived from +/[2]V∘.=V) to average the

ranks when tied observations occur. The function RANK is thus defined by

```
      ∇R ← RANK V;W
[1]   W ← V ∘.= V
[2]   R ← (W +.× ⍋⍋V) ÷ +/[2] W∇
```

From an APL point of view the function RANK may be defined in a more effective and possibly more elegant way [Snyder, 1973]

```
0.5×(⍋⍋V)+⍒⍋⌽V
```

but the writer of this article considers the RANK function shown more descriptive, and at the same time it illustrates the use of the so-called 'composite functions': reduction (+/), inner product (+.×) and outer 'product' (∘.=).

Spearman's formula is in traditional notation

$$r_s = 1 - \frac{6 \cdot \sum_{i=1}^{N} d_i^2}{N^3 - N}$$

where d_i is the vector of the N pairs of rank differences. In APL notation SPEARMAN may be defined as

```
      ∇R ← X SPEARMAN Y;D;N
[1]   D ← (RANK X) - RANK Y
[2]   N ← ⍴D
[3]   R ← 1 - (6×SUM D*2) ÷ (N*3) - N ∇
```

5.3 Solution of Linear Least Squares Problems

The primitive mixed function ⌹ is used monadically for inversion of matrices, and if used dyadically it will solve a system of linear equations:

```
             X ← B⌹A
so that      (A+.×X) ∧.= B      is fulfilled.
```

If A contains more rows than columns

```
+/+/((A+.×X)-B)*2
```

is minimized, that is, solved as a linear least squares problem.

Let us use this to define a 'polynomial fit', degree N, in which we create the powers 0, 1, .., N, which is done directly with ∘.* . The observations vectors are called X (independent variable) and Y (dependent variable):

```
M ← X∘.*0,⍳N
C ← Y ⌹ M
R ← +/((M+.×C)-Y)*2
```

where C constitutes the coefficients of X^0, X^1, X^2,.. X^N, and R is the sum of the residuals squared.

6. CONCLUDING REMARKS

In the opinion of the author there are several reasons why APL lends itself so well to statistics. The main reason may be the way in which the primitive scalar functions of APL are extended to arrays, but also that the descriptive notation itself may be executed directly from an APL time-sharing terminal in one of the implemented APL-systems. It follows that APL like all other notation forms, are best acquired in step with the requirements.

REFERENCES

Anscombe, F.J. Use of Iverson's Language APL for Statistical Computing. Technical Report No. 4, Department of Statistics, Yale University, New Haven, Conn. 1968.

IBM. APL\360-OS and APL\360-DOS User's Manual. Form. No. GH20-0906-1. International Business Machines Corp. 1973a.

IBM. APLSV User's Manual. Form. No. SH20-1460. International Business Machines Corp. 1973b.

Rosenkrands, B. Graphics by APL. APL Congress 1971. IRIA Colloque APL, Paris 1971.

Smillie, K.W. STATPACK2: An APL Statistical Package. Department of Computing Science. Publication No. 17, University of Alberta, Edmonton, Alberta, Canada. 1969a.

Smillie, K.W. Some APL Algorithms for Orthogonal Factorial Experiments. Department of Computing Science. Publication No. 18, University of Alberta, Edmonton, Alberta, Canada. 1969b.

Smillie, K.W. An Introduction to APL\360 with Some Statistical Examples. Department of Computing Science. Publication No. 19, University of Alberta, Edmonton, Alberta, Canada. 1970.

Smillie, K.W. APL and Statistics. Programs or Insight? APL Congress 1971. IRIA Colloque APL. Paris. 1971.

Snyder, M. Interactive Data Analysis and Non-parametric Statistics. APL Congress 1973. P. Gjerlov, H.J. Helms and Johs Nielsen. North-Holland Elsevier. Amsterdam 1973.

- - * - -

Econometric Modeling in APL

Franz Schober, Stuttgart

1. INTRODUCTION

The system EPLAN (Econometric Planning) is an experimental, interactive computer language for econometric modeling and forecasting. It was developed while the author was with the IBM Philadelphia Scientific Center. The system contains various techniques for data analysis and transformation, tabular and graphical display, parameter estimation in single and simultaneous equation models, and model solution. EPLAN provides the econometric researcher with a tool for specification and execution of his models in an interactive mode without much knowledge of the basic computer principles in the background.

In this paper the design principles of a central part of EPLAN, the model language, will be described. As the system is written in the programming language APLSV, see Falkoff and Iverson [1973], the presentation emphasizes the application of APL as a notational language. Iverson [1962, 1969, 1972a,b], Berry, Falkoff and Iverson [1970], Berry, Bartoli, Dell'Aquila and Spadavecchia [1973] have used APL for the presentation of the concepts of a formal discipline, such as mathematics, physics and computer science. By that means, once defined concepts can be immediately experienced on a computer terminal. For several reasons APL may serve as a conceptual notation, namely because of its rich set of primitive functions and its compact data handling, as well as the simple way in which the user can expand the set of primitive functions by his own functions and thus expand the APL notation into his discipline of interest. Here, we shall concentrate on the discipline of econometric modeling, i.e. the formulation, estimation and solution of certain economic models. We will show, how APL concepts may be directly used for the design of a model language.

The design of the model language observes a hierarchical structure. The basic elements of this structure are defined by the economic variables (a). Subsequent stages comprise (b) operations between variables, (c) the formulation of equations and (d) of models. The paragraphs of this paper are arranged accordingly. The style of presentation is inevitably brief with the main emphasis put on the model language itself. Other features of EPLAN, such as display or file routines, are omitted, but may be referenced in Schober [1974b], or partly in an earlier version in Schober [1974a].

2. ECONOMIC VARIABLES

By applying the characteristic "time", two types of variables may be distinguished: time series and cross-sectional variables. Let us first consider time series and assume, that they are equidistant. A time series is defined in EPLAN by means of the APL-function *DF*, e.g.

```
      PRICE←12 1973 5 DF 4.1 5.2 5.9 6.3 7.1 6.8

      PRICE
2368112 4.1 5.2 5.9 6.3 7.1 6.8
```

The right argument of *DF* contains the period values, ordered with increasing time (here we assume for each period only one value, although series with multiple classification may be considered in EPLAN, too). The three-component left argument of *DF* contains all information about the time characteristic: the first component denotes the "periodicity" of the series ; here 12 indicates monthly series. In general, the periodicity describes the length of one period and is defined as the number of periods corresponding to one full year, e.g. 1=yearly, 4=quarterly. The two other components define the "origin" of the series, which is the period of the first value of the right argument. Above, 1973 5, specifies the origin May 1973; the corresponding period value is 4.1. For internal convenience, the left argument is transformed by *DF* into a scalar "header" via a simple, unique transformation, and catenated in front of the value array. The explicit result with an optional name (*PRICE*) defines a global APL-variable. Thus, time series are also available for manipulations outside of the system EPLAN.

Cross-sectional variables are viewed as "degenerate time series" with a zero header, which serves merely as an indicator. The value array may be ordered to some user-chosen criterion. Non- equidistant time series can be artificially represented by two cross-sectional variables, one containing the values and one the period lengths or end points. In the following we will restrict ourselves to the treatment of equidistant time series as the most important variable type in econometric modeling. But it should be mentioned that most concepts work in a very similar way also for cross-sectional variables and non-equidistant time series.

3. OPERATIONS BETWEEN TIME SERIES

In order to define a diadic operation *Q* between two time series, *X*1 and *X*2:

```
R←X1 Q X2
```

both series must have same periodicity. However, they still may have different origin and number of periods. Therefore, an element by element application of *Q* can only take place after bringing the series into a conformal format. A natural way for doing this is to consider for the operation only the common periods of both series, i.e. to truncate them to the largest common intersection in time. In this case the result *R* can be a time series with less periods than the operands *X*1 and *X*2. If the intersection is empty, *R* is an empty array.

Accordingly, a diadic operation is defined in EPLAN by a diadic APL-function with a mnemonic name. For example

```
      UNITS←12 1973 2 DF 800 850 930 1020 1050 1100 1200

      REVENUE←UNITS T PRICE

      REVENUE
2368112 4182 5460 6490 7560

      EC REVENUE[1]
12 1973 5
```

shows the multiplication (*T* for "times") of the series *UNITS* and *PRICE* (see paragraph 2). The result *REVENUE* extends over the largest common intersection of both series (*EC* is an auxiliary function to encode the header). In its main body the function *T* builds the intersection and then performs the multiplication. Thus *T*, like all other operators in EPLAN, is in effect an extension of the the APL-primitive operators to the data type "time series".

One of the operands may be a scalar constant:

```
      NEWPRICE←1.15 T PRICE

      NEWPRICE
2368112 4.715 5.98 6.785 7.245 8.165 7.82
```

In this case the operation is performed for all periods. Similarly, a monadic operation, such as the logarithm

```
      LPRICE←LOG PRICE

      LPRICE
2368112 1.410987 1.6486586 1.7749524 1.8405496
      1.9600948 1.9169226
```

does not involve an intersection process. Here, the EPLAN operator function is almost identical with the corresponding APL-primitive, with the exception of the exclusion of the header from the operation.

In this manner all important arithmetic and logical operators are represented by mnemonic APL-functions. The following table contains a summary.

Summary of operators defined in EPLAN

	Operation	EPLAN Function Name
Diadic	+	P
	-	M
	×	T
	÷	D
	*	PW
	<	L
	≤	LE
	=	E
	≥	GE
	>	G
	≠	NE
	∧	AND
	∨	OR
Monadic	log	LOG
	exp	EXP
	sin	SIN
	cos	COS
	tan	TAN
	abs value	ABS
	cumul value	CUM
	~	NOT
Spec. Oper.	shifts	LAG
	first diff.	DEL
	first quot.	RTO
	rect.distr.	RECT
	norm.distr.	NORM
	change period.	CHANGE

Included are also some special operators which are frequently used in econometric models, such as time shifts (leads and lags), first differences and quotients, the generation of random variates, and techniques to change the periodicity of a time series. Compound expressions may be formulated, for example

```
      DREVENUE←DEL LOG UNITS T PRICE

      DREVENUE
2368212 0.26665919 0.17281374 0.15260866
```

(*DEL X* defines the first differences of a series *X*). According to APL syntax rules the evaluation of a compound expression is done strictly from right to left, if not differently specified by the use of parentheses.

It should be mentioned that EPLAN contains a general system command, which provides the user with an explicit control over the intersection process between two (or a list of) time series. By that means he may reduce the number of periods involved in an operation. Besides that application, the command may in general be used for the controlled limitation of the econometric analysis to only a subset of the available data.

4. EQUATIONS

4.1. Definitorial Equations (Identities)

In the previous paragraph operations were executed immediately after input, defining a new, transformed time series. The relations between the operand series were lost after execution. According to econometric terminology we will refer to this case as a variables transformation and distinguish it from equations, which are stored relations for later execution.

In EPLAN an equation is stored as an APL-character vector:

```
      REVEQU←'DREVENUE←DEL LOG UNITS T PRICE'

      REVEQU
DREVENUE←DEL LOG UNITS T PRICE
```

An equation can be executed at any time, provided the right-hand side is completely specified, by means of the APL-primitive ⍎

```
      ⍎REVEQU

      DREVENUE
2368212 0.26665919 0.17281374 0.15260866
```

4.2. Structural Equations (Stochastic Equations)

EPLAN includes various regression techniques to estimate the parameters of a structural equation: ordinary least squares, generalized least squares, instrumental variables substitution (two-stage least squares), principal components substitution, and polynomial distributed lags. The equation must be linear in the unknown parameters. Several regression techniques may be combined within the same estimation (if meaningful). The following example, taken from Schober [1974b], illustrates the estimation of an equation with the name *MEQU*3

```
MEQU3←'W1←B0 P (B1 T S1 D S2) P B2 T 1 LAG S1 D S2'
```

where *B*0, *B*1, *B*2 are the unknown parameters. (1 LAG *X* defines a lag of one period of the time series *X*).

```
      INVAR←'1,1 LAG S1,1 LAG S2,1 LAG S1 D S2,DI'


      MEQU3←'W1' REGRESS '1,S1 D S2,1 LAG S1 D S2'
WITH: A1 ι0
RHO: ¯0.15594204
WITH: INVAR INST 'S1 D S2'
WITH:


COEF/VALUE/ST ERR/T-STAT.....

 1    .30321    .06374   4.75700
 2    .33815    .28317   1.19414
 3   ¯.15191    .23604   ¯.64357

NO OF VARIABLES.........    3.00000
NO OF OBSERVATIONS......   10.00000
SS DUE TO REGRESSION....     .03232
SS DUE TO RESIDUALS.....     .00506
F-STATISTIC.............   12.77891
STANDARD ERROR..........     .02903
R*2 -STATISTIC..........     .86467
R*2 CORRECTED...........     .82601
DURBIN WATSON STATISTIC.    2.02001



      MEQU3
W1← ( 0.303213 T 1 ) P ( 0.338148 T S1 D S2 ) P
      ( ¯0.15191 T 1 LAG S1 D S2)
```

The function *REGRESS* responds with an inquiry *WITH*, expecting the selection of the desired regression technique. The user response *A*1 ι0 in the example performs a first-order autoregressive correction of the error term, the second response ...*INST*... an instrumental variables substitution for the variable *S*1 *D* *S*2, using the left argument list of time series as instruments. (For a description of these techniques, see e.g. Theil, 1971). *REGRESS* continues to respond by means of the request *WITH*, until a blank input finishes the selection phase. The techniques are executed in the sequence of selection. At the end, ordinary least squares is performed for the modified observation series (ordinary least squares alone is indicated by a blank input after the first *WITH* request).

The explicit result of *REGRESS* is a well-defined equation, in the same representation as for definitorial equations. It contains already numereric parameter estimates (the precision can be specified by the user). Together with the definitorial equations the structural equations can in this form be immediately used for the assembly of the complete model.

5. MODELS

5.1. Definition

Models are a set of definitorial and structural equations. A complete model has as many dependent (endogenous) variables as there are equations. In EPLAN a model is represented by an APL-character matrix, where each row corresponds to one equation. The model may by assembled by means of the function *DMODEL*

```
      EQULIST←'EQU1,MEQU2,MEQU3,EQU4,D1,D2,D3,D4,D5'

      MODEL1←DMODEL EQULIST


      MODEL1
S1← ( 227.177 T WR ) P ( 0.546535 T SL1) P ( 0.120248 T DI )
S2← ( ¯80.4977 T WR ) P ( 0.830654 T SL2) P ( 0.117227 T DI )
W1← ( 0.303213 T 1 ) P ( 0.338148 T SR ) P ( ¯0.15191 T SRL)
W2← ( 0.385065 T 1 ) P ( ¯0.291008 T SR ) P ( 0.198681 T SRL)
SR←S1 D S2
SL1←1 LAG S1
SL2←1 LAG S2
SRL←SL1 D SL2
WR←W1 D W2
```

The right argument is a list of equation names; the sequence of these names indicates the sequence of the equations in the model (the example is again taken from Schober, 1974b). The equations have to be written in normalized form, so that each endogenous variable occurs in exactly one equation of the model on the left-hand side. The same variable may not occur on the right-hand side of that equation, except in lagged form. However, it may occur on the right-hand side of other equations.

5.2. Solution

Models are solved in EPLAN by means of a multi-dimensional variant of the well-known GAUSS- SEIDEL technique. Essentially, this technique successively substitutes estimates for the endogenous variables into the right-hand side of the model, starting from "good" initial estimates (see e.g. Ortega and Rheinboldt, 1970). Note, that our model representation in character matrix form can be interpreted as the canonical representation of a valid APL-function. Thus, each iteration of the GAUSS-SEIDEL technique corresponds to one execution of that function. The solution process via the function *SOLVE* makes direct use of that fact. Before, it is mostly desirable to reorder the sequence of equations to obtain an almost recursive model. This is done by the function *ORDER*, using VAN DER GIESSEN's substitution algorithm [1970]. Convergence may be considerably accelerated by that procedure. Before executing *SOLVE*, initial estimates for only those endogenous variables must be supplied which are not recursively evaluated within the first iteration (in the example below *S*1 and *S*2 as indicated by the execution of *ORDER*). Also various convergence parameters may be set (not shown). After successful execution of *SOLVE* the solutions can be called by name.

```
      MODEL2←' ' ORDER MODEL1
SEQUENCE OF EQUS:
SL1,SR ,SL2,SRL,W1  ,W2  ,WR  ,S1  ,S2

GIVE INIT APPR FOR:
S1  ,S2


      MODEL2
SL1←1 LAG S1
SR←S1 D S2
SL2←1 LAG S2
SRL←SL1 D SL2
W1← ( 0.303213 T 1 ) P ( 0.338148 T SR ) P ( ¯0.15191 T SRL)
W2← ( 0.385065 T 1 ) P ( ¯0.291008 T SR ) P ( 0.198681 T SRL)
WR←W1 D W2
S1← ( 227.177 T WR ) P ( 0.546535 T SL1) P ( 0.120248 T DI )
S2← ( ¯80.4977 T WR ) P ( 0.830654 T SL2) P ( 0.117227 T DI )
```

```
      ⍝ INITIAL ESTIMATES FOR S1 AND S2:

      S1
195401 250.328 398.658 690.081 952.423 1232.111 1414.718
       1526.195 1571.303 1613.578 1657.487 1731.53 1672.207
       1702.119

      S2
195401 3196.318 2856.406 2668.824 2453.522 2178.211 2008.075
       2006.827 1991.59 1971.696 1900.684 1808.007 1590.523
       1633.202


      'S1,S2,W1,W2' SOLVE MODEL2
CONV AT IT: 9


      ⍝ SOLUTIONS, E.G.

      S1
195401 250.328 711.40429 1001.2352 1189.0217 1302.9291 1372.3554
       1449.049 1521.7438 1587.4465 1653.2687 1709.2758 1764.6163
       1818.5691

      S2
195401 3196.318 2870.6459 2612.7354 2405.3939 2233.213 2088.7515
       1978.954 1873.5413 1776.1248 1711.8389 1663.6987 1648.1199
       1651.1821
```

6. SUMMARY

The paper concentrated on the design of the model language, a central part of the econometric system EPLAN. The design employs APL in a very direct manner, thus avoiding on the one hand a language interpreter (with a minor exception within the body of the *ORDER* and *SOLVE* functions). On the other hand it is believed that the resulting model language syntax is convenient enough for the application to practical econometric problems.

REFERENCES

Berry, P.C., A.D. Falkoff and K.E. Iverson, Using the Computer to Compute: a Direct but Neglected Approach to Teaching Mathematics, IBM Philadelphia Scientific Center Technical Report No. 320-2988, 1970.

Berry, P.C., G. Bartoli, C. Dell'Aquila and V. Spadavecchia, APL and Insight: The Use of Programs to Represent Concepts in Teaching, IBM Philadelphia Scientific Center Technical Report No. 320-3020, 1973.

Falkoff, A.D. and K.E. Iverson, APLSV User's Manual. APL Shared Variable System, IBM Corporation, 1973.

Iverson, K.E., A Programming Language, Wiley, New York 1962.

Iverson, K.E., The Use of APL in Teaching, IBM Corporation, 1969.

Iverson, K.E., APL in Exposition, IBM Philadelphia Scientific Center Technical Report No. 320-3010, 1972a.

Iverson, K.E., Algebra: An Algorithmic Treatment, Addison-Wesley, Menlo Park 1972b.

Ortega, J. and W. Rheinboldt, Iterative Solution of Nonlinear Equations in Several Variables, Academic Press, New York 1970.

Schober, F., Eine interaktive oekonometrische Modellsprache in APL, in: Hansen, H.R. (Hsg.), Computergestuetzte Marketing-Planung, Verlag Moderne Industrie, Muenchen (erscheint 1974a).

Schober, F., EPLAN. An APL-based Language for Econometric Modeling and Forecasting, IBM Philadelphia Scientific Center Technical Report, 1974b.

Theil, H., Principles of Econometrics, Wiley, New York 1971.

Van der Giessen, A.A., Solving Non-linear Systems by Computers; a New Method, Statistica Neerlandia 24, p. 1-10, 1970.

A Project for Standardization and Coordination of Statistical Programs in the BRD*

N. Victor, F. Tanzer, H. J. Trampisch, R. Zentgraf, P. Schaumann,
and W. Strudthoff, Gießen and Munich

1. INTRODUCTION

Within the scope of the 2nd EDP-Project (2. DV-Förderungsprogramm) of the government of the BRD the Ministry of Research and Technology (Bundesministerium für Forschung und Technologie, BMFT) finances a project for standardization and coordination of statistical programs with main attention to the analysis of medical data bases.

The preparations for this project are running since July 1972. In July 1973 the first part of the funds had been granted, for the purpose of supplying a pilot study. In January 1974 the phase of actual execution has been started.

During the phase of planning and development, the responsibility for the project is at the Department of Biomathematics of Gießen University. To guarantee an optimal utility of the project for a long time, after completion of the test phase, the project will be handed over to a permanent installation of the government (the Institute of Medical Data Processing of the GSF (IMD), Munich). An intensive cooperation of the IMD during the current phase of

---

* Project supported by a BMFT-grant with Project No. DVM 1o7

the project is supposed to guarantee a jointless transition.
At first we want to make some introductory remarks about the aims of the project; these aims will be specified in the next paragraphs and finally we will characterize the approaches chosen.

First section

The construction of a software-information-service (Software-Informationsdienst, SID) is supposed to make information about existing programs and systems available to all institutions with need of statistical programs as complete as possible. Aim is the promotion of the 'bilateral' program-exchange and the 'computing out of home' with the aim of avoiding multiple developments.

Second section

The next step has been the determination of standards for the programming language, the program structure and the data interfaces. This is supposed to simplify the exchangeability of new programs and to supply the base for an open programsystem. Especially we aspire a reduction of the amount of work necessary for adaptation.

Third section

After the completion of the preparatory work the construction of this open system should be fulfilled. The system is modular and consists of subprograms and segments linked by data interfaces. With the funds granted for the project we do not want to establish a new system; purpose of the project is the construction of the system with the contributions of several institutions by use of existing modules.

Fourth section

Finally we plan to examine the possibilities of reducing the amount of work necessary when evaluating medical data bases by using interactive data-manipulation and evaluation programs.
The project-management has close touch with several institutions and study groups from the fields of software-information and statistical programs; e.g. the project is affiliated with the software-information-center which is supported by the BMFT, also. A coordination on international level is aspired.
Profiteers of the project are mainly those institutions, which are supported within the 2nd EDP-Project; furthermore the services of the project are available to all institutions interested, here as well as abroad.

2. ORGANIZATION OF THE PROJECT

Core of the project is the coordination-group (Koordinations-gruppe, KG), which is assissted in its work by statistical teams (Arbeitskreise, AK) and by adaptation-groups (AG). A schematic representation of the project is shown in figure 1.
The adaptation-groups will guarantee the transferability of all programs developed to the installations of the most important manufacturers. During the first sections their duty will be to help specifying the standardizing principles in consideration of the special peculiarities of their installations. The main work for the AGs will start with the second phase of the project as soon as adaptations of programs will become necessary.

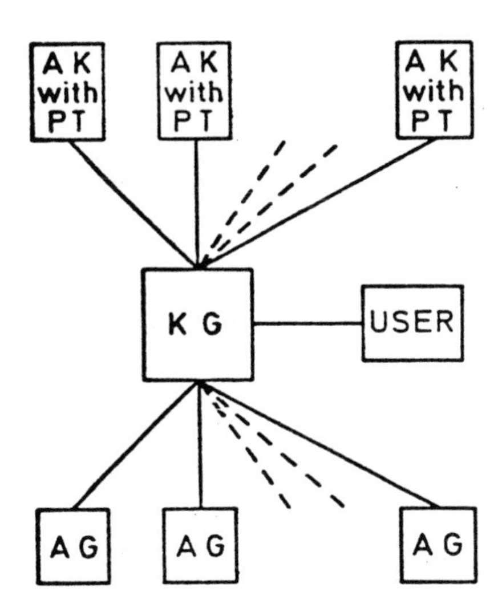

Fig.1: Scheme of the Organisation of the Projekt (see Text).

The statistical teams have been devided up according to their statistical areas, have specialized knowledge on certain fiels of statistics and should have a general view of the programs existing in their field, most of them have a programming team (PT) at their disposal. Their tasks will be the valuation of the programs for the software information service, the selection of program modules for the system to be established and the supply of new programs, where gaps in their special statistical field make this necessary. To solve the third of the tasks mentioned, the statistical teams dispose of the programming teams.

The coordination-group provides the outlines, coordinates the work of the AKs and the AGs, contributes on one field as statistical team, realizes the system on one installation type (i.e. works as an AG, also), provides programs for the linking of the

planned system and is crossing for the users of the services supplied by the project.
The outlines provided by the KG are discussed, modified and passed together with delegates of the AKs during work-shops. At present the coordination-group is equipped with four positions and shall be extended to seven subsequently. Up to now ten statistical teams have been established; so far the installation types of IBM, CDC, PDP, AEG-Telefunken and Siemens are covered. Additional statistical teams and adaptation-groups are scheduled.

3. SOFTWARE-INFORMATION-SERVICE

For the software-informations-service (SID) the coordination group supplies a data base on the existing statistical programs. This data base is organized according to the statistical methods. The index of this data base is due to a scheme of classification, which divides the field of statistics into sections (like in Statistical Theory and Method Abstracts).
Details about the programs are reported by use of information sheets, which are filled out by the AKs. The first part of this information sheet consists, in addition to biographical data, of information about language, programming techniques and program structure. In addition there is information about the amount of hardware and software necessary for executing the program. Using these specifications the user will be able to estimate the amount of work that will result, if the program is adapted to his peculiar installation. Finally the information sheet supplies a critical examination of the program, covering the statis-

tical and numerical methods, the programming techniques used and the suitability of the program for the evaluation of medical studies.

The user may ask the coordination group for information about programs about certain statistical topics. As an answer he will receive a list of all programs present and will get all information sheets which he desires. Then he will be able to decide, which program will be the best for his evaluation problem. After this he may contact the institution using this program and may order the program from there (bilateral exchange). On the other hand, if the user, on the basis of the information sheets, will have the impression that none of the programs suitable could be adapted to his installation within a reasonable amount of time, he may evaluate his data at an institution, where a program suitable is running (computing 'out of home'). A list of all those institutions will be sent to the user if desired. Usually the coordination group does not participate in this exchange; it only assists if eventual difficulties occur.

4. STANDARDIZATION

For the sake of the standardization,the programming language used, the manner of documentation and the data interfaces have been marked out. Thus shall be ensured that all programs supplied by any statistical team may be used with any installation without large scale adaptation.

As programming language because of its wide distribution FORTRAN has been chosen. Basing on the USASI-standards, programming-rules have been developed, which in general are restrictions (e.g. COMMON is not used); as an extension the use of random access I/O for external storage is permitted. Since, despite of the programming-rules,the adaptation of I/O-statements may cause difficulties, the use of special I/O-modules has been determined. Thus there will be no adaptation necessary for the actual statistical program.

As an additional aspect of standardization, principles for the internal and external documentation of the programs have been established. In addition to the user manual for the user who is not familiar with the system, there should be a detailed external documentation which includes information about the statistical methods used. This is supposed to help the experienced user in linking segments and in inserting own programs. The internal documentation (with Comment-cards) will ease the adaptation of the programs for the programmer.

The third aspect of the standardization concerns the establishment of uniform data interfaces. The logical structure and the physical realization of the most important interfaces have been fixed. At time additional interfaces may be defined. The physical realization of the logical interfaces will be done in special subroutines.

One of the interfaces - called So - is of special importance. It contains the details about experimental design, labels and data. Usually it will be located between data manipulation and evaluation. Because of the interfaces, the statistican will be able to concentrate on the evaluation programs, while the more complicated data manipulation programs will be supplied by EDP specialists.

5. SET-UP OF AN OPEN SYSTEM

Major aim of the project is the establishment of a flexible tool to solve various statistical problems; as such a tool we will develop an open modular system, which is supposed to satisfy the following conditions:

(i) The system should consist of programs developed by various teams; thus shall be ensured that all programs are written by specialists only.

(ii) An easy expansion of the system by the implementation of new programs has to be assured.

(iii) The system should be designed in a way, that every single part can be used efficently.

(iv) A great number of standard-subroutines should be available to the programmer.

(v) The experienced user should have the possibility of choosing his own sequence of evaluation by linking segments.

(vi) The user not familiar with EDP should be able to get complete evaluation programs of superior quality.

The program-system conceived is fixed by standards and interfaces only. Because of this flexible frame a continuous expansibility even by user programs will be guaranteed.

As a first step the creation of a common subroutine-pool (Unterprogramm-Pool, UP-Pool) has been arranged. In this there are subroutines for various problems, numerical, statistical, data-technical as well as I/O-problems. The subroutines should be designed in a way that they may be used in different segments. Because of the standardization the use of the UP-Pool is possible with many installations and wise even outside the system. While the accept of programs for the software-information-service will occur uncontrolled, programs for the UP-Pool are accepted only after passing a close examination. To do this for example a syntax-checker will be employed.

One segment stands for a meaningly step of statistical evaluation. To build up the segments the subroutines of the UP-Pool are used. The structure of the system is given in figure 2.

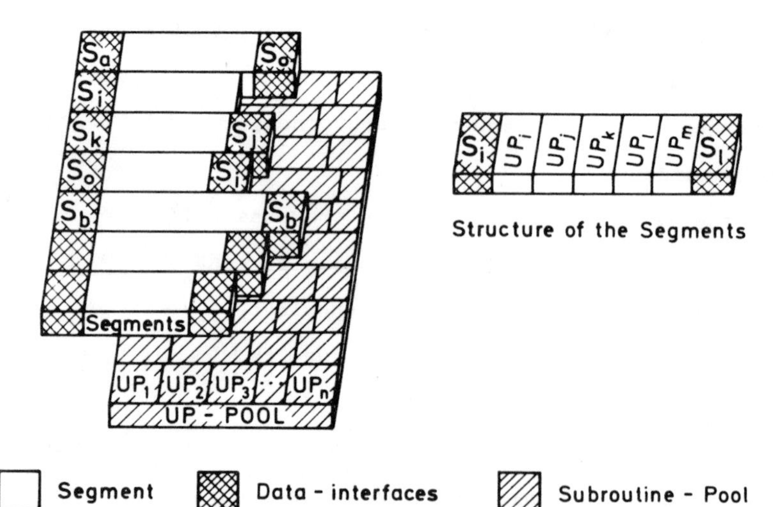

Subroutines are used for several Segments

Fig. 2: STRUCTURE OF THE SYSTEM.

Several segments may be put together to reach a complete evaluation program. The data exchange between the different segments will occur by use of the standardized interfaces. The evaluation programs, consisting of segments linked by a specialist, may easily be applied by a user not familiar with EDP. In this way there will be easy utilization as well as high flexibility.

PROSPECTIVES

It is planned to have the software-information-service available in the latter part of 1974. The collection of the standardized subroutines for the UP-Pool is scheduled for 1975. Subsequently the entire system will be made available.

LIST OF AUTHORS

BEAUCHAMP, K. G., Cranfield Inst. of Technology, Bedford, England, MK 43 OAL Cranfield, GB
BEDALL, Fritz K., Dr., Staatsinstitut f. Bildungsforschung, Arabella 1, D-8000 Munich, GFR
BELLACICCO, Antonio, Dept. of Statistics, Viale Ippocrate 92, I-Rome, Italy
BJÖRCK, Ake, Prof., Department of Mathematics, Linkoeping University, S-58183 Linköping, Sweden
BOOSTER, P., Ing., Hoodgovens B. V., Centraal Laboratorium, NL-Ijmuiden, Netherlands
BRUNS, Hermann, Dr., Bayerische Motoren Werke AG., Petuelring 130, D-8000 Munich 40, GFR
BRYCE, Gale Rex, Brigham Young University, 210 MSCB, 84602 Provo, Utah, USA
BUHLER, Roald, Princeton University, Computer Center, 87 Prospect Avenue, 08540 Princeton, New Jersey, USA
CALZOLARI, Giorgio, IBM Italia, S. Maria 67, I-56100 Pisa, Italy
CEHAK, Konrad, Univ. Doz. Dr., Zentralanst. f. Meteorologie und Geodynamik, Hohe Warte 38, A-1190 Vienna, Austria
CHORIN, Alexandre J., Univ. of California, California, 94720 Berkeley, USA
CHRISTELLER, S., F. Hoffmann-La Roche, Grenzackerstrasse, CH-4000 Basle, Switzerland
COHEN, Vidal, Université de Paris, Place de Lattre, F-750/6 Paris, France
EL-GENDI, Salah E., Dr., Faculty of Economics, Benghasi, Libya
ELSHOUT, I. I., University of Utrecht, Psychological Lab. Transitorium II. (R. 317), Heidelberglaan 2, Uithof/ NL-Utrecht, Netherlands
ERBER, Georg, Institut f. angewandte Statistik, FU Berlin, c/o Peter Naeve, Correnspatz 2, D-1000 Berlin 33, GFR
ESCOUFIER, Yves, Dept. Informatique, I. U. T., Av. d'Occitanie, F-3400 Montpellier, France
GASSER, Theo, Dr., ETH, Math. Statistik, Clausiusstr. 55, CH-8006 Zurich, Switzerland
GORDESCH, Johannes, Computing Centre, University of Vienna, Universitätsstrasse 7, A-1010 Vienna, Austria
GORENSTEIN, Samuel, IBM Thomas J. Watson Research Center, Yorktown Heights, New York, USA
GREENWOOD, Arthur J., Institute of Oceanography, City University New York, Box 2324 Grand Central Station, N. Y. 10017, USA
HABBEMA, J. D. F., Afdeling Medische Statistiek, Rijksuniversiteit te Leide, Wassenaarseweg 80, P. O. Box 2060, NL-Leiden, Netherlands
HELM, M., Dr., Siemens AG, Leopoldstr. 208, D-8000 Munich, GFR

HUBER, Peter J., Prof.Dr., ETH Zurich, Clausiusstrasse 5
CH-8006 Zurich, Switzerland
ITZINGER, Oskar, Dr., Inst.f.Höhere Studien, Stumpergasse
A-1060 Vienna, Austria
JACKSON, David M., Dr., Dept.of Combinatorics and Optimization, University of Waterloo, CDN-Waterloo, Ontario, Canada
KARASEK, Mirek, Econometric Research Branch, 56 Wellesley St. W., CDN-Toronto, Ontario, Canada
KEMPF, Wilhelm F., Dr., IPN-Kiel, Olshausenstrasse 40-60, D-2300 Kiel, GFR
KOBAYASHI, H., Computer Sciences Deptm., IBM Thomas J.Watson Research Center, Yorktown Heights, N.Y., USA
KOHLAS, J., Prof.Dr., Inst.f.Automation, 1.Route du Jura, CH-1700 Fribourg, Switzerland
LAVENBERG, Stephen S., IBM, Monterey and Cottle Roads, 95193 San Jose, California, USA
LEBART, Ludovic, CREDOC-CNRS, Boulevard de la Gare 45, F-750013 Paris-13, France
LECLERC, Annette, INSERM, Chemin de Ronde 44, F-78110 Le Vesinet France
LENZ, Hans-J., Dr., Inst.f.quant.Ökonomik und Statistik der FU Berlin, Garystr.21, D-1000 Berlin 33, GFR
MAIGNAN, M.F., Inst.Stat.u.P.Swiss Alum. Ltd, Feldegg Str.4, CH-8034 Zurich, Switzerland
NAEVE, Peter, Prof.Dr., Inst.f.quant.Ökonomik und Statistik der FU Berlin, Corrensplatz 2, D-1000 Berlin 33, GFR
NELDER, J.A., Rothamsted Experimental Sta., Harpendgn Herts England, GB
REISINGER, Leo, Institut für Statistik, Universitätsstrasse 7, A-1010 Vienna, Austria
ROSENKRANDS, B., IBM Headquarters, 91 Nymollevej, DK-2800 Lyngby, Denmark
SALEEB, Shafeek I., Prof., Inst.Stat.Stud.Res.Cairo Uni., Tharwat Str.5, ET-1017 Cairo, Egypt
SALFI, Robert, Institute of Oceanography, City University of New York, 15 Decker Avenue, Elizabeth NJ 07208, USA
SCHOBER, Franz, Dr., IBM Philadelphia Scientific Center, 3401 Market Street, PA 19104, Philadelphia, USA
STAHLIE, T.J., Nationaal Lucht- en Ruimevaartlaboratorium, Sloterweg 145, NL-Amsterdam, Netherlands
STAUDER, Ernoe, Dr., Central Stat.Office, 1.Szabo Ilonka 75/77, H-1015 Budapest, Hungary
VAN DRIEL, Otto P., Philips Research Laboratories, NL-Eindhoven, Netherlands
VAN HEMERT, M., University of Utrecht, Psychological Lab. Transitorium II (R.317), Heidelberglaan 2 Uithof, NL-Utrecht, Netherlands
VAN HEMERT, N.A., Univ.of Amsterdam, Psychological Lab., Weesperplein 8 (R.743), NL-Amsterdam, Netherlands.

VENKATACHARYA, K., Dr., Univ. of Libya, Faculty of Economics
Benghazi, LY-Benghazi, Libya
VICTOR, N., Prof. Dr., Abt. Biom. Inst. Bioch. Endok. FB Vet. Med.,
Frankfurter-Str. 112, D-6300 Giessen, GFR
WALL, M., Dr., F. Hoffmann La Roche & Co., CH-Basle, Switzerland
WALLIS, James R., Dr., Centro Scientifico 1, Via Santa Maria 67,
I-56100 Pisa, Italy
ZIELIŃSKI, Ryszard, Inst. of Mathem. Pol. Aca. Scienc., Bruna 30 M. 73,
PL-02-594 Warszawa, Poland

ERRATA

(COMPSTAT 1974 – Proceedings in Computational Statistics)

Page	
6	*Last line is missing* GASSER, TH.: Spline Smoothing of Spectra 323
7	*Line 16* Read: VAN HEMERT instead of VON HEMERT
22	*Line 7 from below* Read: $\Gamma(p)$ instead of $\Gamma(x)$ *Line 2 from below* Add: "$-\log_e \Gamma(p)$" to r.h.s.
23	*Line 3* Add: "$-\log_e \Gamma(p)$" to r.h.s.
31	*Section 4, line 14* For esceeds read exceeds
32	*Section 5, line 3* For]969 read 1969
33	*Table 2, heading* For 2 read χ^2
36	*Line 4* Read: $V = \{r \in R: t(r) > t_\epsilon\}$ ... instead of $V = \{r \in R: t(r) < t_\epsilon\}$ ...
38	*Line after eq. (3)* Read: if $t_1, t_2 \ldots t_m$ instead of $t_1, t_2 \ldots t_s$
60	*Add to References* Wallis, J.R., N.C.Matalas, and J.R. Slack, "Just a moment!" Water Resources Research, Vol.10. 2 April 1974.

Page	
65	*Eq. (1) and Eq. (2)* Read: $= \sum_{r=0}^{\infty} \sum_{\alpha_1 + \ldots + \alpha_k = r} \frac{y_1^{\alpha_1} \ldots y_k^{\alpha_k}}{r!} \frac{\partial^{(r)} f(x)}{\partial x_1^{\alpha_1} \ldots \partial x_k^{\alpha_k}}$ instead of $= \sum_{r=0}^{\infty} \sum_{\alpha_1 + \ldots + \alpha_k = r} \frac{y_1^{\alpha_1} \ldots y_k^{\alpha_k}}{r!} \cdot \frac{\partial^{(r)} f(x)}{x_1^{\alpha_1} \ldots x_k^{\alpha_k}}$
82	*Line 6 from below* Read: $A, B, C \in \underline{C}$ instead of $A, B, C\ \underline{C}$
83	*Line 7 from below* Read: $\zeta : A \rightarrow E$ instead of $: A\ E$ *Line 2 from below* Read: $f : G \rightarrow G'$ instead of $f : G\ G'$
84	*Line 5 from above* Read: form instead of from
85	*Line 3* Read: $h : X \rightarrow Y$ instead of $h : X\ Y$ *Line 4 (diagram)* Read: $\underline{g}(h)\ t_X = t_Y \underline{f}(h)$ instead of $\underline{g}(h)\ t_X = t_{Y\underline{f}(h)}$

Page

86 *Line 17*
Read: $\underline{T}$ [set of all $\underline{P}_n$; mor ($\underline{P}_n, \underline{P}_{n'}$)], instead of $\underline{T}$ set of all $\underline{P}_n$; mor ($\underline{P}_n, \underline{P}_{n'}$),
Line 24
Read: ($\underline{P}_n, \underline{P}_{n'}$) instead ($\underline{P}_n, \underline{P}_n$)

87 *Line 3*
Read: interval [0,1]
instead of interval 0,1

88 *Line 13*
Read: nodes and edges instead of nodes and / or edges

89 *Line 2*
Read: determine a similarity
instead of determine similarity

102 *Section 2 line 2-7*
Read: The form of this estimator, as used by us, is for a sample x_{j1}, ..., x_{jNj} given by

$$f_j(x) = \frac{1}{N_j} \sum_{r=1}^{N_j} K\{(x-x_{jr})' \Sigma_j^{-1} (x-x_{jr})\} \quad j = 1, \ldots k \quad (1)$$

Here Σ j is a diagonal matrix of smoothing parameters. A well known difficulty is the choice of Σ_j.
Instead of: The general form ... the selection of a value for σ_j.

103 *Line 3*
Read: $\Sigma_j = \sigma_j^2 I$ instead of $\Sigma_j = I$

111 *Line 1*
Read: classification instead of classiciation

115 *Line 2 from below*
Read: $C_i : | x | = \rho_i$
instead of $C_i : | x_u | = \rho_i$

135 *Line 8 from below*
Read: $\bar{X}_2$ instead of X_2
Line 7 and 6 from below
Read always: $\bar{X}_2'$ instead of X_2'
Line 4 from below
Read: $P_3' \tilde{y}_2$ instead of $P_3\, y_2$

136 *Line 8*
Read: $P_4' \tilde{q}_2$ instead of $P_4\, \tilde{q}_2$

201 *Eq (4.6) rhs*

Read: $= \sum_{\substack{s=1 \\ \beta \leq k-s}}^{k-1} N_{k-s} \cdot \ldots$

instead of $\sum_{s=1}^{k-1} N_{k-s} \cdot \ldots$

228 *Eq. (12)*
Read: $P_k = (\psi_k^T \; \psi_k)^{-1}$ (see (7))
instead of $P_k = (\underline{\psi}_k^T \; \underline{\psi}_k)^{-1}$
Eq. (13)
Read: $\hat{\Theta}_k = P_k \, \psi_k^T \, \underline{y}_k$
instead of $\hat{\underline{\Theta}}_k = P_k \, \underline{\psi}_k^T \, \underline{y}_k$

230 *Insert Eq. (25) and Eq. (26)*

$$\hat{\underline{\Theta}}_{k+1} = \hat{\underline{\Theta}}_k + \Delta \hat{\underline{\Theta}}_{k+1} \quad (25)$$

$$\hat{\underline{d}}_{k+1} = \hat{\underline{d}}_k + \Delta \hat{\underline{d}}_{k+1} \quad (26)$$

257 *Line 7*
Read: solution instead of selection

258 *Eq. (14)*
Read: $i \leqslant k$ instead of $i \geqslant k$

324 *Line 16*
Read: $I_q = \frac{1}{N} | Y_q |^2$
instead of $Y_q = \frac{1}{N} | Y_q |^2$

Page		Page	

393 *Line 1*
Read: modelling instead of modellig

Line 2 from below
Read: interpretation instead of interpretatron

394 *Point 8*
Read: with instead of ith

395 *Section 2, line 6*
Read: request instead of reques

Line 7, section 2
Read: provided if instead of provided

400 *Line 24*
Read: assumptions instead of assumtions

405 *Line 11*
Read: forty-two instead of fourty-two

413 *Reference Caroll*
Read: contemporary Psychology 17 instead of contemporary Psychology 6

448 There are *cut off the last letters in the following lines:*

1	two
8	following
9	factorization
14	Versions
17	fact-
25	opera-
32	other
37	distribu-

449 *Line 1*
A vertical line should preceed $|R_r$

450 *Equation (5)*
Φ_j in the last term should be Φ'_j

454 *Line 5 from below*
Read: $\underline{y} - \hat{\underline{y}} = \underline{y} - \underline{x}\hat{\underline{\beta}}$ instead of $\underline{y} - \hat{\underline{y}} = \underline{\bar{y}} - \underline{x}\hat{\underline{\beta}}$

455 *Table 1 model C*
The error sum of squares should be 1204.0

456 *The following reference should be added*
Searle, S.R. "Another Look at Henderson's Methods of Estimating Variance Components" Biometrics 24 : 749 - 787; (1968).

461 *Line 24*
Read: ENCODE (26, 60, QQED) instead of ENCODE (26, 60 QQED)

464 *Line 1*
Read: memory instead of memo

Line 12
Read: either instead of eithe

504 *Line 3 from below*
Read: factor rotation instead of factor notation

537 The *name* ELSHOUT, I.I., must be ELSHOUT, J.J. and the *address* must be: Univ.of Amsterdam, Psychological lab., Weesperplein 8 (R.751), NL-Amsterdam, Netherlands